CAN THEORIES BE REFUTED?

SYNTHESE LIBRARY

MONOGRAPHS ON EPISTEMOLOGY,

LOGIC, METHODOLOGY, PHILOSOPHY OF SCIENCE,

SOCIOLOGY OF SCIENCE AND OF KNOWLEDGE,

AND ON THE MATHEMATICAL METHODS OF

SOCIAL AND BEHAVIORAL SCIENCES

VOLUME 81

CAN THEORIES BE REFUTED?

Essays on the Duhem–Quine Thesis

Edited by

SANDRA G. HARDING

State University of New York at Albany

D. REIDEL PUBLISHING COMPANY

DORDRECHT-HOLLAND / BOSTON-U.S.A.

Q
175
.C238

Library of Congress Cataloging in Publication Data

Main entry under title:

Can theories be refuted?

(Synthese library ; 81)
Includes bibliographies and index.
1. Science—Philosophy. 2. Science—Methodology.
3. Duhem, Pierre Maurice Marie, 1861–1916. 4. Quine, Willard
Van Orman. I. Harding, Sandra G. II. Title: Duhem–Quine
thesis.
Q175.C238   501   75-28339
ISBN 90–277–0629–8
ISBN 90–277–0630–1 pbk.

Published by D. Reidel Publishing Company,
P.O. Box 17, Dordrecht, Holland

Sold and distributed in the U.S.A., Canada, and Mexico
by D. Reidel Publishing Company, Inc.
Lincoln Building, 160 Old Derby Street, Hingham,
Mass. 02043, U.S.A.

Printed in The Netherlands by D. Reidel, Dordrecht

To Dorian and Emily

TABLE OF CONTENTS

INTRODUCTION

According to a view assumed by many scientists and philosophers of science and standardly found in science textbooks, it is controlled experience which provides the basis for distinguishing between acceptable and unacceptable theories in science: acceptable theories are those which can pass empirical tests. It has often been thought that a certain sort of test is particularly significant: 'crucial experiments' provide supporting empirical evidence for one theory while providing conclusive evidence against another. However, in 1906 Pierre Duhem argued that the falsification of a theory is necessarily ambiguous and therefore that there are no crucial experiments; one can never be sure that it is a given theory rather than auxiliary or background hypotheses which experiment has falsified. W. V. Quine has concurred in this judgment, arguing that "our statements about the external world face the tribunal of sense experience not individually but only as a corporate body".

Some philosophers have thought that the Duhem–Quine thesis gratuitously raises perplexities. Others see it as doubly significant; these philosophers think that it provides a base for criticism of the foundational view of knowledge which has dominated much of western thought since Descartes, and they think that it opens the door to a new and fruitful way to conceive of scientific progress in particular and of the nature and growth of knowledge in general.

In this introductory essay, I shall indicate what considerations led Duhem and Quine to their views, and how some other leading philosophers and historians of science independently have arrived at similar conclusions. Then the major criticisms of the Duhem–Quine thesis will be presented. Finally I shall sketch the outlines of the rich and wide-ranging discussion of the implications of the Duhem–Quine thesis – a discussion which has occurred mainly in the last decade.

In his great book, *The Aim and Structure of Physical Theory*, Duhem was concerned with the way scientific theories were discussed by most scientists and philosophers of science of the late nineteenth century.

While these scientists and philosophers recognized that theories about nature could not be proved true, they did believe that by eliminating rival hypotheses through prescribed methods, science could finally reveal the residual, single, true description of nature. One kind of experiment was thought to be ideal for the purpose: 'crucial experiments' simultaneously refuted one hypothesis while verifying another hypothesis which was presumed to be the only logical alternative to the target hypothesis. Crucial experiments were thus thought to play a central role in science's project of searching for the truth.

In the selection presented here, Chapter VI of his book, Duhem argues against this view. He shows that two conditions must be satisfied if simultaneous falsification and verification are to take place, and that neither of these conditions can, as a matter of fact, be fulfilled. In the first place, an unambiguous falsification procedure must exist. *Modus tollens* arguments are usually taken to represent the appropriate falsification procedure[1], but Duhem argues that *modus tollens* is rarely, if ever, the structure of argument in the sciences since a scientist's predictions are in fact based not on any single hypothesis but, instead, on at least several assumptions and rules of inference, some of which are often only tacitly held. It is the target hypothesis plus a set of auxiliary hypotheses from which predictions are deduced. "The physicist can never subject an isolated hypothesis to experimental test, but only a whole group of hypotheses". Thus there is no reason to single out any particular hypothesis as the guilty one for isolated hypotheses are immune from refutation: Duhem denies that unambiguous falsification procedures do exist in science.

Secondly, even if it were possible to refute a particular hypothesis, one would not be justified in presuming that one had thereby shown any alternative hypothesis to be true, or as the claim has more recently been stated, shown any alternative hypothesis to be closer to the truth. In order to make this stronger truth claim – or truth-like claim – one must be able to show that *reductio ad absurdum* methods are applicable to scientific inference. First of all, Duhem points out that if it were possible to falsify any single hypothesis, then it might prove possible in the future to falsify any hypothesis to date unrefuted. But furthermore, in the future it might also be the case that some alternative explanation more satisfactory than any now known might be produced or discovered. We can see that this second problem arises because the concepts involved in our hypotheses change as knowledge grows; the plausibility of a description of some characteristic of nature is as much a consequence of the adequacy of one's concepts as it is due to the truth of one's claims. Thus, even if one could falsify a given hypothesis, the only truth established by such a falsification would be the denial of the hypothesis. But the denial of the hypothesis is not itself a single hypothesis but, given conceptual creativity, a potentially infinite disjunction of hypotheses. Because the physicist – unlike the Greek geometer – cannot enumerate all the possible alternative hypotheses which would explain an event, *reductio* methods are not applicable to scientific inference: we can't "assimilate experimental contradiction to reduction to absurdity", as Duhem says. So neither condition required for experiments to be crucial can, in fact, be satisfied, according to Duhem.

In his well-known essay, 'Two Dogmas of Empiricism', Quine refers approvingly to Duhem when he argues that only science as a whole, including the laws of logic, is empirically testable. Many have seen this as a radical conventionalist thesis. A great deal of critical attention has been focused on Quine's attack on the analytic/synthetic distinction; but that may well turn out to be the less important claim Quine makes in this essay. In the first four sections of the essay, Quine criticizes several arguments which might be given in defense of the analytic/synthetic distinction. He then goes on to consider a way of defending the distinction which relies on the verification theory of meaning. Perhaps a statement can be taken to be analytic if it is confirmed by anything whatever that happens in the world. Surely, if we can take some particular statements to be verified by particular experiences, we can also take other statements to be verified 'come what may'. But to this line of argument Quine objects that in fact no individual statement can be verified. Well, one might think, perhaps a statement is analytic if it is disconfirmed by nothing whatever that happens in the world. But, Quine notes, "any statement can be held true come what may, if we make drastic enough adjustments elsewhere in the system". And, by the same token, "no statement is immune to revision". He says, "the unit of empirical significance is the whole of science". If Quine could provide the philosophic underpinnings to defend this point, he would have shown not only that the analytic/synthetic distinction is untenable, but, more importantly, that the true/false distinction is not defensible except as applied to science as a whole, and that we cannot defend on epistemological grounds the distinction between physics and logic. However, Quine does not really provide the philosophic underpinnings needed to support these broad claims.

Quine's thesis is stronger than Duhem's, for where Duhem claimed that the physicist can never be sure that no saving set of auxiliary assumptions exists which, together with the target hypothesis, would entail the actual observational results, Quine seems to hold that saving hypotheses always exist: "Any statement can be held true come what may". Quine's thesis is also more general than Duhem's, for Quine extends Duhem's claim for conventionalism in physics to include the truths of logic as well as *all* of the laws of science.

Starting from somewhat different problems, both Carl Hempel and Thomas S. Kuhn have arrived at conclusions similar to Duhem's and Quine's. Hempel began with the problem of defining theoretical terms. He argued that the positivists were wrong to think that the theoretical terms of science can be explicitly or operationally defined using only observation terms. Instead, the theoretical terms must be introduced into science by the theories themselves, and this means that the theoretical term is, in effect, implicitly defined not by observation terms but by the theory. However, because statements are deducible from the theory which do not contain the theoretical terms in question, the theory as a whole can be said to have empirical significance, and, Hempel thinks, can be confirmed or falsified. He reminds that

It is not correct to speak, as is often done, of 'the experiential meaning' of a term or a sentence in isolation. ... A single sentence in a scientific theory does not, as a rule, entail any observation sentences; consequences asserting the occurrence of certain observable phenomena can be derived from it only by conjoining it with a set of other, subsidiary, hypotheses. Of the latter, some will usually be observation sentences, others will be previously accepted theoretical statements.[2]

Thus for Hempel, the unit of empirical significance – the unit which is tested – must be only the theory as a whole, where this is evidently taken to include all the possible statements of any kind required for the derivation of observation sentences. Hempel has in effect given a defense of the Duhem–Quine thesis for that part of science which includes theories.

Kuhn's project was to give an account of the nature of the scientific enterprise and the reasons for its special success – an account which would fit the history and practice of science better than what he claims are the obviously inadequate accounts standardly found in science textbooks and in many discussions in the philosophy of science. Kuhn proposed a fundamental change in our perception and evaluation of familiar episodes from the history and practice of science. He suggested that the everyday practice of science – 'normal science' – consists for the most part of a very important kind of puzzle solving in which the nature of the puzzle and the terms set for its solution are not themselves regarded critically or subjected to test by scientists in any significant way. Normal science takes place within holistic 'paradigms' – extremely broad theoretical, metaphysical and methodological models of nature and of how to discover her secrets. In Chapter X of *The Structure of Scientific Revolutions*, the selection included here, Kuhn argues that the theories which are part of scientific paradigms are not refutable by observations at all. This is because, on Kuhn's view, sensory experience is not fixed and neutral and theories are not simply man-made interpretations of given data: "What occurs during a scientific revolution is not fully reducible to a reinterpretation of individual and stable data.... Paradigms are not corrigible by normal science at all". Instead, it seems that the paradigmatic theories in fact define what is to count as a relevant observation; they define the 'world' within which the scientist works. So, a theory is not refuted by experience but instead simply abandoned when there are a large number of problems with it and when a possibly more fruitful way of perceiving the world is at hand. Kuhn claims that his account of science directly challenges the Cartesian paradigm which has guided epistemology and accounts of science for three-hundred years.

The two strongest attacks on the Duhem–Quine thesis have come from Adolf Grünbaum explicitly, and, mainly implicitly, from Karl Popper. In his extraordinarily influential *The Logic of Scientific Discovery*, published in German in 1935, but translated into English only 24 years later, Popper begins by identifying what he takes to be the fundamental problems for philosophy of science and for epistemology. Virtually all problems in epistemology come down to two, Popper says: the problem of induction ('Hume's problem'), and the problem of finding an acceptable criterion of demarcation between science and metaphysics ('Kant's problem'). With respect to Hume's problem, Popper says that the inductivists have simply misconceived the nature of scientific method and of human knowledge. Scientists do not try to prove universal statements true by demonstrating them to be reducible to true singular statements; what scientists do is deduce singular statements from a theory and compare these with the results of experiments. If the comparison is favorable, the theory has temporarily passed its test; if the conclusions of the deductions have been falsified, then their falsification also falsifies the theory from which they were deduced. Therefore, it is falsifiability, not verifiability, which is at the heart of scientific method. Popper claims to dissolve Hume's problem of 'inductivism' by substituting for it a theory of 'deductivism' – a theory of falsifiability.

And it is his theory of falsifiability which provides the demarcation line between science and metaphysics, Popper says, thereby solving Kant's problem. For the positivists at one time, both the universal laws of science and also metaphysical speculations were assigned the status of 'non-statements' – they were 'meaningless', unable to be reduced to elementary statements of experience. But for Popper, while the universal laws of science cannot be verified, they, unlike metaphysical speculations, can be falsified by experience. Thus a theory of the falsifiability of singular statements and of the universal laws from which they can be deduced lies at the heart of Popper's solutions to what he takes to be the two leading problems in epistemology.

In the sixth section of the first selection here – Chapter I of *The Logic of Scientific Discovery* – Popper considers the objection that no theoretical system is ever conclusively falsified for it can always be saved by the logically admissible procedures of "introducing *ad hoc* auxiliary hypotheses, by changing *ad hoc* a definition, or by simply refusing to acknowledge any falsifying experience whatever". Popper proposes that "the empirical method shall be characterized as a method that excludes precisely those ways of evading falsification which... are logically admissible". The method of science is precisely a method insuring that the singular statements deduced from theories can be falsified by experience, thereby transmitting their falsity back up to the theory from which they were deduced. Appended here to Chapter I of *The Logic of Scientific Discovery* are two brief selections from later pages of Popper's book which stress his reasons for arguing that empirical science should be characterized by its methods.

More recently, in the short excerpt from *Conjectures and Refutations* which is reprinted here, Popper argues specifically that the Quine–Duhem thesis, "the holistic view of tests", both "does not create a serious difficulty for the fallibilist and falsificationist" and also that "on the other hand... the holistic argument goes much too far". Evidently he thinks it does not create a difficulty since while the falsificationist does indeed take for granted a vast amount of traditional knowledge,

He does not accept this background knowledge; neither as established nor as fairly certain, nor yet as probable. He knows that even its tentative acceptance is risky, and stresses that every bit of it is open to criticism even though only in a piecemeal way. We can never be certain that we shall challenge the right bit; but since our quest is not for certainty, this does not matter.... Now it has to be admitted that we can often test only a large chunk of a theoretical system, and sometimes perhaps only the whole system, and that, in these cases, it is sheer guesswork which of its ingredients should be held responsible for any falsification.[3]

However, the holistic argument goes too far since it is possible in "quite a few cases to find which hypothesis is responsible for the refutation; or in other words, which part or group of hypotheses was necessary for the derivation of the refuted prediction". Does the fact that in some cases scientists reach intersubjective agreement, at least temporarily, as to which part of their theories to revise save Popper's falsificationism from the Duhem–Quine thesis? The latter would appear to challenge not this uncontroversial sociological fact but the notion that it is tests which determine *which* part of our web of hypotheses and beliefs should be counted as refuted. We must still ask how Popper has succeeded in deflecting the challenge posed to his falsificationism by "the holistic view of tests".

In several publications Adolf Grünbaum has challenged Duhem's thesis that the falsifiability of an isolated empirical hypothesis as an explanans is unavoidably inconclusive. He regards the Duhemian thesis as a conventionalist ploy to be found also in Einstein, Poincaré, and Quine. In his well-known 1960 essay reprinted in this collection, Grünbaum argues that the Duhem–Quine thesis is both a logical non-sequitur and, furthermore, false. First he argues that conclusive falsifying experiments are possible. To deny this; Grünbaum says, Duhem would have to prove on general logical grounds that for any empirical finding whatever (e.g., $\sim O$), there is a set of non-trivial auxiliary hypotheses from which, together with the target hypothesis in question, the findings could be deduced. But this Duhem cannot guarantee. Thus Duhem made a logical error in taking the conventionalist thesis to follow from the occasional inconclusiveness of supposedly crucial experiments. But, furthermore, Grünbaum argues that as a matter of fact crucial falsifying experiments have actually occurred in physics, and he discusses what he thinks is an example of such a case. In his response, Quine himself suggests that he finds the Duhem–Quine thesis as challenged by Grünbaum tenable only if taken trivially. Grünbaum has since discussed this thesis in several places, most recently in the second of his essays reprinted here.

Recently a number of philosophers have thought further about the Duhem–Quine thesis, in large part due to Grünbaum's arguments. Laurens Laudan, Carlo Giannoni, and Gary Wedeking all point out that there are two versions of the Duhem–Quine thesis. There is a stronger one held not by Duhem but probably by Quine; the weaker one, actually held by Duhem, is untouched by Grünbaum's attack.

Grünbaum presumes, Laudan points out, that the burden of proof is on the scientist who refuses to call a refuted hypothesis false to show that his hypothesis can be saved by some suitable auxiliary hypothesis. But Duhem did not make this strong claim but only the weaker one that those who deny the target hypothesis must show that there does not exist an auxiliary hypothesis which would make the target hypothesis compatible with the unforeseen experimental results. Unless such a proof is forthcoming, Laudan points out, "a scientist is logically justified in seeking some rapprochement between his hypothesis and the uncooperative data". Duhem does not make the purported logical blunder ascribed to him by Grünbaum. Laudan goes on to argue that Grünbaum's counter-example purporting to make out a case for conclusive falsification does not hold up either. Grünbaum says that in the history of science one can find cases where a set of hypotheses has been falsified. But Duhem would not disagree with that holistic view, but only with a claim that isolated hypotheses are falsifiable. Furthermore, Grünbaum assumes that the fact that in a particular case the auxiliary hypotheses are highly probable forces a scientist to relinquish the target hypothesis. But 'highly probable' is not 'known to be true'; "the demands of prudence do not carry logical weight", Laudan notes.

After criticizing the Grünbaum objections to the Duhemian thesis, Giannoni goes on to defend that balance of the Quinean thesis which is both the same as the original Duhemian thesis and which is not defended by Hempel. Giannoni defends the Duhem–Quine thesis for the part of science depending on the use of operational definitions which refer to measuring instruments, and, in particular, where the measurement is derivative rather than fundamental. In the last part of his essay, Giannoni points to some broader implications of the Duhemian thesis for our conception of scientific knowledge. The Duhemian thesis is not an epistemological thesis regarding our knowledge of the world, he says, but a semantical thesis regarding the meaning of scientific words and of scientific language. But to say this is not to trivialize the thesis, he thinks, for the thesis is required by the very notion of scientific discovery. This and other considerations lead Giannoni to suggest that the question about the nature of Duhemian conventionalism in physics comes down to a choice between a realistic or a nominalistic approach to the symbols of physics. And this in turn leads him to conclude that the Duhem–Quine thesis shows that the distinction between descriptive simplicity and inductive simplicity is not a logical distinction but an ontological one. Where a realist searches for inductive simplicity, a nominalist searches for descriptive simplicity.

Wedeking argues that Popper as well as Grünbaum are examples of the "empirically-minded philosophers" who think that there are empirical theories which are 'falsifiable' in the sense that each such theory unambiguously separates out those basic statements which it prohibits and those which it permits. After discussing Grünbaum's position, Wedeking tries to formulate the Quinean version of the Duhem–Quine thesis in such a way that its validity can be tested. But, he can turn up only the pragmatic formulation that the new theoretical system in which the target hypothesis is held true in light of the unforeseen observations must be one which is 'adequate' to the facts of experience. Thus Wedeking doubts the truth of the Quinean thesis in any but its most trivial form. Does this mean, he asks, that Quine's theory is worthless? No, for it leads us to observe that those cases in which Quine's thesis holds only trivially and those in which we are confronted with a real and significant alternative as to which sentences we should reject as false and which we should retain as true "differ only in a matter of degree".

The next two selections are extraordinarily rich and complex essays which, along with Feyerabend's essay, set the Duhem–Quine thesis and its criticisms in the broadest philosophic context. Hesse shows how

Duhem introduces two important modifications into classical empiricism: a new theory of correspondence and a new theory of coherence. She argues that Quine has taken up both aspects of Duhem's new empiricism; and, after countering both Popper's and Grünbaum's objections to the Duhem–Quine thesis, she points to what she thinks are two residual and significant problems with Duhem's and Quine's 'network theory'. First, some theory of relative empirical confirmation must be given, and this must be done without identifying any statement of the system which expresses the evidence incorrigibly. Secondly, a way must be provided for analysing stability and change of meaning in the network. Hesse proposes solutions to these problems, and then marks five features which distinguish the new Duhem–Quine empiricism, as she has now developed it, from the older, classical empiricism. First, she points out how the difference between theoretical and observational aspects of science is not espistemological but pragmatic and causal. Second, on the Duhem–Quine theory, empirical applications of observation predicates are not incorrigible, and the empirical laws taken to hold between them are not infallible. Third, in order to avoid the meaning variance paradoxes, the majority of descriptive predicates must be intersubjectively stable. However, we cannot know which observation predicates will retain stability of meaning in subsequent theories, just as we cannot know ahead of time which observation statements will be retained as true in later theories. Fourth, she thinks that we must assume some prior principles of selection for picking well-confirmed theories, and some criteria for shifts of applicability of some observation predicates if we are to avoid complete arbitrariness in adoption of a new 'best' theory. Finally, while some aspects of the epistemological problem are systematically conflated with causal mechanisms in the new empiricism, this does not result in a fatal circularity for the network model. She concludes that while the new Duhem–Quine view of science can provide no guarantee that our knowledge of this world is firmly based, that demand of the old empiricists was evidently unreasonable anyway.

Lakatos thinks that Kuhn, in particular, faced with the problems with both the older empiricist views and with what he calls 'naive falsificationism', abandons efforts to give a rational explanation of the success of science. Kuhn settles, he says, for merely trying to explain changes in paradigms in terms of social psychology. Lakatos proposes, instead, to follow Popper's lead and find a workable criterion of rational progress in the growth of knowledge; he presents a theory of 'progressive problem shifts'. On Lakatos' theory, the history of science is not the history of scientific theories but the history of series of theories – the history of 'research programmes'. A research programme consists of methodological rules: a negative heuristic tells us what paths of research to avoid, a positive heuristic tells us what paths to pursue. In particular, the negative heuristic forbids us to direct our *modus tollens* at the 'hard core' of a series of theories. The positive heuristic consists of a partially articulated set of suggestions or hints on how to develop the 'refutable variants' of the research programme, how to modify and sophisticate the refutable protective belt of auxiliary hypotheses. Lakatos claims that his demarcation between progressive and degenerating problem shifts is almost identical with Popper's celebrated demarcation criterion between science and metaphysics.

Like Popper, Lakatos does not think that the 'weak form' of the Duhem–Quine thesis conflicts with methodological falsificationism (though it does with 'naive falsificationism'). But a strong form, which he thinks Quine probably holds, is inconsistent with all forms of methodological falsificationism because, as he sees it, it excludes any rational selection rules among the alternative theories. Do Lakatos and Popper take the conventionalism of Duhem and Quine to involve a greater component of arbitrariness in the construction and selection of theories than would be espoused by either Duhem or Quine?

In the next section, Grünbaum in part modifies his original critique of the Duhem–Quine thesis. At least in some cases, he now holds, we can ascertain the falsity of a component hypothesis to all scientific intents and purposes, although we cannot falsify it beyond any and all possibility of subsequent rehabilitation. Grünbaum here replies to the discussions of his earlier position by Laudan, Giannoni, Hesse, Lakatos and others.

The essays above have shown how the Duhem–Quine thesis provides the cutting edge for an attack on the empiricist reliance on the authority of the senses and, more generally, for a criticism of Cartesian foundationalism, or authoritarianism, in epistemology. Many would take Feyerabend to be the most extreme anti-Cartesian today, and in the last essay in this collection, one can detect attacks on two forms of what might be called the 'methodological authoritarianism' which characterizes much of epistemology since Descartes. In its historically extremely influential form, the 'naive inductivist' holds that there are indeed rational principles for the discovery of nature's regularities, and that these are identical with the rational principles for justifying the accepted theories of science at any particular time. The naive inductivist has the idea that there exist definite principles and rules of a quasi-logical sort for inducing theories from phenomena. But recent philosophers of science as different as Hempel and Popper have retreated to a second position: science definitely can be reconstructed as a rational enterprise, but there may well be no principles for a prospective methodology of science. However, Feyerabend holds that there are no rationally defensible rules for either a prospective methodology of science or for a retrospective assessment of science.[4] In the selection presented here, the last five sections of his long (110 pp) essay 'Against Method', Feyerabend presents four objections to the first two positions, and suggests that the persistent attempt of epistemologists to identify a rational 'authority' – a firm source of knowledge or a sure method for discriminating between reliable and unreliable hypotheses – is not entirely separable from authoritarianism in values.

Against the naive inductivist, Feyerabend argues that since the history of science reveals no rationally justifiable rules for a prospective methodology, it should be granted that there are no rationally justifiable rules for retrospective assessments of science either. Second, Feyerabend points out that any proposed reconstructions of science which seem to us to be reasonably rational actually end up being useless, for they only succeed in eliminating all of science as irrational. Third, Feyerabend repeats his well-known criticisms of the purported asymmetry between observation statements and theoretical statements. It is not the case that observations are irrefutable while theories are refutable. While we often do test our theories by experience, we equally often assess experience in light of new or newer theories, changing our assessment of experience accordingly. Thus, for Feyerabend, research is an interaction between new theories explicitly stated and older beliefs which lie deep in the observation language. Finally, Feyerabend presents his well-known arguments for the incommensurability of theories.

Throughout Feyerabend's writing, and here in particular, can be found a conscious and conscientious attempt to push philosophy of science into an ethical dimension. Feyerabend sees epistemological and ethical preferences as tightly linked. For one thing, someone, or a culture, which prefers or is led to prefer a secure, routine, entirely predictable way of life will naturally be led to accept an epistemology characterized by unchanging, 'objective', guaranteed facts and by a theory of how, using just the right methods, we can apprehend and collect them. Ways of life and theories of knowledge are mutually supporting, Feyerabend suggests. And, the choice between alternative theories of knowledge becomes a normative act, the conscious selection of a certain way of understanding the world.

What exactly are the ideals of rationality which Feyerabend is here attacking? Is he in fact appealing to other, hidden ideals of rationality in his attack on Cartesianism? Is he proposing that there is an ideal of rationality which lies wholly or in part outside the forces which have shaped the history of science, outside the forces which produce maximum 'progress' in science? What account can be given of *this* notion of rationality?

With this concluding essay, the consequences of the Duhem–Quine thesis have been extended not only to challenge Cartesian epistemology, but also to open the possibility of a reconsideration of the link between epistemology, on the one hand, and ethics and political theory, on the other hand. The Duhem–Quine thesis may well take its place in the history of ideas as signaling a radical change in our understanding of the nature of both human knowledge and human knowers.

SANDRA G. HARDING

NOTES

[1] *Modus tollens* is represented by the schema $[(H \rightarrow O) \cdot \sim O] \rightarrow \sim H$. If, from a given hypothesis, H, we predict a certain observation, O, then, if the prediction turns out to be false, $\sim O$, that would serve to refute the hypothesis from which the prediction was deduced, $\sim H$.

[2] Carl G. Hempel, 'Empiricist Criteria of Cognitive Significance: Problems and Changes', *below*, p. 75.

[3] Karl R. Popper, 'Truth, Rationality, and the Growth of Scientific Knowledge', *Conjectures and Refutations*, Routledge and Kegan Paul, London, 1963, p. 238–239, *below* p. 114.

[4] C. A. Hooker has also set Feyerabend's philosophy of science in this context. Cf. his interesting review of 'Against Method': 'Critical Notice of Minnesota Studies in the Philosophy of Science', Part II, *Canadian Journal of Philosophy* 1 (1972).

PIERRE DUHEM

PHYSICAL THEORY AND EXPERIMENT*

1. The Experimental Testing of a Theory Does Not Have the Same Logical Simplicity in Physics as in Physiology

The sole purpose of physical theory is to provide a representation and classification of experimental laws; the only test permitting us to judge a physical theory and pronounce it good or bad is the comparison between the consequences of this theory and the experimental laws it has to represent and classify. Now that we have minutely analyzed the characteristics of a physical experiment and of a physical law, we can establish the principles that should govern the comparison between experiment and theory; we can tell how we shall recognize whether a theory is confirmed or weakened by facts.

When many philosophers talk about experimental sciences, they think only of sciences still close to their origins, e.g., physiology or certain branches of chemistry where the experimenter reasons directly on the facts by a method which is only common sense brought to greater attentiveness but where mathematical theory has not yet introduced its symbolic representations. In such sciences the comparison between the deductions of a theory and the facts of experiment is subject to very simple rules. These rules were formulated in a particularly forceful manner by Claude Bernard, who would condense them into a single principle, as follows:

"The experimenter should suspect and stay away from fixed ideas, and always preserve his freedom of mind".

"The first condition that has to be fulfilled by a scientist who is devoted to the investigation of natural phenomena is to preserve a complete freedom of mind based on philosophical doubt."[1]

If a theory suggests experiments to be done, so much the better: "... we can follow our judgment and our thought, give free rein to our imagination provided that all our ideas are only pretexts for instituting new experiments that may furnish us probative facts or unexpected and fruitful ones".[2] Once the experiment is done and the results clearly established, if a theory takes them over in order to generalize them, coordinate them, and draw from them new subjects for experiment, still so much the better: "... if one is imbued with the principles of experimental method, there is nothing to fear; for so long as the idea is a right one, it will go on being developed; when it is an erroneous idea, experiment is there to correct it". [3] But so long as the experiment lasts, the theory should remain waiting, under strict orders to stay outside the door of the laboratory; it should keep silent and leave the scientist without disturbing him while he faces the facts directly; the facts must be observed without a preconceived idea and gathered with the same scrupulous impartiality, whether they confirm or contradict the predictions of the theory. The report that the observer will give us of his experiment should be a faithful and scrupulously exact reproduction of the phenomena, and should not let us even guess what system the scientist places his confidence in or distrusts.

Men who have an excessive faith in their theories or in their ideas are not only poorly disposed to make discoveries but they also make very poor observations. They necessarily observe with a preconceived idea and, when they have begun an experiment, they want to see in its results only a confirmation of their theory. Thus they distort observation and often neglect very important facts because they go counter to their goal. That is what made us say elsewhere that we must never do experiments in order to confirm our ideas but merely to check them.... But it quite naturally happens that those who believe too much in their own theories do not sufficiently believe in the theories of others. Then the dominant idea of these condemners of others is to find fault with the theories of the latter and to seek to contradict them. The setback for science remains the same. They are doing experiments only in order to destroy a theory instead of doing them in order to look for the truth. They also make poor observations because they take into the results of their experiments only what fits their purpose, by neglecting what is unrelated to it, and by very carefully avoiding whatever might go in the direction of the idea they wish to combat. Thus one is led by two parallel paths to the same result, that is to say, to falsifying science and the facts.

The conclusion of all this is that it is necessary to obliterate one's opinion as well as that of others when faced with the decisions of the experiment; ... we must accept the results of experiment just as they present themselves with all that is unforeseen and accidental in them.[4]

Here, for example, is a physiologist who admits that the anterior roots of the spinal nerve contain the motor nerve-fibers and the posterior roots the sensory fibers. The theory he accepts leads him to imagine an experiment: if he cuts a certain anterior root, he ought to be suppressing the mobility of a certain part of the body without destroying its sensibility; after making the section of this root, when he observes the consequences of his operation and when he makes a report of it, he must put aside all his ideas concerning the physiology of the spinal nerve; his report must be a raw description of the facts; he is not permitted to overlook or fail to mention any movement or quiver contrary to his predictions or to attribute it to some secondary cause unless some special experiment has given evidence of this cause; he must, if he does not wish to be accused of scientific bad faith, establish an absolute separation or watertight compartment between the consequences of his theoretical deductions and the establishing of the facts shown by his experiments.

Such a rule is not by any means easily followed; it requires of the scientist an absolute detachment from his own thought and a complete absence of animosity when confronted with the opinion of another person; neither vanity nor envy ought to be countenanced by him. As Bacon put it, he should never show eyes lustrous with human passions. Freedom of mind, which constitutes the sole principle of experimental method, according to Claude Bernard, does not depend merely on intellectual conditions, but also on moral conditions, making its practice rarer and more meritorious.

But if experimental method as just described is difficult to practice, the logical analysis of it is very simple. This is no longer the case when the theory to be subjected to test by the facts is not a theory of physiology but a theory of physics. In the latter case, in fact, it is impossible to leave outside the laboratory door the theory that we wish to test, for without theory it is impossible to regulate a single instrument or to interpret a single reading. We have seen that in the mind of the physicist there are contantly present two sorts of apparatus: one is the concrete apparatus in glass and metal, manipulated by him, the other is the schematic and abstract apparatus which theory substitutes for the concrete apparatus and on which the physicist does his reasoning. For these two ideas are indissolubly connected in his intelligence, and each necessarily calls on the other; the physicist can no sooner conceive the concrete apparatus without associating with it the idea of the schematic apparatus than a Frenchman can conceive an idea without associating it with the French word expressing it. This radical impossibility, preventing one from dissociating physical theories from the experimental procedures appropriate for testing these theories, complicates this test in a singular way, and obliges us to examine the logical meaning of it carefully.

Of course, the physicist is not the only one who appeals to theories at the very time he is experimenting or reporting the results of his experiments. The chemist and the physiologist when they make use of physical instruments, e.g., the thermometer, the manometer, the calorimeter, the galvanometer, and the saccharimeter, implicitly admit the accuracy of the theories justifying the use of these pieces of apparatus as well as of the theories giving meaning to the abstract ideas of temperature, pressure, quantity of heat, intensity of current, and polarized light, by means of which the concrete indications of these instruments are translated. But the theories used, as well as the instruments employed, belong to the domain of physics; by accepting with these instruments the theories without which their readings would be devoid of meaning, the chemist and the physiologist show their confidence in the physicist, whom they suppose to be infallible. The physicist, on the other hand, is obliged to trust his own theoretical ideas or those of his fellow-physicists. From the standpoint of logic, the difference is of little importance; for the physiologist and chemist as well as for the physicist, the statement of the result of an experiment implies, in general, an act of faith in a whole group of theories.

2. AN EXPERIMENT IN PHYSICS CAN NEVER CONDEMN AN ISOLATED HYPOTHESIS BUT ONLY A WHOLE THEORETICAL GROUP

The physicist who carries out an experiment, or gives a report of one, implicitly recognizes the accuracy of a whole group of theories. Let us accept this principle and see what consequences we may deduce from it when we seek to estimate the role and logical import of a physical experiment.

In order to avoid any confusion we shall distinguish two sorts of experiments: experiments of *application*, which we shall first just mention, and experiments of *testing*, which will be our chief concern.

You are confronted with a problem in physics to be solved practically; in order to produce a certain effect you wish to make use of knowledge acquired by physicists; you wish to light an incandescent bulb; accepted theories indicate to you the means for solving the problem; but to make use of these means you have to secure certain information; you ought, I suppose, to determine the electromotive force of the battery of generators at your disposal; you measure this electromotive force: that is what I call an experiment of application. This experiment does not aim at discovering whether accepted theories are accurate or not; it merely intends to draw on these theories. In order to carry it out, you make use of instruments that these same theories legitimize; there is nothing to shock logic in this procedure.

But experiments of application are not the only ones the physicist has to perform; only with their aid can science aid practice, but it is not through them that science creates and develops itself; besides experiments of application, we have experiments of testing.

A physicist disputes a certain law; he calls into doubt a certain theoretical point. How will he justify these doubts? How will he demonstrate the inaccuracy of the law? From the proposition under indictment he will derive the prediction of an experimental fact; he will bring into existence the conditions under which this fact should be produced; if the predicted fact is not produced, the proposition which served as the basis of the prediction will be irremediably condemned.

F. E. Neumann assumed that in a ray of polarized light the vibration is parallel to the plane of polarization, and many physicists have doubted this proposition. How did O. Wiener undertake to transform this doubt into a certainty in order to condemn Neumann's proposition? He deduced from this proposition the following consequence: If we cause a light beam reflected at 45° from a plate of glass to interfere with the incident beam polarized perpendicularly to the plane of incidence, there ought to appear alternately dark and light interference bands parallel to the reflecting surface; he brought about the conditions under which these bands should have been produced and showed that the predicted phenomenon did not appear, from which he concluded that Neumann's proposition is false, viz., that in a polarized ray of light the vibration is not parallel to the plane of polarization.

Such a mode of demonstration seems as convincing and as irrefutable as the proof by reduction to absurdity customary among mathematicians; moreover, this demonstration is copied from the reduction to absurdity, experimental contradiction playing the same role in one as logical contradiction plays in the other.

Indeed, the demonstrative value of experimental method is far from being so rigorous or absolute: the conditions under which it functions are much more complicated than is supposed in what we have just said; the evaluation of results is much more delicate and subject to caution.

A physicist decides to demonstrate the inaccuracy of a proposition; in order to deduce from this proposition the prediction of a phenomenon and institute the experiment which is to show whether this phenomenon is or is not produced, in order to interpret the results of this experiment and establish that the predicted phenomenon is not produced, he does not confine himself to making use of the proposition in question; he makes use also of a whole group of theories accepted by him as beyond dispute. The prediction of the phenomenon, whose nonproduction is to cut off debate, does not derive from the proposition challenged if taken by itself, but from the proposition at issue joined to that whole group of theories; if the predicted phenomenon is not produced, not only is the proposition questioned at fault, but so is the whole theoretical scaffolding used by the physicist. The only thing the experiment teaches us is that among the propositions used to predict the phenomenon and to establish whether it would be produced, there is at least one error; but where this error lies is just what it does not tell us. The physicist may declare that this error is contained in exactly the proposition he wishes to refute, but is he sure it is not in another proposition? If he is, he accepts implicitly the accuracy of all the other propositions he has used, and the validity of his conclusion is as great as the validity of his confidence.

Let us take as an example the experiment imagined by Zenker and carried out by O. Wiener. In order to predict the formation of bands in certain circumstances and to show that these did not appear, Wiener did not make use merely of the famous proposition of F. E. Neumann, the proposition which he wished to refute; he did not merely admit that in a polarized ray vibrations are parallel to the plane of polarization; but he used, besides this, propositions, laws, and hypotheses constituting the optics commonly accepted: he admitted that light consists in simple periodic vibrations, that these vibrations are normal to the light ray, that at each point the mean kinetic energy of the vibratory motion is a measure of the intensity of light, that the more or less complete attack of the gelatine coating on a photographic plate indicates the various degrees of this intensity. By joining these propositions, and many others that would take too long to enumerate, to Neumann's proposition, Wiener was able to formulate a forecast and establish that the experiment belied it. If he attributed this solely to Neumann's proposition, if it alone bears the responsibility for the error this negative result has put in evidence, then

Wiener was taking all the other propositions he invoked as beyond doubt. But this assurance is not imposed as a matter of logical necessity; nothing stops us from taking Neumann's proposition as accurate and shifting the weight of the experimental contradiction to some other proposition of the commonly accepted optics; as H. Poincaré has shown, we can very easily rescue Neumann's hypothesis from the grip of Wiener's experiment on the condition that we abandon in exchange the hypothesis which takes the mean kinetic energy as the measure of the light intensity; we may, without being contradicted by the experiment, let the vibration be parallel to the plane of polarization, provided that we measure the light intensity by the mean potential energy of the medium deforming the vibratory motion.

These principles are so important that it will be useful to apply them to another example; again we choose an experiment regarded as one of the most decisive ones in optics.

We know that Newton conceived the emission theory for optical phenomena. The emission theory supposes light to be formed of extremely thin projectiles, thrown out with very great speed by the sun and other sources of light; these projectiles penetrate all transparent bodies; on account of the various parts of the media through which they move, they undergo attractions and repulsions; when the distance separating the acting particles is very small these actions are very powerful, and they vanish when the masses between which they act are appreciably far from each other. These essential hypotheses joined to several others, which we pass over without mention, lead to the formulation of a complete theory of reflection and refraction of light; in particular, they imply the following proposition: The index of refraction of light passing from one medium into another is equal to the velocity of the light projectile within the medium it penetrates, divided by the velocity of the same projectile in the medium it leaves behind.

This is the proposition that Arago chose in order to show that the theory of emission is in contradiction with the facts. From this proposition a second follows: Light travels faster in water than in air. Now Arago had indicated an appropriate procedure for comparing the velocity of light in air with the velocity of light in water; the procedure, it is true, was inapplicable, but Foucault modified the experiment in such a way that it could be carried out; he found that the light was propagated less rapidly in water than in air. We may conclude from this, with Foucault, that the system of emission is incompatible with the facts.

I say the *system* of emission and not the *hypothesis* of emission; in fact, what the experiment declares stained with error is the whole group of propositions accepted by Newton, and after him by Laplace and Biot, that is, the whole theory from which we deduce the relation between the index of refraction and the velocity of light in various media. But in condemning this system as a whole by declaring it stained with error, the experiment does not tell us where the error lies. Is it in the fundamental hypothesis that light consists in projectiles thrown out with great speed by luminous bodies? Is it in some other assumption concerning the actions experienced by light corpuscles due to the media through which they move? We know nothing about that. It would be rash to believe, as Arago seems to have thought, that Foucault's experiment condemns once and for all the very hypothesis of emission, i.e., the assimilation of a ray of light to a swarm of projectiles. If physicists had attached some value to this task, they would undoubtedly have succeeded in founding on this assumption a system of optics that would agree with Foucault's experiment.

In sum, the physicist can never subject an isolated hypothesis to experimental test, but only a whole group of hypotheses; when the experiment is in disagreement with his predictions, what he learns is that at least one of the hypotheses constituting this group is unacceptable and ought to be modified; but the experiment does not designate which one should be changed.

We have gone a long way from the conception of the experimental method arbitrarily held by persons unfamiliar with its actual functioning. People generally think that each one of the hypotheses employed in physics can be taken in isolation, checked by experiment, and then, when many varied tests have established its validity, given a definitive place in the system of physics. In reality, this is not the case. Physics is not a machine which lets itself be taken apart; we cannot try each piece in isolation and, in order to adjust it, wait until its solidity has been carefully checked. Physical science is a system that must be taken as a whole; it is an organism in which one part cannot be made to function except when the parts that are most remote from it are called into play, some more so than others, but all to some degree. If something goes wrong, if some discomfort is felt in the functioning of the organism, the physicist will have to ferret out through its effect on the entire system which organ needs to be remedied or modified without the possibility of isolating this organ and examining it apart. The watchmaker to whom you give a watch that has stopped separates all the wheelworks and examines them one by one until he finds the part that is defective or broken. The doctor to whom a patient appears cannot dissect him in order to establish his diagnosis; he has to guess the seat and cause of the ailment solely by inspecting disorders affecting the whole body. Now, the physicist concerned with remedying a limping theory resembles the doctor and not the watchmaker.

3. A "CRUCIAL EXPERIMENT" IS IMPOSSIBLE IN PHYSICS

Let us press this point further, for we are touching on one of the essential features of experimental method, as it is employed in physics.

Reduction to absurdity seems to be merely a means of refutation, but it may become a method of demonstration: in order to demonstrate the truth of a proposition it suffices to corner anyone who would admit the contradictory of the given proposition into admitting an absurd consequence. We know to what extent the Greek geometers drew heavily on this mode of demonstration.

Those who assimilate experimental contradiction to reduction to absurdity imagine that in physics we may use a line of argument similar to the one Euclid employed so frequently in geometry. Do you wish to obtain from a group of phenomena a theoretically certain and indisputable explanation? Enumerate all the hypotheses that can be made to account for this group of phenomena; then, by experimental contradiction eliminate all except one; the latter will no longer be a hypothesis, but will become a certainty.

Suppose, for instance, we are confronted with only two hypotheses. Seek experimental conditions such that one of the hypotheses forecasts the production of one phenomenon and the other the production of quite a different effect; bring these conditions into existence and observe what happens; depending on whether you observe the first or the second of the predicted phenomena, you will condemn the second or the first hypothesis; the hypothesis not condemned will be henceforth indisputable; debate will be cut off, and a new truth will be acquired by science. Such is the experimental test that the author of the *Novum Organum* called the "*fact of the cross*, borrowing this expression from the crosses which at an intersection indicate the various roads".

We are confronted with two hypotheses concerning the nature of light; for Newton, Laplace, or Biot light consisted of projectiles hurled with extreme speed, but for Huygens, Young, or Fresnel light consisted of vibrations whose waves are propagated within an ether. These are the only two possible hypotheses as far as one can see: either the motion is carried away by the body it excites and remains attached to it, or else it passes from one body to another. Let us pursue the first hypothesis; it declares that light travels more quickly in water than in air; but if we follow the second, it declares that light travels more quickly in air than in water. Let us set up Foucault's apparatus; we set into motion the turning mirror; we see two luminous spots formed before us, one colorless, the other greenish. If the greenish band is to the left of the colorless one, it means that light travels faster in water than in air, and that the hypothesis of vibrating waves is false. If, on the contrary, the greenish band is to the right of the colorless one, that means that light travels faster in air than in water, and that the hypothesis of emissions is condemned. We look through the magnifying glass used to examine the two luminous spots, and we notice that the greenish spot is to the right of the colorless one; the debate is over; light is not a body, but a vibratory wave motion propagated by the ether; the emission hypothesis has had its day; the wave hypothesis has been put beyond doubt, and the crucial experiment has made it a new article of the scientific credo.

What we have said in the foregoing paragraph shows how mistaken we should be to attribute to Foucault's experiment so simple a meaning and so decisive an importance; for it is not between two hypotheses, the emission and wave hypotheses, that Foucault's experiment judges trenchantly; it decides rather between two sets of theories each of which has to be taken as a whole, i.e., between two entire systems, Newton's optics and Huygens' optics.

But let us admit for a moment that in each of these systems everything is compelled to be necessary by strict logic, except a single hypothesis; consequently, let us admit that the facts, in condemning one of the two systems, condemn once and for all the single doubtful assumption it contains. Does it follow that we can find in the 'crucial experiment' an irrefutable procedure for transforming one of the two hypotheses before us into a demonstrated truth? Between two contradictory theorems of geometry there is no room for a third judgment; if one is false, the other is necessarily true. Do two hypotheses in physics ever constitute such a strict dilemma? Shall we ever dare to assert that no other hypothesis is imaginable? Light may be a swarm of projectiles, or it may be a vibratory motion whose waves are propagated in a medium; is it forbidden to be anything else at all? Arago undoubtedly thought so when he formulated this incisive alternative: Does light move more quickly in water than in air? "Light is a body. If the contrary is the case, then light is a wave". But it would be difficult for us to take such a decisive stand; Maxwell, in fact, showed that we might just as well attribute light to a periodical electrical disturbance that is propagated within a dielectric medium.

Unlike the reduction to absurdity employed by geometers, experimental contradiction does not have the power to transform a physical hypothesis into an indisputable truth; in order to confer this power on it, it would be necessary to enumerate completely the various hypotheses which may cover a determinate group of phenomena; but the physicist is never sure he has exhausted all the imaginable assumptions. The truth of a physical theory is not decided by heads or tails.

4. CRITICISM OF THE NEWTONIAN METHOD. FIRST EXAMPLE: CELESTIAL MECHANICS

It is illusory to seek to construct by means of experimental contradiction a line of argument in imitation of the reduction to absurdity; but the geometer is acquainted with other methods for attaining certainty than the method of reducing to an absurdity; the direct demonstration in which the truth of a proposition is established by itself and not by the refutation of the contradictory proposition seems to him the most perfect of arguments. Perhaps physical theory would be more fortunate in its attempts if it sought to imitate direct demonstration. The hypotheses from which it starts and develops its conclusions would then be tested one by one; none would have to be accepted until it presented all the certainty that experimental method can confer on an abstract and general proposition; that is to say, each would necessarily be either a law drawn from observation by the sole use of those two intellectual operations called induction and generalization, or else a corollary mathematically deduced from such laws. A theory based on such hypotheses would then not present anything arbitrary or doubtful; it would deserve all the confidence merited by the faculties which serve us in formulating natural laws.

It was this sort of physical theory that Newton had in mind when, in the 'General Scholium' which crowns his *Principia*, he rejected so vigorously as outside of natural philosophy any hypothesis that induction did not extract from experiment; when he asserted that in a sound physics every proposition should be drawn from phenomena and generalized by induction.

The ideal method we have just described therefore deserves to be named the Newtonian method. Besides, did not Newton follow this method when he established the system of universal attraction, thus adding to his precepts the most magnificent of examples? Is not his theory of gravitation derived entirely from the laws which were revealed to Kepler by observation, laws which problematic reasoning transforms and whose consequences induction generalizes?

This first law of Kepler's, "The radial vector from the sun to a planet sweeps out an area proportional to the time during which the planet's motion is observed", did, in fact, teach Newton that each planet is constantly subjected to a force directed toward the sun.

The second law of Kepler's, "The orbit of each planet is an ellipse having the sun at one focus", taught him that the force attracting a given planet varies with the distance of this planet from the sun, and that it is in an inverse ratio to the square of this distance.

The third law of Kepler's, "The squares of the periods of revolution of the various planets are proportional to the cubes of the major axes of their orbits", showed him that different planets would, if they were brought to the same distance from the sun, undergo in relation to it attractions proportional to their respective masses.

The experimental laws established by Kepler and transformed by geometric reasoning yield all the characteristics present in the action exerted by the sun on a planet; by induction Newton generalized the result obtained; he allowed this result to express the law according to which any portion of matter acts on any other portion whatsoever, and he formulated this great principle: "Any two bodies whatsoever attract each other with a force which is proportional to the product of their masses and in inverse ratio to the square of the distance between them". The principle of universal gravitation was found, and it was obtained, without any use having been made of any fictive hypothesis, by the inductive method the plan of which Newton outlined.

Let us again examine this application of the Newtonian method, this time more closely; let us see if a somewhat strict logical analysis will leave intact the appearance of rigor and simplicity that this very summary exposition attributes to it.

In order to assure this discussion of all the clarity it needs, let us begin by recalling the following principle, familiar to all those who deal with mechanics: We cannot speak of the force which attracts a body in given circumstances before we have designated the supposedly fixed term of reference to which we relate the motion of all bodies; when we change this point of reference or term of comparison, the force representing the effect produced on the observed body by the other bodies surrounding it changes in direction and magnitude according to the rules stated by mechanics with precision.

That posited, let us follow Newton's reasoning.

Newton first took the sun as the fixed point of reference; he considered the motions affecting the different planets by reference to the sun; he admitted Kepler's laws as governing these motions, and derived the following proposition: If the sun is the point of reference in relation to which all forces are compared, each planet is subjected to a force directed toward the sun, a force proportional to the mass of the planet and to the inverse square of its distance from the sun. Since the latter is taken as the reference point, it is not subject to any force.

In an analogous manner Newton studied the motion of the satellites and for each of these he chose as a fixed reference point the planet which the satellite accompanies, the earth in the case of the moon, Jupiter in the case of the masses moving around Jupiter. Laws just like Kepler's were taken as governing these motions, from which it follows that we can formulate the following proposition: If we take as a fixed reference point the planet accompanied by a satellite, this satellite is subject to a force directed toward the planet varying inversely with the square of the distance. If, as happens with Jupiter, the same planet possesses several satellites, these satellites, were they at the same distance from the planet, would be acted on by the latter with forces proportional to their respective masses. The planet is itself not acted on by the satellite.

Such, in very precise form, are the propositions which Kepler's laws of planetary motion and the extension of these laws to the motions of satellites authorize us to formulate. For these propositions Newton substituted another which may be stated as follows: Any two celestial bodies whatsoever exert on each other a force of attraction in the direction of the straight line joining them, a force proportional to the product of their masses and to the inverse square of the distance between them. This statement presupposes all motions and forces to be related to the same reference point; the latter is an ideal standard of reference which may well be conceived by the geometer but which does not characterize in an exact and concrete manner the position in the sky of any body.

Is this principle of universal gravitation merely a generalization of the two statements provided by Kepler's laws and their extension to the motion of satellites? Can induction derive it from these two statements? Not at all. In fact, not only is it more general than these two statements and unlike them, but it contradicts them. The student of mechanics who accepts the principle of universal attraction can calculate the magnitude and direction of the forces between the various planets and the sun when the latter is taken as the reference point, and if he does he finds that these forces are not what our first statement would require. He can determine the magnitude and direction of each of the forces between Jupiter and its satellites when we refer all the motions to the planet, assumed to be fixed, and if he does he notices that these forces are not what our second statement would require.

The principle of universal gravity, very far from being derivable by generalization and induction from the observational laws of Kepler, formally contradicts these laws. If Newton's theory is correct, Kepler's laws are necessarily false.

Kepler's laws based on the observation of celestial motions do not transfer their immediate experimental certainty to the principle of universal weight, since if, on the contrary, we admit the absolute exactness of Kepler's laws, we are compelled to reject the proposition on which Newton based his celestial mechanics. Far from adhering to Kepler's laws, the physicist who claims to justify the theory of universal gravitation finds that he has, first of all, to resolve a difficulty in these laws: he has to prove that his theory, incompatible with the exactness of Kepler's laws, subjects the motions of the planets and satellites to other laws scarcely different enough from the first laws for Tycho Brahé, Kepler, and their contemporaries to have been able to discern the deviations between the Keplerian and Newtonian orbits. This proof derives from the circumstances that the sun's mass is very large in relation to the masses of the various planets and the mass of a planet is very large in relation to the masses of its satellites.

Therefore, if the certainty of Newton's theory does not emanate from the certainty of Kepler's laws, how will this theory prove its validity? It will calculate, with all the high degree of approximation that the constantly perfected methods of algebra involve, the perturbations which at each instant remove every heavenly body from the orbit assigned to it by Kepler's laws; then it will compare the calculated perturbations with the perturbations observed by means of the most precise instruments and the most scrupulous methods. Such a comparison will not only bear on this or that part of the Newtonian principle, but will involve all its parts at the same time; with those it will also involve all the principles of dynamics; besides, it will call in the aid of all the propositions of optics, the statics of gases, and the theory of heat, which are necessary to justify the properties of telescopes in their construction, regulation, and correction, and in the elimination of the errors caused by diurnal or annual aberration and by atmospheric refraction. It is no longer a matter of taking, one by one, laws justified by observation, and raising each of them by induction and generalization to the rank of a principle; it is a matter of comparing the corollaries of a whole group of hypotheses to a whole group of facts.

Now, if we seek out the causes which have made the Newtonian method fail in this case for which it was imagined and which seemed to be the most perfect application for it, we shall find them in that double character of any law made use of by theoretical physics: This law is symbolic and approximate.

Undoubtedly, Kepler's laws bear quite directly on the very objects of astronomical observation; they are as little symbolic as possible. But in this purely experimental form they remain inappropriate for suggesting the principle of universal gravitation; in order to acquire this fecundity they must be transformed and must yield the characters of the forces by which the sun attracts the various planets.

Now this new form of Kepler's laws is a symbolic form; only dynamics gives meanings to the words 'force' and 'mass', which serve to state it, and only dynamics permits us to substitute the new symbolic formulas for the old realistic formulas, to substitute statements relative to 'forces' and 'masses' for laws relative to orbits. The legitimacy of such a substitution implies full confidence in the laws of dynamics.

And in order to justify this confidence let us not proceed to claim that the laws of dynamics were beyond doubt at the time Newton made use of them in symbolically translating Kepler's laws; that they had received enough empirical confirmation to warrant the support of reason. In fact, the laws of dynamics had been subjected up to that time to only very limited and very crude tests. Even their enunciations had remained very vague and involved; only in Newton's *Principia* had they been for the first time formulated in a precise manner. It was in the agreement of the facts with the celestial mechanics which Newton's labors gave birth to that they received their first convincing verification.

Thus the translation of Kepler's laws into symbolic laws, the only kind useful for a theory, presupposed the prior adherence of the physicist to a whole group of hypotheses. But, in addition, Kepler's laws being only approximate laws, dynamics permitted giving them an infinity of different symbolic translations. Among these various forms, infinite in number, there is one and only one which agrees with Newton's principle. The observations of Tycho Brahé, so felicitously reduced to laws by Kepler, permit the theorist to choose this form, but they do not constrain him to do so, for there is an infinity of others they permit him to choose.

The theorist cannot, therefore, be content to invoke Kepler's laws in order to justify his choice. If he wishes to prove that the principle he has adopted is truly a principle of natural classification for celestial motions, he must show that the observed perturbations are in agreement with those which had been calculated in advance; he has to show how from the course of Uranus he can deduce the existence and position of a new planet, and find Neptune in an assigned direction at the end of his telescope.

5. CRITICISM OF THE NEWTONIAN METHOD (CONTINUED). SECOND EXAMPLE: ELECTRODYNAMICS

Nobody after Newton except Ampère has more clearly declared that all physical theory should be derived from experience by induction only; no work has been more closely modelled after Newton's *Philosophiae naturalis Principia mathematica* than Ampère's *Théorie mathématique des phénomènes électrodynamiques uniquement déduite de l'expérience.*

The epoch marked by the works of Newton in the history of the sciences is not only one of the most important discoveries that man has made concerning the causes of the great phenomena of nature, but it is also the epoch in which the human mind opened a new route in the sciences whose object is the study of these phenomena.

These are the lines with which Ampère began the exposition of his *Théorie mathématique*; he continued in the following terms:

"Newton was far from thinking" that the law of universal weight

could be discovered by starting from more or less plausible abstract considerations. He established the fact that it had to be deduced from observed facts, or rather from those empirical laws which, like those of Kepler, are but results generalized from a great number of facts.

To observe the facts first, to vary their circumstances as far as possible, to make precise measurements along with this first task in order to deduce from them general laws based only on experience, and to deduce from these laws, independently of any hypothesis about the nature of the forces producing the phenomena, the mathematical value of these forces, i.e., the formula representing them – that is the course Newton followed. It has been generally adopted in France by the scientists to whom physics owes the enormous progress it has made in recent times, and it has served me as a guide in all my research on electrodynamic phenomena. I have consulted only experience in order to establish the laws of these phenomena, and I have deduced from them the formula which can only represent the forces to which they are due; I have made no investigation about the cause itself assignable to these forces, well convinced that any investigation of this kind should be preceded simply by experimental knowledge of the laws and of the determination, deduced solely from these laws, of the value of the elementary force.

Neither very close scrutiny nor great perspicacity is needed in order to recognize that the *Théorie mathématique des phénomènes électrodynamiques* does not in any way proceed according to the method prescribed by Ampère and to see that it is not "deduced only from experience" (*uniquement déduite de l'expérience*). The facts of experience taken in their primitive rawness cannot serve mathematical reasoning; in order to feed this reasoning they have to be transformed and put into a symbolic form. This transformation Ampère did make them undergo. He was not content merely with reducing the metal apparatus in which currents flow to simple geometric figures; such an assimilation imposes itself too naturally to give way to any serious doubt. Neither was he content merely to use the notion of force, borrowed from mechanics, and various theorems constituting this science; at the time he wrote, these theorems might be considered as beyond dispute. Besides all this, he appealed to a whole set of entirely new hypotheses which are entirely gratuitous and sometimes even rather surprising. Foremost among these hypotheses it is appropriate to mention the intellectual operation by which he decomposed into infinitely small elements the electric current, which, in reality, cannot be broken without ceasing to exist; then the supposition that all real electrodynamic actions are resolved into fictive actions involving the pairs that the elements of current form, one pair at a time; then the postulate that the mutual actions of two elements are reduced to two forces applied to the elements in the direction of the straight line joining them, forces equal and opposite in direction; then the postulate that the distance between two elements enters simply into the formula of their mutual action by the inverse of a certain power.

These diverse assumptions are so little self-evident and so little necessary that several of them have been criticized or rejected by Ampère's successors; other hypotheses equally capable of translating symbolically the fundamental experiments of electrodynamics have been proposed by other physicists, but none of them has succeeded in giving this translation without formulating some new postulate, and it would be absurd to claim to do so.

The necessity which leads the physicist to translate experimental facts symbolically before introducing them into his reasoning, renders the purely inductive path Ampère drew impracticable; this path is also forbidden to him because each of the observed laws is not exact but merely approximate.

Ampère's experiments have the grossest degree of approximation. He gave a symbolic translation of the facts observed in a form appropriate for the success of his theory, but how easily he might have taken advantage of the uncertainty of the observations in order to give quite a different translation! Let us listen to Wilhelm Weber:

Ampère made a point of expressly indicating in the title of his memoir that his mathematical theory of electrodynamic phenomena is *deduced only from experiment*, and indeed in his book we find expounded in detail the simple as well as ingenious method which led him to his goal. There we find, presented with all the precision and scope desirable, the exposition of his experiments, the deductions that he draws from them for theory, and the description of the instruments he employs. But in fundamental experiments, such as we have here, it is not enough to indicate the general meaning of an experiment, to describe the instruments used in performing it, and to tell in a general way that it has yielded the result expected; it is indispensable to go into the details of the experiment itself, to say how often it has been repeated, how the conditions were modified, and what the effect of these modifications has been; in a word, to compose a sort of brief of all the circumstances permitting the reader to sit in judgment on the degree of reliability and certainty of the result. Ampère does *not* give these precise details concerning his experiments, and the demonstration of the fundamental law of electrodynamics still awaits this indispensable supplementation. The fact of the mutual attraction of two conducting wires has been verified over an over again and is beyond all dispute; but these verifications have always been made under conditions and by such means that no *quantitative* measurement was possible and these measurements are far from having reached the degree of precision required for considering the law of these phenomena demonstrated.

More than once, Ampère has drawn from the *absence* of any electrodynamic action the same consequences as from a measurement that would have given him a result equal to *zero*, and by this artifice, with great sagacity and with even greater skill, he has succeeded in bringing together the data necessary for the establishment and demonstration of his theory; but these *negative* experiments with which we must be content in the absence of direct *positive* measurements,

those experiments in which all passive resistances, all friction, all causes of error tend precisely to produce the effect we wish to observe,

cannot have all the value or demonstrative force of those positive measurements, especially when they are not obtained with the procedures and under the conditions of true measurement, which are moreover impossible to obtain with the instruments Ampère has employed.[5]

Experiments with so little precision leave the physicist with the problem of choosing between an infinity of equally possible symbolic translations, and confer no certainty on a choice they do not impose; only intuition, guessing the form of theory to be established, directs this choice. This role of intuition is particularly important in the work of Ampère; it suffices to run through the writings of this great geometer in order to recognize that his fundamental formula of electrodynamics was found quite completely by a sort of divination, that his experiments were thought up by him as afterthoughts and quite purposefully combined so that he might be able to expound according to the Newtonian method a theory that he had constructed by a series of postulates.

Besides, Ampère had too much candor to dissimulate very learnedly that what was artificial in his exposition was *entirely deduced from experi-*

ment; at the end of his *Théorie mathématique des phénomènes électro-dynamiques* he wrote the following lines: "I think I ought to remark in finishing this memoir that I have not yet had the time to construct the instruments represented in Diagram 4 of the first plate and in Diagram 20 of the second plate. The experiments for which they were intended have not yet been done." Now the first of the two sets of apparatus in question aimed to bring into existence the last of the four fundamental cases of equilibrium which are like columns in the edifice constructed by Ampère: it is with the aid of the experiment for which this apparatus was intended that we were to determine the power of the distance according to which electrodynamic actions proceed. Very far from its being the case that Ampère's electrodynamic theory was *entirely deduced from experiment*, experiment played a very feeble role in its formation: it was merely the occasion which awakened the intuition of this physicist of genius, and his intuition did the rest.

It was through the research of Wilhelm Weber that the very intuitive theory of Ampère was first subjected to a detailed comparison with the facts; but this comparison was not guided by the Newtonian method. Weber deduced from Ampère's theory, taken as a whole, certain effects capable of being calculated; the theorems of statics and of dynamics, and also even certain propositions of optics, permitted him to conceive an apparatus, the electrodynamometer, by means of which these same effects may be subjected to precise measurements; the agreement of the calculated predictions with the results of the measurements no longer, then, confirms this or that isolated proposition of Ampère's theory, but the whole set of electrodynamical, mechanical, and optical hypotheses that must be invoked in order to interpret each of Weber's experiments.

Hence, where Newton had failed, Ampère in his turn just stumbled. That is because two inevitable rocky reefs make the purely inductive course impracticable for the physicist. In the first place, no experimental law can serve the theorist before it has undergone an interpretation transforming it into a symbolic law; and this interpretation implies adherence to a whole set of theories. In the second place, no experimental law is exact but only approximate, and is therefore susceptible to an infinity of distinct symbolic translations; and among all these translations the physicist has to choose one which will provide him with a fruitful hypothesis without his choice being guided by experiment at all.

This criticism of the Newtonian method brings us back to the conclusions to which we have already been led by the criticism of experimental contradiction and of the crucial experiment. These conclusions merit our formulating them with the utmost clarity. Here they are:

To seek to separate each of the hypotheses of theoretical physics from the other assumptions on which this science rests in order to subject it in isolation to observational test is to pursue a chimera; for the realization and interpretation of no matter what experiment in physics imply adherence to a whole set of theoretical propositions.

The only experimental check on a physical theory which is not illogical consists in comparing the *entire system of the physical theory with the whole group of experimental laws,* and in judging whether the latter is represented by the former in a satisfactory manner.

6. CONSEQUENCES RELATIVE TO THE TEACHING OF PHYSICS

Contrary to what we have made every effort to establish, it is generally accepted that each hypothesis of physics may be separated from the group and subjected in isolation to experimental test. Of course, from this erroneous principle false consequences are deduced concerning the method by which physics should be taught. People would like the professor to arrange all the hypotheses of physics in a certain order, to take the first one, enounce it, expound its experimental verifications, and then when the latter have been recognized as sufficient, declare the hypothesis accepted. Better still, people would like him to formulate this first hypothesis by inductive generalization of a purely experimental law; he would begin this operation again on the second hypothesis, on the third, and so on until all of physics was constituted. Physics would be taught as geometry is: hypotheses would follow one another as theorems follow one another; the experimental test of each assumption would replace the demonstration of each proposition; nothing which is not drawn from facts or immediately justified by facts would be promulgated.

Such is the ideal which has been proposed by many teachers, and which several perhaps think they have attained. There is no lack of authoritative voices inviting them to the pursuit of this ideal. M. Poincaré says:

It is important not to multiply hypotheses excessively, but to make them only one after the other. If we construct a theory based on multiple hypotheses, and experiment

condemns the theory, which one among our premises is it necessary to change? It will
be impossible to know. And if, on the other hand, the experiment succeeds, shall we
think we have verified all these hypotheses at the same time? Shall we think we have
determined several unknowns with a single equation?[6]

In particular, the purely inductive method whose laws Newton formulated is given by many physicists as the only method permitting one to
expound rationally the science of nature. Gustave Robin says:

The science we shall make will be only a combination of simple inductions suggested
by experience. As to these inductions, we shall formulate them always in propositions
easy to retain and *susceptible of direct verification*, never losing sight of the fact that *a
hypothesis cannot be verified by its consequences*.[7]

This is the Newtonian method recommended if not prescribed for those
who plan to teach physics in the secondary schools. They are told:

The procedures of mathematical physics are not adequate for secondary-school instruction, for they consist in starting from hypotheses or from definitions posited a
priori in order to deduce from them conclusions which will be subjected to experimental check. This method may be suitable for specialized classes in mathematics, but
it is wrong to apply it at present in our elementary courses in mechanics, hydrostatics,
and optics. Let us replace it by the inductive method.[8]

The arguments we have developed have established more than sufficiently the following truth: It is as impracticable for the physicist to follow the inductive method whose practice is recommended to him as it is
for the mathematician to follow that perfect deductive method which
would consist in defining and demonstrating everything, a method of
inquiry to which certain geometers seem passionately attached, although
Pascal properly and rigorously disposed of it a long time ago. Therefore,
it is clear that those who claim to unfold the series of physical principles
by means of this method are naturally giving an exposition of it that is
faulty at some point.

Among the vulnerable points noticeable in such an exposition, the
most frequent and, at the same time, the most serious, because of the false
ideas it deposits in the minds of students, is the 'fictitious experiment'.
Obliged to invoke a principle which has not really been drawn from facts
or obtained by induction, and averse, moreover, to offering this principle
for what it is, namely, a postulate, the physicist invents an imaginary experiment which, were it carried out with success, would possibly lead to
the principle whose justification is desired.

To invoke such a fictitious experiment is to offer an experiment to be done for an experiment done; this is justifying a principle not by means of facts observed but by means of facts whose existence is predicted, and this prediction has no other foundation than the belief in the principle supported by the alleged experiment. Such a method of demonstration implicates him who trusts it in a vicious circle; and he who teaches it without making it exactly clear that the experiment cited has not been done commits an act of bad faith.

At times the fictitious experiment described by the physicist could not, if we attempted to bring it about, yield a result of any precision; the very indecisive and rough results it would produce could undoubtedly be put into agreement with the proposition claimed to be warranted; but they would agree just as well with certain very different propositions; the demonstrative value of such an experiment would therefore be very weak and subject to caution. The experiment that Ampère imagined in order to prove that electrodynamic actions proceed according to the inverse square of the distance, but which he did not perform, gives us a striking example of such a fictitious experiment.

But there are worse things. Very often the fictitious experiment invoked is not only not realized but incapable of being realized; it presupposes the existence of bodies not encountered in nature and of physical properties which have never been observed. Thus Gustave Robin, in order to give the principles of chemical mechanics the purely inductive exposition that he wishes, creates at will what he calls witnessing bodies (*corps témoins*), bodies which by their presence alone are capable of agitating or stopping a chemical reaction.[9] Observation has never revealed such bodies to chemists.

The unperformed experiment, the experiment which would not be performed with precision, and the absolutely unperformable experiment do not exhaust the diverse forms assumed by the fictitious experiment in the writings of physicists who claim to be following the experimental method; there remains to be pointed out a form more illogical than all the others, namely, the absurd experiment. The latter claims to prove a proposition which is contradictory if regarded as the statement of an experimental fact.

The most subtle physicists have not always known how to guard against the intervention of the absurd experiment in their expositions. Let us quote, for instance, some lines taken from J. Bertrand:

If we accept it as an experimental fact that electricity is carried to the surface of bodies, and as a necessary principle that the action of free electricity on the points of conductors should be null, we can deduce from these two conditions, supposing they are strictly satisfied, that electrical attractions and repulsions are inversely proportional to the square of the distance.[10]

Let us take the proposition "There is no electricity in the interior of a conducting body when electrical equilibrium is established in it", and let us inquire whether it is possible to regard it as the statement of an experimental fact. Let us weigh the exact sense of the words figuring in the statement, and particularly, of the word interior. In the sense we must give this word in this proposition, a point interior to a piece of electrified copper is a point taken within the mass of copper. Consequently, how can we go about establishing whether there is or is not any electricity at this point? It would be necessary to place a testing body there, and to do that it would be necessary to take away beforehand the copper that is there, but then this point would no longer be within the mass of copper; it would be outside that mass. We cannot without falling into a logical contradiction take our proposition as a result of observation.

What, therefore, is the meaning of the experiments by which we claim to prove this proposition? Certainly, something quite different from what we make them say. We hollow out a cavity in a conducting mass and note that the walls of this cavity are not charged. This observation proves nothing concerning the presence or absence of electricity at points deep within the conducting mass. In order to pass from the experimental law noted to the law stated we play on the word interior. Afraid to base electrostatics on a postulate, we base it on a pun.

If we simply turn the pages of the treatises and manuals of physics we can collect any number of fictitious experiments; we should find there abundant illustrations of the various forms that such an experiment can assume, from the merely unperformed experiment to the absurd experiment. Let us not waste time on such a fastidious task. What we have said suffices to warrant the following conclusion: The teaching of physics by the purely inductive method such as Newton defined it is a chimera. Whoever claims to grasp this mirage is deluding himself and deluding his pupils. He is giving them, as facts seen, facts merely foreseen; as precise observations, rough reports; as performable procedures, merely ideal experiments; as experimental laws, propositions whose terms cannot be taken as real without contradiction. The physics he expounds is false and falsified.

Let the teacher of physics give up this ideal inductive method which proceeds from a false idea, and reject this way of conceiving the teaching of experimental science, a way which dissimulates and twists its essential character. If the interpretation of the slightest experiment in physics presupposes the use of a whole set of theories, and if the very description of this experiment requires a great many abstract symbolic expressions whose meaning and correspondence with the facts are indicated only by theories, it will indeed be necessary for the physicist to decide to develop a long chain of hypotheses and deductions before trying the slightest comparison between the theoretical structure and the concrete reality; also, in describing experiments verifying theories already developed, he will very often have to anticipate theories to come. For example, he will not be able to attempt the slightest experimental verification of the principles of dynamics before he has not only developed the chain of propositions of general mechanics but also laid the foundations of celestial mechanics; and he will also have to suppose as known, in reporting the observations verifying this set of theories, the laws of optics which alone warrant the use of astronomical instruments.

Let the teacher therefore develop, in the first place, the essential theories of the science; without doubt, by presenting the hypotheses on which these theories rest, it is necessary for him to prepare their acceptance; it is good for him to point out the data of common sense, the facts gathered by ordinary observation or simple experiments or those scarcely analyzed which have led to formulating these hypotheses. To this point, moreover, we shall insist on returning in the next chapter; but we must proclaim loudly that these facts sufficient for suggesting hypotheses are not sufficient to verify them; it is only after he has constituted an extensive body of doctrine and constructed a complete theory that he will be able to compare the consequences of this theory with experiment.

Instruction ought to get the student to grasp this primary truth: Experimental verifications are not the base of theory but its crown. Physics does not make progress in the way geometry does: the latter grows by the continual contribution of a new theorem demonstrated once and for all and added to theorems already demonstrated; the former is a symbolic painting in which continual retouching gives greater comprehensiveness and unity, and the *whole* of which gives a picture resembling more and more the *whole* of the experimental facts, whereas each detail of the picture cut off and isolated from the whole loses all meaning and no longer represents anything.

To the student who will not have perceived this truth, physics will appear as a monstrous confusion of fallacies of reasoning in circles and begging the question; if he is endowed with a mind of high accuracy, he will repel with disgust these perpetual defiances of logic; if he has a less accurate mind, he will learn by heart here words with inexact meaning, these descriptions of unperformed and unperformable experiments, and lines of reasoning which are sleight-of-hand passes, thus losing in such unreasoned memory work the little correct sense and critical mind he used to possess.

The student who, on the other hand, will have seen clearly the ideas we have just formulated will have done more than learned a certain number of propositions of physics; he will have understood the nature and true method of experimental science.[11]

7. Consequences relative to the mathematical development of physical theory

Through the preceding discussions the exact nature of physical theory and of its relations with experiment emerge more and more clearly and precisely.

The materials with which this theory is constructed are, on the one hand, the mathematical symbols serving to represent the various quantities and qualities of the physical world, and, on the other hand, the general postulates serving as principles. With these materials theory builds a logical structure; in drawing the plan of this structure it is hence bound to respect scrupulously the laws that logic imposes on all deductive reasoning and the rules that algebra prescribes for any mathematical operation.

The mathematical symbols used in theory have meaning only under very definite conditions; to define these symbols is to enumerate these conditions. Theory is forbidden to make use of these signs outside these conditions. Thus, an absolute temperature by definition can be positive only, and by definition the mass of a body is invariable; never will theory in its formulas give a zero or negative value to absolute temperature, and never in its calculations will it make the mass of a given body vary.

Theory is in principle grounded on postulates, that is to say, on propositions that it is at leisure to state as it pleases, provided that no contradiction exists among the terms of the same postulate or between two distinct postulates. But once these postulates are set down it is bound to guard them with jealous rigor. For instance, if it has placed at the base of its system the principle of the conservation of energy, it must forbid any assertion in disagreement with this principle.

These rules bring all their weight to bear on a physical theory that is being constructed; a single default would make the system illogical and would oblige us to upset it in order to reconstruct another; but they are the only limitations imposed. IN THE COURSE OF ITS DEVELOPMENT, *a physical theory is free to choose any path it pleases provided that it avoids any logical contradiction; in particular, it is free not to take account of experimental facts.*

This is no longer the case WHEN THE THEORY HAS REACHED ITS COMPLETE DEVELOPMENT. When the logical structure has reached its highest point it becomes necessary to compare the set of mathematical propositions obtained as conclusions from these long deductions with the set of experimental facts; by employing the adopted procedures of measurement we must be sure that the second set finds in the first a sufficiently similar image, a sufficiently precise and complete symbol. If this agreement between the conclusions of theory and the facts of experiment were not to manifest a satisfactory approximation, the theory might well be logically constructed, but it should nonetheless be rejected because it would be contradicted by observation, because it would be *physically* false.

This comparison between the conclusions of theory and the truths of experiment is therefore indispensable, since only the test of facts can give physical validity to a theory. But this test by facts should bear exclusively on the conclusions of a theory, for only the latter are offered as an image of reality; the postulates serving as points of departure for the theory and the intermediary steps by which we go from the postulates to the conclusions do not have to be subject to this test.

We have in the foregoing pages very thoroughly analyzed the error of those who claim to subject one of the fundamental postulates of physics directly to the test of facts through a procedure such as a crucial experiment; and especially the error of those who accept as principles only "in-

ductions consisting exclusively in erecting into general laws not the interpretation but *the very result of a very large number of experiments*".[12]

There is another error lying very close to this one; it consists in requiring that all the operations performed by the mathematician connecting
postulates with conclusions should have *a physical meaning*, in wishing
"to reason only about *performable operations*", and in "introducing only
magnitudes accessible to experiment".[13]

According to this requirement any magnitude introduced by the
physicist in his formulas should be connected through a process of measurement to a property of a body; any algebraic operation performed on
these magnitudes should be translated into concrete language by the
employment of these processes of measurement; thus translated, it should
express a real or possible fact.

Such a requirement, legitimate when it comes to the final formulas at
the end of a theory, has no justification if applied to the intermediary
formulas and operations establishing the transition from postulates to
conclusions.

Let us take an example.

J. Willard Gibbs studied the theory of the dissociation of a perfect
composite gas into its elements, also regarded as perfect gases. A formula
was obtained expressing the law of chemical equilibrium internal to such
a system. I propose to discuss this formula. For this purpose, keeping
constant the pressure supporting the gaseous mixture, I consider the
absolute temperature appearing in the formula and I make it vary from
0 to $+\infty$.

If we wish to attribute a physical meaning to this mathematical operation, we shall be confronted with a host of objections and difficulties. No
thermometer can reveal temperatures below a certain limit, and none can
determine temperatures high enough; this symbol which we call 'absolute
temperature' cannot be translated through the means of measurement at
our disposal into something having a concrete meaning unless its numerical value remains between a certain minimum and a certain maximum.
Moreover, at temperatures sufficiently low this other symbol which thermodynamics calls 'a perfect gas' is no longer even an approximate image
of any real gas.

These difficulties and many others, which it would take too long to
enumerate, disappear if we heed the remarks we have formulated. In the

construction of the theory, the discussion we have just given is only an intermediary step, and there is no justification for seeking a physical meaning in it. Only when this discussion shall have led us to a series of propositions, shall we have to submit these propositions to the test of facts; then we shall inquire whether, within the limits in which the absolute temperature may be translated into concrete readings of a thermometer and the idea of a perfect gas is approximately embodied in the fluids we observe, the conclusions of our discussion agree with the results of experiment.

By requiring that mathematical operations by which postulates produce their consequences shall always have a physical meaning, we set unjustifiable obstacles before the mathematician and cripple his progress. G. Robin goes so far as to question the use of the differential calculus; if Professor Robin is intent on constantly and scrupulously satisfying this requirement, he would practically be unable to develop any calculation; theoretical deduction would be stopped in its tracks from the start. A more accurate idea of the method of physics and a more exact line of demarcation between the propositions which have to submit to factual test and those which are free to dispense with it would give back to the mathematician all his freedom and permit him to use all the resources of algebra for the greatest development of physical theories.

8. ARE CERTAIN POSTULATES OF PHYSICAL THEORY INCAPABLE OF BEING REFUTED BY EXPERIMENT?

We recognize a correct principle by the facility with which it straightens out the complicated difficulties into which the use of erroneous principles brought us.

If, therefore, the idea we have put forth is correct, namely, that comparison is established necessarily between the *whole* of theory and the *whole* of experimental facts, we ought in the light of this principle to see the disappearance of the obscurities in which we should be lost by thinking that we are subjecting each isolated theoretical hypothesis to the test of facts.

Foremost among the assertions in which we shall aim at eliminating the appearance of paradox, we shall place one that has recently been often formulated and discussed. Stated first by G. Milhaud in connection with the '*pure bodies*' of chemistry,[14] it has been developed at length and forcefully by H. Poincaré with regard to principles of mechanics;[15] Edouard Le Roy has also formulated it with great clarity.[16]

That assertion is as follows: Certain fundamental hypotheses of physical theory cannot be contradicted by any experiment, because they constitute in reality *definitions*, and because certain expressions in the physicist's usage take their meaning only through them.

Let us take one of the examples cited by Le Roy:

When a heavy body falls freely, the acceleration of its fall is constant. Can such a law be contradicted by experiment? No, for it constitutes the very definition of what is meant by "falling freely." If while studying the fall of a heavy body we found that this body does not fall with uniform acceleration, we should conclude not that the stated law is false, but that the body does not fall freely, that some cause obstructs its motion, and that the deviations of the observed facts from the law as stated would serve to discover this cause and to analyze its effects.

Thus, M. Le Roy concludes,

laws are verifiable, taking things strictly ..., because they constitute the very criterion by which we judge appearances as well as the methods that it would be necessary to utilize in order to submit them to an inquiry whose precision is capable of exceeding any assignable limit.

Let us study again in greater detail, in the light of the principles previously set down, what this comparison is between the law of falling bodies and experiment.

Our daily observations have made us acquainted with a whole category of motions which we have brought together under the name of motions of heavy bodies; among these motions is the falling of a heavy body when it is not hindered by any obstacle. The result of this is that the words "free fall of a heavy body" have a meaning for the man who appeals only to the knowledge of common sense and who has no notion of physical theories.

On the other hand, in order to classify the laws of motion in question the physicist has created a theory, the theory of weight, an important application of rational mechanics. In that theory, intended to furnish a symbolic representation of reality, there is also the question of "free fall of a heavy body", and as a consequence of the hypotheses supporting this whole scheme free fall must necessarily be a uniformly accelerated motion.

The words "free fall of a heavy body" now have two distinct meanings. For the man ignorant of physical theories, they have their *real* meaning, and they mean what common sense means in pronouncing them; for the physicist they have a *symbolic* meaning, and mean "uniformly accelerated motion". Theory would not have realized its aim if the second meaning were not the sign of the first, if a fall regarded as free by common sense were not also regarded as uniformly accelerated, or *nearly* uniformly accelerated, since common-sense observations are essentially devoid of precision, according to what we have already said.

This agreement, without which the theory would have been rejected without further examination, is finally arrived at: a fall declared by common sense to be nearly free is also a fall whose acceleration is nearly constant. But noticing this crudely approximate agreement does not satisfy us; we wish to push on and surpass the degree of precision which common sense can claim. With the aid of the theory that we have imagined, we put together apparatus enabling us to recognize with sensitive accuracy whether the fall of a body is or is not uniformly accelerated; this apparatus shows us that a certain fall regarded by common sense as a free fall has a slightly variable acceleration. The proposition which in our theory gives its symbolic meaning to the words "free fall" does not represent with sufficient accuracy the properties of the real and concrete fall that we have observed.

Two alternatives are then open to us.

In the first place, we can declare that we were right in regarding the fall studied as a free fall and in requiring that the theoretical definition of these words agree with our observations. In this case, since our theoretical definition does not satisfy this requirement, it must be rejected; we must construct another mechanics on new hypotheses, a mechanics in which the words "free fall" no longer signify "uniformly accelerated motion", but "fall whose acceleration varies according to a certain law".

In the second alternative, we may declare that we were wrong in establishing a connection between the concrete fall we have observed and the symbolic free fall defined by our theory, that the latter was too simplified a scheme of the former, that in order to represent suitably the fall as our experiments have reported it the theorist should give up imagining a weight falling freely and think in terms of a weight hindered by certain obstacles like the resistance of the air, that in picturing the action of these obstacles by means of appropriate hypotheses he will compose a more complicated scheme than a free weight but one more apt to reproduce the details of the experiment; in short, in accord with the language we have previously established (Ch. IV, Sec. 3), we may seek to eliminate by means of suitable "corrections" the "causes of error", such as air resistance, which influenced our experiment.

M. Le Roy asserts that we shall prefer the second to the first alternative, and he is surely right in this. The reasons dictating this choice are easy to perceive. By taking the first alternative we should be obliged to destroy from top to bottom a very vast theoretical system which represents in a most satisfactory manner a very extensive and complex set of experimental laws. The second alternative, on the other hand, does not make us lose anything of the terrain already conquered by physical theory; in addition, it has succeeded in so large a number of cases that we can bank with interest on a new success. But in this confidence accorded the law of fall of weights, we see nothing analogous to the certainty that a mathematical definition draws from its very essence, that is, to the kind of certainty we have when it would be foolish to doubt that the various points on a circumference are all equidistant from the center.

We have here nothing more than a particular application of the principle set down in Section 2 of this chapter. A disagreement between the concrete facts constituting an experiment and the symbolic representation which theory substitutes for this experiment proves that some part of this symbol is to be rejected. But which part? This the experiment does not tell us; it leaves to our sagacity the burden of guessing. Now among the theoretical elements entering into the composition of this symbol there is always a certain number which the physicists of a certain epoch agree in accepting without test and which they regard as beyond dispute. Hence, the physicist who wishes to modify this symbol will surely bring his modification to bear on elements other than those just mentioned.

But what impels the physicist to act thus is *not* logical necessity. It would be awkward and ill inspired for him to do otherwise, but it would not be doing something logically absurd; he would not for all that be walking in the footsteps of the mathematician mad enough to contradict his own definitions. More than this, perhaps some day by acting differently, by refusing to invoke causes of error and take recourse to corrections in order to reestablish agreement between the theoretical scheme and the fact, and by resolutely carrying out a reform among the propositions declared untouchable by common consent, he will accomplish the work of a genius who opens a new career for a theory.

Indeed, we must really guard ourselves against believing forever warranted those hypotheses which have become universally adopted conventions, and whose certainty seems to break through experimental contradiction by throwing the latter back on more doubtful assumptions. The history of physics shows us that very often the human mind has been led to overthrow such principles completely, though they have been regarded by common consent for centuries as inviolable axioms, and to rebuild its physical theories on new hypotheses.

Was there, for instance, a clearer or more certain principle for thousands of years than this one: In a homogeneous medium, light is propagated in a straight line? Not only did this hypothesis carry all former optics, catoptrics, and dioptrics, whose elegant geometric deductions represented at will an enormous number of facts, but it had become, so to speak, the physical definition of a straight line. It is to this hypothesis that any man wishing to make a straight line appeals, the carpenter who verifies the straightness of a piece of wood, the surveyor who lines up his sights, the geodetic surveyor who obtains a direction with the help of the pinholes of his alidade, the astronomer who defines the position of stars by the optical axis of his telescope. However, the day came when physicists tired of attributing to some cause of error the diffraction effects observed by Grimaldi, when they resolved to reject the law of the rectilinear propagation of light and to give optics entirely new foundations; and this bold resolution was the signal of remarkable progress for physical theory.

9. ON HYPOTHESES WHOSE STATEMENT HAS NO EXPERIMENTAL MEANING

This example, as well as others we could add from the history of science, should show that it would be very imprudent for us to say concerning a hypothesis commonly accepted today: "We are certain that we shall never be led to abandon it because of a new experiment, no matter how precise it is". Yet M. Poincaré does not hesitate to enunciate it concerning the principles of mechanics.[17]

To the reasons already given to prove that these principles cannot be reached by experimental refutation, M. Poincaré adds one which seems even more convincing: Not only can these principles not be refuted by experiment because they are the universally accepted rules serving to discover in our theories the weak spots indicated by these refutations, but also, they cannot be refuted by experiment because *the operation which would claim to compare them with the facts would have no meaning.*

Let us explain that by an illustration.

The principle of inertia teaches us that a material point removed from the action of any other body moves in a straight line with uniform motion. Now, we can observe only relative motions; we cannot, therefore, give an experimental meaning to this principle unless we assume a certain point chosen or a certain geometric solid taken as a fixed reference point to which the motion of the material point is related. The fixation of this reference frame constitutes an integral part of the statement of the law, for if we omitted it, this statement would be devoid of meaning. There are as many different laws as there are distinct frames of reference. We shall be stating one law of inertia when we say that the motion of an isolated point assumed to be seen from the earth is rectilinear and uniform, and another when we repeat the same sentence in referring the motion to the sun, and still another if the frame of reference chosen is the totality of fixed stars. But then, one thing is indeed certain, namely, that whatever the motion of a material point is, when seen from a first frame of reference, we can always and in infinite ways choose a second frame of reference such that seen from the latter our material point appears to move in a straight line with uniform motion. We cannot, therefore, attempt an experimental verification of the principle of inertia; false when we refer the motions to one frame of reference, it will become true when selection is made of another term of comparison, and we shall always be free to choose the latter. If the law of inertia stated by taking the earth as a frame of reference is contradicted by an observation, we shall substitute for it the law of inertia whose statement refers the motion to the sun; if the latter in its turn is contraverted, we shall replace the sun in the statement of the law by the system of fixed stars, and so forth. It is impossible to stop this loophole.

The principle of the equality of action and reaction, analyzed at length by M. Poincaré,[18] provides room for analogous remarks. This principle may be stated thus: "The center of gravity of an isolated system can have only a uniform rectilinear motion".

This is the principle that we propose to verify by experiment.

Can we make this verification? For that it would be necessary for isolated systems to exist. Now, these systems do not exist; the only isolated system is the whole universe.

But we can observe only relative motions; the absolute motion of the center of the universe will therefore be forever unknown. We shall never be able to know if it is rectilinear and uniform or, better still, the question has no meaning. Whatever facts we may observe, we shall hence always be free to assume our principle is true.

Thus many a principle of mechanics has a form such that it is absurd to ask one's self: "Is this principle in agreement with experiment or not?" This strange character is not peculiar to the principles of mechanics; it also marks certain fundamental hypotheses of our physical or chemical theories.[19]

For example, chemical theory rests entirely on the "law of multiple proportions"; here is the exact statement of this law:

Simple bodies A, B, and C may by uniting in various proportions form various compounds M, M',.... The masses of the bodies A, B, and C combining to form the compound M are to one another as the three numbers a, b, and c. Then the masses of the elements A, B, and C combining to form the compound M' will be to one another as the numbers xa, yb, and zc (x, y, and z being three whole numbers).

Is this law perhaps subject to experimental test? Chemical analysis will make us acquainted with the chemical composition of the body M' not exactly but with a certain approximation. The uncertainty of the results obtained can be extremely small; it will never be strictly zero. Now, in whatever relations the elements A, B, and C are combined within the compound M', we can always represent these relations, with as close an approximation as you please, by the mutual relations of three products xa, yb, and zc, where x, y, and z are whole numbers; in other words, whatever the results given by the chemical analysis of the compound M', we are always sure to find three integers x, y, and z thanks to which the law of multiple proportions will be verified with a precision greater than that of the experiment. Therefore, no chemical analysis, no matter how refined, will ever be able to show the law of multiple proportions to be wrong.

In like manner, all crystallography rests entirely on the "law of rational indices" which is formulated in the following way:

A trihedral being formed by three faces of a crystal, a fourth face cuts the three edges of this trihedral at distances from the summit which are proportional to one another as three given numbers, the parameters of the crystal. Any other face whatsoever should cut these same edges at distances from the summit which are to one another as xa, yb, and zc, where x, y, and z are three integers, the indices of the new face of the crystal.

The most perfect protractor determines the direction of a crystal's face only with a certain degree of approximation; the relations among the three segments that such a face makes on the edges of the fundamental trihedral are always able to get by with a certain error; now, however small this error is, we can always choose three numbers x, y, and z such that the mutual relations of these segments are represented with the least amount of error by the mutual relations of the three numbers xa, yb, and zc; the crystallographer who would claim that the law of rational indices is made justifiable by his protractor would surely not have understood the very meaning of the words he is employing.

The law of multiple proportions and the law of rational indices are mathematical statements deprived of all physical meaning. A mathematical statement has physical meaning only if it retains a meaning when we introduce the word 'nearly' or 'approximately'. This is not the case with the statements we have just alluded to. Their object really is to assert that certain relations are *commensurable* numbers. They would degenerate into mere truisms if they were made to declare that these relations are approximately commensurable, for any incommensurable relation whatever is always approximately commensurable; it is even as near as you please to being commensurable.

Therefore, it would be absurd to wish to subject certain principles of mechanics to *direct* experimental test; it would be absurd to subject the law of multiple proportions or the law of rational indices to this *direct* test.

Does it follow that these hypotheses placed beyond the reach of direct experimental refutation have nothing more to fear from experiment? That they are guaranteed to remain immutable no matter what discoveries observation has in store for us? To pretend so would be a serious error.

Taken in isolation these different hypotheses have no experimental meaning; there can be no question of either confirming or contradicting them by experiment. But these hypotheses enter as essential foundations into the construction of certain theories of rational mechanics, of chemical theory, of crystallography. The object of these theories is to represent experimental laws; they are schematisms intended essentially to be compared with facts.

Now this comparison might some day very well show us that one of our representations is ill adjusted to the realities it should picture, that the corrections which come and complicate our schematism do not produce sufficient concordance between this schematism and the facts, that the theory accepted for a long time without dispute should be rejected, and that an entirely different theory should be constructed on entirely different or new hypotheses. On that day some one of our hypotheses, which taken in isolation defied direct experimental refutation, will crumble with the system it supported under the weight of the contradictions inflicted by reality on the consequences of this system taken as a whole.[20]

In truth, hypotheses which by themselves have no physical meaning undergo experimental testing in exactly the same manner as other hypotheses. Whatever the nature of the hypothesis is, we have seen at the beginning of this chapter that it is never in isolation contradicted by experiment; experimental contradiction always bears as a whole on the entire group constituting a theory without any possibility of designating which proposition in this group should be rejected.

There thus disappears what might have seemed paradoxical in the following assertion: Certain physical theories rest on hypotheses which do not by themselves have any physical meaning.

10. GOOD SENSE IS THE JUDGE OF HYPOTHESES WHICH OUGHT TO BE ABANDONED

When certain consequences of a theory are struck by experimental contradiction, we learn that this theory should be modified but we are not told by the experiment what must be changed. It leaves to the physicist the task of finding out the weak spot that impairs the whole system. No absolute principle directs this inquiry, which different physicists may conduct in very different ways without having the right to accuse one another of illogicality. For instance, one may be obliged to safeguard certain fundamental hypotheses while he tries to reestablish harmony between the consequences of the theory and the facts by complicating the schematism in which these hypotheses are applied, by invoking various causes of error, and by multiplying corrections. The next physicist, disdainful of these complicated artificial procedures, may decide to change some one of the essential assumptions supporting the entire system. The first physicist does not have the right to condemn in advance the boldness of the second one, nor does the latter have the right to treat the timidity of the first physicist as absurd. The methods they follow are justifiable only by experiment, and if they both succeed in satisfying the requirements of experiment each is logically permitted to declare himself content with the work that he has accomplished.

That does not mean that we cannot very properly prefer the work of one of the two to that of the other. Pure logic is not the only rule for our judgments; certain opinions which do not fall under the hammer of the principle of contradiction are in any case perfectly unreasonable. These motives which do not proceed from logic and yet direct our choices, these "reasons which reason does not know" and which speak to the ample "mind of finesse" but not to the "geometric mind", constitute what is appropriately called good sense.

Now, it may be good sense that permits us to decide between two physicists. It may be that we do not approve of the haste with which the second one upsets the principles of a vast and harmoniously constructed theory whereas a modification of detail, a slight correction, would have sufficed to put these theories in accord with the facts. On the other hand, it may be that we may find it childish and unreasonable for the first physicist to maintain obstinately at any cost, at the price of continual repairs and many tangled-up stays, the worm-eaten columns of a building tottering in every part, when by razing these columns it would be possible to construct a simple, elegant, and solid system.

But these reasons of good sense do not impose themselves with the same implacable rigor that the prescriptions of logic do. There is something vague and uncertain about them; they do not reveal themselves at the same time with the same degree of clarity to all minds. Hence, the possibility of lengthy quarrels between the adherents of an old system and the partisans of a new doctrine, each camp claiming to have good sense on its side, each party finding the reasons of the adversary inadequate. The history of physics would furnish us with innumerable illustrations of these quarrels at all times and in all domains. Let us confine ourselves to the tenacity and ingenuity with which Biot by a continual bestowal of corrections and accessory hypotheses maintained the emissionist doctrine in optics, while Fresnel opposed this doctrine constantly with new experiments favoring the wave theory.

In any event this state of indecision does not last forever. The day arrives when good sense comes out so clearly in favor of one of the two sides that the other side gives up the struggle even though pure logic would not forbid its continuation. After Foucault's experiment had shown that light traveled faster in air than in water, Biot gave up supporting the emission hypothesis; strictly, pure logic would not have compelled him to give it up, for Foucault's experiment was *not* the crucial experiment that Arago thought he saw in it, but by resisting wave optics for a longer time Biot would have been lacking in good sense.

Since logic does not determine with strict precision the time when an inadequate hypothesis should give way to a more fruitful assumption, and since recognizing this moment belongs to good sense, physicists may hasten this judgment and increase the rapidity of scientific progress by trying consciously to make good sense within themselves more lucid and more vigilant. Now nothing contributes more to entangle good sense and to disturb its insight than passions and interests. Therefore, nothing will delay the decision which should determine a fortunate reform in a physical theory more than the vanity which makes a physicist too indulgent towards his own system and too severe towards the system of another. We are thus led to the conclusion so clearly expressed by Claude Bernard: The sound experimental criticism of a hypothesis is subordinated to certain moral conditions; in order to estimate correctly the agreement of a physical theory with the facts, it is not enough to be a good mathematician and skillful experimenter; one must also be an impartial and faithful judge.

NOTES

* Chapter VI of *The Aim and Structure of Physical Theory*, translated by Philip Wiener (copyright © 1954 by Princeton University Press), pp. 180–218. Reprinted by permission of Princeton University Press. Originally published in French in 1906.
1 Claude Bernard, *Introduction à la Médecine expérimentale*, Paris, 1865, p. 63. (Translator's note: Translated into English by H. C. Greene, *An Introduction to Experimental Medicine*, Henry Schuman, New York, 1949.)

[2] Claude Bernard, *Introduction à la Médecine expérimentale*, Paris, 1865, p. 64.

[3] *Ibid.*, p. 70.

[4] *Ibid.*, p. 67.

[5] Wilhelm Weber, *Electrodynamische Maassbestimmungen*, Leipzig, 1846. Translated into French in *Collection de Mémoires relatifs à la Physique* (Société française de Physique), Vol. III: *Mémoires sur l'Electrodynamique*.

[6] H. Poincaré, *Science et Hypothèse*, p. 179.

[7] G. Robin, *Oeuvres scientifiques, Thermodynamique générale*, Paris, 1901, Introduction, p. xii.

[8] Note on a lecture of M. Joubert, inspector-general of secondary-school instruction, *L'Enseignement secondaire*, April 15, 1903.

[9] G. Robin, *op. cit.*, p. ii.

[10] J. Bertrand, *Leçons sur la Théorie mathématique de l'Electricité*, Paris, 1890, p. 71.

[11] It will be objected undoubtedly that such teaching of physics would be hardly accessible to young minds; the answer is simple: Do not teach physics to minds not yet ready to assimilate it. Mme. de Sévigné used to say, speaking of young children: "Before you give them the food of a truckdriver, find out if they have the stomach of a truckdriver".

[12] G. Robin, *op. cit.*, p. xiv.

[13] *loc. cit.*

[14] G. Milhaud, 'La Science rationnelle', *Revue de Métaphysique et de Morale*, IV, 1896, p. 280. Reprinted in *Le Rationnel*, Paris, 1898, p. 45.

[15] H. Poincaré, 'Sur les Principes de la Mécanique', *Bibliothèque du Congrès International de Philosophie*, III: *Logique et Histoire des Sciences*, Paris, 1901, p. 457; 'Sur la Valeur objective des Théories physiques', *Revue de Métaphysique et de Morale*, X, 1902, p. 263; *La Science et l'Hypothèse*, p. 110.

[16] E. Le Roy, 'Un Positivisme Nouveau', *Revue de Métaphysique et de Morale*, IX, 1901, pp. 143–144.

[17] H. Poincaré, 'Sur les Principes de la Mécanique', *Bibliothèque du Congrès international de Philosophie*, Sec. III: 'Logique et Histoire des Sciences', Paris, 1901, pp. 475, 491.

[18] *Ibid.*, pp. 472ff.

[19] P. Duhem, *Le Mixte et la Combinaison chimique: Essai sur l'évolution d'une idée*, Paris, 1902, pp. 159–161.

[20] At the International Congress of Philosophy held in Paris in 1900, M. Poincaré developed this conclusion: "Thus is explained how experiment may have been able to edify (or suggest) the principles of mechanics, but will never be able to overthrow them". Against this conclusion, M. Hadamard offered various remarks, among them the following: "Moreover, in conformity with a remark of M. Duhem, it is not *an* isolated hypothesis but the whole group of the hypotheses of mechanics that we can try to verify experimentally". *Revue de Métaphysique et de Morale*, VIII, 1900, p. 559.

WILLARD VAN ORMAN QUINE

TWO DOGMAS OF EMPIRICISM*

0. INTRODUCTION

Modern empiricism has been conditioned in large part by two dogmas. One is a belief in some fundamental cleavage between truths which are *analytic*, or grounded in meanings independently of matters of fact, and truths which are *synthetic*, or grounded in fact. The other dogma is *reductionism*: the belief that each meaningful statement is equivalent to some logical construct upon terms which refer to immediate experience. Both dogmas, I shall argue, are ill-founded. One effect of abandoning them is, as we shall see, a blurring of the supposed boundary between speculative metaphysics and natural science. Another effect is a shift toward pragmatism.

1. BACKGROUND FOR ANALYTICITY

Kant's cleavage between analytic and synthetic truths was foreshadowed in Hume's distinction between relations of ideas and matters of fact, and in Leibniz's distinction between truths of reason and truths of fact. Leibniz spoke of the truths of reason as true in all possible worlds. Picturesqueness aside, this is to say that the truths of reason are those which could not possibly be false. In the same vein we hear analytic statements defined as statements whose denials are self-contradictory. But this definition has small explanatory value; for the notion of self-contradictoriness, in the quite broad sense needed for this definition of analyticity, stands in exactly the same need of clarification as does the notion of analyticity itself. The two notions are the two sides of a single dubious coin.

Kant conceived of an analytic statement as one that attributes to its subject no more than is already conceptually contained in the subject. This formulation has two shortcomings: it limits itself to statements of subject-predicate form, and it appeals to a notion of containment which is left at a metaphorical level. But Kant's intent, evident more from the use he makes of the notion of analyticity than from his definition of it, can be restated thus: a statement is analytic when it is true by virtue of meanings and independently of fact. Pursuing this line, let us examine the concept of *meaning* which is presupposed.

Meaning, let us remember, is not to be identified with naming.[1] Frege's example of 'Evening Star' and 'Morning Star', and Russell's of 'Scott' and 'the author of *Waverley*', illustrate that terms can name the same thing but differ in meaning. The distinction between meaning and naming is no less important at the level of abstract terms. The terms '9' and 'the number of the planets' name one and the same abstract entity but presumably must be regarded as unlike in meaning; for astronomical observation was needed, and not mere reflection on meanings, to determine the sameness of the entity in question.

The above examples consist of singular terms, concrete and abstract. With general terms, or predicates, the situation is somewhat different but parallel. Whereas a singular term purports to name an entity, abstract or concrete, a general term does not; but a general term is *true* of an entity, or of each of many, or of none.[2] The class of all entities of which a general term is true is called the *extension* of the term. Now paralleling the contrast between the meaning of a singular term and the entity named, we must distinguish equally between the meaning of a general term and its extension. The general terms 'creature with a heart' and 'creature with kidneys', for example, are perhaps alike in extension but unlike in meaning.

Confusion of meaning with extension, in the case of general terms, is less common than confusion of meaning with naming in the case of singular terms. It is indeed a commonplace in philosophy to oppose intension (or meaning) to extension, or, in a variant vocabulary, connotation to denotation.

The Aristotelian notion of essence was the forerunner, no doubt, of the modern notion of intension or meaning. For Aristotle it was essential in men to be rational, accidental to be two-legged. But there is an important difference between this attitude and the doctrine of meaning. From the latter point of view it may indeed be conceded (if only for the sake of argument) that rationality is involved in the meaning of the word 'man' while two-leggedness is not; but two-leggedness may at the same time be viewed as involved in the meaning of 'biped' while rationality is not. Thus from the point of view of the doctrine of meaning it makes no sense to say of the actual individual, who is at once a man and a biped, that his rationality is essential and his two-leggedness accidental or vice versa. Things had essences, for Aristotle, but only linguistic forms have meaning. Meaning is what essence becomes when it is divorced from the object of reference and wedded to the word.

For the theory of meaning a conspicuous question is the nature of its objects: what sort of things are meanings? A felt need for meant entities may derive from an earlier failure to appreciate that meaning and reference are distinct. Once the theory of meaning is sharply separated from the theory of reference, it is a short step to recognizing as the primary business of the theory of meaning simply the synonymy of linguistic forms and the analyticity of statements; meanings themselves, as obscure intermediary entities, may well be abandoned.[3]

The problem of analyticity then confronts us anew. Statements which are analytic by general philosophical acclaim are not, indeed, far to seek. They fall into two classes. Those of the first class, which may be called *logically true*, are typified by:

(1)   No unmarried man is married.

The relevant feature of this example is that it not merely is true as it stands, but remains true under any and all reinterpretations of 'man' and 'married'. If we suppose a prior inventory of *logical* particles, comprising 'no', 'un-', 'not', 'if', 'then', 'and', etc., then in general a logical truth is a statement which is true and remains true under all reinterpretations of its components other than the logical particles.

But there is also a second class of analytic statements, typified by:

(2)   No bachelor is married.

The characteristic of such a statement is that it can be turned into a logical truth by putting synonyms for synonyms; thus (2) can be turned into (1) by putting 'unmarried man' for its synonym 'bachelor'. We still lack a proper characterization of this second class of analytic statements, and therewith of analyticity generally, inasmuch as we have had in the above description to lean on a notion of "synonymy" which is no less in need of clarification than analyticity itself.

In recent years Carnap has tended to explain analyticity by appeal to what he calls state-descriptions.[4] A state-description is any exhaustive assignment of truth values to the atomic, or noncompound, statements of the language. All other statements of the language are, Carnap assumes, built up of their component clauses by means of the familiar logical devices, in such a way that the truth value of any complex statement is fixed for each state-description by specifiable logical laws. A statement is then explained as analytic when it comes out true under every state-description. This account is an adaptation of Leibniz's "true in all possible worlds". But note that this version of analyticity serves its purpose only if the atomic statements of the language are, unlike 'John is a bachelor' and 'John is married', mutually independent. Otherwise there would be a state-description which assigned truth to 'John is a bachelor' and to 'John is married', and consequently 'No bachelors are married' would turn out synthetic rather than analytic under the proposed criterion. Thus the criterion of analyticity in terms of state-descriptions serves only for languages devoid of extralogical synonym-pairs, such as 'bachelor' and 'unmarried man' – synonym-pairs of the type which give rise to the "second class" of analytic statements. The criterion in terms of state-descriptions is a reconstruction at best of logical truth, not of analyticity.

I do not mean to suggest that Carnap is under any illusions on this point. His simplified model language with its state-descriptions is aimed primarily not at the general problem of analyticity but at another purpose, the clarification of probability and induction. Our problem, however, is analyticity; and here the major difficulty lies not in the first class of analytic statements, the logical truths, but rather in the second class, which depends on the notion of synonymy.

2. Definition

There are those who find it soothing to say that the analytic statements of the second class reduce to those of the first class, the logical truths, by *definition*; 'bachelor', for example, is *defined* as 'unmarried man'. But how do we find that 'bachelor' is defined as 'unmarried man'? Who defined it thus, and when? Are we to appeal to the nearest dictionary, and accept the lexicographer's formulation as law? Clearly this would be to put the cart before the horse. The lexicographer is an empirical scientist, whose business is the recording of antecedent facts; and if he glosses 'bachelor' as 'unmarried man' it is because of his belief that there is a relation of synonymy between those forms, implicit in general or preferred usage prior to his own work. The notion of synonymy presupposed here has still to be clarified, presumably in terms relating to linguistic behavior. Certainly the "definition" which is the lexicographer's report of an observed synonymy cannot be taken as the ground of the synonymy.

Definition is not, indeed, an activity exclusively of philologists. Philosophers and scientists frequently have occasion to "define" a recondite term by paraphrasing it into terms of a more familiar vocabulary. But ordinarily such a definition, like the philologist's, is pure lexicography, affirming a relation of synonymy antecedent to the exposition in hand.

Just what it means to affirm synonymy, just what the interconnections may be which are necessary and sufficient in order that two linguistic forms be properly describable as synonymous, is far from clear; but, whatever these interconnections may be, ordinarily they are grounded in usage. Definitions reporting selected instances of synonymy come then as reports upon usage.

There is also, however, a variant type of definitional activity which does not limit itself to the reporting of preëxisting synonymies. I have in mind what Carnap calls *explication* – an activity to which philosophers are given, and scientists also in their more philosophical moments. In explication the purpose is not merely to paraphrase the definiendum into an outright synonym, but actually to improve upon the definiendum by refining or supplementing its meaning. But even explication, though not merely reporting a preëxisting synonymy between definiendum and definiens, does rest nevertheless on *other* preëxisting synonymies. The matter may be viewed as follows. Any word worth explicating has some contexts which, as wholes, are clear and precise enough to be useful; and the purpose of explication is to preserve the usage of these favored contexts while sharpening the usage of other contexts. In order that a given definition be suitable for purposes of explication, therefore, what is required is not that the definiendum in its antecedent usage be synonymous with the definiens, but just that each of these favored contexts of the definiendum, taken as a whole in its antecedent usage, be synonymous with the corresponding context of the definiens.

Two alternative definientia may be equally appropriate for the purposes of a given task of explication and yet not be synonymous with each other; for they may serve interchangeably within the favored contexts but diverge elsewhere. By cleaving to one of these definientia rather than the other, a definition of explicative kind generates, by fiat, a relation of synonymy between definiendum and definiens which did not hold before. But such a definition still owes its explicative function, as seen, to preëxisting synonymies.

There does, however, remain still an extreme sort of definition which does not hark back to prior synonymies at all: namely, the explicitly conventional introduction of novel notations for purposes of sheer abbreviation. Here the definiendum becomes synonymous with the definiens simply because it has been created expressly for the purpose of being synonymous with the definiens. Here we have a really transparent case of synonymy created by definition; would that all species of synonymy were as intelligible. For the rest, definition rests on synonymy rather than explaining it.

The word 'definition' has come to have a dangerously reassuring sound, owing no doubt to its frequent occurrence in logical and mathematical writings. We shall do well to digress now into a brief appraisal of the role of definition in formal work.

In logical and mathematical systems either of two mutually antagonistic types of economy may be striven for, and each has its peculiar practical utility. On the one hand we may seek economy of practical expression – ease and brevity in the statement of multifarious relations. This sort of economy calls usually for distinctive concise notations for a wealth of concepts. Second, however, and oppositely, we may seek economy in grammar and vocabulary; we may try to find a minimum of basic concepts such that, once a distinctive notation has been appropriated to each of them, it becomes possible to express any desired further concept by mere combination and iteration of our basic notations. This second sort of economy is impractical in one way, since a poverty in basic idioms tends to a necessary lengthening of discourse. But it is practical in another way: it greatly simplifies theoretical discourse *about* the language, through minimizing the terms and the forms of construction wherein the language consists.

Both sorts of economy, though prima facie incompatible, are valuable in their separate ways. The custom has consequently arisen of combining both sorts of economy by forging in effect two languages, the one a part of the other. The inclusive language, though redundant in grammar and vocabulary, is economical in message lengths, while the part, called primitive notation, is economical in grammar and vocabulary. Whole and part are correlated by rules of translation whereby each idiom not in primitive notation is equated to some complex built up of primitive notation. These rules of translation are the so-called *definitions* which appear in formalized systems. They are best viewed not as adjuncts to one language but as correlations between two languages, the one a part of the other.

But these correlations are not arbitrary. They are supposed to show how the primitive notations can accomplish all purposes, save brevity and convenience, of the redundant language. Hence the definiendum and its definiens may be expected, in each case, to be related in one or another of the three ways lately noted. The definiens may be a faithful paraphrase of the definiendum into the narrower notation, preserving a direct synonymy[5] as of antecedent usage; or the definiens may, in the spirit of explication, improve upon the antecedent usage of the definiendum; or finally, the definiendum may be a newly created notation, newly endowed with meaning here and now.

In formal and informal work alike, thus we find that definition – except in the extreme case of the explicitly conventional introduction of new notations – hinges on prior relations of synonymy. Recognizing then that the notion of definition does not hold the key to synonymy and analyticity, let us look further into synonymy and say no more of definition.

3. INTERCHANGEABILITY

A natural suggestion, deserving close examination, is that the synonymy of two linguistic forms consists simply in their interchangeability in all contexts without change of truth value – interchangeability, in Leibniz's phrase, *salva veritate*.[6] Note that synonyms so conceived need not even be free from vagueness, as long as the vaguenesses match.

But it is not quite true that the synonyms 'bachelor' and 'unmarried man' are everywhere interchangeable *salva veritate*. Truths which become false under substitution of 'ummarried man' for 'bachelor' are easily constructed with the help of 'bachelor of arts' or 'bachelor's buttons'; also with the help of quotation, thus:

'Bachelor' has less than ten letters.

Such counterinstances can, however, perhaps be set aside by treating the phrases 'bachelor of arts' and 'bachelor's buttons' and the quotation "bachelor" each as a single indivisible word and then stipulating that the interchangeability *salva veritate* which is to be the touchstone of synonymy is not supposed to apply to fragmentary occurrences inside of a word. This account of synonymy, supposing it acceptable on other counts, has indeed the drawback of appealing to a prior conception of "word" which can be counted on to present difficulties of formulation in its turn. Nevertheless some progress might be claimed in having reduced the problem of synonymy to a problem of wordhood. Let us pursue this line a bit, taking "word" for granted.

The question remains whether interchangeability *salva veritate* (apart from occurrences within words) is a strong enough condition for synonymy, or whether, on the contrary, some heteronymous expressions might be thus interchangeable. Now let us be clear that we are not concerned here with synonymy in the sense of complete identity in psychological associations or poetic quality; indeed no two expressions are synonymous in such a sense. We are concerned only with what may be called *cognitive* synonymy. Just what this is cannot be said without successfully finishing the present study; but we know something about it from the need which arose for it in connection with analyticity in Section 1. The sort of synonymy needed there was merely such that any analytic statement could be turned into a logical truth by putting synonyms for synonyms. Turning the tables and assuming analyticity, indeed, we could explain cognitive synonymy of terms as follows (keeping to the familiar example): to say that 'bachelor' and 'unmarried man' are cognitively synonymous is to say no more nor less than that the statement:

(3)   All and only bachelors are unmarried men

is analytic.[7]

What we need is an account of cognitive synonymy not presupposing analyticity – if we are to explain analyticity conversely with help of cognitive synonymy as undertaken in Section 1. And indeed such an independent account of cognitive synonymy is at present up for consideration, namely, interchangeability *salva veritate* everywhere except within words. The question before us, to resume the thread at last, is whether such interchangeability is a sufficient condition for cognitive synonymy.

We can quickly assure ourselves that it is, by examples of the following sort. The statement:

(4)   Necessarily all and only bachelors are bachelors

is evidently true, even supposing 'necessarily' so narrowly construed as to be truly applicable only to analytic statements. Then, if 'bachelor' and 'unmarried man' are interchangeable *salva veritate*, the result:

(5)   Necessarily all and only bachelors are unmarried men

of putting 'unmarried man' for an occurrence of 'bachelor' in (4) must, like (4), be true. But to say that (5) is true is to say that (3) is analytic, and hence that 'bachelor' and 'unmarried man' are cognitively synonymous.

Let us see what there is about the above argument that gives it its air of hocus-pocus. The condition of interchangeability *salva veritate* varies in its force with variations in the richness of the language at hand. The above argument supposes we are working with a language rich enough to contain the adverb 'necessarily', this adverb being so construed as to yield truth when and only when applied to an analytic statement. But can we condone a language which contains such an adverb? Does the adverb really make sense? To suppose that it does is to suppose that we have already made satisfactory sense of 'analytic'. Then what are we so hard at work on right now?

Our argument is not flatly circular, but something like it. It has the form, figuratively speaking, of a closed curve in space.

Interchangeability *salva veritate* is meaningless until relativized to a language whose extent is specified in relevant respects. Suppose now we consider a language containing just the following materials. There is an indefinitely large stock of one-place predicates (for example, 'F' where 'Fx' means that x is a man) and many-place predicates (for example, 'G' where 'Gxy' means that x loves y), mostly having to do with extralogical subject matter. The rest of the language is logical. The atomic sentences consist each of a predicate followed by one or more variables 'x', 'y', etc.; and the complex sentences are built up of the atomic ones by truth functions ('not', 'and', 'or', etc.) and quantification.[8] In effect such a language enjoys the benefits also of descriptions and indeed singular terms generally, these being contextually definable in known ways.[9] Even abstract singular terms naming classes, classes of classes, etc., are contextually definable in case the assumed stock of predicates includes the two-place predicate of class membership.[10] Such a language can be adequate to classical mathematics and indeed to scientific discourse generally, except insofar as the latter involves debatable devices such as contrary-to-fact conditionals or modal adverbs like 'necessarily'.[11] Now a language of this type is extensional, in this sense: any two predicates which agree extensionally (that is, are true of the same objects) are interchangeable *salva veritate*.[12]

In an extensional language, therefore, interchangeability *salva veritate* is no assurance of cognitive synonymy of the desired type. That 'bachelor' and 'unmarried man' are interchangeable *salva veritate* in an extensional language assures us of no more than that (3) is true. There is no assurance here that the extensional agreement of 'bachelor' and 'unmarried man' rests on meaning rather than merely on accidental matters of fact, as does the extensional agreement of 'creature with a heart' and 'creature with kidneys'.

For most purposes extensional agreement is the nearest approximation to synonymy we need care about. But the fact remains that extensional agreement falls far short of cognitive synonymy of the type required for explaining analyticity in the manner of Section 1. The type of cognitive synonymy required there is such as to equate the synonymy of 'bachelor' and 'unmarried man' with the analyticity of (3), not merely with the truth of (3).

So we must recognize that interchangeability *salva veritate*, if construed in relation to an extensional language, is not a sufficient condition of cognitive synonymy in the sense needed for deriving analyticity in the manner of Section 1. If a language contains an intensional adverb 'necessarily' in the sense lately noted, or other particles to the same effect, then interchangeability *salva veritate* in such a language does afford a sufficient condition of cognitive synonymy; but such a language is intelligible only insofar as the notion of analyticity is already understood in advance.

The effort to explain cognitive synonymy first, for the sake of deriving analyticity from it afterward as in Section 1, is perhaps the wrong approach. Instead we might try explaining analyticity somehow without appeal to cognitive synonymy. Afterward we could doubtless derive cognitive synonymy from analyticity satisfactorily enough if desired. We have seen that cognitive synonymy of 'bachelor' and 'unmarried man' can be explained as analyticity of (3). The same explanation works for any pair of one-place predicates, of course, and it can be extended in obvious fashion to many-place predicates. Other syntactical categories can also be accommodated in fairly parallel fashion. Singular terms may be said to be cognitively synonymous when the statement of identity formed by putting '=' between them is analytic. Statements may be said simply to be cognitively synonymous when their biconditional (the result of joining them by 'if and only if') is analytic.[13] If we care to lump all categories into a single formulation, at the expense of assuming again the notion of "word" which was appealed to early in this section, we can describe any two linguistic forms as cognitively synonymous when the two forms are interchangeable (apart from occurrences within "words") *salva* (no longer *veritate* but) *analyticitate*. Certain technical questions arise, indeed, over cases of ambiguity or homonymy; let us not pause for them, however, for we are already digressing. Let us rather turn our backs on the problem of synonymy and address ourselves anew to that of analyticity.

4. SEMANTICAL RULES

Analyticity at first seemed most naturally definable by appeal to a realm of meanings. On refinement, the appeal to meanings gave way to an appeal to synonymy or definition. But definition turned out to be a will-o'-the-wisp, and synonymy turned out to be best understood only by dint of a prior appeal to analyticity itself. So we are back at the problem of analyticity.

I do not know whether the statement 'Everything green is extended' is analytic. Now does my indecision over this example really betray an incomplete understanding, an incomplete grasp of the "meanings" of 'green' and 'extended'? I think not. The trouble is not with 'green' or 'extended', but with 'analytic'.

It is often hinted that the difficulty in separating analytic statements from synthetic ones in ordinary language is due to the vagueness of ordinary language and that the distinction is clear when we have a precise artificial language with explicit "semantical rules". This, however, as I shall now attempt to show, is a confusion.

The notion of analyticity about which we are worrying is a purported relation between statements and languages: a statement S is said to be *analytic for* a language L, and the problem is to make sense of this relation generally, that is, for variable 'S' and 'L'. The gravity of this problem is not perceptibly less for artificial languages than for natural ones. The problem of making sense of the idiom 'S is analytic for L', with variable 'S' and 'L', retains its stubbornness even if we limit the range of the variable 'L' to artificial languages. Let me now try to make this point evident.

For artificial languages and semantical rules we look naturally to the writings of Carnap. His semantical rules take various forms, and to make my point I shall have to distinguish certain of the forms. Let us suppose, to begin with, an artificial language L_0 whose semantical rules have the form explicitly of a specification, by recursion or otherwise, of all the analytic statements of L_0. The rules tell us that such and such statements, and only those, are the analytic statements of L_0. Now here the difficulty is simply that the rules contain the word 'analytic', which we do not understand! We understand what expressions the rules attribute analyticity to, but we do not understand what the rules attribute to those expressions. In short, before we can understand a rule which begins 'A statement S is analytic for language L_0 if and only if...', we must understand the general relative term 'analytic for'; we must understand 'S is analytic for L' where 'S' and 'L' are variables.

Alternatively we may, indeed, view the so-called rule as a conventional definition of a new simple symbol 'analytic-for-L_0', which might better be written untendentiously as 'K' so as not to seem to throw light on the interesting word 'analytic'. Obviously any number of classes K, M, N, etc. of statements of L_0 can be specified for various purposes or for no purpose; what does it mean to say that K, as against M, N, etc., is the class of the "analytic" statements of L_0?

By saying what statements are analytic for L_0 we explain 'analytic-for-L_0' but not 'analytic', not 'analytic for'. We do not begin to explain the idiom 'S is analytic for L' with variable 'S' and 'L', even if we are content to limit the range of 'L' to the realm of artificial languages.

Actually we do know enough about the intended significance of 'analytic' to know that analytic statements are supposed to be true. Let us then turn to a second form of semantical rule, which says not that such and such statements are analytic but simply that such and such statements are included among the truths. Such a rule is not subject to the criticism of containing the un-understood word 'analytic'; and we may grant for the sake of argument that there is no difficulty over the broader term 'true'. A semantical rule of this second type, a rule of truth, is not supposed to specify all the truths of the language; it merely stipulates, recursively or otherwise, a certain multitude of statements which, along with others unspecified, are to count as true. Such a rule may be conceded to be quite clear. Derivatively, afterward, analyticity can be demarcated thus: a statement is analytic if it is (not merely true but) true according to the semantical rule.

Still there is really no progress. Instead of appealing to an unexplained word 'analytic', we are now appealing to an unexplained phrase 'semantical rule'. Not every true statement which says that the statements of some class are true can count as a semantical rule – otherwise *all* truths would be "analytic" in the sense of being true according to semantical rules. Semantical rules are distinguishable, apparently, only by the fact of appearing on a page under the heading 'Semantical Rules'; and this heading is itself then meaningless.

We can say indeed that a statement is *analytic-for-L_0* if and only if it is true according to such and such specifically appended "semantical rules", but then we find ourselves back at essentially the same case which was originally discussed: 'S is analytic-for-L_0 if and only if...'. Once we seek to explain 'S is analytic for L' generally for variable 'L' (even allowing limitation of 'L' to artificial languages), the explanation 'true according to the semantical rules of L' is unavailing; for the relative term 'semantical rule of' is as much in need of clarification, at least, as 'analytic for'.

It may be instructive to compare the notion of semantical rule with that of postulate. Relative to a given set of postulates, it is easy to say what a postulate is: it is a member of the set. Relative to a given set of semantical rules, it is equally easy to say what a semantical rule is. But given simply a notation, mathematical or otherwise, and indeed as thoroughly understood a notation as you please in point of the translations or truth conditions of its statements, who can say which of its true statements rank as postulates? Obviously the question is meaningless – as meaningless as asking which points in Ohio are starting points. Any finite (or effectively specifiable infinite) selection of statements (preferably true ones, perhaps) is as much a set of postulates as any other. The word 'postulate' is significant only relative to an act of inquiry; we apply the word to a set of statements just insofar as we happen, for the year or the moment, to be thinking of those statements in relation to the statements which can be reached from them by some set of transformations to which we have seen fit to direct our attention. Now the notion of semantical rule is as sensible and meaningful as that of postulate, if conceived in a similarly relative spirit – relative, this time, to one or another particular enterprise of schooling unconversant persons in sufficient conditions for truth of statements of some natural or artificial language L. But from this point of view no one signalization of a subclass of the truths of L is intrinsically more a semantical rule than another; and, if 'analytic' means 'true by semantical rules', no one truth of L is analytic to the exclusion of another.[14]

It might conceivably be protested that an artificial language L (unlike a natural one) is a language in the ordinary sense *plus* a set of explicit semantical rules – the whole constituting, let us say, an ordered pair; and that the semantical rules of L then are specifiable simply as the second component of the pair L. But, by the same token and more simply, we might construe an artificial language L outright as an ordered pair whose second component is the class of its analytic statements; and then the analytic statements of L become specifiable simply as the statements in the second component of L. Or better still, we might just stop tugging at our bootstraps altogether.

Not all the explanations of analyticity known to Carnap and his readers have been covered explicitly in the above considerations, but the extension to other forms is not hard to see. Just one additional factor should be mentioned which sometimes enters: sometimes the semantical rules are in effect rules of translation into ordinary language, in which case the analytic statements of the artificial language are in effect recognized as such from the analyticity of their specified translations in ordinary language. Here certainly there can be no thought of an illumination of the problem of analyticity from the side of the artificial language.

From the point of view of the problem of analyticity the notion of an artificial language with semantical rules is a *feu follet par excellence*. Semantical rules determining the analytic statements of an artificial language are of interest only insofar as we already understand the notion of analyticity; they are of no help in gaining this understanding.

Appeal to hypothetical languages of an artificially simple kind could conceivably be useful in clarifying analyticity, if the mental or behavioral or cultural factors relevant to analyticity – whatever they may be – were somehow sketched into the simplified model. But a model which takes analyticity merely as an irreducible character is unlikely to throw light on the problem of explicating analyticity.

It is obvious that truth in general depends on both language and extralinguistic fact. The statement 'Brutus killed Caesar' would be false if the world had been different in certain ways, but it would also be false if the word 'killed' happened rather to have the sense of 'begat'. Thus one is tempted to suppose in general that the truth of a statement is somehow analyzable into a linguistic component and a factual component. Given this supposition, it next seems reasonable that in some statements the factual component should be null; and these are the analytic statements. But, for all its a priori reasonableness, a boundary between analytic and synthetic statements simply has not been drawn. That there is such a distinction to be drawn at all is an unempirical dogma of empiricists, a metaphysical article of faith.

5. The Verification Theory and Reductionism

In the course of these somber reflections we have taken a dim view first of the notion of meaning, then of the notion of cognitive synonymy, and finally of the notion of analyticity. But what, it may be asked, of the verification theory of meaning? This phrase has established itself so firmly as a catchword of empiricism that we should be very unscientific indeed not to look beneath it for a possible key to the problem of meaning and the associated problems.

The verification theory of meaning, which has been conspicuous in the literature from Peirce onward, is that the meaning of a statement is the method of empirically confirming or infirming it. An analytic statement is that limiting case which is confirmed no matter what.

As urged in Section 1, we can as well pass over the question of meanings as entities and move straight to sameness of meaning, or synonymy. Then what the verification theory says is that statements are synonymous if and only if they are alike in point of method of empirical confirmation or infirmation.

This is an account of cognitive synonymy not of linguistic forms generally, but of statements.[15] However, from the concept of synonymy of statements we could derive the concept of synonymy for other linguistic forms, by considerations somewhat similar to those at the end of Section 3. Assuming the notion of "word", indeed, we could explain any two forms as synonymous when the putting of the one form for an occurrence of the other in any statement (apart from occurrences within "words") yields a synonymous statement. Finally, given the concept of synonymy thus for linguistic forms generally, we could define analyticity in terms of synonymy and logical truth as in Section 1. For that matter, we could define analyticity more simply in terms of just synonymy of statements together with logical truth; it is not necessary to appeal to synonymy of linguistic forms other than statements. For a statement may be described as analytic simply when it is synonymous with a logically true statement.

So, if the verification theory can be accepted as an adequate account of statement synonymy, the notion of analyticity is saved after all. However, let us reflect. Statement synonymy is said to be likeness of method of empirical confirmation or infirmation. Just what are these methods which are to be compared for likeness? What, in other words, is the nature of the relation between a statement and the experiences which contribute to or detract from its confirmation?

The most naive view of the relation is that it is one of direct report. This is *radical reductionism*. Every meaningful statement is held to be translatable into a statement (true or false) about immediate experience. Radical reductionism, in one form or another, well antedates the verification theory of meaning explicitly so called. Thus Locke and Hume held that every idea must either originate directly in sense experience or else be compounded of ideas thus originating; and taking a hint from Tooke we might rephrase this doctrine in semantical jargon by saying that a term, to be significant at all, must be either a name of a sense datum or a compound of such names or an abbreviation of such a compound. So stated, the doctrine remains ambiguous as between sense data as sensory events and sense data as sensory qualities; and it remains vague as to the admissible ways of compounding. Moreover, the doctrine is unnecessarily and intolerably restrictive in the term-by-term critique which it imposes. More reasonably, and without yet exceeding the limits of what I have called radical reductionism, we may take full statements as our significant units – thus demanding that our statements as wholes be translatable into sense-datum language, but not that they be translatable term by term.

This emendation would unquestionably have been welcome to Locke and Hume and Tooke, but historically it had to await an important re-orientation in semantics – the reorientation whereby the primary vehicle of meaning came to be seen no longer in the term but in the statement. This reorientation, seen in Bentham and Frege, underlies Russell's concept of incomplete symbols defined in use,[16] also it is implicit in the verification theory of meaning, since the objects of verification are statements.

Radical reductionism, conceived now with statements as units, set it-self the task of specifying a sense-datum language and showing how to translate the rest of significant discourse, statement by statement, into it. Carnap embarked on this project in the *Aufbau*.

The language which Carnap adopted as his starting point was not a sense-datum language in the narrowest conceivable sense, for it included also the notations of logic, up through higher set theory. In effect it in-cluded the whole language of pure mathematics. The ontology implicit in it (that is, the range of values of its variables) embraced not only sensory events but classes, classes of classes, and so on. Empiricists there are who would boggle at such prodigality. Carnap's starting point is very parsi-monious, however, in its extralogical or sensory part. In a series of con-structions in which he exploits the resources of modern logic with much ingenuity, Carnap succeeds in defining a wide array of important addi-tional sensory concepts which, but for his constructions, one would not have dreamed were definable on so slender a basis. He was the first em-piricist who, not content with asserting the reducibility of science to terms of immediate experience, took serious steps toward carrying out the reduction.

If Carnap's starting point is satisfactory, still his constructions were, as he himself stressed, only a fragment of the full program. The construction of even the simplest statements about the physical world was left in a sketchy state. Carnap's suggestions on this subject were, despite their sketchiness, very suggestive. He explained spatio-temporal point-instants as quadruples of real numbers and envisaged assignment of sense qualities to point-instants according to certain canons. Roughly summarized, the plan was that qualities should be assigned to point-instants in such a way as to achieve the laziest world compatible with our experience. The prin-ciple of least action was to be our guide in constructing a world from experience.

Carnap did not seem to recognize, however, that his treatment of physical objects fell short of reduction not merely through sketchiness, but in principle. Statements of the form 'Quality q is at point-instant x; y; z; t' were, according to his canons, to be apportioned truth values in such a way as to maximize and minimize certain over-all features, and with growth of experience the truth values were to be progressively revised in the same spirit. I think this is a good schematization (deliberately over-simplified, to be sure) of what science really does; but it provides no indication, not even the sketchiest, of how a statement of the form 'Quality q is at x; y; z; t' could ever be translated into Carnap's initial language of sense data and logic. The connective 'is at' remains an added undefined connective; the canons counsel us in its use but not in its elimination.

Carnap seems to have appreciated this point afterward; for in his later writings he abandoned all notion of the translatability of statements about the physical world into statements about immediate experience. Reductionism in its radical form has long since ceased to figure in Carnap's philosophy.

But the dogma of reductionism has, in a subtler and more tenuous form, continued to influence the thought of empiricists. The notion lingers that to each statement, or each synthetic statement, there is associated a unique range of possible sensory events such that the occurrence of any of them would add to the likelihood of truth of the statement, and that there is associated also another unique range of possible sensory events whose occurrence would detract from that likelihood. This notion is of course implicit in the verification theory of meaning.

The dogma of reductionism survives in the supposition that each statement, taken in isolation from its fellows, can admit of confirmation or infirmation at all. My countersuggestion, issuing essentially from Carnap's doctrine of the physical world in the *Aufbau*, is that our statements about the external world face the tribunal of sense experience not individually but only as a corporate body.[17]

The dogma of reductionism, even in its attenuated form, is intimately connected with the other dogma – that there is a cleavage between the analytic and the synthetic. We have found oursleves led, indeed, from the latter problem to the former through the verification theory of meaning. More directly, the one dogma clearly supports the other in this way: as long as it is taken to be significant in general to speak of the confirmation and infirmation of a statement, it seems significant to speak also of a limiting kind of statement which is vacuously confirmed, *ipso facto*, come what may; and such a statement is analytic.

The two dogmas are, indeed, at root identical. We lately reflected that in general the truth of statements does obviously depend both upon language and upon extralinguistic fact; and we noted that this obvious circumstance carries in its train, not logically but all too naturally, a feeling that the truth of a statement is somehow analyzable into a linguistic component and a factual component. The factual component must, if we are empiricists, boil down to a range of confirmatory experiences, in the extreme case where the linguistic component is all that matters, a true statement is analytic. But I hope we are now impressed with how stubbornly the dsitinction between analytic and synthetic has resisted any straightforward drawing. I am impressed also, apart from prefabricated examples of black and white balls in an urn, with how baffling the problem has always been of arriving at any explicit theory of the empirical confirmation of a synthetic statement. My present suggestion is that it is nonsense, and the root of much nonsense, to speak of a linguistic component and a factual component in the truth of any individual statement. Taken collectively, science has its double dependence upon language and experience; but this duality is not significantly traceable into the statements of science taken one by one.

The idea of defining a symbol in use was, as remarked, an advance over the impossible term-by-term empiricism of Locke and Hume. The statement, rather than the term, came with Bentham to be recognized as the unit accountable to an empiricist critique. But what I am now urging is that even in taking the statement as unit we have drawn our grid too finely. The unit of empirical significance is the whole of science.

6. EMPIRICISM WITHOUT THE DOGMAS

The totality of our so-called knowledge or beliefs, from the most casual matters of geography and history to the profoundest laws of atomic physics or even of pure mathematics and logic, is a man-made fabric which impinges on experience only along the edges. Or, to change the figure, total science is like a field of force whose boundary conditions are experience. A conflict with experience at the periphery occasions readjustments in the interior of the field. Truth values have to be redistributed over some of our statements. Reëvaluation of some statements entails reëvaluation of others, because of their logical interconnections – the logical laws being in turn simply certain further statements ot the system, certain further elements of the field. Having reëvaluated one statement we must reëvaluate some others, which may be statements logically connected with the first or may be the statements of logical connections themselves. But the total field is so underdetermined by its boundary conditions, experience, that there is much latitude of choice as to what statements to reëvaluate in the light of any single contrary experience. No particular experiences are linked with any particular statements in the interior of the field, except indirectly through considerations of equilibrium affecting the field as a whole.

If this view is right, it is misleading to speak of the empirical content of an individual statement – especially if it is a statement at all remote from the experiential periphery of the field. Furthermore it becomes folly to seek a boundary between synthetic statements, which hold contingently on experience, and analytic statements, which hold come what may. Any statement can be held true come what may, if we make drastic enough adjustments elsewhere in the system. Even a statement very close to the periphery can be held true in the face of recalcitrant experience by pleading hallucination or by amending certain statements of the kind called logical laws. Conversely, by the same token, no statement is immune to revision. Revision even of the logical law of the excluded middle has been proposed as a means of simplifying quantum mechanics; and what difference is there in principle between such a shift and the shift whereby Kepler superseded Ptolemy, or Einstein Newton, or Darwin Aristotle?

For vividness I have been speaking in terms of varying distances from a sensory periphery. Let me try now to clarify this notion without metaphor. Certain statements, though *about* physical objects and not sense experience, seem peculiarly germane to sense experience – and in a selective way: some statements to some experiences, others to others. Such statements, especially germane to particular experiences, I picture as near the periphery. But in this relation of "germaneness" I envisage nothing more than a loose association reflecting the relative likelihood, in practice, of our choosing one statement rather than another for revision in the event of recalcitrant experience. For example, we can imagine recalcitrant experiences to which we would surely be inclined to accommodate our system by reëvaluating just the statement that there are brick houses on Elm Street, together with related statements on the same topic. We can imagine other recalcitrant experiences to which we would be inclined to accommodate our system by reëvaluating just the statement that there are no centaurs, along with kindred statements. A recalcitrant experience can, I have urged, be accommodated by any of various alternative reëvaluations in various alternative quarters of the total system; but, in the cases which we are now imagining, our natural tendency to disturb the total system as little as possible would lead us to focus our revisions upon these specific statements concerning brick houses or centaurs. These statements are felt, therefore, to have a sharper empirical reference than highly theoretical statements of physics or logic or ontology. The latter statements may be thought of as relatively centrally located within the total network, meaning merely that little preferential connection with any particular sense data obtrudes itself.

As an empiricist I continue to think of the conceptual scheme of science as a tool, ultimately, for predicting future experience in the light of past experience. Physical objects are conceptually imported into the situation as convenient intermediaries – not by definition in terms of experience, but simply as irreducible posits[18] comparable, epistemologically, to the gods of Homer. For my part I do, qua lay physicist, believe in physical objects and not in Homer's gods; and I consider it a scientific error to believe otherwise. But in point of epistemological footing the physical objects and the gods differ only in degree and not in kind. Both sorts of entities enter our conception only as cultural posits. The myth of physical objects is epistemologically superior to most in that it has proved more efficacious than other myths as a device for working a manageable structure into the flux of experience.

Positing does not stop with macroscopic physical objects. Objects at the atomic level are posited to make the laws of macroscopic objects, and ultimately the laws of experience, simpler and more manageable; and we need not expect or demand full definition of atomic and subatomic entities in terms of macroscopic ones, any more than definition of macroscopic things in terms of sense data. Science is continuation of common sense, and it continues the common-sense expedient of swelling ontology to simplify theory.

Physical objects, small and large, are not the only posits. Forces are another example; and indeed we are told nowadays that the boundary between energy and matter is obsolete. Moreover, the abstract entities which are the substance of mathematics – ultimately classes and classes of classes and so on up – are another posit in the same spirit. Epistemologically these are myths on the same footing with physical objects and gods, neither better nor worse except for differences in the degree to which they expedite our dealings with sense experiences.

The over-all algebra of rational and irrational numbers is underdetermined by the algebra of rational numbers, but is smoother and more convenient; and it includes the algebra of rational numbers as a jagged or gerrymandered part.[19] Total science, mathematical and natural and human, is similarly but more extremely underdetermined by experience. The edge of the system must be kept squared with experience; the rest, with all its elaborate myths or fictions, has as its objective the simplicity of laws.

Ontological questions, under this view, are on a par with questions of natural science.[20] Consider the question whether to countenance classes as entities. This, as I have argued elsewhere,[21] is the question whether to quantify with respect to variables which take classes as values. Now Carnap (1950a) has maintained that is a question not of matters of fact but of choosing a convenient language form, a convenient conceptual scheme or framework for science. With this I agree, but only on the proviso that the same be conceded regarding scientific hypotheses generally. Carnap (1950a, p. 32n) has recognized that he is able to preserve a double standard for ontological questions and scientific hypotheses only by assuming an absolute distinction between the analytic and the synthetic; and I need not say again that this is a distinction which I reject.[22]

The issue over there being classes seems more a question of convenient conceptual scheme; the issue over there being centaurs, or brick houses on Elm Street, seems more a question of fact. But I have been urging that this difference is only one of degree, and that it turns upon our vaguely pragmatic inclination to adjust one strand of the fabric of science rather than another in accommodating some particular recalcitrant experience. Conservatism figures in such choices, and so does the quest for simplicity.

Carnap, Lewis, and others take a pragmatic stand on the question of choosing between language forms, scientific frameworks; but their pragmatism leaves off at the imagined boundary between the analytic and the synthetic. In repudiating such a boundary I espouse a more thorough pragmatism. Each man is given a scientific heritage plus a continuing barrage of sensory stimulation; and the considerations which guide him in warping his scientific heritage to fit his continuing sensory promptings are, where rational, pragmatic.

NOTES

* Reprinted by permission of the publishers from Willard Van Orman Quine, *From a Logical Point of View*, Harvard University Press, Cambridge, Mass., Copyright © 1953, 1961 by the President and Fellows of Harvard College. First published in *The Philosophical Review* **60** (1951).

[1] See *From a Logical Point of View (FLPV)*, p. 9.

[2] See *FLPV*, p. 10, and pp. 107–115.

[3] See *FLPV*, pp. 11f, and pp. 48f.

[4] Carnap (1947), pp. 9ff; (1950b), pp. 70ff.

[5] According to an important variant sense of 'definition', the relation preserved may be the weaker relation of mere agreement in reference; see *FLPV*, p. 132. But definition in this sense is better ignored in the present connection, being irrelevant to the question of synonymy.

[6] Cf. Lewis (1918), p. 373.

[7] This is cognitive synonymy in a primary, broad sense. Carnap (1947, pp. 56ff) and Lewis (1946, pp. 83ff) have suggested how, once this notion is at hand, a narrower sense of cognitive synonymy which is preferable for some purposes can in turn be derived. But this special ramification of concept-building lies aside from the present purposes and must not be confused with the broad sort of cognitive synonymy here concerned.

[8] Pp. 81ff, *FLPV*, contain a description of just such a language except that there happens there to be just one predicate, the two-place predicate 'e'.

[9] See *FLPV*, pp. 5–8; also pp. 85f, 166f.

[10] See *FLPV*, p. 87.

[11] On such devices see also Essay VIII, 'Reference and Modality', in *FLPV*.

[12] This is the substance of Quine, *121.

[13] The 'if and only if' itself is intended in the truth functional sense. See Carnap (1947), p. 14.

[14] The foregoing paragraph was not part of the present essay as originally published. It was prompted by Martin (see Bibliography), as was the end of Essay VII, 'Notes on the Theory of Reference', in *FLPV*.

[15] The doctrine can indeed be formulated with terms rather than statements as the units. Thus Lewis describes the meaning of a term as "*a criterion in mind*, by reference to which one is able to apply or refuse to apply the expression in question in the case of presented, or imagined, things or situations" (1946, p. 133). – For an instructive account of the vicissitudes of the verification theory of meaning, centered however on the question of meaning*fulness* rather than synonymy and analyticity, see Hempel.

[16] See *FLPV*, p. 6.
[17] This doctrine was well argued by Duhem, pp. 303–328. Or see Lowinger, pp. 132–140.
[18] Cf. *FLPV*, pp. 17f.
[19] Cf. p. 18, *FLPV*.
[20] "L'ontologie fait corps avec la science elle-même et ne peut en être separée". Meyerson, p. 439.
[21] *FLPV*, pp. 12f; pp. 102ff.
[22] For an effective expression of further misgivings over this distinction, see White.

BIBLIOGRAPHY

Carnap, Rudolf: *Meaning and Necessity*, University of Chicago Press, Chicago, 1947.
Carnap, Rudolf: 'Empiricism, Semantics, and Ontology', *Revue Internationale de Philosophie* 4 (1950a), 20–40.
Carnap, Rudolf: *Logical Foundations of Probability*, University of Chicago Press, Chicago, 1950b.
Duhem, Pierre: *La Théorie Physique: Son Objet et Sa Structure*, Paris, 1906.
Hempel, C. G.: 'Problems and Changes in the Empiricist Criterion of Meaning', *Revue Internationale de Philosophie* 4 (1950), 41–63.
Hempel, C. G.: 'The Concept of Cognitive Significance: A Reconsideration', *Proceedings of American Academy of Arts and Sciences* 80 (1951), 61–77.
Lewis, C. I.: *A Survey of Symbolic Logic*, Berkeley, 1918.
Lewis, C. I.: *An Analysis of Knowledge and Valuation*, Open Court, Le Salle, Ill., 1946.
Lowinger, Armand: *The Methodology of Pierre Duhem*, Columbia University Press, New York, 1941.
Martin, R. M.: 'On "Analytic"', *Philosophical Studies* 3 (1952), 42–47.
Meyerson, Émile: *Identité et Realité*, Paris, 1908; 4th ed., 1932.
Quine, W. V.: *Mathematical Logic*, Norton, New York, 1940; Harvard University Press, Cambridge, 1947; rev. ed., Harvard University Press, Cambridge, 1951.
White, Morton: 'The Analytic and the Synthetic: An Untenable Dualism', in Sidney Hook (ed.), *John Dewey: Philosopher of Science and Freedom*, Dial Press, New York, 1950, pp. 316–330.

CARL G. HEMPEL

EMPIRICIST CRITERIA OF COGNITIVE SIGNIFICANCE: PROBLEMS AND CHANGES*

1. THE GENERAL EMPIRICIST CONCEPTION OF COGNITIVE AND EMPIRICAL SIGNIFICANCE**

It is a basic principle of contemporary empiricism that a sentence makes a cognitively significant assertion, and thus can be said to be either true or false, if and only if either (1) it is analytic or contradictory – in which case it is said to have purely logical meaning or significance – or else (2) it is capable, at least potentially, of test by experiential evidence – in which case it is said to have empirical meaning or significance. The basic tenet of this principle, and especially of its second part, the so-called testability criterion of empirical meaning (or better: meaningfulness), is not peculiar to empiricism alone: it is characteristic also of contemporary operationism, and in a sense of pragmatism as well; for the pragmatist maxim that a difference must make a difference to be a difference may well be construed as insisting that a verbal difference between two sentences must make a difference in experiential implications if it is to reflect a difference in meaning.

How this general conception of cognitively significant discourse led to the rejection, as devoid of logical and empirical meaning, of various formulations in speculative metaphysics, and even of certain hypotheses offered within empirical science, is too well known to require recounting. I think that the general intent of the empiricist criterion of meaning is basically sound, and that notwithstanding much oversimplification in its use, its critical application has been, on the whole, enlightening and salutary. I feel less confident, however, about the possibility of restating the general idea in the form of precise and general criteria which establish sharp dividing lines (a) between statements of purely logical and statements of empirical significance, and (b) between those sentences which do have cognitive significance and those which do not.

In the present paper, I propose to reconsider these distinctions as conceived in recent empiricism, and to point out some of the difficulties they present. The discussion will concern mainly the second of the two distinctions; in regard to the first, I shall limit myself to a few brief remarks.

2. THE EARLIER TESTABILITY CRITERIA OF MEANING AND THEIR SHORTCOMINGS

Let us note first that any general criterion of cognitive significance will have to meet certain requirements if it is to be at all acceptable. Of these, we note one, which we shall consider here as expressing a necessary, though by no means sufficient, *condition of adequacy* for criteria of cognitive significance.

(A)   If under a given criterion of cognitive significance, a sentence N is non-significant, then so must be all truth-functional compound sentences in which N occurs nonvacuously as a component. For if N cannot be significantly assigned a truth value, then it is impossible to assign truth values to the compound sentences containing N; hence, they should be qualified as nonsignificant as well.

We note two corollaries of requirement (A):

(A1)   If under a given criterion of cognitive significance, a sentence S is nonsignificant, then so must be its negation, $\sim S$.

(A2)   If under a given criterion of cognitive significance, a sentence N is nonsignificant, then so must be any conjunction $N \cdot S$ and any disjunction $N \vee S$, no matter whether S is significant under the given criterion or not.

We now turn to the initial attempts made in recent empiricism to establish general criteria of cognitive significance. Those attempts were governed by the consideration that a sentence, to make an empirical assertion must be capable of being borne out by, or conflicting with, phenomena which are potentially capable of being directly observed. Sentences describing such potentially observable phenomena – no matter whether the latter do actually occur or not – may be called observation sentences. More specifically, an *observation sentence* might be construed as a sentence – no matter whether true or false – which asserts or denies that a specified object, or group of objects, of macroscopic size has a particular *observable characteristic*, i.e., a characteristic whose presence or absence can, under favorable circumstances, be ascertained by direct observation.[1]

The task of setting up criteria of empirical significance is thus transformed into the problem of characterizing in a precise manner the relationship which obtains between a hypothesis and one or more observation sentences whenever the phenomena described by the latter either confirm or disconfirm the hypothesis in question. The ability of a given sentence to enter into that relationship to some set of observation sentences would then characterize its testability-in-principle, and thus its empirical significance. Let us now briefly examine the major attempts that have been made to obtain criteria of significance in this manner.

One of the earliest criteria is expressed in the so-called *verifiability requirement*. According to it, a sentence is empirically significant if and only if it is not analytic and is capable, at least in principle, of complete verification by observational evidence; i.e., if observational evidence can be described which, if actually obtained, would conclusively establish the truth of the sentence.[2] With the help of the concept of observation sentence, we can restate this requirement as follows: A sentence S has empirical meaning if and only if it is possible to indicate a finite set of observation sentences, $O_1, O_2 ..., O_n$, such that if these are true, then S is necessarily true, too. As stated, however, this condition is satisfied also if S is an analytic sentence or if the given observation sentences are logically incompatible with each other. By the following formulation, we rule these cases out and at the same time express the intended criterion more precisely:

2.1. REQUIREMENT OF COMPLETE VERIFIABILITY IN PRINCIPLE. A sentence has empirical meaning if and only if it is not analytic and follows logically from some finite and logically consistent class of observation sentences.[3] These observation sentences need not be true, for what the criterion is to explicate is testability by "potentially observable phenomena", or testability "in principle".

In accordance with the general conception of cognitive significance outlined earlier, a sentence will now be classified as cognitively significant if either it is analytic or contradictory, or it satisfies the verifiability requirement.

This criterion, however, has several serious defects. One of them has been noted by several writers:

(a) Let us assume that the properties of being a stork and of being red-legged are both observable characteristics, and that the former does not logically entail the latter. Then the sentence

(S1)   All storks are red-legged,

is neither analytic nor contradictory; and clearly, it is not deducible from a finite set of observation sentences. Hence, under the contemplated criterion, (S1) is devoid of empirical significance; and so are all other sentences purporting to express universal regularities or general laws. And since sentences of this type constitute an integral part of scientific theories, the verifiability requirement must be regarded as overly restrictive in this respect.

Similarly, the criterion disqualifies all sentences such as 'For any substance there exists some solvent', which contain both universal and existential quantifiers (i.e., occurrences of the terms 'all' and 'some' or their equivalents); for no sentences of this kind can be logically deduced from any finite set of observation sentences.

Two further defects of the verifiability requirement do not seem to have been widely noticed:

(b) As is readily seen, the negation of (S1)

($\sim S$1)   There exists at least one stork that is not red-legged

is deducible from any two observation sentences of the type 'a is a stork' and 'a is not red-legged'. Hence, ($\sim S$1) is cognitively significant under our criterion, but (S1) is not, and this constitutes a violation of condition (A1).

(c) Let S be a sentence which does, and N a sentence which does not satisfy the verifiability requirement. Then S is deducible from some set of observation sentences; hence, by a familiar rule of logic, $S \vee N$ is deducible from the same set, and therefore cognitively significant according to our criterion. This violates condition (A2) above.[4]

Strictly analogous considerations apply to an alternative criterion, which makes complete falsifiability in principle the defining characteristic of empirical significance. Let us formulate this criterion as follows:

2.2. REQUIREMENT OF COMPLETE FALSIFIABILITY IN PRINCIPLE. A sentence has empirical meaning if and only if its negation is not analytic and follows logically from some finite logically consistent class of observation sentences.

This criterion qualifies a sentence as empirically meaningful if its negation satisfies the requirement of complete verifiability; as it is to be expected, it is therefore inadequate on similar grounds as the latter:

(a) It denies cognitive significance to purely existential hypotheses, such as 'There exists at least one unicorn', and all sentences whose formulation calls for mixed – i.e., universal and existential – quantification, such as 'For every compound there exists some solvent', for none of these can possibly be conclusively falsified by a finite number of observation sentences.

(b) If 'P' is an observation predicate, then the assertion that all things have the property P is qualified as significant, but its negation, being equivalent to a purely existential hypothesis, is disqualified [cf. (a)]. Hence, criterion (2.2.) gives rise to the same dilemma as (2.1.).

(c) If a sentence S is completely falsifiable whereas N is a sentence which is not, then their conjunction, $S \cdot N$ (i.e., the expression obtained by connecting the two sentences by the word 'and') is completely falsifiable; for if the negation of S is entailed by a class of observation sentences, then the negation of $S \cdot N$ is, *a fortiori*, entailed by the same class. Thus, the criterion allows empirical significance to many sentences which an adequate empiricist criterion should rule out, such as 'All swans are white and the absolute is perfect'.

In sum, then, interpretations of the testability criterion in terms of complete verifiability or of complete falsifiability are inadequate because they are overly restrictive in one direction and overly inclusive in another, and because both of them violate the fundamental requirement A.

Several attempts have been made to avoid these difficulties by construing the testability criterion as demanding merely a partial and possibly indirect confirmability of empirical hypotheses by observational evidence.

A formulation suggested by Ayer[5] is characteristic of these attempts to set up a clear and sufficiently comprehensive criterion of confirmability. It states, in effect, that a sentence S has empirical import if from S in conjunction with suitable subsidiary hypotheses it is possible to derive observation sentences which are not derivable from the subsidiary hypotheses alone.

This condition is suggested by a closer consideration of the logical structure of scientific testing; but it is much too liberal as it stands. Indeed, as Ayer himself has pointed out in the second edition of his book, *Language, Truth, and Logic*,[6] his criterion allows empirical import to any sentence whatever. Thus, e.g., if S is the sentence 'The absolute is perfect', it suffices to choose as a subsidiary hypothesis the sentence 'If the absolute is perfect then this apple is red' in order to make possible the deduction of the observation sentence 'This apple is red', which clearly does not follow from the subsidiary hypothesis alone.

To meet this objection, Ayer proposed a modified version of his testability criterion. In effect, the modification restricts the subsidiary hypotheses mentioned in the previous version to sentences which either are analytic or can independently be shown to be testable in the sense of the modified criterion.[7]

But it can readily be shown that this new criterion, like the requirement of complete falsifiability, allows empirical significance to any conjunction $S \cdot N$, where S satisfies Ayer's criterion while N is a sentence such as 'The absolute is perfect', which is to be disqualified by that criterion. Indeed, whatever consequences can be deduced from S with the help of permissible subsidiary hypotheses can also be deduced from $S \cdot N$ by means of the same subsidiary hypotheses; and as Ayer's new criterion is formulated essentially in terms of the deducibility of a certain type of consequence from the given sentence, it countenances $S \cdot N$ together with S. Another difficulty has been pointed out by Church, who has shown[8] that if there are any three observation sentences none of which alone entails any of the others, then it follows for any sentence S whatsoever that either it or its denial has empirical import according to Ayer's revised criterion.

All the criteria considered so far attempt to explicate the concept of empirical significance by specifying certain logical connections which must obtain between a significant sentence and suitable observation sentences. It seems now that this type of approach offers little hope for the attainment of precise criteria of meaningfulness: this conclusion is suggested by the preceding survey of some representative attempts, and it receives additional support from certain further considerations, some of which will be presented in the following sections.

3. CHARACTERIZATION OF SIGNIFICANT SENTENCES BY CRITERIA FOR THEIR CONSTITUENT TERMS

An alternative procedure suggests itself which again seems to reflect well the general viewpoint of empiricism: It might be possible to characterize cognitively significant sentences by certain conditions which their constituent terms have to satisfy. Specifically, it would seem reasonable to say that all extralogical terms[9] in a significant sentence must have experiential reference, and that therefore their meanings must be capable of explication by reference to observables exclusively.[10] In order to exhibit certain analogies between this approach and the previous one, we adopt the following terminological conventions:

Any term that may occur in a cognitively significant sentence will be called a *cognitively significant term*. Furthermore, we shall understand by an *observation term* any term which either (a) is an *observation predicate*, i.e., signifies some observable characteristic (as do the terms 'blue', 'warm', 'soft', 'coincident with', 'of greater apparent brightness than') or (b) names some physical object of macroscopic size (as do the terms 'the needle of this instrument', 'the Moon', 'Krakatoa Volcano', 'Greenwich, England', 'Julius Caesar').

Now while the testability criteria of meaning aimed at characterizing the cognitively significant sentences by means of certain inferential connections in which they must stand to some observation sentences, the alternative approach under consideration would instead try to specify the vocabulary that may be used in forming significant sentences. This vocabulary, the class of significant terms, would be characterized by the condition that each of its elements is either a logical term or else a term with empirical significance; in the latter case, it has to stand in certain definitional or explicative connections to some observation terms. This approach certainly avoids any violations of our earlier conditions of adequacy. Thus, e.g., if S is a significant sentence, i.e., contains cognitively significant terms only, then so is its denial, since the denial sign, and its verbal equivalents, belong to the vocabulary of logic and are thus significant. Again, if N is a sentence containing a non-significant term, then so is any compound sentence which contains N.

But this is not sufficient, of course. Rather, we shall now have to consider a crucial question analogous to that raised by the previous approach:

Precisely how are the logical connections between empirically significant terms and observation terms to be construed if an adequate criterion of cognitive significance is to result? Let us consider some possibilities.

3.1. The simplest criterion that suggests itself might be called the *requirement of definability*. It would demand that any term with empirical significance must be explicitly definable by means of observation terms.

This criterion would seem to accord well with the maxim of operationism that all significant terms of empirical science must be introduced by operational definitions. However, the requirement of definability is vastly too restrictive, for many important terms of scientific and even prescientific discourse cannot be explicitly defined by means of observation terms.

In fact, as Carnap[11] has pointed out, an attempt to provide explicit definitions in terms of observables encounters serious difficulties as soon as disposition terms, such as 'soluble', 'malleable', 'electric conductor', etc., have to be accounted for; and many of these occur even on the prescientific level of discourse.

Consider, for example, the word 'fragile'. One might try to define it by saying that an object x is fragile if and only if it satisfies the following condition: If at any time t the object is sharply struck, then it breaks at that time. But if the statement connectives in this phrasing are construed truth-functionally, so that the definition can be symbolized by

(D) $Fx \equiv (t)(Sxt \supset Bxt),$

then the predicate 'F' thus defined does not have the intended meaning. For let a be any object which is not fragile (e.g., a raindrop or a rubber band), but which happens not to be sharply struck at any time throughout its existence. Then 'Sat' is false and hence '$Sat \supset Bat$' is true for all values of 't'; consequently, 'Fa' is true though a is not fragile.

To remedy this defect, one might construe the phrase 'if...then...' in the original definiens as having a more restrictive meaning than the truth-functional conditional. This meaning might be suggested by the subjunctive phrasing 'If x were to be sharply struck at any time t, then x would break at t.' But a satisfactory elaboration of this construal would require a clarification of the meaning and the logic of counterfactual and subjunctive conditionals, which is a thorny problem.[12]

An alternative procedure was suggested by Carnap in his theory of reduction sentences.[13] These are sentences which, unlike definitions, specify the meaning of a term only conditionally or partially. The term 'fragile', for example, might be introduced by the following reduction sentence:

(R) $(x)\,(t)\,[Sxt \supset (Fx \equiv Bxt)]$

which specifies that if x is sharply struck at any time t, then x is fragile if and only if x breaks at t.

Our earlier difficulty is now avoided, for if a is a nonfragile object that is never sharply struck, then that expression in R which follows the quantifiers is true of a; but this does not imply that 'Fa' is true. But the reduction sentence R specifies the meaning of 'F' only for application to those objects which meet the 'test condition' of being sharply struck at some time; for these it states that fragility then amounts to breaking. For objects that fail to meet the test condition, the meaning of 'F' is left undetermined. In this sense, reduction sentences have the character of partial or conditional definitions.

Reduction sentences provide a satisfactory interpretation of the experiential import of a large class of disposition terms and permit a more adequate formulation of so-called operational definitions, which, in general, are not complete definitions at all. These considerations suggest a greatly liberalized alternative to the requirement of definability:

3.2. *The Requirement of Reducibility*

Every term with empirical significance must be capable of introduction, on the basis of observation terms, through chains of reduction sentences.

This requirement is characteristic of the liberalized versions of positivism and physicalism which, since about 1936, have superseded the older, overly narrow conception of a full definability of all terms of empirical science by means of observables,[14] and it avoids many of the shortcomings of the latter. Yet, reduction sentences do not seem to offer an adequate means for the introduction of the central terms of advanced scientific theories, often referred to as theoretical constructs. This is indicated by the following considerations: A chain of reduction sentences provides a necessary and a sufficient condition for the applicability of the term it introduces. (When the two conditions coincide, the chain is tantamount to an explicit definition.) But now take, for example, the concept of length as used in classical physical theory. Here, the length in centimeters of the distance between two points may assume any positive real number as its value; yet it is clearly impossible to formulate, by means of observation terms, a sufficient condition for the applicability of such expressions as 'having a length of $\sqrt{2}$ cm' and 'having a length of $\sqrt{2}+10^{-100}$ cm'; for such conditions would provide a possibility for discrimination, in observational terms, between two lengths which differ by only 10^{-100} cm.[15]

It would be ill-advised to argue that for this reason, we ought to permit only such values of the magnitude, length, as permit the statement of sufficient conditions in terms of observables. For this would rule out, among others, all irrational numbers and would prevent us from assigning, to the diagonal of a square with sides of length 1, the length $\sqrt{2}$, which is required by Euclidean geometry. Hence, the principles of Euclidean geometry would not be universally applicable in physics. Similarly, the principles of the calculus would become inapplicable, and the system of scientific theory as we know it today would be reduced to a clumsy, unmanageable torso. This, then, is no way of meeting the difficulty. Rather, we shall have to analyze more closely the function of constructs in scientific theories, with a view to obtaining through such an analysis a more adequate characterization of cognitively significant terms.

Theoretical constructs occur in the formulation of scientific theories. These may be conceived of, in their advanced stages, as being stated in the form of deductively developed axiomatized systems. Classical mechanics, or Euclidean or some Non-Euclidean form of geometry in physical interpretation, present examples of such systems. The extralogical terms used in a theory of this kind may be divided, in familiar manner, into primitive or basic terms, which are not defined within the theory, and defined terms, which are explicitly defined by means of the primitives. Thus, e.g., in Hilbert's axiomatization of Euclidean geometry, the terms 'point', 'straight line', 'between' are among the primitives, while 'line segment', 'angle', 'triangle', 'length' are among the defined terms. The basic and the defined terms together with the terms of logic constitute the vocabulary out of which all the sentences of the theory are constructed. The latter are divided, in an axiomatic presentation, into primitive statements (also called postulates or basic statements) which, in the theory, are not derived from any other statements, and derived ones, which are obtained by logical deduction from the primitive statements.

From its primitive terms and sentences, an axiomatized theory can be developed by means of purely formal principles of definition and deduction, without any consideration of the empirical significance of its extralogical terms. Indeed, this is the standard procedure employed in the axiomatic development of uninterpreted mathematical theories such as those of abstract groups or rings or lattices, or any form of pure (i.e., noninterpreted) geometry.

However, a deductively developed system of this sort can constitute a scientific theory only if it has received an empirical interpretation[16] which renders it relevant to the phenomena of our experience. Such interpretation is given by assigning a meaning, in terms of observables, to certain terms or sentences of the formalized theory. Frequently, an interpretation is given not for the primitive terms or statements but rather for some of the terms definable by means of the primitives, or for some of the sentences deducible from the postulates.[17] Furthermore, interpretation may amount to only a partial assignment of meaning. Thus, e.g., the rules for the measurement of length by means of a standard rod may be considered as providing a *partial* empirical interpretation for the term 'the length, in centimeters, of interval i', or alternatively, for some sentences of the form 'the length of interval i is r centimeters'. For the method is applicable only to intervals of a certain medium size, and even for the latter it does not constitute a full interpretation since the use of a standard rod does not constitute the only way of determining length: various alternative procedures are available involving the measurement of other magnitudes which are connected, by general laws, with the length that is to be determined.

This last observation, concerning the possibility of an indirect measurement of length by virtue of certain laws, suggests an important reminder. It is not correct to speak, as is often done, of 'the experiential meaning' of a term or a sentence in isolation. In the language of science, and for similar reasons even in pre-scientific discourse, a single statement usually has no experiential implications. A single sentence in a scientific theory does not, as a rule, entail any observation sentences; consequences asserting the occurrence of certain observable phenomena can be derived from it only by conjoining it with a set of other, subsidiary, hypotheses. Of the latter, some will usually be observation sentences, others will be previously accepted theoretical statements. Thus, e.g., the relativistic theory of the deflection of light rays in the gravitational field of the sun entails assertions about observable phenomena only if it is conjoined with a considerable body of astronomical and optical theory as well as a large number of specific statements about the instruments used in those observations of solar eclipses which serve to test the hypothesis in question.

Hence, the phrase, 'the experiential meaning of expression E' is elliptical: What a given expression 'means' in regard to potential empirical data is relative to two factors, namely:

(I)   *the linguistic framework L* to which the expression belongs. Its rules determine, in particular, what sentences – observational or otherwise – may be inferred from a given statement or class of statements;

(II)   the theoretical context in which the expression occurs, i.e., the class of those statements in L which are available as subsidiary hypotheses.

Thus, the sentence formulating Newton's law of gravitation has no experiential meaning by itself; but when used in a language whose logical apparatus permits the development of the calculus, and when combined with a suitable system of other hypotheses – including sentences which connect some of the theoretical terms with observation terms and thus establish a partial interpretation – then it has a bearing on observable phenomena in a large variety of fields. Analogous considerations are applicable to the term 'gravitational field', for example. It can be considered as having experiential meaning only within the context of a theory, which must be at least partially interpreted; and the experiential meaning of the term – as expressed, say, in the form of operational criteria for its application – will depend again on the theoretical system at hand, and on the logical characteristics of the language within which it is formulated.

4. COGNITIVE SIGNIFICANCE AS A CHARACTERISTIC OF INTERPRETED SYSTEMS

The preceding considerations point to the conclusion that a satisfactory criterion of cognitive significance cannot be reached through the second avenue of approach here considered, namely by means of specific requirements for the terms which make up significant sentences. This result accords with a general characteristic of scientific (and, in principle, even pre-scientific) theorizing: Theory formation and concept formation go hand in hand; neither can be carried on successfully in isolation from the other.

If, therefore, cognitive significance can be attributed to anything, then only to entire theoretical systems formulated in a language with a well-determined structure. And the decisive mark of cognitive significance in such a system appears to be the existence of an interpretation for it in terms of observables. Such an interpretation might be formulated, for example, by means of conditional or biconditional sentences connecting nonobservational terms of the system with observation terms in the given language; the latter as well as the connecting sentences may or may not belong to the theoretical system.

But the requirement of partial interpretation is extremely liberal; it is satisfied, for example, by the system consisting of contemporary physical theory combined with some set of principles of speculative metaphysics, even if the latter have no empirical interpretation at all. Within the total system, these metaphysical principles play the role of what K. Reach and also O. Neurath liked to call *isolated sentences*: They are neither purely formal truths or falsehoods, demonstrable or refutable by means of the logical rules of the given language system; nor do they have any experiential bearing; i.e., their omission from the theoretical system would have no effect on its explanatory and predictive power in regard to potentially observable phenomena (i.e., the kind of phenomena described by observation sentences). Should we not, therefore, require that a cognitively significant system contain no isolated sentences? The following criterion suggests itself:

(4.1.) A theoretical system is cognitively significant if and only if it is partially interpreted to at least such an extent that none of its primitive sentences is isolated.

But this requirement may bar from a theoretical system certain sentences which might well be viewed as permissible and indeed desirable. By way of a simple illustration, let us assume that our theoretical system T contains the primitive sentence

(S1) $(x) [P_1 x \supset (Qx \equiv P_2 x)]$,

where 'P_1' and 'P_2' are observation predicates in the given language L, while 'Q' functions in T somewhat in the manner of a theoretical construct and occurs in only one primitive sentence of T, namely (S1). Now (S1) is not a truth or falsehood of formal logic; and furthermore, if (S1) is omitted from the set of primitive sentences of T, then the resulting system, T', possesses exactly the same systematic, i.e., explanatory and predictive, power as T. Our contemplated criterion would therefore qualify (S1) as an isolated sentence which has to be eliminated – excised by means of Occam's razor, as it were – if the theoretical system at hand is to be cognitively significant.

But it is possible to take a much more liberal view of (S1) by treating it as a partial definition for the theoretical term 'Q'. Thus conceived, (S1) specifies that in all cases where the observable characteristic P_1 is present, 'Q' is applicable if and only if the observable characteristic P_2 is present as well. In fact, (S1) is an instance of those partial, or conditional, definitions which Carnap calls bilateral reduction sentences. These sentences are explicitly qualified by Carnap as analytic (though not, of course, as truths of formal logic), essentially on the ground that all their consequences which are expressible by means of observation predicates (and logical terms) alone are truths of formal logic.[18]

Let us pursue this line of thought a little further. This will lead us to some observations on analytic sentences and then back to the question of the adequacy of (4.1.).

Suppose that we add to our system T the further sentence

(S2) $(x) [P_3x \supset (Qx \equiv P_4x)],$

where 'P_3', 'P_4' are additional observation predicates. Then, on the view that "every bilateral reduction sentence is analytic",[19] (S2) would be analytic as well as (S1). Yet, the two sentences jointly entail non-analytic consequences which are expressible in terms of observation predicates alone, such as[20]

(Q) $(x) [\sim (P_1x \cdot P_2x \cdot Px_3 \cdot \sim P_4x) \cdot \sim (P_1x \cdot \sim P_2x \cdot P_3x \cdot P_4x)].$

But one would hardly want to admit the consequence that the conjunction of two analytic sentences may be synthetic. Hence if the concept of analyticity can be applied at all to the sentences of interpreted deductive systems, then it will have to be relativized with respect to the theoretical context at hand. Thus, e.g., (S1) might be qualified as analytic relative to the system T, whose remaining postulates do not contain the term 'Q', but as synthetic relative to the system T enriched by (S2). Strictly speaking, the concept of analyticity has to be relativized also in regard to the rules of the language at hand, for the latter determine what observational or other consequences are entailed by a given sentence. This need for at least a twofold relativization of the concept of analyticity was almost to be expected in view of those considerations which required the same twofold relativization for the concept of experiential meaning of a sentence.

If, on the other hand, we decide not to permit (S1) in the role of a partial definition and instead reject it as an isolated sentence, then we are led to an analogous conclusion: Whether a sentence is isolated or not will depend on the linguistic frame and on the theoretical context at hand: While (S1) is isolated relative to T (and the language in which both are formulated), it acquires definite experiential implications when T is enlarged by (S2).

Thus we find, on the level of interpreted theoretical systems, a peculiar rapprochement, and partial fusion, of some of the problems pertaining to the concepts of cognitive significance and of analyticity: Both concepts need to be relativized; and a large class of sentences may be viewed, apparently with equal right, as analytic in a given context, or as isolated, or nonsignificant, in respect to it.

In addition to barring, as isolated in a given context, certain sentences which could just as well be construed as partial definitions, the criterion (4.1.) has another serious defect. Of two logically equivalent formulations of a theoretical system it may qualify one as significant while barring the other as containing an isolated sentence among its primitives. For assume that a certain theoretical system T1 contains among its primitive sentences S', S'',... exactly one, S', which is isolated. Then T1 is not significant under (4.1.). But now consider the theoretical system T2 obtained from T1 by replacing the two first primitive sentences, S', S'', by one, namely their conjunction. Then, under our assumptions, none of the primitive sentences of T2 is isolated, and T2, though equivalent to T1, is qualified as significant by (4.1.). In order to do justice to the intent of (4.1.), we would therefore have to lay down the following stricter requirement:

(4.2.)   A theoretical system is cognitively significant if and only if it is partially interpreted to such an extent that in no system equivalent to it at least one primitive sentence is isolated.

Let us apply this requirement to some theoretical system whose postulates include the two sentences $(S1)$ and $(S2)$ considered before, and whose other postulates do not contain 'Q' at all. Since the sentences $(S1)$ and $(S2)$ together entail the sentence O, the set consisting of $(S1)$ and $(S2)$ is logically equivalent to the set consisting of $(S1)$, $(S2)$ and O. Hence, if we replace the former set by the latter, we obtain a theoretical system equivalent to the given one. In this new system, both $(S1)$ and $(S2)$ are isolated since, as can be shown, their removal does not affect the explanatory and predictive power of the system in reference to observable phenomena. To put it intuitively, the systematic power of $(S1)$ and $(S2)$ is the same as that of O. Hence, the original system is disqualified by (4.2.). From the viewpoint of a strictly sensationalist positivism as perhaps envisaged by Mach, this result might be hailed as a sound repudiation of theories making reference to fictitious entities, and as a strict insistence on theories couched exclusively in terms of observables. But from a contemporary vantage point, we shall have to say that such a procedure overlooks or misjudges the important function of constructs in scientific theory: The history of scientific endeavor shows that if we wish to arrive at precise, comprehensive, and well-confirmed general laws, we have to rise above the level of direct observation. The phenomena directly accessible to our experience are not connected by general laws of great scope and rigor. Theoretical constructs are needed for the formulation of such higher-level laws. One of the most important functions of a well-chosen construct is its potential ability to serve as a constituent in ever new general connections that may be discovered; and to such connections we would blind ourselves if we insisted on banning from scientific theories all those terms and sentences which could be 'dispensed with' in the sense indicated in (4.2.). In following such a narrowly phenomenalistic or positivistic course, we would deprive ourselves of the tremendous fertility of theoretical constructs, and we would often render the formal structure of the expurgated theory clumsy and inefficient.

Criterion (4.2.), then, must be abandoned, and considerations such as those outlined in this paper seem to lend strong support to the conjecture that no adequate alternative to it can be found; i.e., that it is not possible to formulate general and precise criteria which would separate those partially interpreted systems whose isolated sentences might be said to have a significant function from those in which the isolated sentences are, so to speak, mere useless appendages.

We concluded earlier that cognitive significance in the sense intended by recent empiricism and operationism can at best be attributed to sentences forming a theoretical system, and perhaps rather to such systems as wholes. Now, rather than try to replace (4.2.) by some alternative, we will have to recognize further that cognitive significance in a system is a matter of degree: Significant systems range from those whose entire extralogical vocabulary consists of observation terms, through theories whose formulation relies heavily on theoretical constructs, on to systems with hardly any bearing on potential empirical findings. Instead of dichotomizing this array into significant and non-significant systems it would seem less arbitrary and more promising to appraise or compare different theoretical systems in regard to such characteristics as these:

(a) the clarity and precision with which the theories are formulated, and with which the logical relationships of their elements to each other and to expressions couched in observational terms have been made explicit;

(b) the systematic, i.e., explanatory and predictive, power of the systems in regard to observable phenomena;

(c) the formal simplicity of the theoretical system with which a certain systematic power is attained;

(d) the extent to which the theories have been confirmed by experiential evidence.

Many of the speculative philosophical approaches to cosmology, biology, or history, for example, would make a poor showing on practically all of these counts and would thus prove no matches to available rival theories, or would be recognized as so unpromising as not to warrant further study or development.

If the procedure here suggested is to be carried out in detail, so as to become applicable also in less obvious cases, then it will be necessary, of course, to develop general standards, and theories pertaining to them, for the appraisal and comparison of theoretical systems in the various respects just mentioned. To what extent this can be done with rigor and precision cannot well be judged in advance. In recent years, a considerable amount of work has been done towards a definition and theory of the concept of degree of confirmation, or logical probability, of a theoretical system;[21] and several contributions have been made towards the clarification of some of the other ideas referred to above.[22] The continuation of this research represents a challenge for further constructive work in the logical and methodological analysis of scientific knowledge.

NOTES

* From *Aspects of Scientific Explanation*. Copyright © 1965 by The Free Press. Reprinted by permission.

** This essay combines, with certain omissions and some other changes, the contents of two articles: 'Problems and Changes in the Empiricist Criterion of Meaning', *Revue Internationale de Philosophie* (1950), 41–63; and 'The Concept of Cognitive Significance: A Reconsideration', *Proceedings of the American Academy of Arts and Sciences* **80** (1951), 61–77. This material is reprinted with kind permission of the Director of *Revue Internationale de Philosophie* and of the American Academy of Arts and Sciences.

[1] Observation sentences of this kind belong to what Carnap has called the thing-language, cf., e.g. (1938), pp. 52–53. That they are adequate to formulate the data which serve as the basis for empirical tests is clear in particular for the intersubjective testing procedures used in science as well as in large areas of empirical inquiry on the common-sense level. In epistemological discussions, it is frequently assumed that the ultimate evidence for beliefs about empirical matters consists in perceptions and sensations whose description calls for a phenomenalistic type of language. The specific problems connected with the phenomenalistic approach cannot be discussed here; but it should be mentioned that at any rate all the critical considerations presented in this article in regard to the testability criterion are applicable, *mutatis mutandis*, to the case of a phenomenalistic basis as well.

[2] Originally, the permissible evidence was meant to be restricted to what is observable by the speaker and perhaps his fellow beings during their life times. Thus construed, the criterion rules out, as cognitively meaningless, all statements about the distant future or the remote past, as has been pointed out, among others, by Ayer (1946), Chapter I; by Pap (1949), Chapter 13, esp. pp. 333ff.; and by Russell (1948), pp. 445–47. This difficulty is avoided, however, if we permit the evidence to consist of any finite set of "logically possible observation data", each of them formulated in an observation sentence. Thus, e.g., the sentence S_1, "The tongue of the largest dinosaur in New York's Museum of Natural History was blue or black" is completely verifiable in our sense; for it is a logical consequence of the sentence S_2, "The tongue of the largest dinosaur in New York's Museum of Natural History was blue"; and this is an observation sentence, in the sense just indicated.

And if the concept of *verifiability in principle* and the more general concept of *confirmability in principle*, which will be considered later, are construed as referring to *logically possible evidence* as expressed by observation sentences, then it follows similarly that the class of statements which are verifiable, or at least confirmable, in principle include such assertions as that the planet Neptune and the Antarctic Continent existed before they were discovered, and that atomic warfare, if not checked, will lead to the extermination of this planet. The objections which Russell (1948), pp. 445 and

447, raises against the verifiability criterion by reference to those examples do not apply therefore if the criterion is understood in the manner here suggested. Incidentally, statements of the kind mentioned by Russell, which are not actually verifiable by any human being, were explicitly recognized as cognitively significant already by Schlick (1936), Part V, who argued that the impossibility of verifying them was 'merely empirical'. The characterization of verifiability with the help of the concept of observation sentence as suggested here might serve as a more explicit and rigorous statement of that conception.

[3] As has frequently been emphasized in the empiricist literature, the term 'verifiability' is to indicate, of course, the conceivability, or better, the logical possibility, of evidence of an observational kind which, if actually encountered, would constitute conclusive evidence for the given sentence; it is not intended to mean the technical possibility of performing the tests needed to obtain such evidence, and even less the possibility of actually finding directly observable phenomena which constitute conclusive evidence for that sentence – which would be tantamount to the actual existence of such evidence and would thus imply the truth of the given sentence. Analogous remarks apply to the terms 'falsifiability' and 'confirmability'. This point has clearly been disregarded in some critical discussions of the verifiability criterion. Thus, e.g., Russell (1948), p. 448 construes verifiability as the actual existence of a set of conclusively verifying occurrences. This conception, which has never been advocated by any logical empiricist, must naturally turn out to be inadequate since according to it the empirical meaningfulness of a sentence could not be established without gathering empirical evidence, and moreover enough of it to permit a conclusive proof of the sentence in question! It is not surprising, therefore, that his extraordinary interpretation of verifiability leads Russell to the conclusion: "In fact, that a proposition is verifiable is itself not verifiable" (*l.c.*). Actually, under the empiricist interpretation of complete verifiability, any statement asserting the verifiability of some sentence S whose text is quoted, is either analytic or contradictory; for the decision whether there exists a class of observation sentences which entail S, i.e., whether such observation sentences can be formulated, no matter whether they are true or false – that decision is a purely logical matter.

[4] The arguments here adduced against the verifiability criterion also prove the inadequacy of a view closely related to it, namely that two sentences have the same cognitive significance if any set of observation sentences which would verify one of them would also verify the other, and conversely. Thus, e.g., under this criterion, any two general laws would have to be assigned the same cognitive significance, for no general law is verified by any set of observation sentences. The view just referred to must be clearly distinguished from a position which Russell examines in his critical discussion of the positivistic meaning criterion. It is "the theory that two propositions whose verified consequences are identical have the same significance" (1948), p. 448. This view is untenable indeed, for what consequences of a statement have actually been verified at a given time is obviously a matter of historical accident which cannot possibly serve to establish identity of cognitive significance. But I am not aware that any logical empiricist ever subscribed to that 'theory'.

[5] (1936, 1946), Chapter I. The case against the requirements of verifiability and of falsifiability, and in favor of a requirement of partial confirmability and disconfirmability, is very clearly presented also by Pap (1949), Chapter 13.

[6] (1946), 2nd ed., pp. 11–12.

[7] This restriction is expressed in recursive form and involves no vicious circle. For the full statement of Ayer's criterion, see Ayer (1946), p. 13.

[8] Church (1949). An alternative criterion recently suggested by O'Connor (1950) as a revision of Ayer's formulations is subject to a slight variant of Church's stricture: It can be shown that if there are three observation sentences none of which entails any of the others, and if S is any noncompound sentence, then either S or $\sim S$ is significant under O'Connor's criterion.

[9] An extralogical term is one that does not belong to the specific vocabulary of logic. The following phrases, and those definable by means of them, are typical examples of logical terms: 'not', 'or', 'if... then', 'all', 'some', '... is an element of class...'. Whether it is possible to make a sharp theoretical distinction between logical and extra-logical terms is a controversial issue related to the problem of discriminating between analytic and synthetic sentences. For the purpose at hand, we may simply assume that the logical vocabulary is given by enumeration.

[10] For a detailed exposition and critical discussion of this idea, see H. Feigl's stimulating and enlightening article (1950).

[11] Cf. (1936–37), especially Section 7.

[12] On this subject, see for example Langford (1941); Lewis (1946), pp. 210–30; Chisholm (1946); Goodman (1947); Reichenbach (1947), Chapter VIII; Hempel and Oppenheim (1948), Part III; Popper (1949); and especially Goodman's further analysis (1955).

[13] Cf. Carnap, *loc. cit.* Note 11. For a brief elementary presentation of the main idea, see Carnap (1938), Part III. The sentence R here formulated for the predicate 'F' illustrates only the simplest type of reduction sentence, the so-called bilateral reduction sentence.

[14] Cf. the analysis in Carnap (1936–37), especially Section 15; also see the briefer presentation of the liberalized point of view in Carnap (1938).

[15] (Added in 1964.) This is not strictly correct. For a more circumspect statement, see Note 12 in 'A Logical Appraisal of Operationism' and the fuller discussion in Section 7 of the essay 'The Theoretician's Dilemma'. Both of these pieces are reprinted in *Aspects of Scientific Explanation*.

[16] The interpretation of formal theories has been studied extensively by Reichenbach, especially in his pioneer analyses of space and time in classical and in relativistic physics. He describes such interpretation as the establishment of *coordinating definitions* (Zuordnungsdefinitionen) for certain terms of the formal theory. See, for example, Reichenbach (1928). More recently, Northrop [cf.(1947), Chapters VII, and also the detailed study of the use of deductively formulated theories in science, *ibid.*, Chapters IV, V, VI] and H. Margenau [cf., for example (1935)] have discussed certain aspects of this process under the title of *epistemic correlation*.

[17] A somewhat fuller account of this type of interpretation may be found in Carnap (1939), Section 24. The articles by Spence (1944) and by MacCorquodale and Meehl (1948) provide enlightening illustrations of the use of theoretical constructs in a field outside that of the physical sciences, and of the difficulties encountered in an attempt to analyze in detail their function and interpretation.

[18] Cf. Carnap (1936–37), especially Sections 8 and 10.

[19] Carnap (1936–37), p. 452.

[20] The sentence O is what Carnap calls the *representative sentence* of the couple consisting of the sentences $(S1)$ and $(S2)$; see (1936–37), pp. 450–53.

[21] Cf., for example, Carnap (1945) 1 and (1945) 2, and especially (1950). Also see Helmer and Oppenheim (1945).

[22] On simplicity, cf. especially Popper (1935), Chapter V; Reichenbach (1938), Section 42; Goodman (1949)1, (1949)2, (1950); on explanatory and predictive power, cf. Hempel and Oppenheim (1948), Part IV.

BIBLIOGRAPHY

Ayer, A. J.: *Language, Truth and Logic*, London, 1936; 2nd ed. 1946.

Carnap, R.: 'Testability and Meaning', *Philosophy of Science* 3 (1936) and 4 (1937).

Carnap, R.: 'Logical Foundations of the Unity of Science', in *International Encyclopedia of Unified Science* I, 1; Chicago, 1938.

Carnap, R.: *Foundations of Logic and Mathematics*, Chicago, 1939.

Carnap, R.: 'On Inductive Logic', *Philosophy of Science* 12 (1945). Referred to as (1945) 1 in this article.

Carnap, R.: 'The Two Concepts of Probability', *Philosophy and Phenomenological Research* 5 (1945). Referred to as (1945) 2 in this article.

Carnap, R.: *Logical Foundations of Probability*, Chicago, 1950.

Chisholm, R. M.: 'The Contrary-to-Fact Conditional', *Mind* 55 (1946).

Church, A.: 'Review of Ayer (1946)', *The Journal of Symbolic Logic* 14 (1949), 52–53.

Feigl, H.: 'Existential Hypotheses: Realistic vs. Phenomenalistic Interpretations', *Philosophy of Science* 17 (1950).

Goodman, N.: 'The Problem of Counterfactual Conditionals', *The Journal of Philosophy* 44 (1947).

Goodman, N.: 'The Logical Simplicity of Predicates', *The Journal of Symbolic Logic* 14 (1949). Referred to as (1949) 1 in this article.

Goodman, N.: 'Some Reflections on the Theory of Systems', *Philosophy and Phenomenological Research* 9 (1949). Referred to as (1949) 2 in this article.

Goodman, N.: 'An Improvement in the Theory of Simplicity', *The Journal of Symbolic Logic* 15 (1950).

Goodman, N.: *Fact, Fiction, and Forecast*, Cambridge, Massachusetts, 1955.

Helmer, O. and Oppenheim, P.: 'A Syntactical Definition of Probability and of Degree of Confirmation', *The Journal of Symbolic Logic* 10 (1945).

Hempel, C. G. and Oppenheim, P.: 'Studies in the Logic of Explanation', *Philosophy of Science* 15 (1948).

Langford, C. H.: Review in *The Journal of Symbolic Logic* 6 (1941), 67–68.

Lewis, C. I.: *An Analysis of Knowledge and Valuation*, La Salle, Ill., 1946.

MacCorquodale, K. and Meehl, P. E.: 'On a Distinction Between Hypothetical Constructs and Intervening Variables', *Psychological Review* 55 (1948).

Margenau, H.: 'Methodology of Modern Physics', *Philosophy of Science* 2 (1935).

Northrop, F. S. C.: *The Logic of the Sciences and the Humanities*, New York, 1947.

O'Connor, D. J.: 'Some Consequences of Professor A. J. Ayer's Verification Principle', *Analysis* 10 (1950).

Pap, A.: *Elements of Analytic Philosophy*, New York, 1949.

Popper, K.: *Logik der Forschung*, Wien, 1935.

Popper, K.: 'A Note on Natural Laws and So-Called "Contrary-to-Fact Conditionals"', *Mind* 58 (1949).

Reichenbach, H.: *Philosophie der Raum-Zeit-Lehre*, Berlin, 1928.

Reichenbach, H.: *Elements of Symbolic Logic*, New York, 1947.

Russell, B.: *Human Knowledge*, New York, 1948.

Schlick, M.: 'Meaning and Verification', *Philosophical Review* 45 (1936). Also reprinted in Feigl, H. and W. Sellars (eds.), *Readings in Philosophical Analysis*, New York, 1949.

Spence, Kenneth W.: 'The Nature of Theory Construction in Contemporary Psychology', *Psychological Review* 51 (1944).

COGNITIVE SIGNIFICANCE

The preceding essay is a conflation of two articles: 'Problems and Changes in the Empiricist Criterion of Meaning', *Revue Internationale de Philosophie*, No. 11 (1950), and 'The Concept of Cognitive Significance: A Reconsideration', *Proceedings of the American Academy of Arts and Sciences* **80** (1951). In combining the two, I omitted particularly some parts of the first article, which had been largely superseded by the second one;[1] I also made a few minor changes in the remaining text. Some of the general problems raised in the combined essay are pursued further in *Aspects of Scientific Explanation*, especially in 'The Theoretician's Dilemma'. In this Postscript, I propose simply to note some second thoughts concerning particular points in the preceding essay.

(i) The objections 2.1(c) and 2.2(c) against the requirements of complete verifiability and of complete falsifiability are, I think, of questionable force. For $S \vee N$ can properly be said to be entailed by S, and S in turn by $S \cdot N$, only if N as well as S is a declarative sentence and thus is either true or false. But if the criterion of cognitive significance is understood to delimit the class of sentences which make significant assertions, and which are thus either true or false, then the sentence N invoked in the objections is not declarative, and neither are $S \vee N$ or $S \cdot N$; hence the alleged inferences from $S \cdot N$ to S and from S to $S \vee N$ are inadmissible.[2]

My objection retains its force, however, against the use of falsifiability, not as a criterion of significance, but as a "criterion of demarcation". This use would draw a dividing line "between the statements, or systems of statements, of the empirical sciences, and all other statements – whether they are of a religious or of a metaphysical character, or simply pseudo-scientific".[3] For the argument 2.2(c) shows that the conjunction of a scientific statement S with a nonscientific statement N is falsifiable and thus qualifies as a scientific statement; and this would defeat the intended purpose of the criterion of demarcation.

(ii) My assertion, in 2.1(a) and 2.2(a), that the requirements of verifiability and of falsifiability would rule out *all* hypotheses of mixed quantificational form is false. Consider the hypothesis 'All ravens are black and something is white', or, in symbolic notation

$$(x) (Rx \supset Bx) \cdot (\exists y) \, Wy,$$

which is equivalent to

$$(x) (\exists y) [(Rx \supset Bx) \cdot Wy].$$

This sentence satisfies the falsifiability requirement because it implies the purely universal hypothesis '$(x) (Rx \supset Bx)$', which would be falsified, for example, by the following set of observation sentences: $\{'Ra', '\sim Ba'\}$. Similarly, the sentence

$$(\exists x) (y) (Rx \vee Wy)$$

is verifiable since it is implied, for example, by 'Ra'.

The essential point of the objection remains unaffected, however: Many scientific hypotheses of mixed quantificational form are neither verifiable nor falsifiable; these would therefore be disqualified by the requirement of verifiability as well as by that of falsifiability; and if the latter is used as a criterion of demarcation rather than of significance, it excludes those hypotheses from the class of scientific statements. These consequences are unacceptable.

(iii) An even stronger criticism of the criteria of verifiability and of falsifiability results from condition (A1), which is stated early in Section 2, and which demands in effect that any acceptable criterion of significance which admits a sentence as significant must also admit its negation. That this condition must be met is clear, for since a significant sentence is one that is either true or false, its negation can be held nonsignificant only on pain of violating a fundamental principle of logic. And even if the falsifiability criterion is used as a criterion of demarcation rather than of cognitive significance, satisfaction of (A1) seems imperative. Otherwise, a scientist reporting that he had succeeded in refuting a scientific hypothesis S of universal form would be making a nonscientific statement if he were to say: "Hence, it is not the case that S holds", for this statement would not be falsifiable. More generally, formally valid deductive logical inference would often lead from scientific premises to nonscientific conclusions – e.g., from '$Ra \cdot \sim Ba$' to '$(\exists x) (Rx \cdot \sim Bx)$'; and, surely, this is intolerable.

But when the requirement of verifiability, or that of falsifiability, is combined with condition (A1), then a sentence qualifies as cognitively significant just in case it and its negation are verifiable, or just in case it and its negation are falsifiable. These two criteria now demand the same thing of a significant sentence, namely, that it be both verifiable and falsifiable. This characterization admits, besides all truth-functional compounds of observation sentences, also certain sentences containing quantifiers. For example, '$Pa \vee (x) Qx$' is verifiable by 'Pa' and falsifiable by $\{$'$\sim Pa$', '$\sim Qb$'$\}$; and as is readily seen, '$Pa \cdot (\exists x) Qx$' equally meets the combined requirement. But this requirement excludes all strictly general hypotheses, i.e., those containing essential occurrences of quantifiers but not of individual constants; such as '$(x) (Rx \supset Bx)$', '$(x) (\exists y) (Rxy \supset Sxy)$', and so forth. Again, this consequence is surely unacceptable, no matter whether the criterion is meant to delimit the class of significant sentences or the class of statements of empirical science.

NOTES

[1] The basic ideas presented in the earlier articles and in the present conflated version are penetratingly examined by I. Scheffler in *The Anatomy of Inquiry*, New York, 1963. Part II of his book deals in detail with the concept of cognitive significance.

[2] I owe this correction to graduate students who put forth the above criticism in one of my seminars. The same point has recently been stated very clearly by D. Rynin in 'Vindication of L*G*C*L*P*S*T*V*M' *Proceedings and Addresses of the American Philosophical Association* **30** (1957); see especially pp, 57–58.

[3] K. R. Popper, 'Philosophy of Science: A Personal Report', In C. A. Mace (ed.), *British Philosophy in the Mid-Century*, London, 1957, pp. 155–91; quotations from pp. 163, 162.

KARL R. POPPER

SOME FUNDAMENTAL PROBLEMS IN THE LOGIC OF SCIENTIFIC DISCOVERY*

A scientist, whether theorist or experimenter, puts forward statements, or systems of statements, and tests them step by step. In the field of the empirical sciences, more particularly, he constructs hypotheses, or systems of theories, and tests them against experience by observation and experiment.

I suggest that it is the task of the logic of scientific discovery, or the logic of knowledge, to give a logical analysis of this procedure; that is, to analyse the method of the empirical sciences.

But what are these 'methods of the empirical sciences'? And what do we call 'empirical science'?

1. THE PROBLEM OF INDUCTION

According to a widely accepted view – to be opposed in this book – the empirical sciences can be characterized by the fact that they use '*inductive methods*', as they are called. According to this view, the logic of scientific discovery would be identical with inductive logic, i.e. with the logical analysis of these inductive methods.

It is usual to call an inference 'inductive' if it passes from *singular statements* (sometimes also called 'particular' statements), such as accounts of the results of observations or experiments, to *universal statements*, such as hypotheses or theories.

Now it is far from obvious, from a logical point of view, that we are justified in inferring universal statements from singular ones, no matter how numerous; for any conclusion drawn in this way may always turn out to be false: no matter how many instances of white swans we may have observed, this does not justify the conclusion that *all* swans are white.

The question whether inductive inferences are justified, or under what conditions, is known as *the problem of induction*.

The problem of induction may also be formulated as the question of how to establish the truth of universal statements which are based on experience, such as the hypotheses and theoretical systems of the empirical sciences. For many people believe that the truth of these universal statements is *'known by experience'*; yet it is clear that an account of an experience – of an observation or the result of an experiment – can in the first place be only a singular statement and not a universal one. Accordingly, people who say of a universal statement that we know its truth from experience usually mean that the truth of this universal statement can somehow be reduced to the truth of singular ones, and that these singular ones are known by experience to be true; which amounts to saying that the universal statement is based on inductive inference. Thus to ask whether there are natural laws known to be true appears to be only another way of asking whether inductive inferences are logically justified.

Yet if we want to find a way of justifying inductive inferences, we must first of all try to establish a *principle of induction*. A principle of induction would be a statement with the help of which we could put inductive inferences into a logically acceptable form. In the eyes of the upholders of inductive logic, a principle of induction is of supreme importance for scientific method: "...this principle", says Reichenbach,

determines the truth of scientific theories. To eliminate it from science would mean nothing less than to deprive science of the power to decide the truth or falsity of its theories. Without it, clearly, science would no longer have the right to distinguish its theories from the fanciful and arbitrary creations of the poet's mind.[1]

Now this principle of induction cannot be a purely logical truth like a tautology or an analytic statement. Indeed, if there were such a thing as a purely logical principle of induction, there would be no problem of induction; for in this case, all inductive inferences would have to be regarded as purely logical or tautological transformations, just like inferences in deductive logic. Thus the principle of induction must be a synthetic statement; that is, a statement whose negation is not self-contradictory but logically possible. So the question arises why such a principle should be accepted at all, and how we can justify its acceptance on rational grounds.

Some who believe in inductive logic are anxious to point out, with Reichenbach, that "the principle of induction is unreservedly accepted by the whole of science and that no man can seriously doubt this principle in everyday life either".[2] Yet even supposing this were the case – for after all, 'the whole of science' might err – I should still contend that a principle of induction is superfluous, and that it must lead to logical inconsistencies.

That inconsistencies may easily arise in connection with the principle of induction should have been clear from the work of Hume;*[1] also, that they can be avoided, if at all, only with difficulty. For the principle of induction must be a universal statement in its turn. Thus if we try to regard its truth as known from experience, then the very same problems which occasioned its introduction will arise all over again. To justify it, we should have to employ inductive inferences; and to justify these we should have to assume an inductive principle of a higher order; and so on. Thus the attempt to base the principle of induction on experience breaks down, since it must lead to an infinite regress.

Kant tried to force his way out of this difficulty by taking the principle of induction (which he formulated as the 'principle of universal causation') to be 'a priori valid'. But I do not think that his ingenious attempt to provide an a priori justification for synthetic statements was successful.

My own view is that the various difficulties of inductive logic here sketched are insurmountable. So also, I fear, are those inherent in the doctrine, so widely current today, that inductive inference, although not 'strictly valid', *can attain some degree of 'reliability' or of 'probability'*. According to this doctrine, inductive inferences are 'probable inferences'.[3] "We have described", says Reichenbach,

the principle of induction as the means whereby science decides upon truth. To be more exact, we should say that it serves to decide upon probability. For it is not given to science to reach either truth or falsity… but scientific statements can only attain continuous degrees of probability whose unattainable upper and lower limits are truth and falsity.[4]

At this stage I can disregard the fact that the believers in inductive logic entertain an idea of probability that I shall later reject as highly unsuitable for their own purposes (see Section 80, below). I can do so because the difficulties mentioned are not even touched by an appeal to probability. For if a certain degree of probability is to be assigned to statements based on inductive inference, then this will have to be justified by invoking a new principle of induction, appropriately modified. And this new principle in its turn will have to be justified, and so on. Nothing is gained, moreover, if the principle of induction, in its turn, is taken not as 'true' but only as 'probable'. In short, like every other form of inductive logic, the logic of probable inference, or 'probability logic', leads either to an infinite regress, or to the doctrine of *apriorism*.*²

The theory to be developed in the following pages stands directly opposed to all attempts to operate with the ideas of inductive logic. It might be described as the theory of *the deductive method of testing*, or as the view that a hypothesis can only be empirically *tested* – and only *after* it has been advanced.

Before I can elaborate this view (which might be called 'deductivism', in contrast to 'inductivism'⁵) I must first make clear the distinction between the *psychology of knowledge* which deals with empirical facts, and the *logic of knowledge* which is concerned only with logical relations. For the belief in inductive logic is largely due to a confusion of psychological problems with epistemological ones. It may be worth noticing, by the way, that this confusion spells trouble not only for the logic of knowledge but for its psychology as well.

2. ELIMINATION OF PSYCHOLOGISM

I said above that the work of the scientist consists in putting forward and testing theories.

The initial stage, the act of conceiving or inventing a theory, seems to me neither to call for logical analysis nor to be susceptible of it. The question how it happens that a new idea occurs to a man – whether it is a musical theme, a dramatic conflict, or a scientific theory – may be of great interest to empirical psychology; but it is irrelevant to the logical analysis of scientific knowledge. This latter is concerned not with *questions of fact* (Kant's *quid facti?*), but only with questions of *justification or validity* (Kant's *quid juris?*). Its questions are of the following kind. Can a statement be justified? And if so, how? Is it testable? Is it logically dependent on certain other statements? Or does it perhaps contradict them? In order that a statement may be logically examined in this way, it must already have been presented to us. Someone must have formulated it, and submitted it to logical examination.

Accordingly I shall distinguish sharply between the process of conceiving a new idea, and the methods and results of examining it logically. As to the task of the logic of knowledge – in contradistinction to the psychology of knowledge – I shall proceed on the assumption that it consists solely in investigating the methods employed in those systematic tests to which every new idea must be subjected if it is to be seriously entertained.

Some might object that it would be more to the purpose to regard it as the business of epistemology to produce what has been called a *'rational reconstruction'* of the steps that have led the scientist to a discovery – to the finding of some new truth. But the question is: what, precisely, do we want to reconstruct? If it is the processes involved in the stimulation and release of an inspiration which are to be reconstructed, then I should refuse to take it as the task of the logic of knowledge. Such processes are the concern of empirical psychology but hardly of logic. It is another matter if we want to reconstruct rationally the *subsequent tests* whereby the inspiration may be discovered to be a discovery, or become known to be knowledge. In so far as the scientist critically judges, alters, or rejects his own inspiration we may, if we like, regard the methodological analysis undertaken here as a kind of 'rational reconstruction' of the corresponding thought-processes. But this reconstruction would not describe these processes as they actually happen: it can give only a logical skeleton of the procedure of testing. Still, this is perhaps all that is meant by those who speak of a 'rational reconstruction' of the ways in which we gain knowledge.

It so happens that my arguments in this book are quite independent of this problem. However, my view of the matter, for what it is worth, is that there is no such thing as a logical method of having new ideas, or a logical reconstruction of this process. My view may be expressed by saying that every discovery contains 'an irrational element', or 'a creative intuition', in Bergson's sense. In a similar way Einstein speaks of "...the search for those highly universal...laws from which a picture of the world can be obtained by pure deduction. There is no logical path", he says, "leading to these...laws. They can only be reached by intuition, based upon something like an intellectual love (*'Einfühlung'*) of the objects of experience".[1]

3. DEDUCTIVE TESTING OF THEORIES

According to the view that will be put forward here, the method of critically testing theories, and selecting them according to the results of tests, always proceeds on the following lines. From a new idea, put up tentatively, and not yet justified in any way – an anticipation, a hypothesis, a theoretical system, or what you will – conclusions are drawn by means of logical deduction. These conclusions are then compared with one another and with other relevant statements, so as to find what logical relations (such as equivalence, derivability, compatibility, or incompatibility) exist between them.

We may if we like distinguish four different lines along which the testing of a theory could be carried out. First there is the logical comparison of the conclusions among themselves, by which the internal consistency of the system is tested. Secondly, there is the investigation of the logical form of the theory, with the object of determing whether it has the character of an empirical or scientific theory, or whether it is, for example, tautological. Thirdly, there is the comparison with other theories, chiefly with the aim of determining whether the theory would constitute a scientific advance should it survive our various tests. And finally, there is the testing of the theory by way of empirical applications of the conclusions which can be derived from it.

The purpose of this last kind of test is to find out how far the new consequences of the theory – whatever may be new in what it asserts – stand up to the demands of practice, whether raised by purely scientific experiments, or by practical technological applications. Here too the procedure of testing turns out to be deductive. With the help of other statements, previously accepted, certain singular statements – which we may call 'predictions' – are deduced from the theory; especially predictions that are easily testable or applicable. From among these statements, those are selected which are not derivable from the current theory, and more especially those which the current theory contradicts. Next we seek a decision as regards these (and other) derived statements by comparing them with the results of practical applications and experiments. If this decision is positive, that is, if the singular conclusions turn out to be acceptable, or *verified*, then the theory has, for the time being, passed its test: we have found no reason to discard it. But if the decision is negative, or in other words, if the conclusions have been *falsified*, then their falsification also falsifies the theory from which they were logically deduced.

It should be noticed that a positive decision can only temporarily support the theory, for subsequent negative decisions may always overthrow it. So long as a theory withstands detailed and severe tests and is not superseded by another theory in the course of scientific progress, we may say that it has 'proved its mettle' or that it is '*corroborated*'.*[1]

Nothing resembling inductive logic appears in the procedure here outlined. I never assume that we can argue from the truth of singular statements to the truth of theories. I never assume that by force of 'verified' conclusions, theories can be established as 'true', or even as merely 'probable'.

In this book I intend to give a more detailed analysis of the methods of deductive testing. And I shall attempt to show that, within the framework of this analysis, all the problems can be dealt with that are usually called '*epistemological*'. Those problems, more especially, to which inductive logic gives rise, can be eliminated without creating new ones in their place.

4. THE PROBLEM OF DEMARCATION

Of the many objections which are likely to be raised against the view here advanced, the most serious is perhaps the following. In rejecting the method of induction, it may be said, I deprive empirical science of what appears to be its most important characteristic; and this means that I remove the barriers which separate science from metaphysical speculation. My reply to this objection is that my main reason for rejecting inductive logic is precisely that *it does not provide a suitable distinguishing mark* of the empirical, non-metaphysical, character of a theoretical system; or in other words, that *it does not provide a suitable 'criterion of demarcation'*.

The problem of finding a criterion which would enable us to distinguish between the empirical sciences on the one hand, and mathematics and logic as well as 'metaphysical' systems on the other, I call the *problem of demarcation*.[1]

This problem was known to Hume who attempted to solve it.[2] With Kant it became the central problem of the theory of knowledge. If, following Kant, we call the problem of induction 'Hume's problem', we might call the problem of demarcation 'Kant's problem'.

Of these two problems – the source of nearly all the other problems of the theory of knowledge – the problem of demarcation is, I think, the more fundamental. Indeed, the main reason why epistemologists with empiricist leanings tend to pin their faith to the 'method of induction' seems to be their belief that this method alone can provide a suitable criterion of demarcation. This applies especially to those empricists who follow the flag of 'positivism'.

The older positivists wished to admit, as scientific or legitimate, only those *concepts* (or notions or ideas) which were, as they put it, 'derived from experience'; those concepts, that is, which they believed to be logically reducible to elements of sense-experience, such as sensations (or sense-data), impressions, perceptions, visual or auditory memories, and so forth. Modern positivists are apt to see more clearly that science is not a system of concepts but rather a system of *statements*.[*1] Accordingly, they wish to admit, as scientific or legitimate, only those statements which are reducible to elementary (or 'atomic') statements of experience – to 'judgments of perception' or 'atomic propositions' or 'protocol-sentences' or what not.[*2] It is clear that the implied criterion of demarcation is identical with the demand for an inductive logic.

Since I reject inductive logic I must also reject all these attempts to solve the problem of demarcation. With this rejection, the problem of demarcation gains in importance for the present inquiry. Finding an acceptable criterion of demarcation must be a crucial task for any epistemology which does not accept inductive logic.

Positivists usually interpret the problem of demarcation in a *naturalistic* way; they interpret it as if it were a problem of natural science. Instead of taking it as their task to propose a suitable convention, they believe they have to discover a difference, existing in the nature of things, as it were, between empirical science on the one hand and metaphysics on the other. They are constantly trying to prove that metaphysics by its very nature is nothing but nonsensical twaddle – 'sophistry and illusion', as Hume says, which we should 'commit to the flames'.[*3]

If by the words 'nonsensical' or 'meaningless' we wish to express no more, by definition, than 'not belonging to empirical science', then the characterization of metaphysics as meaningless nonsense would be trivial; for metaphysics has usually been defined as non-empirical. But of course, the positivists believe they can say much more about metaphysics than that some of its statements are non-empirical. The words 'meaningless' or 'nonsensical' convey, and are meant to convey, a derogatory evaluation; and there is no doubt that what the positivists really want to achieve is not so much a successful demarcation as the final overthrow[3] and the annihilation of metaphysics. However this may be, we find that each time the positivists tried to say more clearly what 'meaningful' meant, the attempt led to the same result – to a definition of 'meaningful sentence' (in contradistinction to 'meaningless pseudo-sentence') which simply reiterated the criterion of demarcation of their *inductive logic*.

This 'shows itself' very clearly in the case of Wittgenstein, according to whom every meaningful proposition must be *logically reducible*[4] to elementary (or atomic) propositions, which he characterizes as descriptions or 'pictures of reality'[5] (a characterization, by the way, which is to cover all meaningful propositions). We may see from this that Wittgenstein's criterion of meaningfulness coincides with the inductivists' criterion of demarcation, provided we replace their words 'scientific' or 'legitimate' by 'meaningful'. And it is precisely over the problem of induction that this attempt to solve the problem of demarcation comes to grief: positivists, in their anxiety to annihilate metaphysics, annihilate natural science along with it. For scientific laws, too, cannot be logically reduced to elementary statements of experience. If consistently applied, Wittgenstein's criterion of meaningfulness rejects as meaningless those natural laws the search for which, as Einstein says,[6] is "the supreme task of the physicist": they can never be accepted as genuine or legitimate statements. This view, which tries to unmask the problem of induction as an empty pseudoproblem, has been expressed by Schlick[*4] in the following words:

The problem of induction consists in asking for a logical justification of *universal statements* about reality... We recognize, with Hume, that there is no such logical justification: there can be none, simply because *they are not genuine* statements.[7]

This shows how the inductivist criterion of demarcation fails to draw a dividing line between scientific and metaphysical systems, and why it must accord them equal status; for the verdict of the positivist dogma of meaning is that both are systems of meaningless pseudo-statements. Thus instead of eradicating metaphysics from the empirical sciences, positivism leads to the invasion of metaphysics into the scientific realm.[8]

In contrast to these anti-metaphysical stratagems – anti-metaphysical in intention, that is – my business, as I see it, is not to bring about the overthrow of metaphysics. It is, rather, to formulate a suitable characterization of empirical science, or to define the concepts 'empirical science' and 'metaphysics' in such a way that we shall be able to say of a given system of statements whether or not its closer study is the concern of empirical science.

My criterion of demarcation will accordingly have to be regarded as a *proposal for an agreement or convention*. As to the suitability of any such convention opinions may differ; and a reasonable discussion of these questions is only possible between parties having some purpose in common. The choice of that purpose must, of course, be ultimately a matter of decision, going beyond rational argument.*[5]

Thus anyone who envisages a system of absolutely certain, irrevocably true statements[9] as the end and purpose of science will certainly reject the proposals I shall make here. And so will those who see 'the essence of science … in its dignity', which they think resides in its 'wholeness' and its 'real truth and essentiality'.[10] They will hardly be ready to grant this dignity to modern theoretical physics in which I and others see the most complete realization to date of what I call 'empirical science'.

The aims of science which I have in mind are different. I do not try to justify them, however, by representing them as the true or the essential aims of science. This would only distort the issue, and it would mean a relapse into positivist dogmatism. There is only *one* way, as far as I can see, of arguing rationally in support of my proposals. This is to analyse their logical consequences: to point out their fertility – their power to elucidate the problems of the theory of knowledge.

Thus I freely admit that in arriving at my proposals I have been guided, in the last analysis, by value judgments and predilections. But I hope that my proposals may be acceptable to those who value not only logical rigour but also freedom from dogmatism; who seek practical applicability, but are even more attracted by the adventure of science, and by discoveries which again and again confront us with new and unexpected questions, challenging us to try out new and hitherto undreamed-of answers.

The fact that value judgments influence my proposals does not mean that I am making the mistake of which I have accused the positivists –

that of trying to kill metaphysics by calling it names. I do not even go so far as to assert that metaphysics has no value for empirical science. For it cannot be denied that along with metaphysical ideas which have obstructed the advance of science there have been others – such as speculative atomism – which have aided it. And looking at the matter from the psychological angle, I am inclined to think that scientific discovery is impossible without faith in ideas which are of a purely speculative kind, and sometimes even quite hazy; a faith which is completely unwarranted from the point of view of science, and which, to that extent, is 'metaphysical'.[11]

Yet having issued all these warnings, I still take it to be the first task of the logic of knowledge to put forward a *concept of empirical science*, in order to make linguistic usage, now somewhat uncertain, as definite as possible, and in order to draw a clear line of demarcation between science and metaphysical ideas – even though these ideas may have furthered the advance of science throughout its history.

5. EXPERIENCE AS A METHOD

The task of formulating an acceptable definition of the idea of empirical science is not without its difficulties. Some of these arise from *the fact that there must be many theoretical systems* with a logical structure very similar to the one which at any particular time is the accepted system of empirical science. This situation is sometimes described by saying that there are a great many – presumably an infinite number – of 'logically possible worlds'. Yet the system called 'empirical science' is intended to represent only *one* world: the 'real world' or the 'world of our experience'.[*1]

In order to make this idea a little more precise, we may distinguish three requirements which our empirical theoretical system will have to satisfy. First, it must be *synthetic*, so that it may represent a noncontradictory, a *possible* world. Secondly, it must satisfy the criterion of demarcation (cf. Sections 6 and 21), i.e. it must not be metaphysical, but must represent a world of possible *experience*. Thirdly, it must be a system distinguished in some way from other such systems as the one which represents *our* world of experience.

But how is the system that represents our world of experience to be distinguished? The answer is: by the fact that it has been submitted to tests, and has stood up to tests. This means that it is to be distinguished by applying to it that deductive method which it is my aim to analyse, and to describe.

'Experience', on this view, appears as a distinctive *method* whereby one theoretical system may be distinguished from others; so that empirical science seems to be characterized not only by its logical form but, in addition, by its distinctive *method*. (This, of course, is also the view of the inductivists, who try to characterize empirical science by its use of the inductive method.)

The theory of knowledge whose task is the analysis of the method or procedure peculiar to empirical science, may accordingly be described as a theory of the empirical method – *a theory of what is usually called experience*.

6. FALSIFIABILITY AS A CRITERION OF DEMARCATION

The criterion of demarcation inherent in inductive logic – that is, the positivistic dogma of meaning – is equivalent to the requirement that all the statements of empirical science (or all 'meaningful' statements) must be capable of being finally decided, with respect to their truth *and* falsity; we shall say that they must be *'conclusively decidable'*. This means that their form must be such that *to verify them and to falsify them* must both be logically possible. Thus Schlick says: "... a genuine statement must be capable of *conclusive verification*"[1]; and Waismann says still more clearly: "If there is no possible way to *determine whether a statement is true* then that statement has no meaning whatsoever. For the meaning of a statement is the method of its verification."[2]

Now in my view there is no such thing as induction.[*1] Thus inference to theories, from singular statements which are 'verified by experience' (whatever that may mean), is logically inadmissible. Theories are, therefore, *never* empirically verifiable. If we wish to avoid the positivist's mistake of eliminating, by our criterion of demarcation, the theoretical systems of natural science,[*2] then we must choose a criterion which allows us to admit to the domain of empirical science even statements which cannot be verified.

But I shall certainly admit a system as empirical or scientific only if it is capable of being *tested* by experience. These considerations suggest that not the *verifiability* but the *falsifiability* of a system is to be taken as a criterion of demarcation.*³ In other words: I shall not require of a scientific system that it shall be capable of being singled out, once and for all, in a positive sense; but I shall require that its logical form shall be such that it can be singled out, by means of empirical tests, in a negative sense: *it must be possible for an empirical scientific system to be refuted by experience.*³

(Thus the statement, 'It will rain or not rain here tomorrow' will not be regarded as empirical, simply because it cannot be refuted; whereas the statement, 'It will rain here tomorrow' will be regarded as empirical.)

Various objections might be raised against the criterion of demarcation here proposed. In the first place, it may well seem somewhat wrong-headed to suggest that science, which is supposed to give us positive information, should be characterized as satisfying a negative requirement such as refutability. However, I shall show, in Sections 31 to 46, that this objection has little weight, since the amount of positive information about the world which is conveyed by a scientific statement is the greater the more likely it is to clash, because of its logical character, with possible singular statements. (Not for nothing do we call the laws of nature 'laws': the more they prohibit the more they say.)

Again, the attempt might be made to turn against me my own criticism of the inductivist criterion of demarcation; for it might seem that objections can be raised against falsifiability as a criterion of demarcation similar to those which I myself raised against verifiability.

This attack would not disturb me. My proposal is based upon an *asymmetry* between verifiability and falsifiability; an asymmetry which results from the logical form of universal statements.*⁴ For these are never derivable from singular statements, but can be contradicted by singular statements. Consequently it is possible by means of purely deductive inferences (with the help of the *modus tollens* of classical logic) to argue from the truth of singular statements to the falsity of universal statements. Such an argument to the falsity of universal statements is the only strictly deductive kind of inference that proceeds, as it were, in the 'inductive direction'; that is, from singular to universal statements.

A third objection may seem more serious. It might be said that even if the asymmetry is admitted, it is still impossible, for various reasons, that any theoretical system should ever be conclusively falsified. For it is always possible to find some way of evading falsification, for example by introducing *ad hoc* an auxiliary hypothesis, or by changing *ad hoc* a definition. It is even possible without logical inconsistency to adopt the position of simply refusing to acknowledge any falsifying experience whatsoever. Admittedly, scientists do not usually proceed in this way, but logically such procedure is possible; and this fact, it might be claimed, makes the logical value of my proposed criterion of demarcation dubious, to say the least.

I must admit the justice of this criticism; but I need not therefore withdraw my proposal to adopt falsifiability as a criterion of demarcation. For I am going to propose (in Sections 20f.) that the *empirical method* shall be characterized as a method that excludes precisely those ways of evading falsification which, as my imaginary critic rightly insists, are logically admissible. According to my proposal, what characterizes the empirical method is its manner of exposing to falsification, in every conceivable way, the system to be tested. Its aim is not to save the lives of untenable systems but, on the contrary, to select the one which is by comparison the fittest, by exposing them all to the fiercest struggle for survival.

The proposed criterion of demarcation also leads us to a solution of Hume's problem of induction – of the problem of the validity of natural laws. The root of this problem is the apparent contradiction between what may be called 'the fundamental thesis of empiricism' – the thesis that experience alone can decide upon the truth or falsity of scientific statements – and Hume's realization of the inadmissibility of inductive arguments. This contradiction arises only if it is assumed that all empirical scientific statements must be 'conclusively decidable', i.e. that their verification and their falsification must both in principle be possible. If we renounce this requirement and admit as empirical also statements which are decidable in one sense only – unilaterally decidable and, more especially, falsifiable – and which may be tested by systematic attempts to falsify them, the contradiction disappears: the method of falsification presupposes no inductive inference, but only the tautological transformations of deductive logic whose validity is not in dispute.[4]

7. THE PROBLEM OF THE 'EMPIRICAL BASIS'

If falsifiability is to be at all applicable as a criterion of demarcation, then singular statements must be available which can serve as premisses in falsifying inferences. Our criterion therefore appears only to shift the problem – to lead us back from the question of the empirical character of theories to the question of the empirical character of singular statements.

Yet even so, something has been gained. For in the practice of scientific research, demarcation is sometimes of immediate urgency in connection with theoretical systems, whereas in connection with singular statements, doubts as to their empirical character rarely arise. It is true that errors of observation occur and give rise to false singular statements, but the scientist scarcely ever has occasion to describe a singular statement as non-empirical or metaphysical.

Problems of the empirical basis – that is, problems concerning the empirical character of singular statements, and how they are tested – thus play a part within the logic of science that differs somewhat from that played by most of the other problems which will concern us. For most of these stand in close relation to the *practice* of research, whilst the problem of the empirical basis belongs almost exclusively to the *theory* of knowledge. I shall have to deal with them, however, since they have given rise to many obscurities. This is especially true of the relation between *perceptual experiences* and *basic statements*. (What I call a 'basic statement' or a 'basic proposition' is a statement which can serve as a premise in an empirical falsification; in brief, a statement of a singular fact.)

Perceptual experiences have often been regarded as providing a kind of justification for basic statements. It was held that these statements are 'based upon' these experiences; that their truth becomes 'manifest by inspection' through these experiences; or that it is made 'evident' by these experiences, etc. All these expressions exhibit the perfectly sound tendency to emphasize the close connection between the basic statements and our perceptual experiences. Yet it was also rightly felt that *statements can be logically justified only by statements*. Thus the connection between the perceptions and the statements remained obscure, and was described by correspondingly obscure expressions which elucidated nothing, but slurred over the difficulties or, at best, adumbrated them through metaphors.

Here too a solution can be found, I believe, if we clearly separate the psychological from the logical and methodological aspects of the problem. We must distinguish between, on the one hand, *our subjective experiences or our feelings of conviction*, which can never justify any statement (though they can be made the subject of psychological investigation) and, on the other hand, the *objective logical relations* subsisting among the various systems of scientific statements, and within each of them.

The problems of the empirical basis will be discussed in some detail in Sections 25 to 30. For the present I had better turn to the problem of scientific objectivity, since the terms 'objective' and 'subjective' which I have just used are in need of elucidation.

8. SCIENTIFIC OBJECTIVITY AND SUBJECTIVE CONVICTION

The words 'objective' and 'subjective' are philosophical terms heavily burdened with a heritage of contradictory usages and of inconclusive and interminable discussions.

My use of the terms 'objective' and 'subjective' is not unlike Kant's. He uses the word 'objective' to indicate that scientific knowledge should be *justifiable*, independently of anybody's whim: a justification is 'objective' if in principle it can be tested and understood by anybody. "If something is valid", he writes, "for anybody in possession of his reason, then its grounds are objective and sufficient".[1]

Now I hold that scientific theories are never fully justifiable or verifiable, but that they are nevertheless testable. I shall therefore say that the *objectivity* of scientific statements lies in the fact that they can be *intersubjectively tested*.[*1]

The word 'subjective' is applied by Kant to our feelings of conviction (of varying degrees).[2] To examine how these come about is the business of psychology. They may arise, for example, "in accordance with the laws of association".[3] Objective reasons too may serve as "subjective *causes* of judging",[4] in so far as we may reflect upon these reasons, and become convinced of their cogency.

Kant was perhaps the first to realize that the objectivity of scientific statements is closely connected with the construction of theories – with the use of hypotheses and universal statements. Only when certain events recur in accordance with rules or regularities, as is the case with repeatable experiments, can our observations be tested – in principle – by anyone.

We do not take even our own observations quite seriously, or accept them as scientific observations, until we have repeated and tested them. Only by such repetitions can we convince ourselves that we are not dealing with a mere isolated 'coincidence', but with events which, on account of their regularity and reproducibility, are in principle inter-subjectively testable.[5]

Every experimental physicist knows those surprising and inexplicable apparent 'effects' which can perhaps even be reproduced in his laboratory for some time, but which finally disappear without trace. Of course, no physicist would say in such a case that he had made a scientific discovery (though he might try to rearrange his experiments so as to make the effect reproducible). Indeed the scientifically significant *physical effect* may be defined as that which can be regularly reproduced by anyone who carries out the appropriate experiment in the way prescribed. No serious physicist would offer for publication, as a scientific discovery, any such 'occult effect', as I propose to call it – one for whose reproduction he could give no instructions. The 'discovery' would be only too soon rejected as chimerical, simply because attempts to test it would lead to negative results.[6] (It follows that any controversy over the question whether events which are in principle unrepeatable and unique ever do occur cannot be decided by science: it would be a metaphysical controversy.)

We may now return to a point made in the previous section: to my thesis that a subjective experience, or a feeling of conviction, can never justify a scientific statement, and that within science it can play no part but that of the subject of an empirical (a psychological) inquiry. No matter how intense a feeling of conviction it may be, it can never justify a statement. Thus I can be utterly convinced of the truth of a statement; certain of the evidence of my perceptions; overwhelmed by the intensity of my experience: every doubt may seem to me absurd. But does this afford the slightest reason for science to accept my statement? Can any statement be justified by the fact that K.R.P. is utterly convinced of its truth? The answer is, 'No'; and any other answer would be incompatible with the idea of scientific objectivity. Even the fact, for me so firmly established, that I am experiencing this feeling of conviction, cannot appear within the field of objective science except in the form of a *psychological hypothesis* which, of course, calls for inter-subjective testing: from the conjecture that I have this feeling of conviction the psychologist may deduce, with the help of psychological and other theories, certain predictions about my behaviour; and these may be confirmed or refuted in the course of experimental tests. But from the epistemological point of view, it is quite irrelevant whether my feeling of conviction was strong or weak; whether it came from a strong or even irresistible impression of indubitable certainty (or 'self-evidence'), or merely from a doubtful surmise. None of this has any bearing on the question of how scientific statements can be justified.

Considerations like these do not of course provide an answer to the problem of the empirical basis. But at least they help us to see its main difficulty. In demanding objectivity for basic statements as well as for other scientific statements, we deprive ourselves of any logical means by which we might have hoped to reduce the truth of scientific statements to our experiences. Moreover we debar ourselves from granting any favoured status to statements which represent experiences, such as those statements which describe our perceptions (and which are sometimes called 'protocol sentences'). They can occur in science only as psychological statements; and this means, as hypotheses of a kind whose standards of inter-subjective testing (considering the present state of psychology) are certainly not very high.

Whatever may be our eventual answer to the question of the empirical basis, one thing must be clear: if we adhere to our demand that scientific statements must be objective, then those statements which belong to the empirical basis of science must also be objective, i.e. inter-subjectively testable. Yet inter-subjective testability always implies that from the statements which are to be tested, other testable statements can be deduced. Thus if the basic statements in their turn are to be inter-subjectively testable, *there can be no ultimate statements in science*: there can be no statements in science which cannot be tested, and therefore none which cannot in principle be refuted, by falsifying some of the conclusions which can be deduced from them.

We thus arrive at the following view. Systems of theories are tested by deducing from them statements of a lesser level of universality. These statements in their turn, since they are to be inter-subjectively testable, must be testable in like manner – and so *ad infinitum*.

It might be thought that this view leads to an infinite regress, and that it is therefore untenable. In Section I, when criticizing induction, I raised the objection that it may lead to an infinite regress; and it might well appear to the reader now that the very same objection can be urged against that procedure of deductive testing which I myself advocate. However, this is not so. The deductive method of testing cannot establish or justify the statements which are being tested; nor is it intended to do so. Thus there is no danger of an infinite regress. But it must be admitted that the situation to which I have drawn attention – testability *ad infinitum* and the absence of ultimate statements which are not in need of tests – does create a problem. For, clearly, tests cannot in fact be carried on *ad infinitum*: sooner or later we have to stop. Without discussing this problem here in detail, I only wish to point out that the fact that the tests cannot go on for ever does not clash with my demand that every scientific statement must be testable. For I do not demand that every scientific statement must *have in fact been tested* before it is accepted. I only demand that every such statement must be *capable* of being tested; or in other words, I refuse to accept the view that there are statements in science which we have, resignedly, to accept as true merely because it does not seem possible, for logical reasons, to test them.

APPENDIX

... A system such as classical mechanics may be 'scientific' to any degree you like; but those who uphold it dogmatically – believing, perhaps, that it is their business to defend such a successful system against criticism as long as it is not *conclusively disproved* – are adopting the very reverse of that critical attitude which in my view is the proper one for the scientist. In point of fact, no conclusive disproof of a theory can ever be produced; for it is always possible to say that the experimental results are not reliable, or that the discrepancies which are asserted to exist between the experimental results and the theory are only apparent and that they will disappear with the advance of our understanding. (In the struggle against Einstein, both these arguments were often used in support of Newtonian mechanics, and similar arguments abound in the field of the social sciences.) If you insist on strict proof (or strict disproof *1)) in the empirical sciences, you will never benefit from experience, and never learn from it how wrong you are.

If therefore we characterize empirical science merely by the formal or logical structure of its statements, we shall not be able to exclude from it that prevalent form of metaphysics which results from elevating an obsolete scientific theory into an incontrovertible truth.

Such are my reasons for proposing that empirical science should be characterized by its methods: by our manner of dealing with scientific systems: by what we do with them and what we do to them. Thus I shall try to establish the rules, or if you will the norms, by which the scientist is guided when he is engaged in research or in discovery, in the sense here understood.

... The falsifying mode of inference here referred to – the way in which the falsification of a conclusion entails the falsification of the system from which it is derived – is the *modus tollens* of classical logic....

By means of this mode of inference we falsify *the whole system* (the theory as well as the initial conditions) which was required for the deduction of the statement *p*, i.e. of the falsifying statement. Thus it cannot be asserted of any one statement of the system that it is, or is not, specifically upset by the falsification. Only if *p* is *independent* of some part of the system can we say that this part is not involved in the falsification.[1] With this is connected the following possibility: we may, in some cases, perhaps in consideration of the *levels of universality*, attribute the falsification to some definite hypothesis – for instance to a newly introduced hypothesis. This may happen if a well-corroborated theory, and one which continues to be further corroborated, has been deductively explained by a new hypothesis of a higher level. The attempt will have to be made to test this new hypothesis by means of some of its consequences which have not yet been tested. If any of these are falsified, then we may well attribute the falsification to the new hypothesis alone. We shall then seek, in its stead, other high-level generalizations, but we shall not feel obliged to regard the old system, of lesser generality, as having been falsified.

NOTES

* Chapter 1 of *The Logic of Scientific Discovery*, and, as an appendix, selections from pp. 50, 76, 77; copyright © 1959 by Karl Raimund Popper. Published by Basic Books, New York, 1959, seventh English impression Hutchinson of London, 1974. Reprinted by permission. This book is a translation by the author (with the assistance of Julius Freed and Lan Freed) of *Logik der Forschung*, first published in Vienna in 1935. The

starred notes were added to the translation by the author. The various appendices, the Postscript, and the other numbered sections referred to in the footnotes may be found in this translation.

SECTION 1

[1] H. Reichenbach, *Erkenntnis* 1 (1930), 186 (cf. also p. 64 f.).

[2] Reichenbach, *ibid.*, p. 67.

*1 The decisive passages from Hume are quoted in appendix *vii, text to Notes 4, 5, and 6; see also Note 2 to Section 81, below.

[3] Cf. J. M. Keynes, *A Treatise on Probability* (1921); O. Külpe, *Vorlesungen über Logic* (ed. by Selz, 1923); Reichenbach (who uses the term 'probability implications'), 'Axiomatik der Wahrscheinlichkeitrechnung', *Mathem. Zeitschr.* **34** (1932); and in many other places.

[4] Reichenbach, *Erkenntnis* 1 (1930), 186.

*2 See also Chapter x, below, especially Note 2 to Section 81, and Chapter *ii of the *Postscript* for a fuller statement of this criticism.

[5] Liebig (in *Induktion und Deduktion*, 1865) was probably the first to reject the inductive method from the standpoint of natural science; his attack is directed against Bacon. Duhem (in *La Théorie physique, son objet et sa structure*, 1906; English translation by P. P. Wiener: *The Aim and Structure of Physical Theory*, Princeton, 1954) held pronounced deductivist views. (*But there are also inductivist views to be found in Duhem's book, for example in the third chapter, Part One, where we are told that only experiment, induction, and generalization have produced Descartes's law of diffraction; cf. the English translation, p. 455.) See also V. Kraft, *Die Grundformen der Wissenschaftlichen Methoden*, 1925; and Carnap, *Erkenntnis* **2** (1932), 440.

SECTION 2

[1] Address on Max Planck's 60th birthday. The passage quoted begins with the words, "The supreme task of the physicist is to search for those universal laws...", etc. (quoted from A. Einstein, *Mein Weltbild*, 1934, p. 168; English translation by A. Harris: *The World as I see It*, 1935, p. 125). Similar ideas are found earlier in Liebig, *op. cit.*; cf. also Mach, *Principien der Wärmelehre* (1896), p. 443ff. *The German word *'Einfühlung'* is difficult to translate. Harris translates: "sympathetic understanding of experience".

SECTION 3

*1 For this term, see Note *1 before Section 79, and Section *29 of my *Postscript*.

SECTION 4

[1] With this (and also with Sections 1 to 6 and 13 to 24) cf. my note: *Erkenntnis* **3** (1933), 426; *It is now here reprinted, in translation, as appendix*.

[2] Cf. the last sentence of his *Enquiry Concerning Human Understanding*. *With the next paragraph, compare for example the quotation from Reichenbach in the text to Note 1, Section 1.

*1 When I wrote this paragraph I overrated the 'modern positivists', as I now see. I should have remembered that *in this respect* the promising beginning of Wittgenstein's *Tractatus* – "The world is the totality of facts, not of things" – was cancelled by its end which denounced the man who "had given no meaning to certain signs in his propositions". See also my *Open Society and its Enemies*, Chapter II, Section ii, and Chapter *i of my *Postscript*, especially Sections *11 (Note 5), *24 (the last five paragraphs), and *25.

*2 Nothing depends on names, of course. When I invented the new name 'basic statement' (or 'basic proposition'; see below, Sections 7 and 28) I did so only because I needed a term *not* burdened with the connotation of a perception statement. But unfortunately it was soon adopted by others, and used to convey precisely the kind of meaning which I wished to avoid. Cf. also my *Postscript*, *29.

*3 Hume thus condemned his own *Enquiry* on its last page, just as later Wittgenstein condemned his own *Tractatus* on its last page. (See Note 2 to Section 10.)

3 Carnap, *Erkenntnis* 2 (1932), 219 ff. Earlier Mill had used the word 'meaningless' in a similar way, *no doubt under the influence of Comte; cf. Comte's *Early Essays on Social Philosophy*, ed. by H. D. Hutton, 1911, p. 223. See also my *Open Society*, Note 51 to Chapter 11.

4 Wittgenstein, *Tractatus Logico-Philosophicus* (1918 and 1922), Proposition 5. *As this was written in 1934, I am dealing here of course *only* with the *Tractatus*.

5 Wittgenstein, *op. cit.*, Proposition 4.01; 4.03; 2.221.

6 Cf. Note 1 to Section 2.

*4 The idea of treating scientific laws as pseudo-propositions – thus solving the problem of induction – was attributed by Schlick to Wittgenstein. (Cf. my *Open Society*, Notes 46 and 51 f. to Chapter II.) But it is really much older. It is part of the instrumentalist tradition which can be traced back to Berkeley, and further. (See for example my paper 'Three Views Concerning Human Knowledge', in *Contemporary British Philosophy*, 1956; and 'A Note on Berkeley as a Precursor of Mach', in *The British Journal for the Philosophy of Science* iv, 4, (1953), 26 ff., now in my *Conjectures and Refutations*, 1959. Further references in Note *1 before Section 12 (p. 59). The problem is also treated in my *Postscript*, Sections *11 to *14, and *19 to *26.)

7 Schlick, *Naturwissenschaften* 19 (1931), 156. (The italics are mine.) Regarding natural laws Schlick writes (p. 151), "It has often been remarked that, strictly, we can never speak of an absolute verification of a law, since we always, so to speak, tacitly make the reservation that it may be modified in the light of further experience. If I may add, by way of parenthesis", Schlick continues, "a few words on the logical situation, the above-mentioned fact means that a natural law, in principle, does not have the logical character of a statement, but is, rather, a prescription for the formation of statements." *('Formation' no doubt was meant to include transformation or derivation.) Schlick attributed this theory to a personal communication of Wittgenstein's. See also Section *12 of my *Postscript*.

8 Cf. Section 78 (for example Note 1). *See also my *Open Society*, Notes 46, 51, and 52 to Chapter II, and my paper 'The Demarcation between Science and Metaphysics', contributed in January 1955 to the planned Carnap volume of the *Library of Living Philosophers*, ed. by P. A. Schilpp.

*5 I believe that a reasonable discussion is always possible between parties interested in truth, and ready to pay attention to each other. (Cf. my *Open Society*, Chapter 24.)

9 This is Dingler's view; cf. Note 1 to Section 19.

10 This is the view of O. Spaan (*Kategorienlehre*, 1924).

11 Cf. also Planck, *Positivismus und reale Aussenwelt* (1931) and Einstein, 'Die Religiosität der Forschung', in *Mein Weltbild* (1934), p. 43; English translation by A. Harris: *The World as I See It* (1935), p. 23 ff. *See also Section 85, and my *Postscript*.

SECTION 5

*1 Cf. appendix *x.

SECTION 6

¹ Schlick, *Naturwissenschaften* **19** (1931), 150.

² Waismann, *Erkenntnis* **1** (1930), 229.

*¹ I am not, of course, here considering so-called 'mathematical induction'; what I am denying is that there is such a thing as induction in the so-called 'inductive sciences'; that there are either 'inductive procedures' or 'inductive inferences'.

*² In his *Logical Syntax* (1937, p. 321 f.) Carnap admitted that this was a mistake (with a reference to my criticism); and he did so even more fully in 'Testability and Meaning', recognizing the fact that universal laws are not only 'convenient' for science but even 'essential' (*Philosophy of Science* **4** (1937), 27). But in his inductivist *Logical Foundations of Probability* (1950), he returns to a position very like the one here criticized: finding that universal laws have zero probability (p. 511), he is compelled to say (p. 575) that though they need not be expelled from science, science can very well do without them.

*³ Note that I suggest falsifiability as a criterion of demarcation, but *not of meaning*. Note, moreover, that I have already (Section 4) sharply criticized the use of the idea of meaning as a criterion of demarcation, and that I attack the dogma of meaning again, even more sharply, in Section 9. It is therefore a sheer myth (though any number of refutations of my theory have been based upon this myth) that I ever proposed falsifiability as a criterion of meaning. Falsifiability separates two kinds of perfectly meaningful statements: the falsifiable and the non-falsifiable. It draws a line inside meaningful language, not around it. See also Appendix *i, and Chapter *i of my *Postscript*, especially Sections *17 and *19.

³ Related ideas are to be found, for example, in Frank, *Die Kausalität und ihre Grenzen* (1931), Ch. I, Section 10 (p. 15 f); Dubislav, *Die Definition* (3rd edition 1931), p. 100 f. (Cf. also Note 1 to Section 4, above.).

*⁴ This asymmetry is now more fully discussed in Section *22 of my *Postscript*.

⁴ For this see also my paper mentioned in note 1 to section 4, *now here reprinted as appendix *i; and my *Postscript*, esp. Section *2.

SECTION 8

¹ *Kritik der reinen Vernunft*, Methodenlehre, 2. Haupstück, 3. Abschnitt (2nd edition, p. 848; English Translation by N. Kemp Smith, 1933: *Critique of Pure Reason*, The Transcendental Doctrine of Method, Chapter ii, Section 3, p. 645).

*¹ I have since generalized this formulation; for inter-subjective *testing* is merely a very important aspect of the more general idea of inter-subjective *criticism*, or in other words, of the idea of mutual rational control by critical discussion. This more general idea, discussed at some length in my *Open Society and Its Enemies*, Chapters 23 and 24, and in my *Poverty of Historicism*, Section 32, is also discussed in my *Postscript*, especially in Chapters *i, *ii, and *vi.

² *Ibid.*

³ Cf. *Kritik der reinen Vernunft*, Transcendentale Elementarlehre Section 19 (2nd edition, p. 142; English translation by N. Kemp Smith, 1933: *Critique of Pure Reason*, Transcendental Doctrine of Elements, Section 19, p. 159).

⁴ Cf. *Kritik der reinen Vernunft*, Methodenlehre, 2. Haupstück, 3. Abschnitt (2nd edition, p. 849; English translation, Chapter ii, Section 3, p. 646).

⁵ Kant realized that from the required objectivity of scientific statements it follows that they must be at any time inter-subjectively testable, and that they must therefore have the form of universal laws or theories. He formulated this discovery somewhat

obscurely by his "principle of temporal succession according to the law of causality" (which principle he believed that he could prove *a priori* by employing the reasoning here indicated). I do not postulate any such principle (cf. Section 12); but I agree that scientific statements, since they must be inter-subjectively testable, must always have the character of universal hypotheses.* See also Note *1 to Section 22.

[6] In the literature of physics there are to be found some instances of reports, by serious investigators, of the occurrence of effects which could not be reproduced, since further tests led to negative results. A well-known example from recent times is the unexplained positive result of Michelson's experiment observed by Miller (1921–1926) at Mount Wilson, after he himself (as well as Morley) had previously reproduced Michelson's negative result. But since later tests again gave negative results it is now customary to regard these latter as decisive, and to explain Miller's divergent result as "due to unknown sources of error". *See also Section 22, especially Note *1.

APPENDIX

*1 I have now here added in brackets the words 'or strict disproof' to the text (a) because they are clearly implied by what is said immediately before ("no conclusive disproof of a theory can ever be produced"), and (b) because I have been constantly misinterpreted as upholding a criterion (and moreover one of *meaning* rather than of *demarcation*) based upon a doctrine of 'complete' or 'conclusive' falsifiability.

[1] Thus we cannot at first know which among the various statements of the remaining sub-system t' (of which p is not independent) we are to blame for the falsity of p; which of these statements we have to alter, and which we should retain. (I am not here discussing interchangeable statements.) It is often only the scientific instinct of the investigator (influenced, of course, by the results of testing and re-testing) that makes him guess which statements of t' he should regard as innocuous, and which he should regard as being in need of modification. Yet it is worth remembering that it is often the modification of what we are inclined to regard as obviously innocuous (because of its complete agreement with our normal habits of thought) which may produce a decisive advance. A notable example of this is Einstein's modification of the concept of simultaneity.

KARL R. POPPER

BACKGROUND KNOWLEDGE AND SCIENTIFIC GROWTH*

People involved in a fruitful critical discussion of a problem often rely, if only unconsciously, upon two things: the acceptance by all parties of the common aim of getting at the truth, or at least nearer to the truth, and a considerable amount of common background knowledge. This does not mean that either of these two things is an indispensible basis of every discussion, or that these two things are themselves '*a priori*', and cannot be critically discussed in their turn. It only means that criticism never starts from nothing, even though every one of its starting points *may* be challenged, one at a time, in the course of the critical debate.

Yet though every one of our assumptions may be challenged, it is quite impracticable to challenge all of them at the same time. Thus all criticism must be piecemeal (as against the holistic view of Duhem and of Quine); which is only another way of saying that the fundamental maxim of every critical discussion is that we should stick to our problem, and that we should subdivide it, if practicable, and try to solve no more than one problem at a time, although we may, of course, always proceed to a subsidiary problem, or replace our problem by a better one.

While discussing a problem we always accept (if only temporarily) all kinds of things as *unproblematic*: they constitute for the time being, and for the discussion of this particular problem, what I call our *background knowledge*. Few parts of this background knowledge will appear to us in all contexts as absolutely unproblematic, and any particular part of it *may* be challenged at any time, especially if we suspect that its uncritical acceptance may be responsible for some of our difficulties. But almost all of the vast amount of background knowledge which we constantly use in any informal discussion will, for practical reasons, necessarily remain unquestioned; and the misguided attempt to question it all – that is to say, *to start from scratch* – can easily lead to the breakdown of a critical debate. (Were we to start the race where Adam started, I know of no reason why we should get any further than Adam did.)

The fact that, as a rule, we are at any given moment taking a vast amount of traditional knowledge for granted (for almost all our knowledge is traditional) creates no difficulty for the falsificationist or fallibilist. For he does not *accept* this background knowledge; neither as established nor as fairly certain, nor yet as probable. He knows that even its tentative acceptance is risky, and stresses that every bit of it is open to criticism, even though only in a piecemeal way. We can never be certain that we shall challenge the right bit; but since our quest is not for certainty, this does not matter. It will be noticed that this remark contains my answer to Quine's holistic view of empirical tests; a view which Quine formulates (with reference to Duhem), by asserting that our statements about the external world face the tribunal of sense experience not individually but only as a corporate body.[1] Now it has to be admitted that we can often test only a large chunk of a theoretical system, and sometimes perhaps only the whole system, and that, in these cases, it is sheer guesswork which of its ingredients should be held responsible for any falsification; a point which I have tried to emphasize – also with reference to Duhem – for a long time past. Though this argument may turn a verificationist into a sceptic, it does not affect those who hold that all our theories are guesses anyway.

This shows that the holistic view of tests, even if it were true, would not create a serious difficulty for the fallibilist and falsificationist. On the other hand, it should be said that the holistic argument goes much too far. It is possible in quite a few cases to find which hypothesis is responsible for the refutation; or in other words, which part, or group of hypotheses, was necessary for the derivation of the refuted prediction. The fact that such logical dependencies may be discovered is established by the practice of *independence proofs* of axiomatized systems; proofs which show that certain axioms of an axiomatic system cannot be derived from the rest. The more simple of these proofs consist in the construction, or rather in the discovery, of a *model* – a set of things, relations, operations, or functions – which satisfies all the axioms except the *one* whose independence is to be shown: for this one axiom – and therefore for the theory as a whole – the model constitutes a counter example.

Now let us say that we have an axiomatized theoretical system, for example of physics, which allows us to predict that certain things do not happen, and that we discover a counter example. There is no reason whatever why this counter example may not be found to satisfy most of our axioms or even all our axioms except one whose independence would be thus established. This shows that the holistic dogma of the 'global' character of all tests or counter examples is untenable. And it explains why, even without axiomatizing our physical theory, we may well have an inkling of what has gone wrong with our system.

This, incidentally, speaks in favour of operating, in physics, with highly analysed theoretical systems – that is, with systems which, even though they may fuse all the hypotheses into one, allow us to separate various groups of hypotheses, each of which may become an object of refutation by counter examples. (An excellent recent example is the rejection, in atomic theory, of the law of parity; another is the rejection of the law of commutation for conjugate variables, prior to their interpretation as matrices, and to the statistical interpretation of these matrices.)

NOTES

* From 'Truth, Rationality, and the Growth of Knowledge', pp. 238–239, in *Conjectures and Refutations* by Karl R. Popper. Copyright © 1963, 1965, 1969 by Karl R. Popper. Published by Routledge and Kegan Paul, London, and Basic Books, Inc, New York. Reprinted by permission.
[1] See W. V. Quine, *From a Logical Point of View*, 1953, p. 41.

ADOLF GRÜNBAUM*

THE DUHEMIAN ARGUMENT**

ABSTRACT. This paper offers a refutation of P. Duhem's thesis that the *falsifiability* of an isolated empirical hypothesis H as an *explanans* is *unavoidably inconclusive*. Its central contentions are the following:

(1) No general features of the logic of falsifiability can assure, for every isolated empirical hypotheses H and independently of the domain to which it pertains, that H can always be preserved as an *explanans* of any empirical findings O whatever by some modification of the auxiliary assumptions A in conjunction with which H functions as an *explanans*. For Duhem *cannot* guarantee on any general logical grounds the deducibility of O from an *explanans* constituted by the conjunction of H and some revised *non*-trivial version R of A: the existence of the required set R of collateral assumptions must be demonstrated for each particular case.

(2) The categorical form of the Duhemian thesis is not only a *non-sequitur* but actually false. This is shown by adducing the testing of physical geometry as a counterexample to Duhem in the form of a rebuttal to A. Einstein's geometrical articulation of Duhem's thesis.

(3) The possibility of a quasi *a priori* choice of a physical geometry in the sense of Duhem must be clearly *distinguished* from the feasibility of a conventional adoption of such a geometry in the sense of H. Poincaré. And the legitimacy of the latter cannot be invoked to save the Duhemian thesis from refutation by the foregoing considerations.

In a recent paper published in this journal, Herburt [6] expounds and endorses the Duhemian argument but contests its being adduced by Quine to repudiate the distinction between analytic and synthetic statements.

The present paper is intended to report the result of an investigation by the author which shows that the categorical form of Duhem's contention, viz. that the falsification of part of an *explanans* is always unavoidably inconclusive, is untenable. In particular, it turns out that it is one thing to maintain with Herburt that "in every empirical test a certain number of statements of various types is involved" [6, p. 109] but quite another to conclude in Duhemian fashion, as he does, that "in principle, it is possible...to maintain any particular empirical statement, whatever the data of experience, provided we make appropriate changes in the system of hypotheses which is put to test" [6, p. 108].

We must distinguish the following two forms of Duhem's thesis:

(i) the logic of every disconfirmation, no less than of every confirmation, of a presumably empirical hypothesis H is such as to *involve at some stage or other* an entire network of interwoven hypotheses in which H is ingredient rather than the separate testing of the component H,

(ii) *No one constituent hypothesis H* can ever be *extricated* from the ever-present web of collateral assumptions so as to be *open* to *decisive refutation* by the evidence as part of an *explanans* of that evidence, just as no such isolation is achievable for purposes of verification. This conclusion becomes apparent by a consideration of the two parts of the schema of unavoidably inconclusive falsifiability, which are:

(a) it is an elementary fact of *deductive* logic that if certain observational consequences O are entailed by the *conjunction* of H and a set A of auxiliary assumptions, then the *failure* of O to materialize entails *not* the falsity of H by itself but only the weaker conclusion that H and A cannot *both* be true; the falsifiability of H is therefore *inconclusive* in the sense that the *falsity* of H is *not deductively inferable* from the premiss $[(H \cdot A) \to O] \cdot \sim O$,

(b) the actual observational findings O', which are incompatible with O, *allow* that H be true while A is false, because they always permit the theorist to preserve H with impunity as a part of the *explanans* of O' by so modifying A that the *conjunction* of H and the *revised* version A' of A does explain (entail) O'. This preservability of H is to be understood as a retainability *in principle* and does *not* depend on the ability of scientists to propound the required set A' of collateral assumptions at any given time.

Thus, there is an ingression of a kind of *a priori* choice into physical theory: at the price of suitable compensatory modifications in the remainder of the theory, *any one* of its *component* hypotheses H may be retained in the face of seemingly contrary empirical findings as an *explanans* of these very findings. And this quasi *a priori* preservability of H is sanctioned by the far-reaching theoretical *ambiguity* and flexibility of the logical constraints imposed by the observational evidence.[1]

Let us now consider the two parts (a) and (b) of the schema which the *stronger* form (ii) of the Duhemian thesis claims to be the *universal paradigm* of the logic of falsifiability in empirical science. Clearly, part (a) is valid, being a *modus tollens* argument in which the antecedent of the conditional premise is a conjunction. But part (a) utilizes the *de facto* findings O' only to the extent that they are *incompatible* with the observational expectations O derived from the conjunction of H and A. And part (a) is *not at all sufficient to show that the falsifiability of H as part of an explanans of the actual empirical facts O' is unavoidably inconclusive.*

For neither part (a) nor other general logical considerations can *guarantee* the deducibility of O' from an *explanans* constitued by the conjunction of H and some *non-trivial* revised set A' of the auxiliary assumptions which is logically incompatible with A *under the hypothesis H*.[2]

How then does Duhem propose to assure that there exists such a *non-trivial* set A' for any one component hypothesis H *independently* of the domain of empirical science to which H pertains? It would seem that such assurance *cannot* be given on general logical grounds at all but that the existence of the required set A' needs *separate* and *concrete* demonstration for each particular case. In short, even in contexts to which part (a) of the Duhemian schema is applicable – which is *not* true for *all* contexts, as we shall see – neither the premiss

$$[(H \cdot A) \rightarrow O] \cdot \sim O,$$

nor other general logical considerations entail that

$$(\exists A') [(H \cdot A') \rightarrow O'],$$

where A' is non-trivial in the sense of Note 2. And hence Duhem's thesis that the falsifiability of an *explanans* H is unavoidably inconclusive is a *non-sequitur.*

That the Duhemian thesis is not only a *non-sequitur* but actually false is borne out, as we shall now see, by the case of testing the hypothesis that a certain *physical geometry* holds, a case of conclusive falsifiability which yields an important *counterexample* to Duhem's stronger thesis concerning falsifiability but which does justify the *weaker* form (i) of his thesis.

I have given a detailed logical analysis of the highly ramified issue of the *empirical credentials* of physical geometry (and chronometry) in other recent publications [3, 4], and I shall confine the present discussion to summarizing those results of that analysis which serve the following objectives of this paper:

1. To substantiate *geometrically* my claim that

(1) by denying the feasibility of conclusive falsification, the Duhemian schema is a serious *misrepresentation* of the actual logical situation characterizing an important class of cases of falsifiability of a purported explanans, and that

(2) the plausibility of Duhem's thesis derives from the false supposition that part (a) of the schema *is* always applicable *and* that its formal validity guarantees the applicability of part (b) of the schema.

The geometrical substantiation of my claim will make apparent the incorrectness of Herburt's contention that part (a) of the schema "establishes conclusively the cogency of Duhem's claim" [6, p. 107].

2. To clarify the important distinction between the *a priori* choice of a physical geometry in the sense of *Duhem* and the conventional choice of such a geometry in the sense of *Poincaré*. An appreciation of this distinction will counteract the prevalent misunderstanding, apparently shared by Herburt [6, p. 105], that Poincare's conventionalist conception of geometry is to be construed as an espousal of the Duhemian argument as applied to the special case of physical geometry.

1. PHYSICAL GEOMETRY AS A COUNTEREXAMPLE TO THE DUHEMIAN THESIS

Since Duhem's argument was *articulated* and endorsed by Einstein a decade ago in regard to the epistemological status of physical geometry, I shall summarize my critique of Einstein's geometrical version of that argument.

Physical geometry is usually conceived as the system of metric relations exhibited by transported solid bodies *independently of their particular chemical composition*. On this conception, the criterion of congruence can be furnished by a transported solid body for the purpose of determining the geometry by measurement, only if the computational application of suitable 'corrections' (or, ideally, appropriate shielding) has assured rigidity in the sense of essentially eliminating inhomogeneous thermal, elastic, electromagnetic and other perturbational influences. For these influences are 'deforming' in the sense of producing changes of *varying degree* in different kinds of materials. Since the existence of perturbational influences thus issues in a dependence of the coincidence behavior of transported solid rods on the latter's *chemical composition*, and since physical geometry is concerned with the behavior common to all solids apart from their substance-specific idiosyncrasies, the discounting of idiosyncratic distortions is an essential aspect of the logic of physical geometry. The demand for the computational *elimination* of such distortions as a prerequisite to the experimental determination of the geometry has a thermodynamic counterpart: the requirement of a means for measuring temperature which does not yield the discordant results produced by expansion thermometers at other than fixed points when different thermometric substances are employed. This thermometric need is fulfilled successfully by Kelvin's thermodynamic scale of temperature. But attention to the implementation of the corresponding prerequisite of physical geometry has led Einstein to impugn the empirical status of that geometry. He considers the case in which congruence has been defined by the diverse kinds of transported solid measuring rods *as corrected for their respective idiosyncratic distortions* with a view to then making an empirical determination of the prevailing geometry. And Einstein's thesis is that the very logic of computing these corrections precludes that the geometry itself be accessible to experimental ascertainment *in isolation from* other physical regularities. Specifically, he states his case in the form of a dialogue [1, pp. 676–678] in which he attributes his own Duhemian view to Poincaré for reasons that will become clear later on and offers that view in opposition to Hans Reichenbach's conception [10, 11]. But I submit that Poincaré's text will *not* bear Einstein's interpretation. For in speaking of the variations which solids exhibit under distorting influences, Poincaré says "we neglect these variations in laying the foundations of geometry, because, besides their being very slight, they are irregular and consequently seem to us accidental". [8, p. 76] I am therefore taking the liberty of replacing the name 'Poincaré' in Einstein's dialogue by the term 'Duhem and Einstein'. *With this modification*, the dialogue reads as follows [1, pp. 676–678]:

Duhem and Einstein: The empirically given bodies are not rigid, and consequently can not be used for the embodiment of geometric intervals. Therefore, the theorems of geometry are not verifiable.

Reichenbach: I admit that there are no bodies which can be *immediately* adduced for the 'real definition' [i.e. physical definition] of the interval. Nevertheless, this real definition can be achieved by taking the thermal volume-dependence, elasticity, electro- and magneto-striction, etc., into consideration. That this is really and without contradiction possible, classical physics has surely demonstrated.

Duhem and Einstein: In gaining the real definition improved by yourself you have made use of physical laws, the formulation of which presupposes (in this case) Euclidean geometry. The verification, of which you have spoken, refers, therefore, not merely to geometry but to the entire system of physical laws which constitute its foundation. An examination of geometry by itself is consequently not thinkable. – Why should it consequently not be entirely up to me to choose geometry according to my own convenience (i.e., Euclidean) and to fit the remaining (in the usual sense 'physical') laws to this choice in such manner that there can arise no contradiction of the whole with experience?

Einstein is making two major points here:

(1) In obtaining a physical geometry by giving a physical interpretation of the postulates of a formal geometric axiom system, the specification of the physical meaning of such theoretical terms as 'congruent', 'length', or 'distance' is *not* at all simply a matter of giving an operational definition in the strict sense. Instead, what has been variously called a 'rule of correspondence' (Margenau and Carnap), a 'coordinative definition' (Reichenbach), an 'epistemic correlation' (Northrop) or a 'dictionary' (N. R. Campbell) is provided here *through the mediation of hypotheses and laws* which are *collateral* to the geometric theory whose physical meaning is being specified. Einstein's point that the physical meaning of congruence is given by the transported rod *as corrected theoretically* for idiosyncratic distortions is an illuminating one and has an abundance of analogues throughout physical theory, thus showing, incidentally, that strictly operational definitions are a rather simplified and limiting species of rules of correspondence [cf. 4, §2, Section (ii)].

(2) Einstein's second claim, which is the cardinal one for our purposes, is that the role of collateral theory in the physical definition of congruence is such as to issue in the following *circularity*, from which there is no escape, he maintains, short of acknowledging the existence of an *a priori* element *in the sense of the Duhemian ambiguity*: the rigid body is not even defined without first *decreeing* the validity of Euclidean geometry (or of some other particular geometry). For *before* the *corrected* rod can be used to make an empirical determination of the *de facto* geometry, the required corrections must be computed via laws, such as those of elasticity, which involve Euclideanly-calculated areas and volumes. But clearly the warrant for thus introducing Euclidean geometry *at this stage* cannot be empirical.

If Einstein's Duhemian thesis were to prove correct, then it would have to be acknowledged that there is a sense in which physical geometry *itself* does not provide a geometric characterization of physical reality. For by this characterization we understand the articulation of the system of relations obtaining between bodies and transported solid rods quite apart from their substance-specific distortions. And to the extent to which physical geometry is *a priori* in the sense of the Duhemian ambiguity, there is an ingression of *a priori* elements into physical theory to take the place of distinctively geometric gaps in our knowledge of the physical world.

I now wish to set forth my doubts regarding the soundness of Einstein's contention. And I shall do so in two parts the first of which deals with the special case in which effectively no deforming influences are present in a certain region whose geometry is to be ascertained.

(i) If we are confronted with the problem of the falsifiability of the geometry ascribed to a region which is effectively free from deforming influences, then the *correctional* physical laws play no role as auxiliary assumptions, and the latter reduce to the claim that the region in question is, in fact, effectively *free* from deforming influences. And *if* such freedom can be affirmed *without* presupposing collateral theory, then the geometry alone rather than only a wider theory in which it is ingredient will be falsifiable. The question is therefore whether freedom from deforming influences can be asserted and ascertained independently of (sophisticated) collateral theory. My answer to this question is: Yes. For quite independently of the conceptual elaboration of such physical magnitudes as temperature, whose constancy would characterize a region free from deforming influences, the absence of perturbations is certifiable for the region as follows: two solid rods of very different chemical constitution which coincide at one place in the region will also coincide everywhere else in it independently of their paths of transport. Accordingly, the absence of deforming influences is ascertainable *independently* of any assumptions as to the geometry and of other (sophisticated) collateral theory.

Let us now employ our earlier notation and denote the geometry by 'H' and the assertion concerning the freedom from perturbations by 'A'. Then, once we have laid down the congruence definition and the remaining semantical rules, the physical geometry H becomes conclusively falsifiable as an *explanans* of the posited empirical findings O'. For the actual logical situation is characterized *not* by part (a) of the Duhemian schema but instead by the schema

$$[\{(H \cdot A) \to O\} \cdot \sim O \cdot A] \to \, \sim H.$$

It will be noted that we identified the H of the Duhemian schema with the geometry. But since a geometric theory, at least in its synthetic form, can be axiomatized as a conjunction of logically independent postulates, a particular axiomatization of H could be decomposed logically into various sets of component subhypotheses. Thus, for example, the hypothesis of Euclidean geometry could be stated, if we wished, as the conjunction of two parts consisting respectively of the Euclidean parallel postulate and the postulates of absolute geometry. And the hypothesis of hyperbolic geometry could be stated in the form of a conjunction of absolute geometry and the hyperbolic parallel postulate.

In view of the logically-compound character of a geometric hypothesis, Professor Grover Maxwell has suggested that the Duhemian thesis may be tenable in this context if we construe it as pertaining *not* to the falsifiability of a geometry as a whole but to the falsifiability of its component subhypotheses in any given axiomatization. There are two ways in which this proposed interpretation might be understood: (1) as an assertion that *any one component sub*hypothesis eludes conclusive refutation on the grounds that the empirical findings can falsify the set of axioms only as a whole, or (2) in any given axiomatization of a physical geometry there exists *at least one component sub*hypothesis which eludes conclusive refutation.

The first version of the proposed interpretation will not bear examination. For suppose that H is the hypothesis of Euclidean geometry and that we consider absolute geometry as one of its subhypotheses and the Euclidean parallel postulate as the other. If now the empirical findings were to show that the geometry is hyperbolic, then indeed absolute geometry would have eluded refutation. But if, on the other hand, the prevailing geometry were to turn out to be spherical, then the mere replacement of the Euclidean parallel postulate by the spherical one could not possibly save absolute geometry from refutation. For absolute geometry alone is logically incompatible with spherical geometry and hence with the posited empirical findings.

If one were to read Duhem as per the very cautious *second* version of Maxwell's proposed interpretation, then our analysis of the logic of testing the geometry of a *perturbation-free* region could *not* be adduced as having furnished a counter-example to so mild a form of Duhemism. And the question of the validity of this highly attenuated version is thus left open by our analysis without detriment to that analysis.

We now turn to the critique of Einstein's Duhemian argument as applied to the empirical determination of the geometry of a region which *is* subject to deforming influences.

(ii) There can be no question that when deforming influences *are* present, the laws used to make the corrections for deformations involve areas and volumes in a fundamental way (e.g. in the definitions of the elastic stresses and strains) and that this involvement presupposes a geometry, as is evident from the area and volume formulae of differential geometry, which contain the square root of the determinant of the components g_{ik} of the metric tensor. Thus, the empirical determination of the geometry involves the joint assumption of a geometry and of certain collateral hypotheses. But we see already that this assumption cannot be adequately represented by the conjunction $H \cdot A$ of the Duhemian schema where H represents the geometry.

Now suppose that we begin with a set of Euclideanly-formulated physical laws P_0 in correcting for the distortions induced by perturbations and then use the thus Euclideanly-corrected congruence standard for *empirically* exploring the geometry of space by determining the metric tensor. *The initial stipulational affirmation of the Euclidean geometry G_0 in the physical laws P_0 used to compute the corrections in no way assures that the geometry obtained by the corrected rods will be Euclidean.* If it is non-Euclidean, then the question is: what will be involved in Einstein's fitting of the physical laws to preserve Euclideanism and avoid a contradiction of the theoretical system with experience? Will the adjustments in P_0 necessitated by the retention of Euclideanism entail merely a change in the dependence of the length assigned to the transported rod on such *non-positional* parameters as temperature, pressure, magnetic field etc.? Or could the putative empirical findings compel that the length of the transported rod be likewise made a non-constant function of its *position* and *orientation* as independent variables in order to square the coincidence findings with the requirement of Euclideanism? The temporal variability of distorting influences and the possibility of obtaining non-Euclidean results by measurements carried out in a spatial region uniformly characterized by standard conditions of temperature, pressure, electric and magnetic field strength etc. show it to be *quite doubtful* that the preservation of Euclideanism could always be accomplished short of introducing *the dependence of the rod's length on the independent variables of position and orientation.* Thus, in order to retain Euclideanism, it may be necessary to *remetrize* entirely apart from any consideration of idiosyncratic distortions and even after correcting for these in some way or other. But this kind of remetrization, though entirely admissible in *other* contexts, does *not* provide the requisite support for Einstein's Duhemian thesis. For it is the avowed onus of that thesis to show that the geometry *by itself* cannot be held to be empirical even when, with Reichenbach, we have sought to assure its empirical character by choosing and then adhering to the customary (standard) definition of congruence, *which excludes resorting to such remetrization.*

Thus, there may well obtain observational findings O', expressed in terms of a particular definition of congruence (e.g., the *customary* one), which are such that there does *not* exist any *non*-trivial set A' of auxiliary assumptions capable of preserving the Euclidean H in the face of O'. And this result alone suffices to invalidate the Einsteinian version of Duhem's thesis to the effect that any geometry, such as Euclid's, can be preserved in the face of any experimental findings which are expressed in terms of the customary definition of congruence.

But what of the possibility of actually *extricating* the unique underlying geometry (to within experimental accuracy) from the network of hypotheses which enter into the testing procedure? Elsewhere [3, 4], I have given two methods which can determine the unique underlying geometry in the case of a space of constant curvature, it being an open question whether these methods can also be generalized to cover the case of a space of variable curvature, and, if not, whether there is another method which succeeds in that case.[3]

It might appear that my geometric counterexample to the Duhemian thesis of unavoidably inconclusive falsifiability of an *explanans* is vulnerable to the following criticism:

To be sure, Einstein's geometric articulation of that thesis does *not* leave room for saving it by resorting to a remetrization in the sense of making the length of the rod *vary* with position or orientation even *after* it has been corrected for idiosyncratic distortions. But why saddle the Duhemian thesis as such with a restriction peculiar to Einstein's particular version of it? And thus why not allow Duhem to save his thesis by countenancing those *alterations in the congruence definition* which are *remetrizations*?

My reply is that to deny the Duhemian the invocation of such an alteration of the congruence definition *in this context* is *not* a matter of gratuitously requiring him to justify his thesis within the confines of Einstein's particular version of that thesis; instead, the imposition of this restriction is entirely legitimate here, and the Duhemian could hardly wish to reject it as unwarranted. For it is of the essence of Duhem's contention that H (in this case: Euclidean geometry) can always be preserved *not* by tampering with the *semantical rules* (interpretive sentences) linking H to the observational base but rather by availing oneself of the alleged *inductive latitude* afforded by the ambiguity of the experimental evidence to do the following: (a) leave the factual commitments of H *unaltered* by retaining both the statement of H and the semantical rules linking its terms to the observational base, and (b) replace the set A by a set A' of auxiliary assumptions *differing in factual content* from A such that A and A' are logically incompatible under the hypothesis H. Now, the factual content of a geometrical hypothesis can be *changed* either by preserving the original statement of the hypothesis while changing one or more of the semantical rules or by keeping all of the semantical rules intact and suitably changing the statement of the hypothesis [3, pp. 213–214]. We can see, therefore, that the retention of a Euclidean H by the device of changing through remetrization the semantical rule governing the meaning of 'congruent' (for line segments) effects a retention not of the *factual commitments* of the original Euclidean H but only of its *linguistic trappings*. That the thus 'preserved' Euclidean H actually *repudiates* the factual commitments of the *original* one is clear from the following: the *original* Euclidean H had asserted that the coincidence behavior common to all kinds of solid rods is Euclidean, *if* such transported rods are taken as the physical realization of congruent intervals; but the Euclidean H which survived the confrontation with the posited empirical findings only by dint of a *remetrization* is predicated on a *denial* of the very assertion that was made by the original Euclidean H, which it was to 'preserve'.

Hence, the confines within which the Duhemian must make good his claim of the preservability of a Euclidean H do *not* admit of the kind of change in the congruence definition which alone would render his claim tenable under the assumed empirical conditions. Accordingly, the geometrical critique of Duhem's thesis given in this paper does *not* depend for its validity on restrictions peculiar to Einstein's version of it.

Even apart from the fact that Duhem's thesis precludes resorting to an alternative metrization to save it from refutation in our geometrical context, the very feasibility of alternative metrizations is vouchsafed *not* by any general Duhemian considerations pertaining to the logic of falsifiability but by a property peculiar to the subject matter of geometry (and chronometry): the latitude for *convention* in the ascription of the spatial (or temporal) *equality* relation to intervals in the continuous manifolds of physical space (or time) [3, 4].

2. THE DISTINCTION BETWEEN THE CONVENTIONAL ADOPTION OF A PHYSICAL GEOMETRY IN THE SENSE OF H. POINCARÉ AND ITS QUASI 'A PRIORI' CHOICE IN THE SENSE OF P. DUHEM

The key to the difference between the geometric conventionalism of H. Poincaré and the geometrical form of the conventionalism of P. Duhem is furnished by the distinction just used to rebut the objection that Duhem is being unfairly saddled with liabilities incurred by Einstein: the distinction between preserving a particular geometry (e.g. the Euclidean one) by a remetrizational change in the congruence definition, on the one hand, and intending to retain a particular geometry *without* change in that definition (or in other semantical rules) by an alteration of the factual content of the auxiliary assumptions, on the other. More specifically, the Duhemian conception envisions scope for alternative geometric accounts of a given body of evidence only to the extent that these geometries are associated with alternative sets of correctional physical laws. On the other hand, the range of alternative geometric descriptions of given evidence affirmed by Poincaré is far wider and rests on very different grounds: instead of involving the Duhemian *inductive* latitude, Poincaré bases the possibility of giving *either* a Euclidean *or* a non-Euclidean description of the same spatio-physical facts on alternative metrizability. For Poincaré tells us [8, pp. 66–80] *that quite apart from any considerations of substance-specific distorting influences and even after correcting for these in some way or other*, we are at liberty to define congruence – and thereby to fix the geometry appropriate to the given facts – either by calling the solid rod equal to itself everywhere or by making its length vary in a specified way with its position and orientation. Thus, whereas Duhem's affirmation of the retainability of Euclidean geometry in the face of any observational evidence is *inductive*, the preservability of that geometry asserted by Poincaré is *remetrizational*: Poincaré's conventionalist claim regarding geometry is that if the customary definition of congruence on the basis of the coincidence behavior common to all kinds of solid rods does not assure a particular geometric description of the facts, then such a description can be guaranteed remetrizationally, i.e., by merely choosing an appropriately different noncustomary congruence definition which makes the length of every kind of solid rod a specified *non*-constant function of the *independent* variables of position and orientation.

Duhem's and Poincaré's differing conceptions of the retainability of a given metric geometry issue in two correspondingly different views as to the *interdependence* of geometry and of the remainder of physics. The special optical case of Poincaré's much-discussed and widely-misunderstood statement of the possibility of always giving a Euclidean description of any results of stellar parallax measurements [8, p. 81] will serve to articulate the difference between the *linguistic* interdependence of geometry and optics espoused by Poincaré and their *inductive* (epistemological) interdependence as championed by Duhem.

Poincaré's point is that if the paths of light rays are geodesics on a particular definition of congruence, and if these paths are found parallactically to sustain *non*-Euclidean relations on that metrization, then we need only choose a different definition of congruence such that these *same* paths will no longer be geodesics and that the geodesics of the newly chosen congruence are Euclideanly related. From the standpoint of synthetic geometry, the latter choice effects only a *renaming* of optical and other paths and thus is merely a *recasting of the same factual content in Euclidean language rather than a revision of the extra-linguistic content of optical and other laws*. For an alternative metrization affects only the *language* in which the facts of optics and the coincidence behavior of a transported rod are described: the two geometric descriptions respectively associated with two alternative metrizations are *alternative representations of the same factual content*, and so are the two sets of optical laws corresponding to these geometries. Accordingly, Poincaré is affirming a *linguistic* interdependence of the geometric theory of rigid solids and the optical theory of light rays.

On the other hand, the attempt to explain certain parallactic data by different geometries which constitute alternatives in the inductive sense of Duhem would presumably issue in the following alternative between two theoretical systems, each of which comprises a geometry G and an optics O:

(a) G_E: the geometry of the rigid body geodesics is Euclidean, and

O_1: the paths of light rays do *not* coincide with these geodesics but form a non-Euclidean system,

or

(b) $G_{non \cdot E}$: the geodesics of the rigid body congruence are *not* a Euclidean system, and

O_2: the paths of light rays *do* coincide with these geodesics, and thus they form a non-Euclidean system.

We saw that the physically-interpreted alternative geometries associated with two (or more) different metrizations in the sense of Poincaré have precisely the same total factual content, as do the corresponding two sets of optical laws. By contrast, in the Duhemian account, G_E and $G_{non \cdot E}$ not only *differ* in factual content but are logically incompatible, and so are O_1 and O_2. And on the latter conception, there is sameness of factual content *in regard to the assumed parallactic data* only between the *combined* systems formed by the two conjunctions (G_E and O_1) and ($G_{non \cdot E}$ and O_2). Thus, the need for the combined system of G and O to yield the empirical facts, coupled with the avowed epistemological (inductive) *inseparability* of G and O lead the Duhemian to conceive of their *interdependence* as *inductive* (epistemological).

Hence whereas Duhem (and Einstein) construe the interdependence of G and O inductively such that the geometry by itself is *not* accessible to empirical test, Poincaré's conception of their interdependence allows for an empirical determination of G *by itself*, *if* we have *renounced* recourse to an alternative metrization in which the length of the rod is held to vary with its position or orientation.

It would seem that it was Poincaré's discussion of the interdependence of optics and geometry by reference to stellar parallax measurements which led Einstein and others to regard him as a proponent of the Duhemian thesis. But this interpretation appears untenable not only in the light of the immediate context of Poincaré's discussion of his astronomical example, but also, as I have shown elsewhere [2, 4], upon taking account of the remainder of his writings.

NOTES

* The author is indebted to the National Science Foundation for the support of research and wishes to acknowledge the benefit of discussions with Dr Grover Maxwell

and other fellow-participants in the 1959 summer sessions of the Minnesota Center for Philosophy of Science.
** Reprinted from *Philosophy of Science* 27, No. 1, January, 1960. Copyright © 1960, Philosophy of Science Association. Reprinted by permission.
[1] Cf. P. Duhem, *The Aim and Structure of Physical Theory* (tr. by P. P. Wiener), Princeton, 1954, Part II, Chapter VI [Selection 1 in this collection – Ed.], esp. pp. 183–190. Duhem's explicit disavowal of both decisive falsifiability and crucial verifiability of an *explanans* will *not* bear K. R. Popper's reading of him [9, p. 78]: Popper, who is an exponent of decisive falsifiability [9], misinterprets Duhem as allowing that tests of a hypothesis may be decisively *falsifying* and as denying only that they may be crucially *verifying*. Notwithstanding Popper's *exegetical* error, we shall find presently that his thesis of the feasibility of decisively falsifying tests can be buttressed by a telling counterexample to Duhem's categorical denial of that thesis.
A defense of the claim that *isolated* parts of physical theory can be *confirmed* is outlined by H. Feigl in his 'Confirmability and Confirmation', *Revue Internationale de Philosophie* V (1951), 268–279, which is reprinted in P. P. Wiener (ed.), *Readings in Philosophy of Science*, New York, 1953, esp. pp. 528–529.
[2] The requirement of *non-triviality* of A' requires clarification. If one were to allow O' itself, for example, to qualify as a set A', then, of course, O' could be deduced trivially, and H would not even be needed in the *explanans*. Hence a *necessary* condition for the *non-triviality* of A' is that H be required in addition to A' for the deduction of the *explanandum*. But, as N. Rescher has pointed out to me, this necessary condition is not also sufficient. For it *fails* to rule out an A' of the trivial form $\sim H \vee O'$ (or $H \supset O'$) from which O' could *not* be deduced without H.
The unavailability of a formal sufficient condition for non-triviality is not, however, damaging to the critique of Duhem presented in this paper. For surely Duhem's illustrations from the history of physics as well as the whole tenor of his writing indicate that *he intends his thesis to stand or fall on the existence of the kind of A' which we would all recognize as non-trivial in any given case.* Any endeavor to save Duhem's thesis from refutation by invoking the kind of A' which no scientist would accept as admissible would turn Duhem's thesis into a most unenlightening triviality that no one would wish to contest. Thus, I have no intention whatever of denying the following compound formal claim: if H and A jointly entail O, the falsity of O does not entail the falsity of H, *and* there will always be *some kind of A'* which, in conjunction with H, will entail O'.
[3] Whatever the answer to this open question, I have argued in another publication [5] that it is wholly misconceived to suppose with J. Maritain [7] that there are *supra*-scientific philosophical means for ascertaining the underlying geometry, if *scientific* procedures do not succeed in unraveling it.

BIBLIOGRAPHY

[1] Einstein, A.: 'Reply to Criticisms', in *Albert Einstein: Philosopher-Scientist* (ed. by Schilpp, P. A.), Evanston, 1949, pp. 665–688.
[2] Grünbaum, A.: 'Carnap's Views on the Foundations of Geometry', in *The Philosophy of Rudolf Carnap* (ed. by Schilpp, P. A.), New York, 1963.
[3] Grünbaum, A.: 'Conventionalism in Geometry', in *The Axiomatic Method* (ed. by Henkin, L., Suppes, P., and Tarski, A.), Amsterdam, 1959, pp. 204–222.
[4] Grünbaum, A.: 'Geometry, Chronometry and Empiricism', in *Minnesota Studies in the Philosophy of Science* (ed. by Feigl, H. and Maxwell, G.), Vol. III, Minneapolis, 1962.

[5] Grünbaum, A.: 'The *A Priori* in Physical Theory', to appear in the Proceedings of the Symposium on the Nature of Physical Knowledge, held at the summer 1959 meeting of the American Physical Society, Milwaukee, Wisconsin.

[6] Herburt, G. K.: 'The Analytic and the Synthetic', *Philosophy of Science* **26** (1959), 104–113.

[7] Maritain, J.: *The Degrees of Knowledge*, New York, 1959, pp. 165–173.

[8] Poincaré, H.: *The Foundations of Science*, Lancaster, 1946.

[9] Popper, K. R.: *The Logic of Scientific Discovery*, London, 1959.

[10] Reichenbach, H.: 'The Philosophical Significance of the Theory of Relativity', in *Albert Einstein: Philosopher-Scientist* (ed. by Schilpp, P. A.), Evanston, 1949, pp. 287–311.

[11] Reichenbach, H.: *The Philosophy of Space and Time*, New York, 1958, Ch. I, §§ 3–8 incl.

WILLARD VAN ORMAN QUINE

A COMMENT ON GRÜNBAUM'S CLAIM

[*Editor's note*: Quine and Grünbaum have authorized publication of the following letter in which Quine gives his view of Grünbaum's critique as the latter appeared in 'The Falsifiability of Theories: Total or Partial? A Contemporary Evaluation of the Duhem–Quine Thesis', later published in M. Wartofsky (ed.), *Boston Studies in the Philosophy of Science*, Vol. I Reidel, Dordrecht, 1963).]

Professor Adolf Grünbaum   June 1, 1962
Department of Philosophy
University of Pittsburgh
Pittsburgh 13, Pennsylvania

Dear Professor Grünbaum:

I have read your paper on the falsifiability of theories with interest. Your claim that the Duhem–Quine thesis, as you call it, is untenable if taken nontrivially, strikes me as persuasive. Certainly it is carefully argued.

For my own part I would say that the thesis as I have used it *is* probably trivial. I haven't advanced it as an interesting thesis as such. I bring it in only in the course of arguing against such notions as that the empirical content of sentences can in general be sorted out distributively, sentence by sentence, or that the understanding of a term can be segregated from collateral information regarding the object. For such purposes I am not concerned even to avoid the trivial extreme of sustaining a law by changing a meaning; for the cleavage between meaning and fact is part of what, in such contexts, I am questioning.

Actually my holism is not as extreme as those brief vague paragraphs at the end of "Two dogmas of empiricism" are bound to sound. See sections 1–3 and 7–10 of *Word and Object*.

Sincerely yours,
W. V. Quine

THOMAS S. KUHN

SCIENTIFIC REVOLUTIONS AS CHANGES
OF WORLD VIEW*

Examining the record of past research from the vantage of contemporary historiography, the historian of science may be tempted to exclaim that when paradigms change, the world itself changes with them. Led by a new paradigm, scientists adopt new instruments and look in new places. Even more important, during revolutions scientists see new and different things when looking with familiar instruments in places they have looked before. It is rather as if the professional community had been suddenly transported to another planet where familiar objects are seen in a different light and are joined by unfamiliar ones as well. Of course, nothing of quite that sort does occur: there is no geographical transplantation; outside the laboratory everyday affairs usually continue as before. Nevertheless, paradigm changes do cause scientists to see the world of their research-engagement differently. In so far as their only recourse to that world is through what they see and do, we may want to say that after a revolution scientists are responding to a different world.

It is as elementary prototypes for these transformations of the scientist's world that the familiar demonstrations of a switch in visual gestalt prove so suggestive. What were ducks in the scientist's world before the revolution are rabbits afterwards. The man who first saw the exterior of the box from above later sees its interior from below. Transformations like these, though usually more gradual and almost always irreversible, are common concomitants of scientific training. Looking at a contour map, the student sees lines on paper, the cartographer a picture of a terrain. Looking at a bubble-chamber photograph, the student sees confused and broken lines, the physicist a record of familiar subnuclear events. Only after a number of such transformations of vision does the student become an inhabitant of the scientist's world, seeing what the scientist sees and responding as the scientist does. The world that the student then enters is not, however, fixed once and for all by the nature of the environment, on the one hand, and of science, on the other. Rather, it is determined jointly by the environment and the particular normal-scientific tradition

that the student has been trained to pursue. Therefore, at times of revolution, when the normal-scientific tradition changes, the scientist's
perception of his environment must be re-educated – in some familiar
situations he must learn to see a new gestalt. After he has done so the
world of his research will seem, here and there, incommensurable with
the one he had inhabited before. That is another reason why schools
guided by different paradigms are always slightly at cross-purposes.

In their most usual form, of course, gestalt experiments illustrate only
the nature of perceptual transformations. They tell us nothing about the
role of paradigms or of previously assimilated experience in the process of
perception. But on that point there is a rich body of psychological
literature, much of it stemming from the pioneering work of the Hanover
Institute. An experimental subject who puts on goggles fitted with inverting lenses initially sees the entire world upside down. At the start his
perceptual apparatus functions as it had been trained to function in the
absence of the goggles, and the result is extreme disorientation, an acute
personal crisis. But after the subject has begun to learn to deal with his
new world, his entire visual field flips over, usually after an intervening
period in which vision is simply confused. Thereafter, objects are again
seen as they had been before the goggles were put on. The assimilation of a
previously anomalous visual field has reacted upon and changed the
field itself.[1] Literally as well as metaphorically, the man accustomed to
inverting lenses has undergone a revolutionary transformation of vision.

The subjects of the anomalous playing-card experiment discussed in
Section VI experienced a quite similar transformation. Until taught by
prolonged exposure that the universe contained anomalous cards, they
saw only the types of cards for which previous experience had equipped
them. Yet once experience had provided the requisite additional categories, they were able to see all anomalous cards on the first inspection
long enough to permit any identification at all. Still other experiments
demonstrate that the perceived size, color, and so on, of experimentally
displayed objects also varies with the subject's previous training and
experience.[2] Surveying the rich experimental literature from which these
examples are drawn makes one suspect that something like a paradigm
is prerequisite to perception itself. What a man sees depends both upon
what he looks at and also upon what his previous visual-conceptual
experience has taught him to see. In the absence of such training there

can only be, in William James's phrase, "a bloomin' buzzin' confusion".

In recent years several of those concerned with the history of science have found the sorts of experiments described above immensely suggestive. N. R. Hanson, in particular, has used gestalt demonstrations to elaborate some of the same consequences of scientific belief that concern me here.[3] Other colleagues have repeatedly noted that history of science would make better and more coherent sense if one could suppose that scientists occasionally experienced shifts of perception like those described above. Yet, though psychological experiments are suggestive, they cannot, in the nature of the case, be more than that. They do display characteristics of perception that *could* be central to scientific development, but they do not demonstrate that the careful and controlled observation exercised by the research scientist at all partakes of those characteristics. Furthermore, the very nature of these experiments makes any direct demonstration of that point impossible. If historical example is to make these psychological experiments seem relevant, we must first notice the sorts of evidence that we may and may not expect history to provide.

The subject of a gestalt demonstration knows that his perception has shifted because he can make it shift back and forth repeatedly while he holds the same book or piece of paper in his hands. Aware that nothing in his environment has changed, he directs his attention increasingly not to the figure (duck or rabbit) but to the lines on the paper he is looking at. Ultimately he may even learn to see those lines without seeing either of the figures, and he may then say (what he could not legitimately have said earlier) that it is these lines that he really sees but that he sees them alternately *as* a duck and *as* a rabbit. By the same token, the subject of the anomalous card experiment knows (or, more accurately, can be persuaded) that his perception must have shifted because an external authority, the experimenter, assures him that regardless of what he *saw*, he was *looking at* a black five of hearts all the time. In both these cases, as in all similar psychological experiments, the effectiveness of the demonstration depends upon its being analyzable in this way. Unless there were an external standard with respect to which a switch of vision could be demonstrated, no conclusion about alternate perceptual possibilities could be drawn.

With scientific observation, however, the situation is exactly reversed. The scientist can have no recourse above or beyond what he sees with his eyes and instruments. If there were some higher authority by recourse to which his vision might be shown to have shifted, then that authority would itself become the source of his data, and the behavior of his vision would become a source of problems (as that of the experimental subject is for the psychologist). The same sorts of problems would arise if the scientist could switch back and forth like the subject of the gestalt experiments. The period during which light was "sometimes a wave and sometimes a particle" was a period of crisis – a period when something was wrong – and it ended only with the development of wave mechanics and the realization that light was a self-consistent entity different from both waves and particles. In the sciences, therefore, if perceptual switches accompany paradigm changes, we may not expect scientists to attest to these changes directly. Looking at the moon, the convert to Copernicanism does not say, "I used to see a planet, but now I see a satellite". That locution would imply a sense in which the Ptolemaic system had once been correct. Instead, a convert to the new astronomy says, "I once took the moon to be (or saw the moon as) a planet, but I was mistaken". That sort of statement does recur in the aftermath of scientific revolutions. If it ordinarily disguises a shift of scientific vision or some other mental transformation with the same effect, we may not expect direct testimony about that shift. Rather we must look for indirect and behavioral evidence that the scientist with a new paradigm sees differently from the way he had seen before.

Let us then return to the data and ask what sorts of transformations in the scientist's world the historian who believes in such changes can discover. Sir William Herschel's discovery of Uranus provides a first example and one that closely parallels the anomalous card experiment. On at least seventeen different occasions between 1690 and 1781, a number of astronomers, including several of Europe's most eminent observers, had seen a star in positions that we now suppose must have been occupied at the time by Uranus. One of the best observers in this group had actually seen the star on four successive nights in 1769 without noting the motion that could have suggested another identification. Herschel, when he first observed the same object twelve years later, did so with a much improved telescope of his own manufacture. As a result, he was able to notice an apparent disk-size that was at least unusual for stars. Something was awry, and he therefore postponed identification pending futher scrutiny. That scrutiny disclosed Uranus' motion among the stars, and Herschel therefore announced that he had seen a new comet! Only several months later, after fruitless attempts to fit the observed motion to a cometary orbit, did Lexell suggest that the orbit was probably planetary.[4] When that suggestion was accepted, there were several fewer stars and one more planet in the world of the professional astronomer. A celestial body that had been observed off and on for almost a century was seen differently after 1781 because, like an anomalous playing card, it could no longer be fitted to the perceptual categories (star or comet) provided by the paradigm that had previously prevailed.

The shift of vision that enabled astronomers to see Uranus, the planet, does not, however, seem to have affected only the perception of that previously observed object. Its consequences were more far-reaching. Probably, though the evidence is equivocal, the minor paradigm change forced by Herschel helped to prepare astronomers for the rapid discovery, after 1801, of the numerous minor planets or asteroids. Because of their small size, these did not display the anomalous magnification that had alerted Herschel. Nevertheless, astronomers prepared to find additional planets were able, with standard instruments, to identify twenty of them in the first fifty years of the nineteenth century.[5] The history of astronomy provides many other examples of paradigm-induced changes in scientific perception, some of them even less equivocal. Can it conceivably be an accident, for example, that Western astronomers first saw change in the previously immutable heavens during the half-century after Copernicus' new paradigm was first proposed? The Chinese, whose cosmological beliefs did not preclude celestial change, had recorded the appearance of many new stars in the heavens at a much earlier date. Also, even without the aid of a telescope, the Chinese had systematically recorded the appearance of sunspots centuries before these were seen by Galileo and his contemporaries.[6] Nor were sunspots and a new star the only examples of celestial change to emerge in the heavens of Western astronomy immediately after Copernicus. Using traditional instruments, some as simple as a piece of thread, late sixteenth-century astronomers repeatedly discovered that comets wandered at will through the space previously reserved for the immutable planets and stars.[7] The very ease and rapidity with which astronomers saw new things when looking at old objects with old instruments may make us wish to say that, after Copernicus, astronomers lived in a different world. In any case, their research responded as though that were the case.

The preceding examples are selected from astronomy because reports of celestial observation are frequently delivered in a vocabulary consisting of relatively pure observation terms. Only in such reports can we hope to find anything like a full parallelism between the observations of scientists and those of the psychologist's experimental subjects. But we need not insist on so full a parallelism, and we have much to gain by relaxing our standard. If we can be content with the everyday use of the verb 'to see', we may quickly recognize that we have already encountered many other examples of the shifts in scientific perception that accompany paradigm change. The extended use of 'perception' and of 'seeing' will shortly require explicit defense, but let me first illustrate its application in practice.

Look again for a moment at two of our previous examples from the history of electricity. During the seventeenth century, when their research was guided by one or another effluvium theory, electricians repeatedly saw chaff particles rebound from, or fall off, the electrified bodies that had attracted them. At least that is what seventeenth-century observers said they saw, and we have no more reason to doubt their reports of perception than our own. Placed before the same apparatus, a modern observer would see electrostatic repulsion (rather than mechanical or gravitational rebounding), but historically, with one universally ignored exception, electrostatic repulsion was not seen as such until Hauksbee's large-scale apparatus had greatly magnified its effects. Repulsion after contact electrification was, however, only one of many new repulsive effects that Hauksbee saw. Through his researches, rather as in a gestalt switch, repulsion suddenly became *the* fundamental manifestation of electrification, and it was then attraction that needed to be explained.[8] The electrical phenomena visible in the early eighteenth century were both subtler and more varied than those seen by observers in the seventeenth century. Or again, after the assimilation of Franklin's paradigm, the electrician looking at a Leyden jar saw something different from what he had seen before. The device had become a condenser, for which neither the jar shape nor glass was required. Instead, the two conducting coatings – one of which had been no part of the original device – emerged to prominence. As both written discussions and pictorial representations gradually attest, two metal plates with a non-conductor between them had become the prototype for the class.[9] Simultaneously, other inductive effects received new descriptions, and still others were noted for the first time.

Shifts of this sort are not restricted to astronomy and electricity. We have already remarked some of the similar transformations of vision that can be drawn from the history of chemistry. Lavoisier, we said, saw oxygen where Priestley had seen dephlogisticated air and where others had seen nothing at all. In learning to see oxygen, however, Lavoisier also had to change his view of many other more familiar substances. He had, for example, to see a compound ore where Priestley and his contemporaries had seen an elementary earth, and there were other such changes besides. At the very least, as a result of discovering oxygen, Lavoisier saw nature differently. And in the absence of some recourse to that hypothetical fixed nature that he "saw differently", the principle of economy will urge us to say that after discovering oxygen Lavoisier worked in a different world.

I shall inquire in a moment about the possibility of avoiding this strange locution, but first we require an additional example of its use, this one deriving from one of the best known parts of the work of Galileo. Since remote antiquity most people have seen one or another heavy body swinging back and forth on a string or chain until it finally comes to rest. To the Aristotelians, who believed that a heavy body is moved by its own nature from a higher position to a state of natural rest at a lower one, the swinging body was simply falling with difficulty. Constrained by the chain, it could achieve rest at its low point only after a tortuous motion and a considerable time. Galileo, on the other hand, looking at the swinging body, saw a pendulum, a body that almost succeeded in repeating the same motion over and over again ad infinitum. And having seen that much, Galileo observed other properties of the pendulum as well and constructed many of the most significant and original parts of his new dynamics around them. From the properties of the pendulum, for example, Galileo derived his only full and sound arguments for the independence of weight and rate of fall, as well as for the relationship between vertical height and terminal velocity of motions down inclined planes.[10] All these natural phenomena he saw differently from the way they had been seen before.

Why did that shift of vision occur? Through Galileo's individual genius, of course. But note that genius does not here manifest itself in more accurate or objective observation of the swinging body. Descriptively, the Aristotelian perception is just as accurate. When Galileo reported that the pendulum's period was independent of amplitude for amplitudes as great as 90°, his view of the pendulum led him to see far more regularity than we can now discover there.[11] Rather, what seems to have been involved was the exploitation by genius of perceptual possibilities made available by a medieval paradigm shift. Galileo was not raised completely as an Aristotelian. On the contrary, he was trained to analyze motions in terms of the impetus theory, a late medieval paradigm which held that the continuing motion of a heavy body is due to an internal power implanted in it by the projector that initiated its motion. Jean Buridan and Nicole Oresme, the fourteenth-century scholastics who brought the impetus theory to its most perfect formulations, are the first men known to have seen in oscillatory motions any part of what Galileo saw there. Buridan describes the motion of a vibrating string as one in which impetus is first implanted when the string is struck; the impetus is next consumed in displacing the string against the resistance of its tension; tension then carries the string back, implanting increasing impetus until the mid-point of motion is reached; after that the impetus displaces the string in the opposite direction, again against the string's tension, and so on in a symmetric process that may continue indefinitely. Later in the century Oresme sketched a similar analysis of the swinging stone in what now appears as the first discussion of a pendulum.[12] His view is clearly very close to the one with which Galileo first approached the pendulum. At least in Oresme's case, and almost certainly in Galileo's as well, it was a view made possible by the transition from the original Aristotelian to the scholastic impetus paradigm for motion. Until that scholastic paradigm was invented, there were no pendulums, but only swinging stones, for the scientist to see. Pendulums were brought into existence by something very like a paradigm-induced gestalt switch.

Do we, however, really need to describe what separates Galileo from Aristotle, or Lavoisier from Priestley, as a transformation of vision? Did these men really *see* different things when *looking at* the same sorts of objects? Is there any legitimate sense in which we can say that they pursued their research in different worlds? Those questions can no longer be postponed, for there is obviously another and far more usual way to describe all of the historical examples outlined above. Many readers will surely want to say that what changes with a paradigm is only the scientist's interpretation of observations that themselves are fixed once and for all by the nature of the environment and of the perceptual apparatus. On this view, Priestley and Lavoisier both saw oxygen, but they interpreted their observations differently; Aristotle and Galileo both saw pendulums, but they differed in their interpretations of what they both had seen.

Let me say at once that this very usual view of what occurs when scientists change their minds about fundamental matters can be neither all wrong nor a mere mistake. Rather it is an essential part of a philosophical paradigm initiated by Descartes and developed at the same time as Newtonian dynamics. That paradigm has served both science and philosophy well. Its exploitation, like that of dynamics itself, has been fruitful of a fundamental understanding that perhaps could not have been achieved in another way. But as the example of Newtonian dynamics also indicates, even the most striking past success provides no guarantee that crisis can be indefinitely postponed. Today research in parts of philosophy, psychology, linguistics, and even art history, all converge to suggest that the traditional paradigm is somehow askew. That failure to fit is also made increasingly apparent by the historical study of science to which most of our attention is necessarily directed here.

None of these crisis-promoting subjects has yet produced a viable alternate to the traditional epistemological paradigm, but they do begin to suggest what some of that paradigm's characteristics will be. I am, for example, acutely aware of the difficulties created by saying that when Aristotle and Galileo looked at swinging stones, the first saw constrained fall, the second a pendulum. The same difficulties are presented in an even more fundamental form by the opening sentences of this section: though the world does not change with a change of paradigm, the scientist afterward works in a different world. Nevertheless, I am convinced that we must learn to make sense of statements that at least resemble these. What occurs during a scientific revolution is not fully reducible to a reinterpretation of individual and stable data. In the first place, the data are not unequivocally stable. A pendulum is not a falling stone, nor is oxygen dephlogisticated air. Consequently, the data that scientists collect from these diverse objects are, as we shall shortly see, themselves different.

More important, the process by which either the individual or the community makes the transition from constrained fall to the pendulum or from dephlogisticated air to oxygen is not one that resembles interpretation. How could it do so in the absence of fixed data for the scientist to interpret? Rather than being an interpreter, the scientist who embraces a new paradigm is like the man wearing inverting lenses. Confronting the same constellation of objects as before and knowing that he does so, he nevertheless finds them transformed through and through in many of their details.

None of these remarks is intended to indicate that scientists do not characteristically interpret observations and data. On the contrary, Galileo interpreted observations on the pendulum, Aristotle observations on falling stones, Musschenbroek observations on a charge-filled bottle, and Franklin observations on a condenser. But each of these interpretations presupposed a paradigm. They were parts of normal science, an enterprise that, as we have already seen, aims to refine, extend, and articulate a paradigm that is already in existence. Section III provided many examples in which interpretation played a central role. Those examples typify the overwhelming majority of research. In each of them the scientist, by virtue of an accepted paradigm, knew what a datum was, what instruments might be used to retrieve it, and what concepts were relevant to its interpretation. Given a paradigm, interpretation of data is central to the enterprise that explores it.

But that interpretive enterprise – and this was the burden of the paragraph before last – can only articulate a paradigm, not correct it. Paradigms are not corrigible by normal science at all. Instead, as we have already seen, normal science ultimately leads only to the recognition of anomalies and to crises. And these are terminated, not by deliberation and interpretation, but by a relatively sudden and unstructured event like the gestalt switch. Scientists then often speak of the 'scales falling from the eyes' or of the 'lightning flash' that 'inundates' a previously obscure puzzle, enabling its components to be seen in a new way that for the first time permits its solution. On other occasions the relevant illumination comes in sleep.[13] No ordinary sense of the term 'interpretation' fits these flashes of intuition through which a new paradigm is born. Though such intuitions depend upon the experience, both anomalous and congruent, gained with the old paradigm, they are not logically or piecemeal linked to particular items of that experience as an interpretation would be. Instead, they gather up large portions of that experience and transform them to the rather different bundle of experience that will thereafter be linked piecemeal to the new paradigm but not to the old.

To learn more about what these differences in experience can be, return for a moment to Aristotle, Galileo, and the pendulum. What data did the interaction of their different paradigms and their common environment make accessible to each of them? Seeing constrained fall, the Aristotelian would measure (or at least discuss – the Aristotelian seldom measured) the weight of the stone, the vertical height to which it had been raised, and the time required for it to achieve rest. Together with the resistance of the medium, these were the conceptual categories deployed by Aristotelian science when dealing with a falling body.[14] Normal research guided by them could not have produced the laws that Galileo discovered. It could only – and by another route it did – lead to the series of crises from which Galileo's view of the swinging stone emerged. As a result of those crises and of other intellectual changes besides, Galileo saw the swinging stone quite differently. Archimedes' work on floating bodies made the medium non-essential; the impetus theory rendered the motion symmetrical and enduring; and Neoplatonism directed Galileo's attention to the motion's circular form.[15] He therefore measured only weight, radius, angular displacement, and time per swing, which were precisely the data that could be interpreted to yield Galileo's laws for the pendulum. In the event, interpretation proved almost unnecessary. Given Galileo's paradigms, pendulum-like regularities were very nearly accessible to inspection. How else are we to account for Galileo's discovery that the bob's period is entirely independent of amplitude, a discovery that the normal science stemming from Galileo had to eradicate and that we are quite unable to document today. Regularities that could not have existed for an Aristotelian (and that are, in fact, nowhere precisely exemplified by nature) were consequences of immediate experience for the man who saw the swinging stone as Galileo did.

Perhaps that example is too fanciful since the Aristotelians recorded no discussions of swinging stones. On their paradigm it was an extraordinarily complex phenomenon. But the Aristotelians did discuss the simpler case, stones falling without uncommon constraints, and the same differences of vision are apparent there. Contemplating a falling stone,

Aristotle saw a change of state rather than a process. For him the relevant measures of a motion were therefore total distance covered and total time elapsed, parameters which yield what we should now call not speed but average speed.[16] Similarly, because the stone was impelled by its nature to reach its final resting point, Aristotle saw the relevant distance parameter at any instant during the motion as the distance *to* the final end point rather than as that *from* the origin of motion.[17] Those conceptual parameters underlie and give sense to most of his well-known 'laws of motion'. Partly through the impetus paradigm, however, and partly through a doctrine known as the latitude of forms, scholastic criticism changed this way of viewing motion. A stone moved by impetus gained more and more of it while receding from its starting point; distance from rather than distance to therefore became the revelant parameter. In addition, Aristotle's notion of speed was bifurcated by the scholastics into concepts that soon after Galileo became our average speed and instantaneous speed. But when seen through the paradigm of which these conceptions were a part, the falling stone, like the pendulum, exhibited its governing laws almost on inspection. Galileo was not one of the first men to suggest that stones fall with a uniformly accelerated motion.[18] Furthermore, he had developed his theorem on this subject together with many of its consequences before he experimented with an inclined plane. That theorem was another one of the network of new regularities accessible to genius in the world determined jointly by nature and by the paradigms upon which Galileo and his contemporaries had been raised. Living in that world, Galileo could still, when he chose, explain why Aristotle had seen what he did. Nevertheless, the immediate content of Galileo's experience with falling stones was not what Aristotle's had been.

It is, of course, by no means clear that we need be so concerned with 'immediate experience' – that is, with the perceptual features that a paradigm so highlights that they surrender their regularities almost upon inspection. Those features must obviously change with the scientist's commitments to paradigms, but they are far from what we ordinarily have in mind when we speak of the raw data or the brute experience from which scientific research is reputed to proceed. Perhaps immediate experience should be set aside as fluid, and we should discuss instead the concrete operations and measurements that the scientist performs in his laboratory. Or perhaps the analysis should be carried further still from the immediately given. It might, for example, be conducted in terms of some neutral observation-language, perhaps one designed to conform to the retinal imprints that mediate what the scientist sees. Only in one of these ways can we hope to retrive a realm in which experience is again stable once and for all – in which the pendulum and constrained fall are not different perceptions but rather different interpretations of the unequivocal data provided by observation of a swinging stone.

But is sensory experience fixed and neutral? Are theories simply man-made interpretations of given data? The epistemological viewpoint that has most often guided Western philosophy for three centuries dictates an immediate and unequivocal, Yes! In the absence of a developed alternative, I find it impossible to relinquish entirely that viewpoint. Yet it no longer functions effectively, and the attempts to make it do so through the introduction of a neutral language of observations now seem to me hopeless.

The operations and measurements that a scientist undertakes in the laboratory are not 'the given' of experience but rather 'the collected with difficulty'. They are not what the scientist sees – at least not before his research is well advanced and his attention focused. Rather, they are concrete indices to the content of more elementary perceptions, and as such they are selected for the close scrutiny of normal research only because they promise opportunity for the fruitful elaboration of an accepted paradigm. Far more clearly than the immediate experience from which they in part derive, operations and measurements are paradigm-determined. Science does not deal in all possible laboratory manipulations. Instead, it selects those relevant to the juxtaposition of a paradigm with the immediate experience that that paradigm has partially determined. As a result, scientists with different paradigms engage in different concrete laboratory manipulations. The measurements to be performed on a pendulum are not the ones relevant to a case of constrained fall. Nor are the operations relevant for the elucidation of oxygen's properties uniformly the same as those required when investigating the characteristics of dephlogisticated air.

As for a pure observation-language, perhaps one will yet be devised. But three centuries after Descartes our hope for such an eventuality still depends exclusively upon a theory of perception and of the mind. And

modern psychological experimentation is rapidly proliferating phenomena with which that theory can scarcely deal. The duck-rabbit shows
that two men with the same retinal impressions can see different things;
the inverting lenses show that two men with different retinal impressions
can see the same thing. Psychology supplies a great deal of other evidence
to the same effect, and the doubts that derive from it are readily rein-
forced by the history of attempts to exhibit an actual language of observation. No current attempt to achieve that end has yet come close to a
generally applicable language of pure percepts. And those attempts that
come closest share one characteristic that strongly reinforces several of
this essay's main theses. From the start they presuppose a paradigm, taken
either from a current scientific theory or from some fraction of everyday
discourse, and they then try to eliminate from it all non-logical and non-
perceptual terms. In a few realms of discourse this effort has been carried
very far and with fascinating results. There can be no question that efforts
of this sort are worth pursuing. But their result is a language that – like
those employed in the sciences – embodies a host of expectations about
nature and fails to function the moment these expectations are violated.
Nelson Goodman makes exactly this point in describing the aims of his
Structure of Appearance: "It is fortunate that nothing more [than
phenomena known to exist] is in question; for the notion of 'possible'
cases, of cases that do not exist but might have existed, is far from clear".[19]
No language thus restricted to reporting a world fully known in advance
can produce mere neutral and objective reports on 'the given". Philosophical investigation has not yet provided even a hint of what a language
able to do that would be like.

Under these circumstances we may at least suspect that scientists are
right in principle as well as in practice when they treat oxygen and pendulums (and perhaps also atoms and electrons) as the fundamental
ingredients of their immediate experience. As a result of the paradigm-
embodied experience of the race, the culture, and, finally, the profession,
the world of the scientist has come to be populated with planets and
pendulums, condensers and compound ores, and other such bodies besides. Compared with these objects of perception, both meter stick
readings and retinal imprints are elaborate constructs to which experience
has direct access only when the scientist, for the special purposes of his
research, arranges that one or the other should do so. This is not to sug-

gest that pendulums, for example, are the only things a scientist could possibly see when looking at a swinging stone. (We have already noted that members of another scientific community could see constrained fall.) But it is to suggest that the scientist who looks at a swinging stone can have no experience that is in principle more elementary than seeing a pendulum. The alternative is not some hypothetical 'fixed' vision, but vision through another paradigm, one which makes the swinging stone something else.

All of this may seem more reasonable if we again remember that neither scientists nor laymen learn to see the world piecemeal or item by item. Except when all the conceptual and manipulative categories are prepared in advance – e.g., for the discovery of an additional transuranic element or for catching sight of a new house – both scientists and laymen sort out whole areas together from the flux of experience. The child who transfers the word 'mama' from all humans to all females and then to his mother is not just learning what 'mama' means or who his mother is. Simultaneously he is learning some of the differences between males and females as well as something about the ways in which all but one female will behave toward him. His reactions, expectations, and beliefs – indeed, much of his perceived world – change accordingly. By the same token, the Copernicans who denied its traditional title 'planet' to the sun were not only learning what 'planet' meant or what the sun was. Instead, they were changing the meaning of 'planet' so that it could continue to make useful distinctions in a world where all celestial bodies, not just the sun, were seen differently from the way they had been seen before. The same point could be made about any of our earlier examples. To see oxygen instead of dephlogisticated air, the condenser instead of the Leyden jar, or the pendulum instead of constrained fall, was only one part of an integrated shift in the scientist's vision of a great many related chemical, electrical, or dynamical phenomena. Paradigms determine large areas of experience at the same time.

It is, however, only after experience has been thus determined that the search for an operational definition or a pure observation-language can begin. The scientist or philosopher who asks what measurements or retinal imprints make the pendulum what it is must already be able to recognize a pendulum when he sees one. If he saw constrained fall instead, his question could not even be asked. And if he saw a pendulum, but saw it in the same way he saw a tuning fork or an oscillating balance, his question could not be answered. At least it could not be answered in the same way, because it would not be the same question. Therefore, though they are always legitimate and are occasionally extraordinarily fruitful, questions about retinal imprints or about the consequences of particular laboratory manipulations presuppose a world already perceptually and conceptually subdivided in a certain way. In a sense such questions are parts of normal science, for they depend upon the existence of a paradigm and they receive different answers as a result of paradigm change.

To conclude this section, let us henceforth neglect retinal impressions and again restrict attention to the laboratory operations that provide the scientist with concrete though fragmentary indices to what he has already seen. One way in which such laboratory operations change with paradigms has already been observed repeatedly. After a scientific revolution many old measurements and manipulations become irrelevant and are replaced by others instead. One does not apply all the same tests to oxygen as to dephlogisticated air. But changes of this sort are never total. Whatever he may then see, the scientist after a revolution is still looking at the same world. Furthermore, though he may previously have employed them differently, much of his language and most of his laboratory instruments are still the same as they were before. As a result, postrevolutionary science invariably includes many of the same manipulations, performed with the same instruments and described in the same terms, as its prerevolutionary predecessor. If these enduring manipulations have been changed at all, the change must lie either in their relation to the paradigm or in their concrete results. I now suggest, by the introduction of one last new example, that both these sorts of changes occur. Examining the work of Dalton and his contemporaries, we shall discover that one and the same operation, when it attaches to nature through a different paradigm, can become an index to a quite different aspect of nature's regularity. In addition, we shall see that occasionally the old manipulation in its new role will yield different concrete results.

Throughout much of the eighteenth century and into the nineteenth, European chemists almost universally believed that the elementary atoms of which all chemical species consisted were held together by forces of mutual affinity. Thus a lump of silver cohered because of the forces of affinity between silver corpuscles (until after Lavoisier these corpuscles were themselves thought of as compounded from still more elementary particles). On the same theory silver dissolved in acid (or salt in water) because the particles of acid attracted those of silver (or the particles of water attracted those of salt) more strongly than particles of these solutes attracted each other. Or again, copper would dissolve in the silver solution and precipitate silver, because the copper-acid affinity was greater than the affinity of acid for silver. A great many other phenomena were explained in the same way. In the eighteenth century the theory of elective affinity was an admirable chemical paradigm, widely and sometimes fruitfully deployed in the design and analysis of chemical experimentation.[20]

Affinity theory, however, drew the line separating physical mixtures from chemical compounds in a way that has become unfamiliar since the assimilation of Dalton's work. Eighteenth-century chemists did recognize two sorts of processes. When mixing produced heat, light, effervescence or something else of the sort, chemical union was seen to have taken place. If, on the other hand, the particles in the mixture could be distinguished by eye or mechanically separated, there was only physical mixture. But in the very large number of intermediate cases – salt in water, alloys, glass, oxygen in the atmosphere, and so on – these crude criteria were of little use. Guided by their paradigm, most chemists viewed this entire intermediate range as chemical, because the processes of which it consisted were all governed by forces of the same sort. Salt in water or oxygen in nitrogen was just as much an example of chemical combination as was the combination produced by oxidizing copper. The arguments for viewing solutions as compounds were very strong. Affinity theory itself was well attested. Besides, the formation of a compound accounted for a solution's observed homogeneity. If, for example, oxygen and nitrogen were only mixed and not combined in the atmosphere, then the heavier gas, oxygen, should settle to the bottom. Dalton, who took the atmosphere to be a mixture, was never satisfactorily able to explain oxygen's failure to do so. The assimilation of his atomic theory ultimately created an anomaly where there had been none before.[21]

One is tempted to say that the chemists who viewed solutions as compounds differed from their successors only over a matter of definition. In one sense that may have been the case. But that sense is not the one that makes definitions mere conventional conveniences. In the eighteenth century mixtures were not fully distinguished from compounds by operational tests, and perhaps they could not have been. Even if chemists had looked for such tests, they would have sought criteria that made the solution a compound. The mixture-compound distinction was part of their paradigm – part of the way they viewed their whole field of research – and as such it was prior to any particular laboratory test, though not to the accumulated experience of chemistry as a whole.

But while chemistry was viewed in this way, chemical phenomena exemplified laws different from those that emerged with the assimilation of Dalton's new paradigm. In particular, while solutions remained compounds, no amount of chemical experimentation could by itself have produced the law of fixed proportions. At the end of the eighteenth century it was widely known that *some* compounds ordinarily contained fixed proportions by weight of their constituents. For some categories of reactions the German chemist Richter had even noted the further regularities now embraced by the law of chemical equivalents.[22] But no chemist made use of these regularities except in recipes, and no one until almost the end of the century thought of generalizing them. Given the obvious counterinstances, like glass or like salt in water, no generalization was possible without an abandonment of affinity theory and a re-conceptualization of the boundaries of the chemist's domain. That consequence became explicit at the very end of the century in a famous debate between the French chemists Proust and Berthollet. The first claimed that all chemical reactions occurred in fixed proportion, the latter that they did not. Each collected impressive experimental evidence for his view. Nevertheless, the two men necessarily talked through each other, and their debate was entirely inconclusive. Where Berthollet saw a compound that could vary in proportion, Proust saw only a physical mixture.[23] To that issue neither experiment nor a change of definitional convention could be relevant. The two men were as fundamentally at cross-purposes as Galileo and Aristotle had been.

This was the situation during the years when John Dalton undertook the investigations that led finally to his famous chemical atomic theory. But until the very last stages of those investigations, Dalton was neither a chemist nor interested in chemistry. Instead, he was a meteorologist investigating the, for him, physical problems of the absorption of gases by water and of water by the atmosphere. Partly because his training was in a different specialty and partly because of his own work in that specialty, he approached these problems with a paradigm different from that of contemporary chemists. In particular, he viewed the mixture of gases or the absorption of a gas in water as a physical process, one in which forces of affinity played no part. To him, therefore, the observed homogeneity of solutions was a problem, but one which he thought he could solve if he could determine the relative sizes and weights of the various atomic particles in his experimental mixtures. It was to determine these sizes and weights that Dalton finally turned to chemistry, supposing from the start that, in the restricted range of reactions that he took to be chemical, atoms could only combine one-to-one or in some other simple whole-number ratio.[24] That natural assumption did enable him to determine the sizes and weights of elementary particles, but it also made the law of constant proportion a tautology. For Dalton, any reaction in which the ingredients did not enter in fixed proportion was *ipso facto* not a purely chemical process. A law that experiment could not have established before Dalton's work, became, once that work was accepted, a constitutive principle that no single set of chemical measurements could have upset. As a result of what is perhaps our fullest example of a scientific revolution, the same chemical manipulations assumed a relationship to chemical generalization very different from the one they had had before.

Needless to say, Dalton's conclusions were widely attacked when first announced. Berthollet, in particular, was never convinced. Considering the nature of the issue, he need not have been. But to most chemists Dalton's new paradigm proved convincing where Proust's had not been, for it had implications far wider and more important than a new criterion for distinguishing a mixture from a compound. If, for example, atoms could combine chemically only in simple whole-number ratios, then a re-examination of existing chemical data should disclose examples of multiple as well as of fixed proportions. Chemists stopped writing that the two oxides of, say, carbon contained 56 per cent and 72 per cent of oxygen by weight; instead they wrote that one weight of carbon would combine either with 1.3 or with 2.6 weights of oxygen. When the results of old manipulations were recorded in this way, a 2:1 ratio leaped to the eye; and this occurred in the analysis of many well-known reactions and of new ones besides. In addition, Dalton's paradigm made it possible to assimilate Richter's work and to see its full generality. Also, it suggested new experiments, particularly those of Gay-Lussac on combining volumes, and these yielded still other regularities, ones that chemists had not previously dreamed of. What chemists took from Dalton was not new experimental laws but a new way of practicing chemistry (he himself called it the "new system of chemical philosophy"), and this proved so rapidly fruitful that only a few of the older chemists in France and Britain were able to resist it.[25] As a result, chemists came to live in a world where reactions behaved quite differently from the way they had before.

As all this went on, one other typical and very important change occurred. Here and there the very numerical data of chemistry began to shift. When Dalton first searched the chemical literature for data to support his physical theory, he found some records of reactions that fitted, but he can scarcely have avoided finding others that did not. Proust's own measurements on the two oxides of copper yielded, for example, an oxygen weight-ratio of 1.47:1 rather than the 2:1 demanded by the atomic theory; and Proust is just the man who might have been expected to achieve the Daltonian ratio.[26] He was, that is, a fine experimentalist, and his view of the relation between mixtures and compounds was very close to Dalton's. But it is hard to make nature fit a paradigm. That is why the puzzles of normal science are so challenging and also why measurements undertaken without a paradigm so seldom lead to any conclusions at all. Chemists could not, therefore, simply accept Dalton's theory on the evidence, for much of that was still negative. Instead, even after accepting the theory, they had still to beat nature into line, a process which, in the event, took almost another generation. When it was done, even the percentage composition of well-known compounds was different. The data themselves had changed. That is the last of the senses in which we may want to say that after a revolution scientists work in a different world.

NOTES

* Chapter X of *The Structure of Scientific Revolutions*, by Thomas S. Kuhn. Copyright © 1962, 1970 by The University of Chicago. Reprinted by permission.
[1] The original experiments were by George M. Stratton, 'Vision without Inversion of the Retinal Image', *Psychological Review* IV (1897), 341–60, 463–81. A more up-to-date review is provided by Harvey A. Carr, *An Introduction to Space Perception*, New York, 1935, pp. 18–57.

[2] For examples, see Albert H. Hastorf, 'The Influence of Suggestion on the Relationship between Stimulus Size and Perceived Distance', *Journal of Psychology* **XXIX** (1950), 195–217; and Jerome S. Bruner, Leo Postman, and John Rodrigues, 'Expectations and the Perception of Color', *American Journal of Psychology* **LXIV** (1951), 216–27.

[3] N. R. Hanson, *Patterns of Discovery*, Cambridge, 1958, Chapter i.

[4] Peter Doig, *A Concise History of Astronomy*, London, 1950, pp. 115–16.

[5] Rudolph Wolf, *Geschichte der Astronomie*, Munich, 1877, pp. 513–15, 683–93. Notice particularly how difficult Wolf's account makes it to explain these discoveries as a consequence of Bode's Law.

[6] Joseph Needham, *Science and Civilization in China*, III, Cambridge, 1959, 423–29, 434–36.

[7] T. S. Kuhn, *The Copernican Revolution*, Cambridge, Mass., 1957, pp. 206–9.

[8] Duane Roller and Duane H. D. Roller, *The Development of the Concept of Electric Charge*, Cambridge, Mass., 1954, pp. 21–29.

[9] See the discussion in Section VII [Chapter VII of *The Structure of Scientific Revolutions* – ed.] and the literature to which the reference there cited in Note 9 will lead.

[10] Galileo Galilei, *Dialogues concerning Two New Sciences*, trans. by H. Crew and A. de Salvio, Evanston, Ill., 1946, pp. 80–81, 162–66.

[11] *Ibid.*, pp. 91–94, 244.

[12] M. Clagett, *The Science of Mechanics in the Middle Ages*, Madison, Wis., 1959, pp. 537–38, 570.

[13] [Jacques] Hadamard, *Subconscient intuition, et logique dans la recherche scientifique (Conférence faite au Palais de la Découverte le 8 Décembre 1945* [Alençon, n.d.]), pp. 7–8. A much fuller account, though one exclusively restricted to mathematical innovations, is the same author's *The Psychology of Invention in the Mathematical Field*, Princeton, 1949.

[14] T. S. Kuhn, 'A Function for Thought Experiments', in *Mélanges Alexandre Koyré*, ed. R. Taton and I. B. Cohen, to be published by Hermann, Paris, in 1963.

[15] A. Koyré, *Etudes Galiléennes*, Paris, 1939, I, 46–51; and 'Galileo and Plato', *Journal of the History of Ideas* IV (1943), 400–428.

[16] Kuhn, 'A Function for Thought Experiments', in *Mélanges Alexandre Koyré* (see n. 14 for full citation).

[17] Koyré, *Etudes...*, II, 7–11.

[18] Clagett, *op. cit.*, Chaps. iv, vi, and ix.

[19] N. Goodman, *The Structure of Appearance*, Cambridge, Mass., 1951, pp. 4–5. The passage is worth quoting more extensively: "If all and only those residents of Wilmington in 1947 that weigh between 175 and 180 pounds have red hair, then 'red-haired 1947 resident of Wilmington' and '1947 resident of Wilmington weighing between 175 and 180 pounds' may be joined in a constructional definition... The question whether there 'might have been' someone to whom one but not the other of these predicates would apply has no bearing... once we have determined that there is no such person.... It is fortunate that nothing more is in question; for the notion of 'possible' cases, of cases that do not exist but might have existed, is far from clear".

[20] H. Metzger, *Newton, Stahl, Boerhaave et la doctrine chimique*, Paris, 1930, pp. 34–68.

[21] *Ibid.*, pp. 124–29, 139–48. For Dalton, see Leonard K. Nash, *The Atomic-Molecular Theory*, 'Harvard Case Histories in Experimental Science', Case 4; Cambridge, Mass., 1950, pp. 14–21.

[22] J. R. Partington, *A Short History of Chemistry*, 2d ed.; London, 1951, pp. 161–63.

[23] A. N. Meldrum, 'The Development of the Atomic Theory: (1) Berthollet's Doctrine of Variable Proportions', *Manchester Memoirs* **LIV** (1910), 1–16.

[24] L. K. Nash, 'The Origin of Dalton's Chemical Atomic Theory', *Isis* **XLVII** (1956), 101–16.

[25] A. N. Meldrum, 'The Development of the Atomic Theory: (6) The Reception Accorded to the Theory Advocated by Dalton', *Manchester Memoirs* **LV** (1911), 1–10.

[26] For Proust, see Meldrum, 'Berthollet's Doctrine of Variable Proportions', *Manchester Memoirs* **LIV** (1910), 8. The detailed history of the gradual changes in measurements of chemical composition and of atomic weights has yet to be written, but Partington, *op. cit.*, provides many useful leads to it.

LAURENS LAUDAN

GRÜNBAUM ON 'THE DUHEMIAN ARGUMENT'*

In several recent publications[1], Professor Adolf Grünbaum has inveighed against the conventionalism of writers like Einstein, Poincaré, Quine and especially Duhem. Specifically, Grünbaum has assailed the view that a single hypothesis can never be conclusively falsified. Grünbaum claims that the conventionalists' insistence on the immunity of hypotheses from falsification is neither logically valid nor scientifically sound. Directing the weight of his argument against Duhem, Grünbaum launches a two-pronged attack. He insists, first, that conclusive falsifying experiments are possible, suggesting that Duhem's denial of such experiments is a logical non-sequitur. He then proceeds to show that, more than being merely possible, crucial falsifying experiments have occurred in physics. I do not intend to make a logical point against Grünbaum's critique so much as an historical and exegetical one. Put briefly, I believe that he has misconstrued Duhem's views on falsifiability and that the logical blunder which he discussed should not be ascribed to Duhem, but rather to those who have made Duhem's conventionalism into the doctrine which Grünbaum attacks. Whether there are any writers who accept the view he imputes to Duhem, or whether he is exploiting 'straw-men' to give weight to an otherwise trivial argument is an open question. For now, I simply want to suggest that his salvos are wrongly directed against Duhem.

The *locus classicus* for Grünbaum's interpretation is Duhem's discussion of falsifiability in his *Aim and Structure of Physical Theory*.[2] I submit that a careful analysis of the historical and textual context of Duhem's account of crucial experiments will indicate how far Grünbaum's argument misses the mark. In the *Aim and Structure*, Duhem was preoccupied with, and disturbed by, the naive realism with which most scientists of the late nineteenth century discussed their theories. They were firmly convinced that science was a search for truth and though they realized that theories could not be proved by verification, they still believed that by suitably eliminating (i.e., falsifying) rival hypotheses, they could ultimately discover the residual true one. Experiments of this type, which simultaneously refuted one hypothesis and thereby verified another (supposedly its only logical alternative), were called crucial experiments (from Bacon's *'instantiae crucis'*). Though the phrase 'crucial experiment' dates from the seventeenth century, it was really the nineteenth century which adopted it as its own. In nineteenth century chemistry, geology, optics and thermodynamics, much use was made of what were thought to be crucial experiments. The notion of a crucial experiment, and the rigorous empiricism which accompanied it, were particularly common among those scientists who saw themselves in the experimental tradition of Newton, Bacon and Lavoisier. But the phrase 'crucial experiment' enjoyed even wider currency in nineteenth century philosophies of science than in science itself; perhaps because it was so compatible with that century's view of scientific progress as the evolution from 'false' theories to 'true' ones, crucial experiments being the instrument whereby the wrong were separated from the right. It was a dogma of research that some theories were really true and the crucial experiment offered itself as the unerring probe for finding the truth. The doctrine had an appealing, not to say appalling, simplicity.

In his incisive critique of this doctrine, Duhem pointed out that simultaneous falsification and verification required that two conditions be satisfied: (1) that an unambiguous falsification procedure exist and (2) that *reductio ad absurdum* methods be applicable to scientific inference. Duhem argued that neither of these conditions could be fulfilled. Disposing first of conclusive falsification, he pointed out that the alleged refutation of an hypothesis, H, by an observation, $\sim O$, presupposes that scientific reasoning follows the simple schema, $H \rightarrow O$. Falsification could then be represented by a modus tollens argument of the form, $[(H \rightarrow O) \cdot \sim O] \rightarrow \sim H$. Yet this is rarely, if ever, the structure of argument in the sciences. Every prediction the scientist makes is based, not on a single hypothesis, but on several – often tacit – assumptions and rules of inference. Since a falsified prediction, O, is a consequence of the conjunction of several hypotheses (i.e., $(H_1 + H_2 + \cdots + H_n) \rightarrow O$), we have no right to single out one of these hypotheses as the false one. All we are entitled to infer from a successful falsification is that the antecedent conjunction $(H_1 + H_2 + \cdots + H_n)$ is false. Beyond this, we can go no further; isolated hypotheses are immune from refutation. To maintain, as

J. F. Herschel did, that each of the hypotheses of classical mechanics can be independently tested is to misconstrue and oversimplify the relation of theory to experiment.

Duhem's corollary argument further strenghened his denial of crucial experiments. Even if one could, *per impossibile*, falsify a specific hypothesis (e.g., light is corpuscular), one has not thereby proved the truth of any alternative hypothesis (e.g., light is undular). There always remains the possibility of (1) future falsification of the alternative hypothesis and (2) the discovery of presently unknown explanations more satisfactory than the alternative. The only truth established by the falsification of H (assuming, of course, that one could falsify H) is $\sim H$, which is not an hypothesis, but a potentially infinite disjunction of hypotheses. Unlike the mathematician who can often exhaustively enumerate all conceivable cases, the physicist cannot list all the possible alternative hypotheses for explaining an event. If and only if he could would *reductio* methods be applicable. The effect of Duhem's dual attack on cruciality was to convince most philosophers of the impossibility of conclusive falsification. Reichenbach, Frank, Hanson, Quine and Toulmin have been among those who have sought to reinforce Duhem's position.

Recently, however, Professor Grünbaum has asserted that philosophers have been misled by Duhem's denial of crucial falsifying experiments[3] because his doctrine is both logically mistaken and a misrepresentation of the role of experiment in physics. Let us consider Grünbaum's two points in turn.

1. Grünbaum's first argument runs as follows: Duhem's denial of conclusive falsification rests on the assumption that for every hypothesis, H, there always exists a set of auxiliary assumptions, A', such that any observations whatever, O, are compatible with, and deducible from, the conjunction of H and A'. Schematically,

$$(H)(O)(\exists A')(H + A' \to O).$$

But, asks Grünbaum, what assurance have we that such an hypothesis-saver as A' will always exist? While there are occasions when a resourceful and imaginative scientist can save an apparently refuted hypothesis by modifying other assumptions, Duhem has given us no guarantee that such modifications will work for every hypothesis.[4] Unless we have a proof that an appropriate non-trivial A' exists for every H and O (a proof that Duhem never offered), then we need not believe that every falsification is inconclusive. As Grünbaum summarizes his position:

the failure of $\sim O$ to permit the deduction of $\sim H$ does *not* justify the assertion of Duhem's thesis that there always *exists* an A' such that the conjunction of H and A' entails $\sim O$![5]

There is much to be said about, though less to be said for, Grünbaum's argument. Because I think it stems from a misinterpretation of Duhem's text, it is not irrelevant to note that Grünbaum systematically avoids references to, or direct discussions of, the *Aim and Structure*. By conflating the views of Quine, Einstein, Weyl and Duhem, he sets out to refute what he variously calls "Duhem's conception as articulated by W. V. O. Quine" (*Ibid.*, 106), "Einstein's geometrical form of the Duhem thesis" (*Ibid.*, 135) and "Weyl's Duhemian thesis" (*Ibid.*). I question whether it is legitimate to assimilate Duhem's views to those of these other writers. In particular, I want to argue that Duhem did not make the logical blunder of asserting what Grünbaum calls the 'Duhem-thesis'; that is, Duhem does not maintain that

$$(H)\,(O)\,(\exists A')\,(H + A' \to O).$$

Duhem's position is a much milder and more modest one than Grünbaum's formulation of the *D*-thesis would lead us to believe. We must remember that Duhem wanted to show, not that falsification never occurred, but that such falsification was necessarily ambiguous. Given that H and A entail O and that $\sim O$, we can surely infer $\sim(H + A)$, though we have no reason to make an unequivocal denial of H rather than A. But to say that H is not refuted by $\sim O$ is certainly not to make the stronger claim that $(\exists A')\,(H + A' \to O)$. Duhem is not asserting that every hypothesis can be saved, but only that unless one had proved that it cannot be saved, then it is not falsified.

Duhem believed that the apparent asymmetry between verification and falsification (according to which the former was always inconclusive and the latter always conclusive) was chimerical. In fact, reasoned Duhem, we can no more falsify a single hypothesis than we can verify it. Every experiment calls into play a wide assortment of theoretical assumptions, which become tacit premises of the inference from hypothesis to event. The very meaning of the terms in an hypothesis depends upon their use in other hypotheses and a theoretical statement cannot be understood except in the context of the conceptual web of which it forms a strand. When a theorist's predictions are falsified, all he knows is that at least one of the many assumptions he used for making the prediction was incorrect. But the experiment cannot single out the wrong premise(s) from the right one(s). There is a semantical linking between hypotheses which vitiates any attempt at testing them in isolation.[6] To continue to maintain H in the face of $\sim O$ is not necessarily to assert that a suitable A' exists, but simply to allow for the *possibility* that H may still be compatible with $\sim O$, given some suitable A'. The *onus probandi* is not, as Grünbaum supposes, on the scientist who refuses to call a refuted hypothesis false to show that his hypothesis can be saved by some suitable A'. Rather, the burden of proof is on those who deny H to show that there does not exist an A' which would make H compatible with $\sim O$. Schematically, the scientist who claims to have falsified an hypothesis, H, must prove that

$$\sim (\exists A')\,(H + A' \rightarrow \sim O).$$

Unless such a proof is forthcoming, a scientist is logically justified in seeking some sort of rapprochement between his hypothesis and the uncooperative data. It appears, then, that there are two versions of the D-thesis: a stronger one (which Grünbaum attacks) and a weaker (which I believe is Duhem's actual position). They can be formulated as follows:

Stronger D-thesis: For every hypothesis and every observation statement, there exists a set of non-trivial auxiliary assumptions, A', such that H and A' entail O.

Weaker D-thesis: In the absence of a proof that no appropriate hypothesis-saver exists (i.e., unless we prove that $\sim (\exists A')\,(H + A' \rightarrow \sim O)$), then $\sim O$ is not a conclusive refutation of H, even if $H + A' \rightarrow O$.

Insofar as Grünbaum imputes the stronger claim to Duhem, I think he is mistaken. I have already noted that he cites no textual evidence for his reading of Duhem, nor does he give us any reason to believe that Duhem would have accepted the stronger version. It is also worth observing that Grünbaum himself often formulates the D-thesis as if it were simply the claim that hypotheses cannot be conclusively refuted.[7] Surely, the weaker version is a more faithful translation of this claim than the stronger one.[8] By a subtle assimilation of the weaker D-thesis to the stronger and by then refuting the stronger one, Grünbaum talks as if he has thereby refuted the weaker D-thesis as well. But the weaker (and, if I am right, authentic) version of the D-thesis is untouched by Grünbaum's critique.

2. We turn now to consider briefly Grünbaum's second attack on the Duhemian–Quinean–Einsteinean position. Here he seeks to show, by example, that conclusive falsifying experiments do occur. He suggests that we consider the schema

$$[(H + A \rightarrow O)(\bar{O} \cdot A) \rightarrow \bar{H}],$$

where H is a system of geometry, A is a proposition about the perturbation-free characteristics of solid rods, and O is the empirical statement that light rays coincide with rigid-body geodesics. We need not probe into the physics of the problem to understand the substance of Grünbaum's counterexample. He argues, and I think we can safely grant, that we have observed $\sim O$ and, in virtue of independent evidence, can assert with high probability that A is true. On these grounds, Grünbaum claims to have shown that the experiment has falsified the hypothesis H. On the face of it, this is a clear exception to the D-thesis. But there are two factors here which make this argument powerless against Duhem's position. In the first place, a system of geometry (the 'H' in this example) is not the sort of thing which counts for Duhem as an 'isolated hypothesis'. Duhem, we recall, insisted only that isolated hypotheses – not systems of hypotheses such as geometry or classical mechanics – were non-falsifiable. To say that a *set* of hypotheses has been falsified is no refutation of the D-thesis. The second, and more serious flaw in Grünbaum's counterexample, if I understand it correctly, is that A, though probable, is not known to be true: Despite A's high likelihood, a scientist is not forced to relinquish H unless A is known to be true. Since A is subject to some doubt, we cannot necessarily blame the failure of the prediction, O, on H rather than A. To give up H might be more prudent, but the demands of prudence do not carry logical weight. It is perhaps correct to remark that Grünbaum's experiment would cause a rational person to cease to expound H, but the experiment certainly does not provide an unambiguous falsification of H.

In light of the failure of both Grünbaum's logical analysis and his counterexample to make out a case for conclusive falsification, it seems we are still left with Duhem's conventionalism intact.

NOTES

* From *Philosophy of Science*, Vol. 32 (1965). Reprinted by permission.

[1] Cf. Grünbaum's 'The Duhemian Argument', *Philosophy of Science* 27 (1960), 75–87; 'Laws and Conventions in Physical Theory', in *Current Issues in the Philosophy of Science* (ed. Feigl & Maxwell), pp. 140–155 and 161–168; *Philosophical Problems of Space and Time*, pp. 106–152; and 'The Falsifiability of Theories: Total or Partial? A Contemporary Analysis of the Duhem–Quine Thesis', in *Boston Studies in the Philosophy of Science* (ed. Wartofsky).

[2] Cf. Duhem's *Aim and Structure of Physical Theory* (trans. by Wiener), Part II, Chapter vi and *passim*. The translation is based on the 1914 edition of *La Théorie Physique: Son Objet – Sa Structure*.

[3] Grünbaum agrees that crucial experiments cannot verify hypotheses. Like Duhem, he is opposed to the Baconian theory of crucial experiments. But, like Popper and against Duhem, he wants to assert that crucial falsifying experiments occur.

[4] Grünbaum readily grants that by suitable logical gyrations, or by a re-definition of terms, one could always find *ad hoc* some A' such that for a given O, $H+A' \to O$. (The simplest way, of course, would be to take O itself as A'.) But he rightly points out that Duhem's point becomes trivial on such an interpretation. Thus, Grünbaum argues that the Duhemian position requires that there exists a *non-trivial A'* for every H and every O.

[5] *Philosophical Problems of Space and Time*, p. 114.

[6] For example, Duhem writes that "The physicist can never subject an isolated hypothesis to experimental test, but only a whole group of hypotheses; when the experiment is in disagreement with his predictions, what he learns is that at least one of the hypotheses constituting the group is unacceptable and ought to be modified; but the experiment does not indicate which one ought to be changed". *Aim and Structure*, p. 187.

Elsewhere, he notes that "To seek to separate each of the hypotheses of theoretical physics from the other assumptions on which it rests in order to subject it in isolation to observational tests is to pursue a chimera...", *Ibid.*, p. 200.

[7] Thus, Grünbaum writes: "Duhem's thesis is that the falsifiability of the hypothesis H as an explanans of the actual empirical facts O' is always unavoidably inconclusive". 'Laws and Conventions in Physical Theory', in *Current Issues in the Philosophy of Science*, p. 145.

[8] As further evidence for the 'weaker' reading of Duhem, we might consider his discussion in *Aim and Structure*, Part II, Chapter iv, §10. There he explains that when a theory, T, is falsified, a scientist has two options: he can either modify one of the hypotheses in T to make it compatible with the phenomena, or he may discard the whole structure T and opt for an altogether different theory T'. If Duhem accepted the stronger version of the D-thesis, he presumably would have argued that the former alternative is always open. But he does not take this approach. He does say that we have no "right to condemn in advance the boldness" (p. 217) of the physicist who seeks to preserve his theory in the face of embarrassing evidence. But, on the other hand, Duhem admits that *there is no guarantee that he will be successful in finding an A'*. The scientist who takes this ploy is justified only if he succeeds in "satisfying the requirements of experiment" (p. 217), i.e., only if he finds some A' that saves the theory T. Duhem is not asserting that A's will always be found, or even that they exist, if they are not found. He readily admits that if, after a thorough analysis, no A's are forthcoming, then we should give up the entire theoretical system, T. But in giving it up, we must be careful not to say that the individual hypotheses within the theory have been falsified. The theory is no longer fruitful, but that does not mean that all its component hypotheses are false.

CARLO GIANNONI

QUINE, GRÜNBAUM, AND THE DUHEMIAN THESIS*

Quine in his paper 'Two Dogmas of Empiricism'[1] has propounded a radical conventionalist thesis, arguing that only science as a whole, including the laws of logic, is empirically testable. Grünbaum, on the other hand, has in various places including *Philosophical Problems of Space and Time*[2] been critical of an even moderate Duhemian conventionalism, and in particular attempts to show that the geometry of space is testable independently of other physical theory. Between these two extremes lies the Duhemian thesis. We will attempt to describe in this paper the exact nature and extent of Duhemian conventionalism in physics, indicating the manner in which it is a semantical conventionalism while not being trivially so.

Let us begin by looking at Quine's thesis of conventionalism in physics and logic. The point usually derived from the above mentioned paper is that the analytic-synthetic distinction is untenable. It is important to note that Quine in the first section of the paper admits that logically true statements are analytic, and directs his attack towards a second class of analytic statements which are generally called analytic by virtue of the meanings of the nonlogical words, such as 'No bachelor is married'. If synonyms were substituted for synonyms, these statements would be transformed into logical truths. Quine's main attack in the first four sections is on the lack of an adequate notion of synonymy. One might grant this in regard to natural language and say so much the worse for natural languages.

More interesting, more debatable, but less well defended is Quine's discussion in the last two sections of his paper. If the point which Quine makes here is valid for natural language, then it is equally valid for constructed languages. Essentially Quine is proposing an extension of Duhem's thesis of conventionalism in physics to include the truths of logic as well as *all* of the laws of science.

Let us analyze the structure of his argument. Quine suggests several arguments in defense of the analytic-synthetic distinction in the first four sections of the paper, and criticizes each severely. He then goes on to suggest a method of defense based on the verification theory of meaning. Can we not say a statement is analytic if it is confirmed by anything whatever that happens in the world? If we can talk about the verification of particular statements, surely we can also talk about those statements which are verified "come what may". Quine's response to this ploy is to say that no individual statement can be verified. He says, "The unit of empirical significance is the whole of science".[3] Only science as a whole can be verified or falsified. *Ipso facto*, no individual statement can be vacuously confirmed by any experience, and the analytic-synthetic distinction becomes of minor interest in comparison to this far-reaching suggestion of Quine's that *no* individual statement is open to verification. If Quine can make this point, he has not only shown that the analytic-synthetic distinction is untenable, but also and even more important, that the true-false distinction is untenable, except as applied to science as a whole, and that the distinction between physics and logic is completely untenable. We anticipate from Quine at this point a defense as strong and as hard hitting as his critique of the analytic-synthetic distinction. But our expectations are not fulfilled. Quine merely gives examples of cases in which we must conventionally choose to reject one statement or another, for the statements jointly are falsifiable, but not individually. He does not, however, give us a philosophical underpinning to justify his general thesis that only science as a whole can be verified or falsified. Quine gives us this much to work with:

Reevaluation of some statements entails reevaluation of others, because of their logical interconnections – the logical laws being in turn simply certain further statements of the system, certain further elements of the field. Having reevaluated one statement we must reevaluate some others, which may be statements logically connected with the first or may be statements of logical connections themselves.[4]

One wishes that he had spelled out these logical interconnections in some detail. One also wants to know why he is now rejecting the criteria of logical truth of statements which he held at the beginning of the paper.

It is interesting to note here that a conclusion similar to Quine's was arrived at by Hempel and at about the same time (1951).[5] Because of the problem of theoretical terms Hempel arrived at the conclusion that the unit of empirical significance was not the terms or statements of science, but the theories of science. In contradistinction to Quine, however, Hempel gave detailed arguments for this position, possibly because he was reluctant to admit it. The theoretical terms of science cannot be explicitly or operationally defined using only observation terms, but rather must be introduced into science by means of the theories themselves. The theory consists of several statements at least two of which must contain the theoretical term. In effect, the theoretical term is implicitly defined by the theory. Moreover, statements can be deduced from the theory which do not contain theoretical terms, and which, in particular, do not contain the theoretical term in question. Because statements are deducible which do not contain theoretical terms, the theory as a whole can be said to have empirical significance, and can be confirmed or falsified. However, since the theoretical statements themselves contain theoretical terms which are only partially interpreted via the theory, they cannot be individually tested. Therefore, theories, by virtue of containing theoretical terms, must face experience as a whole. The unit of empirical significance and the unit which is to be tested must be the theory as a whole.

We see, therefore, that the Duhem-Quine thesis as applied to *theories* can be defended by purely logical analysis of theories and their terms. To defend the general Quinean thesis that all scientific knowledge is conventional we must be able to apply the same logical analysis throughout science. We will not attempt here to justify Quine's whole thesis, but rather we will attempt to defend that segment which is the same as the original Duhemian thesis. Hempel has already in effect given a defense of the thesis for that segment of science which includes theories. We propose now to defend the thesis for one other segment of science, namely, for that part of science which depends on the use of operational definitions which make reference to measuring instruments, and in particular where the measurement is derivative rather than fundamental.

In defending his thesis, Duhem distinguishes between common sense physical laws and the laws of physics as a developed science, for only to the latter is his thesis applicable. The distinction between these laws arises over the fact that physics uses terms which are not directly observable, such as 'mass', 'temperature', and 'pressure'. In the jargon of philosophy of science these are considered to be operationally defined terms. Duhem does not, however, consider these terms as introduced by definition. While one must certainly use instruments to measure these various quantities, the very use of these instruments implies the belief in certain physical theories, viz., the theories of the instruments themselves.[6] While Duhem would agree with the operationalist that the meaning of these terms depends on the method of measuring them, he would go a step further and say that the meaning of the terms depends also on the theory of the instruments used. In effect, Duhem is treating these terms in the same manner that theoretical terms are currently treated, that is, they are implicitly defined by the theories of the instruments.

Duhem is concerned primarily with quantities which are the subject of derivative measurement rather than fundamental measurement. Of quantities which are derivatively measured we can distinguish three types: (1) Quantities which are *nominally* defined as a mathematical function of other quantities. For example, the density of a substance is defined as the mass of the substance divided by its volume. Thus, in order to measure density one first measures the mass and volume of the substance and then one *calculates* the density. (2) Quantities which are measured by other means than the means originally used in introducing the concept are derivative relative to this method of measurement. For example, length can be fundamentally measured by using meter sticks, but it also can be measured by sending a light beam along the length and back and measuring the time which it takes to complete the round trip. We can then calculate the length by finding the product of the velocity of light (c) and the time. Such a method of measurement is dependent not only on the measurement of time as is the first type of derivative quantity, but also on the law of nature that the velocity of light is a constant equal to c. (3) The third type of derivative quantity, and the type to which the Duhemian thesis is particularly applicable, is exemplified by the ordinary mercury thermometer measurement of temperature. Changes in temperature are correlated with changes in some other quantity, as in the first case of derivative measurement; however, here no significance is given to the absolute length of the strand of mercury. It is rather changes in the length of the strand which are relevant to changes in temperature. Changes in temperature are a function of change in length of the mercury, but temperature itself is not a function of the length. The significant aspect of this type of derivative measurement for the Duhemian thesis is the fact that this type of measurement depends on the assumption of certain laws of nature. While in a certain sense change in temperature is correlated with change in the length of the strand of mercury *by stipulation*, it is also to a certain extent *a law-like connection*, for there are conditions under which one will either disregard or correct a temperature reading as given by a mercury thermometer. For example, scientists believe that the readings of a mercury thermometer placed in a strong magnetic field are incorrect. Therefore, the reading of a mercury thermometer is not taken at face value.

One might object that it is *part of the stipulation* that such perturbing influences as magnetic fields are to be avoided as a condition for *valid* use of the thermometer. Conditions could be added to the stipulation in either of two ways: one might make a blanket statement that the use of a mercury thermometer for temperature measurement is valid only if *all* perturbing influences are absent. This approach is obviously circular, for a perturbing influence is a condition which makes a measurement incorrect. Therefore we know that the measurement is incorrect only if we know that perturbing influences are present, and we know that perturbing influences are present only if we know that the measurement is incorrect. An alternative procedure which avoids the circularity would be to state specifically which conditions are to be considered perturbing. The difficulty with this latter approach is that at the time at which a particular method of derivative measurement is introduced one does not know all of the conditions which one would want to consider perturbing.

We think, rather, that the use of instruments, such as thermometers, involves both stipulative and causal elements. In using an instrument we assume that there is a law-like connection between the property measured and the property by which it is measured. If you will, change in temperature causes change in length of the mercury strand of the thermometer. As with all causal laws, it is assumed that the connection depends on certain conditions, but it is not usually known at first what all of these conditions are. It is stipulative in that the property which is being measured is not directly observable, for there is a difference between the crude concept of temperature of ordinary sensation and the exact concept of temperature of physics. Referring to the concepts of mass, temperature, and pressure Duhem has the following to say: "These ideas are not only abstract; they are, in addition, symbolic, and the symbols assume meaning only by grace of the physical theories"[7]. By an abstract idea Duhem means any idea which is general as opposed to proper names which are particular. Our crude concept of temperature is abstract, but not symbolic. By symbolic, Duhem means theoretical, i.e., unobservable. Since the exact concept of temperature is unobservable, one is free to stipulate what relationship one wants to hold between the length of the mercury strand and the temperature. But in this case, the stipulated connection is nonetheless also a causal connection, for it is a connection which varies with external conditions, at least some of which are unknown. This is the heart of Duhemian conventionalism, that is, that there are *laws of nature* which by the nature of the case must be *stipulated to be true*.

Duhem further substantiates his view that the stipulated connection is causal by noting that when several methods are available for measuring a certain property, no one method is taken as the absolute criterion relative to which the other methods are derivative in the second sense of derivative measurement noted above. Each method is used as a check against the others. Duhem discusses an experiment which consists essentially of an electrical battery with a voltmeter, an incandescent lamp and a coil connected in parallel across it.[8] If the voltmeter indicated a certain reading one would say that the battery was operating. However, even if the voltmeter indicated zero, one might still say that the battery was operating based on the fact that the incandescent lamp was glowing and that the coil was warm. Conversely, if the lamp did not glow, one might still say that the battery was operating because gas bubbles were emerging from the battery or the voltmeter was giving a certain reading or the coil was warm. The operationalist likes to think that one method of measurement is the defining criterion for a certain quantity and that other measurements of the quantity are derivative relative to it, but this does not do justice to the way that scientists in fact proceed. The relationship between method of measurement and quantity measured is not one-one but many-one, and the relationship is many-one because to the physicist the measured quantity is causally related to more than one observable phenomenon. The very construction of measuring instruments oftentimes depends on the prior construction of a theory on the basis of which one can predict a number of observable phenomena, each of which is dependent on a single underlying cause and each of which can operate as a measure of the underlying property.

It is a truism that a scientific law can be tested only if all of the extra-logical terms are interpreted, at least partially. One method of interpreting the symbolic terms of physics, such as temperature, is via measuring instruments, such as thermometers. What is tested then is a physical law together with a particular set of rules of interpretation for its symbolic concepts. If the rules of interpretation are analytic, then any falsification of an interpreted law encountered must be interpreted as a falsification of the law, for an analytic statement cannot be falsified. To insist that any law can be retained by changing the analytic rules of interpretation is to fall into a trivial semantic conventionalism. If Duhem is right, however, in insisting that the rules of interpretation for (some of) the symbolic concepts of physics are law-like, rather than analytic, then a falsification of the interpreted law cannot be taken to be a falsification of the law itself for it may be the interpretation that is false. We then arrive at Duhem's dictum that there are no crucial experiments in physics. This we believe is the extent of Duhem's conventionalism.

Recently Grünbaum has objected to the Duhemian thesis on the grounds that it is an empirical thesis, rather than a logical thesis, and therefore open to falsification, and furthermore that it is in fact false for the case of geometry. *We will attempt to show here that the Duhemian thesis as interpreted by Grünbaum is a logically defensible thesis relative to derivative measurement of the third type*, while agreeing with Grünbaum that it is an *empirical* thesis relative to fundamental measurement. We will furthermore attempt to clarify what we take to be a misunderstanding on the part of Grünbaum of certain aspects of the Duhemian thesis itself.

Let 'H' stand for the hypothesis being tested; 'A' for the interpreting *laws*, (i.e., auxiliary hypotheses); and 'O' for an observational statement. Let us further assume that '$((H \& A) \to O) \& \sim O$' is true. That is, the interpreted hypothesis has been falsified. Grünbaum interprets Duhem as asserting both of the following statements, where 'A'_{nt}' stands for a non-trivial alternative set of interpreting *laws* (i.e., auxiliary hypotheses), and 'H'' for an alternative hypothesis:

$$(\exists A'_{nt}) (H \& A'_{nt}) \to \sim O),$$
$$(\exists H') ((H' \& A) \to \sim O).$$

Duhem would certainly be agreeable to one or the other statement in any particular case *but not both*. His point is not that there is always both an 'A'_{nt}' and an 'H'', but that we cannot determine which it is that exists *independently of the rest of the system*. Within the context of a system, however, he seems to believe that we can determine which it is. Duhem relates physical science to an organism, and he says that "if something goes wrong ... the physicist will have to ferret out through its effect on the entire system which organ needs to be remedied or modified without the possibility of isolating this organ and examining it apart"[9]. Furthermore, Duhem believes that despite the symbolic nature of the theories of physics, we can achieve a 'natural classification' of the world. When we achieve a system which orders and organizes a vast body of previously heterogeneous data of experience, then "it is impossible for us to believe that this order and this organization are not the reflected image of a real order and organization"[10].

Despite the fact that Duhem would not agree to Grünbaum's interpretation of his thesis, we believe that even the stronger thesis, namely, that *both* an 'H'' and an 'A'_{nt}' exist, is defensible. Let us consider the following situation: Charles' Law which holds that temperature times volume is a constant has been relatively well verified in the past. Nevertheless, we run a further test on it and it turns out false. We might say that the law is generally true but that it has exceptions, and that these exceptions can be excluded if we can determine the peculiar conditions operating in this experiment. We might then proceed to determine why the law does not hold in this particular case, that is, to ascertain the perturbing conditions. On the other hand, according to our previous analysis of temperature measurement we know that the result of a thermometer reading is to be accepted only under certain conditions, some of these conditions being unspecified and unknown. Since we might attribute the falsity of Charles' Law in this particular case to a particular perturbing condition, we might also argue that the perturbing condition has not affected the law but rather the measurement of temperature. Why should we say that perturbing conditions affect the law rather than the method of measurement *or* vice-versa? The parallel is complete between Charles' Law and the law of the thermometer. Both are generally reliable, and both are expected to be unreliable under certain conditions, some of which are unknown. When the pair do in fact turn out false we are free to attribute the falsity to either of the laws and to relate the perturbing condition to that law.

While we think that the strong Duhemian thesis is logically defensible with respect to derivative measurement, with respect to fundamental measurement it is at best empirically true. While fundamental measurement includes not only length, but also time, mass, etc., we will concern ourselves only with the former. The fundamental measurement of length depends in the first instance on the existence of a measuring rod. Not any measuring rod will do, however, for most, if not all, measuring rods yield incorrect results because of the influence of perturbing conditions, such as heat. Measurement must therefore be performed with a rigid body where, according to Reichenbach, "Rigid bodies are solid bodies which are not affected by differential forces, or concerning which the influence of differential forces has been eliminated by corrections"[11]. The Duhemian thesis is defensible with respect to geometry if there are as a matter of fact no solid bodies uninfluenced by differential forces.

Grünbaum attempts to prove the falsity of the Duhemian thesis in two ways in regard to geometry. One of the proofs depends on the empirical fact that the material invar has a coefficient of linear expansion which is essentially zero. Invar is, therefore, a rigid material relative to the differential force heat. If one can find for each differential force a material which is rigid relative to it, then one has a standard which can be used to determine the extent to which other materials are influenced by that particular differential force. Whether or not such materials exist is an empirical question which will decide for or against the Duhemian thesis in regard to geometry.

Grünbaum's second proof, which we believe to be invalid, depends on the method of successive approximation. Let us consider the following situation: the length of a rod depends on temperature. In order to correct for the influence of temperature on the rod we use the law of linear expansion. But to determine the law of linear expansion we must use a thermometer. Let us assume that the thermometer is an ordinary mercury thermometer. Since the length of the thermometer itself depends on temperature, the reading of the thermometer itself must be corrected. The very influence that we are measuring affects the measurement of the influence. Have we been caught in a vicious circle? No. The law which is discovered with the mercury thermometer can be corrected by correcting the temperature reading of the thermometer. As we alternatively correct the law of linear expansion and the readings of the mercury thermometer, we will approach a limit, much in the same way that instantaneous velocity and acceleration are determined.

This is the method of successive approximation, but, as Lenzen has noted,[12] this method depends on the fact that the law of linear expansion as discovered by the mercury thermometer is correct to the *first* approximation and incorrect only to the second approximation. The revision of the law which depends on using the corrected thermometer readings will then be correct to the *second* approximation and incorrect only to the *third*, etc. It is for this reason that a limit is reached. Temperature will affect the length of the mercury thermometer only to a small extent relative to its total length, thereby affecting the temperature reading only to a small extent. Thus, the method of successive approximation will give us a law of linear expansion which will differ from the initial law only in the higher orders of approximation.

When we apply these same considerations to the case of geometry, we find that we are concerned on the one hand with a geometry and on the other hand with a set of laws for correcting the measured lengths. If we assume as our congruence standard solid bodies as corrected for differential forces, we know that the geometry of our immediate space has to be in a relatively narrow range centering about Euclidean geometry. Therefore all of the possible geometries will be in a certain sense the same in the first approximation. The corrections which we make for differential forces will then be crucial inasmuch as they will affect our determination of the geometry of space in precisely that order of approximation which makes all the difference. Correspondingly, the laws which we use for correction will differ only in higher orders of approximation. Let us say then that we have two candidates for the geometry of space, G and G', and two candidates for the law of linear expansion, E and E', where one member of each pair differs from the other only in the second order of approximation. Furthermore let us assume that E', but not E, is a function of the curvature of space in the second order of approximation, and that E has been arrived at by the method of successive approximation such that E together with G successfully predicts the observable data. Let us further assume that G' together with E' also successfully predicts the observational data but does so on the basis of a value of E' *which is known to be incorrect* as it assumes a curvature of space other than that of G'.[13] By the method of successive approximation we can alternatively revise E' and G' until we reach in the limit an E" and a G" which successfully predicts the observational data and such that E" assumes the curvature of G″. Is there any guarantee that the G″ and E″ so arrived at will be the same as G and E, respectively? Initially G differed from G′ in the second order of approximation. Since E″ differs from E′ in the second order of approximation, G″ will differ from G′ in the third order of approximation, but will be the same in the second order and therefore *different from G*. It seems then that one would *not* expect G and G′ to converge in the limit, that is, one would expect G and G″ to be different. Therefore, the method of successive approximation which will achieve a unique limit when we are concerned with correcting for the influence of the measured property on the measuring instrument, will not achieve a unique limit where the measured property – in this case the curvature of space – is taken to have an effect on a correctional law which in turn has an effect on the measuring instrument, that is, where a law which is necessary for correcting for differential forces is a function not only of the differential force itself but also of the property which is ultimately being measured. Consequently Duhemian conventionalism can be defended even in the case of geometry which depends on the fundamental measurement of length if it can be empirically shown that there are no bodies rigid relative to each differential force.

In the remainder of the paper we will consider the broad implications of the Duhemian thesis on our conception of scientific knowledge. Specifically we will be concerned with whether the Duhemian thesis is an epistemological thesis regarding our knowledge of the world or a semantical thesis regarding the meaning of scientific words, and scientific language. Let us assume that a hypothesis H together with its interpreting laws A has been falsified, and we determine that the interpreting laws need revision and that the hypothesis is to be retained. Since we have in fact changed the meaning of the symbols of the hypothesis, have we effected nothing more than a trivial semantic change, that is, is not the H which is left after the change in A in fact a new hypothesis with new factual content merely couched in the old symbols? "The symbols assume meaning only by grace of the physical theories"[14]. Certainly in changing the interpreting laws, we have changed the meaning of the symbols, for the symbols are only implicitly defined in the context of theory, but this is equally true if we change the hypothesis, for the symbols depend for their meaning not only on the laws of the measuring instruments but also on the hypotheses in which they occur. A and H together implicitly define the symbols. Duhemian conventionalism is certainly semantical for we are free to change either H or A *because* we are free to change the meaning of the symbols or the theory. The freedom we have in physics, however, is not solely the trivial sort by which we can always change the meaning of a word, for it is also a freedom which is necessitated by the very context of scientific discovery. We could rationally reconstruct physics so as to eliminate this latter freedom by fixing the meaning of the quantities introduced by derivative measurement once and for all by stipulation (definition), but this would be contrary not only to the way physics actually develops, but to the way it should develop. A 'definition' which shifts with the wind, in the light of empirical findings, which is continually modified to protect other laws, is not a definition except in name, and to treat it as a definition is to be intellectually dishonest. If one is going to insist that every statement is either analytic or synthetic, the laws of the instruments should be treated as synthetic; however, we would argue that it is foolish to insist on this distinction in the cases we have discussed above. Let us be quite clear that we are not interested in demolishing the analytic-synthetic distinction entirely. We merely point out that there are cases where it is pointless, and in fact damaging, to insist on it, even in a rational reconstruction of science.

One final problem is yet to be resolved. We know that Duhem holds that the individual statements of physics are not *testable* in isolation from physical theory, but a crucial question is whether or not they are *true or false* in isolation. A group of physical laws together can have factual content, but do the individual laws taken in isolation have factual content? As we interpret Duhem he does believe that the individual statements of physics have factual content and are true or false independently of the rest of physical theory. This belief is based on an 'act of faith' that physical theory does capture the real order in reality, "but while the physicist is powerless to justify this conviction, he is nonetheless powerless to rid his reason of it"[15].

This faith must logically depend on a certain semantical commitment, namely, on the belief that symbols such as temperature have an extra-linguistic meaning which is logically independent of the intra-linguistic meaning which they have by virtue of implicit definition, but which is epistemologically dependent on it. The presupposition then is that there is in reality such a quantity as temperature and that there is a correct measure of it. When a law of the instrument is modified, it may be interpreted as an attempt to bring the *actual* measure of temperature in line with the *correct* measure. When an hypothesis is modified, it may be interpreted as a belief that the actual measure of temperature is also the correct one, and that the difficulty lies with the hypothesis.

If there is such a real quantity in nature as temperature, the question arises as to why we should be at all concerned with it. As far as we can tell this concern in turn depends on the further assumption that nature has in itself a certain simplicity which will be reflected in the laws we discover only if we succeed in using the correct measures. The underlying assumption is a metaphysical belief regarding the nature of the universe. If this assumption is not made, then it does not matter whether or not there is a *real* quantity temperature, and whether or not if there is, we succeed in correctly measuring it.

The question of the nature of conventionalism ultimately resolves itself into whether we are to take a realistic (Platonistic) or a nominalistic approach to the symbols of physics. If we take a realistic position, the individual statements of physics have independent factual content and are true or false; they express propositions. In this case if one were to hold, in contradistinction to Duhem, that one cannot *determine* which statements of science are true, but only whether or not a system of statements adequately predicts the facts, then one would be holding to a merely epistemological thesis regarding the limitations of *our knowledge* of the world.

On the other hand, if one follows a nominalistic approach in regard to these symbols, one is saying that these symbols have extra-linguistic meaning only to the extent that they occur in theories which are ultimately connected with experience. When statements involving these symbols are abstracted from their theoretical context, the meaning of the symbols and *a fortiori* of the statement is stripped from them. Outside of their theoretical context they are meaningless and therefore neither true nor false, while within the theoretical context they are meaningful law-like statements which are either true or false. We would say in this case that a statement 'expresses a proposition' only within a theoretical context.

These considerations illustrate that the distinction between 'descriptive simplicity' and 'inductive simplicity' is not a logical distinction but an ontological one, depending as it does on one's approach to the symbols of physics and the nature of the universe. If one is a realist, one searches for inductive simplicity, for one believes that nature in itself is simple and, therefore, the correct description of nature will be the simplest. If one is a nominalist, one searches for descriptive simplicity, for one believes that the world can be described in simple terms, although not necessarily that it is any truer than a more complex description. The former believes that quantities are to be discovered for they will lead to the correct, that is, the simplest description. The latter believes that quantities are to be created with the view in mind of achieving the simplest description.

NOTES

* From *Nous* **1** (1967). Copyright © Wayne State University Press 1967. Reprinted by permission.

[1] In W. V. O. Quine, *From a Logical Point of View*, Harper and Row, New York, 1963, pp. 20–46.

[2] Alfred A. Knopf, New York, 1963, Chap. 4.

[3] Quine, 'Two Dogmas of Empiricism', p. 42.

[4] *Ibid.*

[5] See his 'The Concept of Cognitive Significance: A Reconsideration', *Proc. Am. Acad. Arts and Sciences* **LXXX** (1951), 61–77, and also his *Aspects of Scientific Explanation*, Free Press, New York, 1965, pp. 101–119.

[6] See Pierre Duhem, *The Aim and Structure of Physical Theory*, trans. by Philip P. Wiener, Princeton University Press, Princeton, 1954, pp. 165–79, for a discussion of the particular characteristics of the laws of physics.

[7] *Ibid.*, p. 166.

[8] *Ibid.*, p. 150.

[9] *Ibid.*, p. 188.

[10] *Ibid.*, p. 26.

[11] *The Philosophy of Space and Time*, Dover, New York, 1958, p. 22.

[12] 'Procedures of Empirical Science', *International Encyclopedia of Unified Science*, University of Chicago Press, Chicago, 1955, Vol. I, pp. 289–95.

[13] The curvature of space assumed by E′ must, of course, be the same as that of G′ if the combination E′ & G′ is to make a claim to truth. We assume in this discussion *as does Grünbaum* that the curvature of space is constant. With this assumption there is no difficulty in manufacturing a second order coefficient of E′ which is a function of curvature and such that the resultant E′ together with *some* G′ yields the same observational consequences as G & E. We are not begging the question for we are not assuming that we can *manufacture* an E′ which can assume the *same* value for the curvature of space as G′ and yield the same observational consequences as E & G.

[14] Duhem, *Aim and Structure of Physical Theory*, p. 166.

[15] *Ibid.*, p. 27. Compare with the following quotation from Albert Einstein, 'Physics and Reality', *Journal of the Franklin Institute* **CCXXL** (1936), reprinted in Edward H. Madden, *The Structure of Scientific Thought*, Houghton Mifflin, Boston, 1960, p. 84: "It is an outcome of faith that nature – as she is perceptible to our five senses – takes the character of such a well formulated puzzle [i.e., has only *one* correct solution]".

GARY WEDEKING

DUHEM, QUINE AND GRÜNBAUM ON FALSIFICATION*

I

A number of empirically-minded philosophers hold that there is a class of 'basic' (or 'protocol') sentences each of which is, in the words of Karl Popper, "a statement of singular fact" ([3], p. 43). Many of these philosophers also maintain, with Popper, that there are empirical theories which are 'falsifiable' in the sense that each such theory

divides the class of all possible basic statements unambiguously into the following two nonempty subclasses. First, the class of all those basic statements with which it is inconsistent (or which it rules out, or prohibits):… secondly, the class of those basic statements which it does not contradict (or which it 'permits') ([3], p. 86).

Duhem has objected to such theories of falsifiability on the grounds that they inadequately represent the complex experimental procedures of theoretical physics. It is not the case that a single physical theory is by itself inconsistent with any experimental findings. For the construction of the physicist's experimental apparatus is based on physical hypotheses other than the one being tested. Thus when the physicist predicts an experimental result O from a hypothesis H, he assumes the truth of all the physical laws that go into the construction of his apparatus and into the calculation of O from H. Thus the ordinary model of the falsification of a hypothesis

$$((H \rightarrow O) \cdot \sim O) \rightarrow \sim H$$

of inductive logic is overly simple. Such a model would represent the testing of a physical hypothesis only if we knew that the other assumptions upon which the prediction was based were true. But inductive logic itself assures us that this is something we cannot know, for it is an accepted inductive maxim that a law (or universally quantified statement) can never be completely verified.[1] An adequate logical model for physical experiments must, then, include in the antecedent of the conditional, of which the consequent is O, not only the hypothesis, but also these other assumptions A. If the result of an experiment is negative, we cannot, therefore, logically conclude that H is false. "The only thing (such an) experiment teaches us is that among the propositions used to predict the phenomenon ... there is at least one error; but where this error lies is just what it does not tell us" ([4], p. 185). Thus for the above simple model of falsification we must substitute

(i) $$(((H \cdot A) \to O) \cdot \sim O) \to (\sim H \vee (\exists p)(p \in A \cdot \sim p)).$$

Grünbaum claims to refute the Duhemian thesis by arguing that that situation is logically characterized, not by (i), but by

(ii) $$(((H \cdot A) \to O) \cdot \sim O \cdot A) \to \sim H.$$

As we noted above, however, A can never be inductively certain. Let p be the probability of A (symbolized $P(A)=p$). Let us also suppose that the antecedent of (i) is true, that $\sim O$ is an experimental result and that $H \cdot A \to O$. Since according to (i), A implies $\sim H$, $\sim H$ must be at least as probable as is A, that is,

$$P(\sim H) \geqslant p.$$

By the Negation Theorem of the probability calculus

$$P(\sim H) = 1 - P(H),$$

which implies that

$$P(H) \leqslant 1 - p.$$

It follows from this that

$$\lim_{p \to 1} P(H) = 0.$$

Thus we see that as the value of p approaches 1 (certainty) the value of $P(H)$ approaches 0 (impossibility). Since A can be, as Grünbaum himself points out, "only more or less highly confirmed" ([2], p. 135), p is always less than 1. Thus the value of $P(H)$ is never $0,$[2] which implies that (ii) does *not* characterize the logical situation.

The result of the above paragraph is, in fact, so obvious from an inspection of (ii), that one might wonder how it is that Grünbaum is led to commit such a blunder. This error can be traced to Grünbaum's failure to understand what Duhem's thesis asserts.[3] Apologizing for the lack of inductive certainty for A, he states that "the inductive risk ... inherent in affirming A does not arise from the alleged inseparability of H and A, and that risk can be made exceedingly small without any involvement of H" ([2], p. 137). It is noteworthy, however, that Duhem nowhere says that the inductive risk of A arises from the inseparability of A and H. Duhem's theory asserts, in fact, the exact converse of this; the inductive inseparability of H and A arises from the inductive risk of A. It is because A is inductively uncertain that (i), rather than (ii), is the correct logical model. Only if A were certain would H be experimentally separable from A. Since it is not certain, the negative result of an experiment (i.e. $\sim O$) will not determine which of the disjuncts of the consequent of (i) is true. Thus any test of H must be a test of both H and A; that is, H and A are experimentally inseparable.

The thesis that Grünbaum seems to have mistaken for Duhem's is that involved in the 'vicious circle' problem propounded by D. M. Y. Sommerville (cf. [2], p. 126). According to Sommerville, there are cases in which an experimental procedure is particularly complicated due to the involvement of H in the determination of A. The paradigmatic case of this situation is when Euclidian measurements are used in experiments designed to determine the geometry of physical space. Such situations would complicate our evaluation of the results of an experiment in an obvious way. The Duhemian theory does not, of course, exclude the possibility of such situations. But neither does its truth depend upon the existence of such situations.

Grünbaum seems also to wish to hold Duhem responsible for a considerably stronger thesis of Quine's. According to this view, "any statement can be held true come what may, if we make drastic enough adjustments elsewhere in the system" ([4], p. 43, cf. [2], p. 108). Applying this theory to the philosophy of physics, Grünbaum reads it as asserting that no matter what the result O' of an experiment designed to test an hypothesis H is, there is always some new set of theoretical assumptions

A' which, together with H, implies O'. Duhem, however, does not hold that an A' having the above property always exists; his is the far weaker thesis that the physicist can never be *sure* that *no* such A' exists.

... experimental contradiction does not have the power to transform a physical hypothesis into an indisputable truth; in order to confer this power on it, it would be necessary to enumerate completely the various hypotheses which may cover a determinate group of phenomena; but the physicist is never sure he has exhausted all the imaginable assumptions ([1], p. 190).

In the interest of historical accuracy, therefore, we will not follow Grünbaum ([2], pp. 110f., etc.) in referring to Quine's theory as 'Duhem's thesis' in the succeeding analysis.

II

According to Grünbaum, Quine's theory asserts that whenever the experimental result O entailed by assumptions A and hypothesis H fails to materialize as expected, there is always some alternative set of assumptions A' such that A' and H entail the actual experimental result O'. That is,

$$\text{If} \quad (H \cdot A \to O) \cdot \sim O, \quad \text{then} \quad (\exists A')(H \cdot A' \to O').$$

Here Grünbaum is substituting H for "any statement" in Quine's assertion that "Any statement can be held true come what may" ([4], p. 43). But that we "can" hold H in the face of O' does not imply, as Grünbaum seems to hold, that H is "needed to deduce O'" ([2], p. 111). At most this only implies that there is some language S' in which H and O' are consistent one with another.[4] There is, as a matter of fact, no reason to assume that Quine must affirm O' to be deducible from any of the true sentences in S' (other than itself). In other words, there is no necessity for Quine to assert of *any* given sentence in S' that it is not independent of the other truths of S'. It might, in such a situation, be objected that no system of physics should contain "observation statements" which cannot be deduced from the "theoretical statements" of that system. But this kind of objection would not be provincial to S', any more than to S in which A is still retained as true. For there is no guarantee that there is *any* consistent language (which is in some sense adequate to the whole of physics) in which any arbitrary statement about an observation can be (nontrivially) deduced. It is true, of course, that we would *prefer* our A' to be such that within the language S' of which H and A' are parts, we could "explain" the result O' of our experiment. But that this is a requirement that we could always impose upon our choice of A' is far from obvious. It is certainly not implied in Quine's thesis.

There is still another way in which Grünbaum's formulation is an inadequate statement of the Quinean thesis. Grünbaum assumes that O', presumably because of its special status as an "observation statement", must count as a true sentence of S'. But when Quine speaks of the possibility of "pleading hallucination" in order to maintain an adequate system "in the face of recalcitrant experience" ([4], p. 43), he is denying that even sentences as relevant to experience as is O' are immune to revision. Quine argues that "it is nonsense ... to speak of a linguistic component and a factual component in the truth of any individual statement" ([4], p. 42). Such nonsense, he writes, has led philosophers to imagine an "extreme case where the linguistic component is all that matters" ([4], p. 41), an analytic statement. Though Quine does not address himself to this issue in 'Two Dogmas', there would seem to be another side to the same coin. The same confusion has led philosophers to believe that there are statements in which all that matters is the factual component. Such "sense-data" statements are, like their polar opposites, analytic statements, thought to be indubitable – immune to revision in any situation whatsoever. Quine's point is that there are *no* statements which are indubitable in this sense. Thus it appears that Grünbaum is mistaken in imposing even the requirement that S' must be *consistent* with O' upon Quine's theory.

But surely there must be *some* restriction or other upon the construction of S'. Grünbaum contends that in order for Quine's theory to be nontrivial, that it is necessary to impose on it the restriction that "*the theoretical language be semantically stable in the relevant respects*" ([2], p. 20). This means, I suppose, that S' must be such that it includes no terms, in particular no terms occurring in H, which have different meanings than they have in S (in which A counts as true). In formulating this requirement, Grünbaum seems to have forgotten that the Quinean thesis under consideration occurs in the final section of an essay in which the analytic-synthetic distinction is denied. In denying this distinction, Quine rejects the opinion "that the truth of a statement is somehow analyzable into a linguistic component and a factual component" ([4], p. 41). Since Grünbaum's requirement of semantical stability (RSS) presupposes that we can always separate these components sufficiently to hold one constant while the other is altered, this requirement makes no sense as applied to the *Quinean* thesis. As Grünbaum offers no argument against the thesis developed in the earlier sections of 'Two Dogmas', we must reject RSS as a legitimate restriction. What, then, is a restriction that we *can* impose upon the construction of S' in order to test the validity of Quine's theory?

A requirement having certain intuitive appeal is that *all true statements be true in S'*. But this requirement is clearly inadequate, for it does not specify the language in which these sentences are true. That is, it does not tell us what language L to compare S' with in order to determine that all the sentences true in L are true in S'. It certainly cannot be construed as requiring that all true statements of any language whatever be true in S'. For it is not in general the case that all true sentences of any given language are even sentences (much less true sentences) of S'.

Perhaps what is needed is some requirement of translatability of truths. A conceivable requirement is that *for every language L there is a truth-preserving translation in S' for each true sentence in L*. But we need *not* require translatability between S' and L for any L whatever. We do not, for example, have to worry about inconsistent languages, nor need we concern ourselves with any language which contains a true sentence such as 'There are centaurs', when there are in fact no centaurs. As a matter of fact, there is no obvious way to specify the class of languages with which we *need* concern ourselves. Certainly we would like each L in this class to be in some sense true to experience. In order to maintain any reasonable translatability requirement, then, we must restrict our choice of L to those languages which are 'adequate' to the facts of experience. But without an independent method of determining this class of languages, this reduces to the requirement that S' *must be 'adequate' to the facts of experience*. As Quine puts it, "The edge of the system must be kept squared with experience" ([4], p. 45).

The Quinean thesis now becomes *for any (noncontradictory) hypothesis H there is a system S' which 'squares' with experience, and in which H is true*. This thesis, it should be noted, is, even without Grünbaum's RSS, far from obvious. For it is by no means clear that there is *any* language (with or without H as a true sentence) which is adequate to experience in the sense required. And this is because it is not clear what sense of "squaring with experience" is required in this connection. Quine contends that our choice *between* the various languages having the required property of squaring with experience is determined by *pragmatic* considerations. This would seem to require a notion of "squaring with experience" which can be rigorously delimited from these pragmatic considerations. But it is not evident that such a rigid delimitation is possible, especially in light of Quine's rejection of the notion of purely factual statements.

The above considerations lead me to doubt the truth of the Quinean thesis in any but its most trivial form, viz. that *a language S' is always possible in which any given sentence H counts as true.* Whenever we attempt to impose additional restrictions on our choice of *S'* in order to prove Quine's theory non-trivial, its meaning begins to elude us. But this is not to say that the theory is worthless, for it results in an observation which is significant indeed: the difference between those situations in which Quine's thesis holds only trivially and those in which we are confronted with a real and significant alternative as to which sentences we should reject as false and which we should retain as true is only a matter of degree.

APPENDIX

Since the above paper was written in the spring of 1965, it does not take into consideration the more recent literature on the controversy. I would like to point out, in particular, Laurens Laudan, 'Grünbaum on "The Duhemian Argument"', *Philosophy of Science* 32 (1965), 295–99, and Carlo Giannoni, 'Quine, Grünbaum, and the Duhemian Thesis', *Nous* 1 (1967), 283–98, both of which give Duhem an interpretation similar to my own. In Grünbaum's more recent writings ('The falsifiability of a Component of a Theoretical System', in Feyerabend and Maxwell (eds.), *Mind, Matter, and Method*, University of Minnesota Press, Minneapolis, 1966, pp. 273–305, and 'Can We Ascertain the Falsity of a Scientific Hypothesis', Chapter 17 of *Philosophical Problems of Space and Time*, Second, enlarged edition, D. Reidel, Dordrecht-Holland, 1973 [excerpted below]), he appears to concede that his was a misinterpretation of Duhem, although he continues to refer to a thesis as 'Duhem's' which is apparently of his own invention, at any rate, certainly not of Duhem's. The issues raised in the second part of my paper could be raised basically unchanged against Grünbaum's newer writings.

NOTES

* From *Philosophy of Science* (1969). Reprinted by permission.

1 This is not to say that *no* universally quantified statement can be known to be true. For those that are logically true can be known to be true. But in these cases the concept of verifications is hardly applicable.

2 We are assuming, of course, that *H* cannot be known to be false on noninductive grounds, i.e. that *H* is not self-contradictory.

3 The extent of Grünbaum's misunderstanding of Duhem can be seen from the fact that he finds no inconsistency in attacking Duhem's denial of falsifiability while holding as "eminently sound" the very basis of this denial, viz. that "the logic of every disconfirmation, no less than of every confirmation of an isolated scientific hypothesis *H* is such as to *involve at some stage or other* an entire network of interwoven hypotheses in which *H* is ingredient rather than in every stage merely the separate hypothesis *H*". ([2], p. 111.)

4 It may be objected that in any system in which both *H* and *O'* are true, '*H* implies *O'*' is also true. But this is true only of the material implication. And Grünbaum clearly intends the relation between *H* and *O'* to consist in the *necessity* of using *H* in the deduction of *A*.

BIBLIOGRAPHY

[1] Duhem, P.: *The Aim and Structure of Physical Theory* (trans. by Philip P. Wiener), Atheneum, New York, 1962.
[2] Grünbaum, A.: *Philosophical Problems of Space and Time*, Alfred Knopf, New York, 1963.
[3] Popper, Karl R.: *The Logic of Scientific Discovery*, Science Editions, New York, 1961.
[4] Quine, W. V. O.: 'Two Dogmas of Empiricism', in *From a Logical Point of View*, Harper & Row, New York, 1963, pp. 20–46.

DUHEM, QUINE AND A NEW EMPIRICISM*

1. THE DUHEM–QUINE THESIS

As in the case of great books in all branches of philosophy, Pierre Duhem's *La Théorie Physique*, first published in 1906, can be looked to as the progenitor of many different and even conflicting currents in subsequent philosophy of science. On a superficial reading, it seems to be an expression of what later came to be called deductivist and instrumentalist analyses of scientific theory. Duhem's very definition of physical theory, put forward early in the book, is the quintessence of instrumentalism:

A physical theory is not an explanation. It is a system of mathematical propositions, deduced from a small number of principles, which aim to represent as simply, as completely, and as exactly as possible a set of experimental laws [p. 19].

The instrumentalist overtones of this become clear from the implications of the denial that theories are explanations. For Duhem an explanation is a metaphysical entity, and science should be independent of metaphysics. But this dictum is not intended, as with the positivists, to dispose of metaphysics as irrational or meaningless; it is rather an assertion of the autonomy and dignity of metaphysics as alone capable of expressing the truth of how things are in the world. Metaphysics according to Duhem is not independent of experience, but its methods are not those of science, and its conclusions stand independently of changing fashions in science. Thus it is for Duhem a grave error to interpret scientific theory as itself providing a metaphysics – a global theory drawn from science such as mechanism is not only false, because science outgrows it by its own methods, but also it is not the kind of theory that could ever be true, because it illegitimately uses the methods of mathematical representation of experimental facts to construct an ontology and to give answers to substantial questions about the nature of the world and of man. But only metaphysics, and in particular a religious metaphysics, can do that. The aim of science must be more modest. A non-interference pact must be established between the domains of science and metaphysics.

Duhem was not the first nor the last philosopher of religion to see the answer to teasing conflicts between science and religion in terms of a complete separation of their spheres of influence, but this is not the aspect of Duhem's thought that I want to discuss here. Indeed, if this were all there were to say about Duhem's philosophy of science it would deserve no more than a minor place in the history of positivism. But his extra-scientific preoccupations did not after all mislead him into so crude an analysis of science itself as his definition of scientific theory would entail. He is saved by a discussion of the observational basis of science that is far subtler than that presupposed by later deductivists and instrumentalists, and paradoxically it is a discussion which can be made to undermine the very foundations of the dichotomy of mathematical theory and explanation, science and metaphysics, that his theory of explanation presupposes.

Most empiricist accounts of science have been based, usually tacitly, on the notion of a comparatively unproblematic observation language. It matters little how this is construed – whether in terms of hard sense data, operational definitions, ordinary language, or what not – the essential point is that there are statements of some kind whose meaning as descriptions of states of affairs is supposed to be transparent, and whose truth-value is supposed to be directly and individually decidable by setting up the appropriate observation situations. It is a long time since anyone seriously claimed that the truth of such statements can be known *incorrigibly*, but most eyes have been averted from the consequences of the significant admission of fallibility of even observation statements, and attention has been concentrated on the way in which meaning and truth-value is conveyed to theories, regarded as in these respects parasitic upon observation statements and clearly distinguishable from them. The consequences for deductivism have been proliferation of a number of insoluble and unnecessary problems regarding the meaning of theoretical statements and the possibility of confirming them, and the result has been a slide into instrumentalism in which, in the end, only observation statements and not theories have empirical interpretation. What that interpretation and its significance is still remains unanalysed.

Duhem introduces two important modifications into this type of classical empiricism. They may be expressed as a new theory of *correspondence* and a new theory of *coherence*.

(i) In his theory of *correspondence*, attention is shifted away from the empirical basis of traditional empiricism to the theoretical *interpretation* of that basis. Duhem sees that what is primarily significant for science is not the precise nature of what we directly observe, which in the end is a *causal* process, itself susceptible of scientific analysis. What is significant is the interpretive expression we give to what is observed, what he calls the *theoretical facts*, as opposed to the 'raw data' represented by *practical facts*. This distinction may best be explained by means of his own example. Consider the theoretical fact 'The temperature is distributed in a certain manner over a certain body' (p. 133). This, says Duhem, is susceptible of precise mathematical formulation with regard to the geometry of the body and the numerical specification of the temperature distribution. Contrast the practical fact. Here geometrical description is at best an idealisation of a more or less rigid body with a more or less indefinite surface. The temperature at a given point cannot be exactly fixed, but is only given as an average value over vaguely defined small volumes. The theoretical fact is an imperfect translation, or interpretation, of the practical fact. Moreover, the relation between them is not one-one, but rather many-many, for an infinity of idealisations may be made to more or less fit the practical fact, and an infinity of practical facts may be expressed by means of one theoretical fact.

Duhem is not careful in his exposition to distinguish *facts* from *linguistic expressions of facts*. Sometimes both practical and theoretical facts seem to be intended as linguistic statements (for instance, where the metaphor of 'translation' is said to be appropriate). But even if this is his intention, it is clear that he does not wish to follow traditional empiricism into a search for forms of expression of practical facts which will constitute the basis of science. Practical facts are not the appropriate place to look for such a basis – they are imprecise, ambiguous, corrigible, and on their own ultimately meaningless. Moreover, there is a sense in which they are literally inexpressible. The absence of distinction between fact and linguistic expression here is not accidental. As soon as we begin to try to capture a practical fact in language, we are committed to some theoretical interpretation. Even to say of the solid body that 'its points are more or less worn down and blunt' is to commit ourselves to the categories of an ideal geometry.

What, then, is the 'basis' of scientific knowledge for Duhem? If we are to use this conception at all, we must say that the basis of science is the set of theoretical facts in terms of which experience is interpreted. But we have just seen that theoretical facts have only a more or less loose and ambiguous relation with experience. How can we be sure that they provide a firm empirical foundation? The answer must be that we cannot be sure. There is no such foundation. It must be admitted that Duhem himself is not consistent on this point, for he sometimes speaks of the persistence of the network of theoretical facts as if this, once established, takes on the privileged character ascribed to observation statements in classical positivism. But this is not the view that emerges from his more careful discussion of examples. For he is quite clear, as in the case of the correction of the 'observational' laws of Kepler by Newton's theory (p. 193), that more comprehensive mathematical representations may show particular theoretical facts to be false.

However, we certainly seem to have a problem here, because if it is admitted that subsets of the theoretical facts may be removed from the corpus of science, and if we yet want to retain empiricism, the decision to remove them can be made only by reference to *other* theoretical facts, whose status is in principle equally insecure. The correspondence with experience, though loose and corrigible, must still be retained, and still remains unanalysed.

(ii) Duhem's theory of *coherence* is indispensable to a satisfactory resolution of this problem. The theory has been much discussed, but unfortunately not always in the context in which Duhem set it, with the result that it has often been misunderstood and even trivialised.

Theoretical facts do not stand on their own, but are bound together, in a network of laws which constitutes the total mathematical representation of experience. The putative theoretical fact that was Kepler's third law of planetary motion, for example, does not fit the network of laws established by Newton's theory. It is therefore modified, and this modification is possible without violating experience because of the many-one relation between the theoretical fact and that practical fact understood as the ultimately inexpressible situation which obtains in regard to the orbits of planets. It follows that neither the truth nor the falsity of a theoretical fact or a lawlike relation connecting such facts can be determined in isolation from the rest of the network. Systems of hypotheses have to come to the test of experience as wholes. Individual hypotheses are not individually falsifiable any more than they are individually verifiable.

Quine, as is well known, has taken up both aspects of Duhem's new empiricism. A bare remnant of empirical correspondence is implied by his dictum that 'our statements about the external world face the tribunal of sense experience not individually but only as a corporate body' – for Quine they do face it; how they face it has come in his recent writings to be a question for a stimulus-response psychology (Quine, 1960, 1968). The coherence of our knowledge is also implied, in the very strong sense (which is never explicitly claimed by Duhem) not only that generally speaking hypotheses cannot individually be shown to be false by experience, but that *no* statement can be; any statement can be maintained true in the face of any evidence: 'Any statement can be held true come what may, if we make drastic enough adjustments elsewhere in the system' (Quine, 1953, p. 43). Because it is doubtful whether we ever want to 'hold a hypothesis *true*' rather than highly confirmed or highly probable, and because I do not here want to beg or examine that question, I shall discuss Quine's claim in a slightly more weakened form than is implied by this quotation. The weaker form, which I shall call the Q-thesis, is that

No descriptive statement can be individually falsified by evidence, whatever the evidence may be, since adjustments in the rest of the system can always be devised to prevent its falsification.

It has seemed to many commentators that to replace the observational basis of science with this shifting network is to open the floodgates to conventionalism, and to a vicious circularity of truth-value and meaning which is in effect an abandonment of empiricism. Popper (1959, p. 78), for example, classes Duhem with Poincaré as a conventionalist. But if by conventionalism is meant, as Poincaré apparently intended in regard to the geometry of physical space, that any given total theoretical system can be imposed upon any logically possible experience, then surely to class Duhem as a conventionalist is a mistake. For neither Duhem nor Quine say anything to imply that a total system is not refutable by experience; indeed that it is so refutable is entailed by their contrast between refutability of individual hypotheses and refutability of the linked system of hypotheses. Once parts of the system have been fixed, perhaps conventionally, there are some extensions of it that are empirically excluded.

But elsewhere Popper (1963, p. 238ff.) demands something more than this:

> We can be reasonably successful in attributing our refutations to definite portions of the theoretical maze. (For we *are* reasonably successful in this – a fact which must remain inexplicable for one who adopts Duhem's or Quine's views on the matter) [p. 243].

The 'holistic argument goes much too far' if it denies that it is ever possible to find out which is the guilty hypothesis. There are, he suggests, three ways in which it may in fact be identified:

(i) We may provisionally take for granted the background knowledge common to two theories for which we design a crucial experiment, and regard the experiment as refuting one or other of the theories rather than the background knowledge. But neither Duhem nor Quine would ever deny this possibility, and it is of course not sufficient to refute Q, since it does not require acceptance of the background knowledge to be anything but *provisional*.

(ii) We may be able to axiomatise the whole theoretical system in such a way as to isolate the effect of a single axiom, which may then be refuted in isolation. But even if we disregard the extreme impracticability of such axiomatisation in the case of most interesting scientific theories, its ideal possibility still does not refute Q, because no axiomatisation can fully account for the empirical applicability of the system, and the correctness of the conditions of application (the so-called 'correspondence rules') might always be called into question to avoid abandonment of any of the axioms.

(iii) Theories need to make successful predictions (to be 'corroborated' in Popper's terminology) as well as being refuted if false. When successful predictions have occurred, Popper seems to suggest, we are more reluctant to abandon those parts of the theory responsible for them, and more willing to locate the responsibility for subsequent refutations in other, less well corroborated parts of the network. Popper's notion of corroboration here as elsewhere is far from clear, but it is difficult to interpret this suggestion in any sense other than in terms of relative inductive *confirmation* of some parts of the system in comparison with others. Some theory of confirmation of the system by experience does indeed seem to be a requirement of the network analysis, and I shall return to this requirement below; but as far as Popper's suggestion goes, he only regards this method of picking out a guilty hypothesis as indicative and not conclusive, and so the method in any case would not refute Q.

The Q-thesis has also recently come under attack from Adolf Grünbaum (1963, Ch. 4; 1966). In a series of articles Grünbaum has sought to show that Q is true only in trivial cases in which 'drastic adjustments elsewhere in the system' are construed as allowing *ad hoc* changes in the rules of English usage. Clearly if a hypothesis predicts that roses are red, and they turn out to be black, Q is not satisfied except trivially by interchanging the uses of 'red' and 'black' in observation reports in English. 'Hence', Grünbaum continues, 'a *necessary* condition for the non-triviality of Duhem's thesis is that *the theoretical language be semantically stable* in the relevant respects' (1966, p. 278). He does not, however, claim to give general sufficient conditions for the non-triviality of auxiliary hypotheses or rules which would preserve the truth of a hypothesis H in the face of apparently contrary evidence, nor does he attempt to spell out in detail what it would be for the theoretical language to be 'semantically stable' or for H to remain the 'same hypothesis', arguing only that Quine's suggestion of resort to a non-standard logic must at least be regarded as trivial, as must *ad hoc* changes in the meanings of descriptive terms.[1]

This criticism suggests, therefore, a second requirement for the Q-thesis to be viable, namely some theory of change and retention of *meaning* within the network, in addition to the first requirement of some theory of confirmation.

To summarise these important and pervasive kinds of doubt about the viability of the Q-thesis, it is convenient to quote further from Grünbaum. Discussing Einstein's assertion that any metric geometry can be preserved in the face of any empirical evidence, he says:

Indeed, if the Duhemian is to maintain, as he does, that a *total theoretical system* is falsifiable by observations, then surely he must assume that the relevant falsifying observations present us with sufficient *relatively* stubborn fact to be falsifying.... And if there were no relatively stubborn fact... how could the Duhemian avoid the following conclusion: "Observational findings are always so unrestrictedly ambiguous as not to permit even the refutation of any given total theoretical system"? But such a result would be tantamount to the absurdity that any total theoretical system can be espoused *a priori* (1966, p. 288).

One could earn a quick point against this passage by remarking a *non sequitur* between 'if there were no *relatively* stubborn fact' and 'observational findings are always so *unrestrictedly* ambiguous ...'. Might there not, one wants to ask, be relatively *un*stubborn facts which were nevertheless not *so* unrestrictedly ambiguous as to warrant the conclusion that any theory might be espoused as true *a priori*? Neither does it obviously follow that if there is no conclusive refutation, any theory goes, for there may be available a theory of relative confirmation. More fundamentally, it should be noted that Grünbaum has almost unwittingly fallen into just the habit of distinguishing the theoretical system from the 'relatively stubborn facts' that are called in question by the Q-thesis. That the facts are only *relatively* stubborn does not save him, for the whole thrust of Q is against the practice of looking in the system for those statements which can, even relatively, form its basis, and upon which the rest of the system is propped up. 'Relatively' in this context is always taken in classical empiricist accounts to imply 'relative to some *more* basic statements which we could uncover if we had time or necessity'. But the only relativity of stubbornness that can be allowed in a Q-system is relativity with respect to the *other theoretical* statements. The structure is mutually supporting. Where the points of external support are applied is a subsidiary matter which cannot be decided independently of the character of the network itself. Grünbaum might well reply that this leaves the theory completely up in the air, and removes it from empirical control in just the way he fears. So we are left with the two requirements of a network theory as constituting problems for explication:

(i) That some relative empirical confirmation should be provided, and that without being able to identify any statement of the system which expresses the evidence incorrigibly.

(ii) That some means of analysing stability and change of meaning in the network should be provided.

2. Criteria of confirmation

When faced with a philosophical tangle which seems to involve logical circularities or contradictions, it is often illuminating to try to conceive of a mechanism which simulates the conditions of the problem, and to see whether a self-consistent model of it is possible. What we need in this case is a machine capable of representing and theorising about its environment according to the conditions just described. We can distinguish in the usual way between the hardware and the software of the machine: the machine has a certain physical constitution, the hardware, which we will assume remains fixed (the machine is not at present regarded as a structurally evolving organism), and its software includes a certain system of coding according to which some of its physical states can be represented in its own 'language', and the representation perhaps printed on an output tape, so that the machine is capable of 'reporting' on its environment. Suppose the machine goes through the following stages of operation:

(i) Physical input from the environment causally modifies part of the machine (its 'receptor').

(ii) The information thus conveyed to the receptor is represented in the machine language according to a code present in the machine. We may assume at this stage that the code is not, at least in practice, infinitely and exactly competent, so that (a) if the input is potentially infinitely various, some information present in the receptor is lost at the coding stage, and (b) the mechanism may make mistakes in a small proportion of translation of the input into code. The product of this stage will be called the *coded input* (C.I.), and corresponds to the set of observation statements produced by a human investigator as the representation in language resulting from experienced sensory input. Notice in particular that C.I. is not necessarily a complete or accurate representation of the input.

(iii) C.I. is examined for repetitions which yield inductive generalisations, and for more complex patterns which yield theories. If the machine is inductivist it may run through all possible systems of generalisations seeking that which is in some specified sense most probable or most simple. If it is deductivist it may have a small stock of patterns to try out on C.I., rejecting those whose fit is too bad, and retaining those whose fit is 'best' or good enough. In either case it is not necessary that the theory arrived at by the machine should be consistent with *every* piece of C.I., only that *most* of it should be consistent. Moreover, *which* parts of C.I. are going to be consistent with the best theory cannot be determined in advance, but only by examining the theory in the light of the complete C.I., and adjusting it to make the best fit. In other words, no single statement of C.I. is incorrigible relative to a good theory, only most of it must be. There are no epistemologically privileged protocol statements, but the element of correspondence which implies an empiricist check on the theory is still present in the whole set of observation statements. Thus the first requirement of the network model of science regarding the possibility of confirmation can be met by providing (a) the empirical check or correspondence element present in the whole of C.I., and (b) some principles of probability or simplicity of theories which are used as the coherence element to choose the best theory and modify and perhaps discard some small part of C.I.

Does the Q-thesis hold for such a machine? No C.I. statement (say C_1) is logically immune from correction by the best theory or theories, and therefore if some given descriptive statement S is contradicted by C_1, S can in principle always be regarded as unfalsified by taking C_1 to be itself false. But what does 'false' mean here? It cannot mean that C_1 is false as a direct representation of the input, because all that can be known about this is that a certain small proportion of such C.I. statements are false, not which ones. It must mean false relative to a 'best' theory constructed in the light of the whole C.I. and the internal coherence criteria for construction of theories. If these criteria are sufficiently modified no doubt C_1 could be made consistent with some theory which satisfied them, but it does not follow that this could be done according to any criteria which would be accepted as reasonable for a good theory. However, this limitation on the applicability of Q only highlights the importance for judging 'truth' and 'falsity' of what are taken to be reasonable criteria, and indicates that these are not immediately determined by the input, but are in a sense *a priori* relative to that input. They may, of course, possibly be regarded as modifiable in a second order sense in the light of the type of theories which seem to be successful for large amounts of input, but then the principles of modification would presuppose yet higher-level criteria, and so on. I shall not pursue here the problem of specifying and justifying criteria for 'best' theories, for in the light of current discussion of various kinds of confirmation theory the only thing that is clear is that the problem is turning out to be unexpectedly deep and difficult, and as yet hardly rewarding. But it does seem important to emphasise that some statement of confirmation criteria for theories seems to be a necessary condition of rebutting the charge that the Q-thesis effectively abandons empiricism.

It may seem that this requirement contradicts the claim that is also integral to the Quinean approach, namely that there is no ultimate distinction between the *a priori* and the *a posteriori*, the analytic and the synthetic. Quine himself (1968) has recently willingly accepted that the possession of some 'innate ideas' is a corollary of his network model of language. This is in reply to objections by Chomsky (1968), who curiously reads Quine, not as abandoning empiricism, but as sticking too closely to it in his analysis of sensory conditioning as the foundation of language learning. There is no empirical evidence, Chomsky claims, for the kind of language learning Quine seems to require, rather all the evidence we have (which incidentally is from syntax rather than semantics, and therefore not clearly relevant to the conditions of applicability of descriptive predicates – but let us take Chomsky for the moment at his own valuation) points to the presence of innate, interlingual dispositions to certain standard linguistic principles. Quine's acceptance of this point seems motivated rather by desire to conform to the present state of empirical linguistics and psychology of perception than by general arguments such as have been put forward here. Innate principles which are understood merely as conditions causally operating on sensory data perhaps need not count for Quine as *a priori* principles which refute his conflation of the prior and the posterior. However, we can hardly be content with this understanding of the principles. To remain so would be like accepting a physico-physiological account of the processes which go on when we do sums, and regarding this as excluding rational discussion of the logically systematic principles involved in doing sums correctly. To adapt the favourite metaphor which Quine takes from Neurath, modifying parts of the network while relying on other parts may be like rebuilding a boat plank by plank while it is afloat, but there are right and wrong ways of doing the rebuilding. To provide a normative inductive logic in which the innate principles are systematically explicated does not *preclude* empirical investigations of the scientific and social facts about inductive reasoning, but it tells us more about them, by showing why and under what conditions they can be regarded as rational. There is a close parallel here with the programme of rational decision theory, which may be assisted, but is not determined, by empirical investigations of practical decision making. In the sense of a rational inductive logic, then, the innate principles would be *a priori* relative to the data which they process, but this sense need not be objectionable to Quine, since no claim is made about the eternal immutability of the principles – different external conditions may cause adaptive organisms to modify these principles too.

3. MEANING

The alternative possibility of saving S from falsification, which has been dismissed by Grünbaum as trivial, is to so 'change the meaning of S' that it no longer contradicts the evidence. How can we understand in terms of the machine model the demand that meanings shall be stable in order to exclude trivial satisfaction of Q? We cannot directly and immediately apply the usual empiricist interpretation of 'the meaning of S' as the empirical conditions necessary and sufficient for S to have the truth-value true, because the only criteria of truth we have are relative to the coherence of the system as well as to its empirical constraints. Indeed, the truth- or probability-value of S *relative to the current best theory* may change as additional evidence replaces best theory by another, and that without direct observation of the empirical conditions of satisfaction of S. So in this sense the meaning of S, like its truth-value, is not invariant to accumulating evidence. Is such instability of meaning an objection to the network model?

It can certainly be interpreted in such a way as to constitute a *reductio ad absurdum* of any model of science which attempts to retain an element of empiricism, including the network model. It has been so interpreted by several recent writers in the guise of what has come to be called the 'meaning variance thesis' (Hanson, 1958; Feyerabend, 1962; Kuhn, 1962), and since I want to distinguish the network model from this thesis in important respects, I shall start by stating and examining the thesis itself.

The original context of the meaning variance thesis was an attack upon the deductive model of theories with its accompanying assumption that there is a comparatively stable and transparent observation language, upon which theoretical language is parasitic. It is pointed out, first, that reliance on deducibility in the deductive account of explanation of observation by theory, and reduction of one theory to another, is vain, because there is always a measure of approximation in such inferences, and hence it is always possible for the same data to be 'explained' in mutually contradictory forms by mutually contradictory theories. For example, Galileo's law is not a logical consequence of Newton's theory; in fact it is contradicted by that theory, because the law asserts that the acceleration of bodies falling along the earth's radii is constant. It was possible to hold Galileo's law to be true only because this discrepancy was concealed by experimental error. And yet Newton's theory is held to *explain* the facts about falling bodies in spite of contradicting the experimental law which had been accepted up to then as a description of those facts. Again, Newtonian mechanics cannot simply be reduced by deduction to the more comprehensive relativity mechanics, because relativity mechanics entails, among other things, that space and time are mutually dependent and inseparable dimensions, and that the mass of a body is not an invariant property, but a function of the body's speed relative to whatever happens to be taken as the rest frame. Such consequences of relativity are strictly *inconsistent* with Newtonian mechanics. Similar objections may be made to the alleged deductive reduction of phenomenological thermodynamics to statistical mechanics, and of quantum to classical electrodynamics. Many of these examples involve something even more radical than mere numerical approximations. It is meaningless, for example, to speak of Newtonian mechanics 'approximating' to relativistic mechanics 'when the constant velocity of light c is taken as infinite', or of quantum theory 'approximating' to classical physics 'when the quantum of action h is taken as zero', because it is of the essence of relativity and quantum theory that the respective constant c and h are *finite constants*, having experimentally specifiable values. Moreover, in passing from one theory to another there are *conceptual* as well as numerical changes in the predicates involved: mass as invariant property becomes variable relation, temperature as property becomes a relational function of velocity, atom as indestructible homogeneous stuff becomes divisible and internally structured.

Such examples as these lead to the second, and more radical, part of the meaning variance thesis, namely that deducibility is impossible not only because numerical fit between theory and observation is at best approximate, but also because the concepts of different theories are governed by rules of syntax and use implicit in the respective theories, and since different theories in a given experimental domain in general conflict, these rules of usage are in general inconsistent. Hence explanation of observation by theory, or reduction of one theory to another, cannot take place by identification of the concepts of one theory with those of observation or of another theory, nor by empirically established relations between them. We cannot even know that different theories are 'about' the same observational subject matter, for if the meaning of the predicates of observation statements are determined by the theoretical beliefs held by their reporters, and if these meanings differ in different theories, then we seem to have an incommensurability between theories which allows no logical comparison between them, and in particular allows no relations of consistency, incompatibility or relative confirmation.

The thrust of the meaning variance thesis is therefore primarily against the notion of a neutral observation language which has meaning invariant to changes of theory. But the thesis becomes impaled on a dilemma. Either there is such an independent observation language, in which case according to the thesis its predicates cannot be related deductively or in any other logical fashion with any theoretical language, or there is no such observation language, in which case every theory provides its own 'theory-laden' observation predicates, and no theory can be logically compared with any other. The consequences of meaning variance can be put in paradoxical form as follows:

(1) The meaning of a term in one theory is not the same as its meaning in another prima facie conflicting theory.

(2) Therefore no statement, and in particular no observation statement, containing the predicate in one theory can contradict a statement containing the predicate in the other.

(3) Therefore no observation statement which belongs to one theory can be used as a test for another theory. There are no crucial experiments between theories.

A similar paradox can be derived from (1) with regard to both explanation and confirmation.

(1a) The meaning of a predicate in the pre-theoretical observation language is different from its meaning in a theory which is said to explain that domain of observation and to be confirmed by it.

(2a) Therefore if the theory entails some observation statement, that statement cannot be the same as any pre-theoretical observation statement, even if it is typographically similar to it.

(3a) Therefore no theory can explain or be confirmed by the statements of the pre-theoretical observation language.

That such paradoxes seem to follow from the meaning variance thesis has been taken to be a strong objection to the thesis, and hence strong support for the view, presupposed in the deductive account, that observation statements have meaning independent of theories. On the other hand there is certainly a prima facie case for item (1) of the meaning variance thesis, and the network model itself is committed to a similar abandonment of the theory-neutral observation language. Must the notion of a theory-laden observation language lead to paradox?

First, it may be wondered whether so radical a departure from deductivism as indicated by (2) is really warranted by the argument for (1). Suppose we grant for the moment (1) in some sense of 'the meaning of a predicate' which could be incorporated into deductivism, for example that the predicates of a theory are 'implicitly defined' by the postulates of that theory, which entails (1). Even so, for the paradoxes to go through, a further step is required. It must be shown either (i) that the sense of 'meaning' required to make (2) true is the same as that required for the truth of (1), or (ii) that another concept of 'meaning' is implicit in (2), that for this concept meaning is also theory-variant, hence that (2) is still true, and the paradoxes follow. (i) can be disposed of very quickly. In order to establish (i) it would be necessary to show that the difference of meaning of 'P' in different theories which is asserted in (1) is such as to preclude substitutivity of 'P' in one theory T for 'P' in the other theory T', so that no relations of consistency, entailment or contradiction could be set up between statements of T and T'. If this were true, however, it would also be impossible to speak of the difference of meaning *of* 'P' in T and T', for this formulation already presupposes some meaning-identity of 'P' which is not theory-variant. Hence (1) would be not just false, but inexpressible. What, then, is the relevant identity of 'P' presupposed by the possibility of asserting (1) which will also make (2) false and hence dissolve the paradoxes? Here typographic similarity will clearly not do. We must appeal somehow to the external empirical reference of T and T' to give the meaning-identity of 'P' that will allow substitutivity of 'P' between the theories.

The suggestion that naturally springs to mind within the deductive framework is to take the class of objects that satisfy P, that is, the extension of 'P', and identify the relevant meaning of 'P' with this extension. In pursuit of this suggestion Israel Scheffler (1967, Ch. 3) proposes to construe 'meaning' in the classic Fregean manner as having two separable components: 'sense' and 'reference'. (1) may be regarded as the assertion that the sense, or definition, or synonymy relations of predicates differ in different theories, but in considering the logical relations of deducibility, consistency, contradiction, and so on, it is sameness of reference or extension that is solely involved. Difference of sense does not imply difference of reference, hence (2) and (3) do not follow from (1). Thus Scheffler claims to reconcile variance of meaning between theories, and between theory and observation, with invariance of reference and hence of logical relations.

Unfortunately this reconciliation does not work even within the deductive framework. Waiving difficulties about construing sense in terms of definitional synonymy relations, the most serious objection is that 'same reference' is neither necessary nor sufficient for the logical comparability that is required of different theories. It is not *sufficient* because the properties ascribed to objects in science are not extensional properties. Suppose two theories T_1 and T_2 are 'about' two quite distinct aspects of a domain of objects: say their colour relations, and their shapes. It may happen that T_1 and T_2 are such that there is an exact one-to-one correspondence between the sets of predicates of T_1 and T_2 respectively, and that as far as is known T_1 is true of any set of objects if and only if T_2 is also true of it. Then the corresponding predicates of the two theories have the same referential meaning. But this does not imply that the theories are the same. So long as no predicates are added to their respective predicate-sets, no development of T_1 can be either consistent with or contradictory to any development of T_2. In other words, because science is about *intensional* properties, sameness of extension does not suffice for logical comparability. Furthermore, sameness of reference is not *necessary* for logical comparability. Two different theories may make use of different categorisations or classifications of objects: thus Dalton's atoms have different extensions from Cannizzaro's atoms, yet we want to be able to say of some of Cannizzaro's statements that they entail or contradict some of Dalton's.

The network model gives promise of resolving the paradoxes by, first, giving a more subtle analysis of the observation language than that presupposed by deductivism, in terms of which what I have called 'intension' of predicates as well as their extension has a place, and second by allowing a distinction to be made between meanings which are internal to a theory, and meanings which are empirically related. Return for a moment to the observing machine described earlier. We have already noticed that the meaning of descriptive statements is internally related to the best theory and its criteria in something like the way the meaning variance theorists describe. It is also the case that no simple account of the meaning of descriptive predicates in terms of their extension is possible in this model, because all we can know about extension is also relative to the state of the evidence. It may be true or highly probable that P applies to a given object according to one best theory, but false or highly improbable according to another theory adopted on different evidence. There is, however, a relation between machine hardware and input that does remain constant during the process of data collection and theory building that has been described. This is the set of physical conditions under which input becomes coded input. These conditions do not demand infinite exactness nor complete freedom from error, but in what has been said so far they have been assumed sufficiently stable to permit the assertion that a high proportion of statements in the C.I. are true, though we don't know which. This stability is sufficient to ensure that trivial changes of meaning are not resorted to to save theories come what may. Translated into terms of human language-users, this stability does not require that they be aware of some transparent empirical relation between observed properties and linguistic predicates, nor even that they always entertain the same theories; it requires only that by learning to apply predicates in an intersubjectively acceptable manner, they have acquired physical dispositions which are invariant to change of evidence.

To express the matter thus is to invite the comment: does not this kind of stability entail undue inflexibility in the use of descriptive predicates? Do not the meanings of our predicates sometimes change even in this respect under pressure of evidence? In other words, does not evidence also educate our dispositions? It seems fairly clear from the history of science that it does. Consider the predicates 'heavy' and 'light' after Newton's theory had been accepted. It then became incorrect to use the word 'light' of air, and correct to use the word 'heavy', because in Newton's theory all material substances are heavy by definition, even if they can be made to cause a balloon to rise. In such cases there is indeed no substitutibility with retention of truth-value of 'heavy' before and after the change, and so the meaning paradoxes seem to arise. But consider the reason why such a change might occur. In machine terms, we might find that certain applications in observational situations of a given predicate to objects of a certain kind were always contradicted by the best theory for a wide variety of evidence. This would not of course *force* on us a change of disposition to apply that predicate to those objects under the appropriate input, because we expect a small proportion of such applications to be in error relative to the best theory. But if these errors seemed to be concentrated in an unexpected way around certain predicates, we might well decide to change the use of these predicates to fit better the best theories as determined by the large proportion of other observation statements which are assumed true. It might even be possible to state explicit rules for such changes of use and disposition, depending for example on the small probability values of the observation statements involved relative to the rest of the evidence. But all this of course depends on any particular occasion on the presence of many predicates which are not so subject to change of use. The solution of the meaning variance paradoxes requires that there are always many stable predicates when one theory gives way to another.

The possibility of some change of use according to empirically control-led rules shows, however, that Grünbaum's requirement of 'stability of meaning' to save the Q-thesis from triviality is too stringent. Allowing the sort of flexibility of meaning which has obviously often occurred in the development of science need not open the floodgates to apriorism.

4. SUMMARY

In summary let me try to state explicitly the main principles of the new Duhem–Quine empiricism in distinction from the old.

(i) There is no need to make a fundamental epistemological distinction between the theoretical and observational aspects of science, either in regard to decidability of truth-value, or transparency of empirical meaning. The network of relatively observational statements can be imagined to be continuous with a network of theoretical relationships. Indeed much of the recent argument in the literature which has been de-signed to show that there is no sharp line between theory and observation has depended upon examples of quasi-direct recognition in some cir-cumstances of the empirical applicability of what are normally called *theoretical* predicates (such as 'particle-pair annihilation', 'glaciation'). The corresponding theoretical properties cannot, of course, be directly observed independently of the surrounding network of theory and empirical laws, but neither can the so-called observable properties. The difference between them is pragmatic and dependent on causal conditions of sense-perception rather than epistemological.

(ii) The corollary is that empirical applications of observation predicates are not incorrigible, and the empirical laws accepted as holding between them are not infallible. A whole theoretical network may force corrections upon empirical laws in any part of it, but not all, or even most, of it can be corrected at once. Moreover, there is no way of telling *a priori* by separating the theoretical from the observational, *which* part may need correction in the light of subsequent evidence and theory.

(iii) Corrections may strongly suggest changes in the conditions of correct intersubjective application of some of the descriptive predicates, and these changes may be made explicitly according to rules which presuppose that other predicates are not subject to change on the same occasion. To save the notion of 'same theory' which is required to avoid the meaning variance paradoxes, there must be some such stability, indeed the majority of descriptive predicates must be stable in this sense, but just as we do not know *a priori* which observation statements will be retained as true in the next theory, neither do we know which observation predicates will retain stability of meaning. Had Aristotle been a Carnapian, 'heavy' would undoubtedly have appeared in his list of primary observation predicates, and he would have held it to be observable that air is not heavy.

(iv) To avoid total arbitrariness in adoption of the 'best' theory on given sensory input, some prior principles of selection of well-confirmed theories, and criteria for shifts of applicability of some observation predicates, must be assumed. This does not seem, however, to be an objectionable apriorism in the context of the new empiricism, since it is always possible that these principles themselves might change under pressure of the evidence in second or higher order network adjustments.

(v) Lurking within many of these elements of the new empiricism is a systematic conflation of certain aspects of the epistemological problem with causal mechanisms. This occurs at the point of what has been called 'coding' of the input into the coded input, and the identification of this process in human observers with the causal process by means of which descriptive language has been learned. Doubtless to the old empiricism this is a fatal cricularity in the network model, because the question will immediately be asked: How do we know anything about the causal coding and the input it processes except in terms of the usual scientific method of observation and theorising? And if this in turn is subject to the conditions of the network model is not the regress irreducibly vicious? Similar objection, it will be recalled, was made to Russell's causal account of the reception of sense data. But there is a crucial difference between the aims of the new empiricism and those of Russell. Russell, in common with most old empiricists, was looking for 'hard data'; new empiricists accept that these are not to be had. This, incidentally, suggests that the approach suggested here to the relatively prior principles of data processing, via a search for a rational inductive logic, is a better reflection of new empiricism than is the purely scientific search for invariants of language which Chomsky favours, or for psychological and machine models of human learning with which some investigators replace the study of inductive logic. Such empirical approaches are always open to the regressive argument, and leave unanswered the question of what prior principles they themselves depend on. The approach via a rational inductive theory, on the other hand, has the merit of exploring possible rational strategies in possible worlds, independently at least of the details of actual learning processes. But it provides no assurance like that sought by old empiricists, that our knowledge of *this* world is firmly based, only that *if* we were given certain interconnected prior conditions, of whose actuality we can never in practice be certain (for example, that the world is not infinitely various), then we could give reasons for our conscious methods of developing science in a world where these conditions obtain. Duhem might hasten to applaud this conclusion as confirming his view that after all scientific knowledge is superficial and transient compared to the revealed truths of a theological metaphysics. We, who do not have this assurance either, must make do with what we have, a poor thing perhaps, but enough.

NOTES

* From The Royal Institute of Philosophy Lectures, *Knowledge and Necessity*, Macmillan and Co., London, Ltd., Vol. III. Copyright © 1970 by The Royal Institute of Philosophy. Reprinted by permission.

[1] *Note added in proof:* Professor Grünbaum has now developed these arguments further in 'Can We Ascertain the Falsity of a Scientific Hypothesis?', Chapter 17 of *Philosophical Problems of Space and Time*, Second, enlarged edition, D. Reidel, Dordrecht-Holland, 1973 [excerpted below – ed.].

BIBLIOGRAPHY

Chomsky, N.: 'Quine's Empirical Assumptions', *Synthese* **19** (1968), 53.

Duhem, P.: *The Aim and Structure of Physical Theory*, Princeton, N.J., 1906; trans. Wiener, Oxford, 1954.

Feyerabend, P. K.: 'Explanation, Reduction and Empiricism', in *Minnesota Studies*, III, ed. by H. Feigl and G. Maxwell, Minneapolis, 1962, p. 28.

Grünbaum, A.: *Philosophical Problems of Space and Time*, New York, 1963.

Grünbaum, A.: 'The Falsifiability of a Component of a Theoretical System', in *Mind, Matter, and Method*, ed. by P. K. Feyerabend and G. Maxwell, Minneapolis, 1966, p. 273.

Hanson, N. R.: *Patterns of Discovery*, Cambridge, 1958.

Kuhn, T. S.: *The Structure of Scientific Revolutions*, Chicago, 1962.

Popper, K. R.: *The Logic of Scientific Discovery*, London, 1959.

Popper, K. R.: *Conjectures and Refutations*, London, 1963.

Quine, W. V. O.: *From a Logical Point of View*, Cambridge, Mass., 1953.

Quine, W. V. O.: *Word and Object*, New York, 1960.

Quine, W. V. O.: 'Replies', *Synthese* **19** (1968), 264.

Scheffler, I.: *Science and Subjectivity*, Indianapolis, 1967.

IMRE LAKATOS

FALSIFICATION AND THE METHODOLOGY OF SCIENTIFIC RESEARCH PROGRAMMES*

1. SCIENCE: REASON OR RELIGION?

For centuries knowledge meant proven knowledge – proven either by the power of the intellect or by the evidence of the senses. Wisdom and intellectual integrity demanded that one must desist from unproven utterances and minimize, even in thought, the gap between speculation and established knowledge. The proving power of the intellect or the senses was questioned by the sceptics more than two thousand years ago; but they were browbeaten into confusion by the glory of Newtonian physics. Einstein's results again turned the tables and now very few philosophers or scientists still think that scientific knowledge is, or can be, proven knowledge. But few realize that with this the whole classical structure of intellectual values falls in ruins and has to be replaced: one cannot simply water down the ideal of proven truth – as some logical empiricists do – to the ideal of 'probable truth'[1] or – as some sociologists of knowledge do – to 'truth by [changing] consensus'.[2]

Popper's distinction lies primarily in his having grasped the full implications of the collapse of the best-corroborated scientific theory of all times: Newtonian mechanics and the Newtonian theory of gravitation. In his view virtue lies not in caution in avoiding errors, but in ruthlessness in eliminating them. Boldness in conjectures on the one hand and austerity in refutations on the other: this is Popper's recipe. Intellectual honesty does not consist in trying to entrench, or establish one's position by proving (or 'probabilifying') it – intellectual honesty consists rather in specifying precisely the conditions under which one is willing to give up one's position. Committed Marxists and Freudians refuse to specify such conditions: this is the hallmark of their intellectual dishonesty. *Belief* may be a regrettably unavoidable biological weakness to be kept under the control of criticism: but *commitment* is for Popper an outright crime.

Kuhn thinks otherwise. He too rejects the idea that science grows by accumulation of eternal truths.[3] He too takes his main inspiration from

Einstein's overthrow of Newtonian physics. His main problem too is *scientific revolution*. But while according to Popper science is 'revolution in permanence', and criticism the heart of the scientific enterprise, according to Kuhn revolution is exceptional and, indeed, extra-scientific, and criticism is, in 'normal' times, anathema. Indeed for Kuhn the transition from criticism to commitment marks the point where progress – and 'normal' science – begins. For him the idea that on 'refutation' one can demand the rejection, the eliminaton of a theory, is 'naive' falsificationism. Criticism of the dominant theory and proposals of new theories are only allowed in the rare moments of 'crisis'. This last Kuhnian thesis has been widely criticized[4] and I shall not discuss it. My concern is rather that Kuhn, having recognized the failure both of justificationism and falsificationism in providing rational accounts of scientific growth, seems now to fall back on irrationalism.

For Popper scientific change is rational or at least rationally reconstructible and falls in the realm of the *logic of discovery*. For Kuhn scientific change – from one 'paradigm' to another – is a mystical conversion which is not and cannot be governed by rules of reason and which falls totally within the realm of the *(social) psychology of discovery*. Scientific change is a kind of religious change.

The clash between Popper and Kuhn is not about a mere technical point in epistemology. It concerns our central intellectual values, and has implications not only for theoretical physics but also for the underdeveloped social sciences and even for moral and political philosophy. If even in science there is no other way of judging a theory but by assessing the number, faith and vocal energy of its supporters, then this must be even more so in the social sciences: truth lies in power. Thus Kuhn's position would vindicate, no doubt, unintentionally, the basic political *credo* of contemporary religious maniacs ('student revolutionaries').

In this paper I shall first show that in Popper's logic of scientific discovery two different positions are conflated. Kuhn understands only one of these, 'naive falsificationism' (I prefer the term 'naive methodological falsificationism'); I think that his criticism of it is correct, and I shall even strengthen it. But Kuhn does not understand a more sophisticated position the rationality of which is not based on 'naive' falsificationism. I shall try to explain – and further strengthen – this stronger Popperian position which, I think, may escape Kuhn's strictures and present scientific revolutions as constituting rational progress rather than as religious conversions.

2. FALLIBILISM VERSUS FALSIFICATIONISM

(a) *Dogmatic (or Naturalistic) Falsificationism. The Empirical Basis*

To see the conflicting theses more clearly, we have to reconstruct the problem situation as it was in philosophy of science after the breakdown of 'justificationism'.

According to the 'justificationists' scientific knowledge consisted of proven propositions. Having recognized that strictly logical deductions enable us only to infer (transmit truth) but not to prove (establish truth), they disagreed about the nature of those propositions (axioms) whose truth can be proved by extra-logical means. *Classical intellectualists* (or 'rationalists' in the narrow sense of the term) admitted very varied – and powerful – sorts of extralogical 'proofs' by revelation, intellectual intuition, experience. These, with the help of logic, enabled them to prove every sort of scientific proposition. *Classical empiricists* accepted as axioms only a relatively small set of 'factual propositions' which expressed the 'hard facts'. Their truth-value was established by experience and they constituted the *empirical basis* of science. In order to prove scientific *theories* from nothing else but the narrow empirical basis, they needed a logic much more powerful than the deductive logic of the classical intellectualists: *'inductive logic'*. All justificationists, whether intellectualists or empiricists, agreed that a singular statement expressing a 'hard fact' may *disprove* a universal theory[5]; but few of them thought that a finite conjunction of factual propositions might be sufficient to *prove* 'inductively' a universal theory.[6]

Justificationism, that is, the identification of knowledge with proven knowledge, was the dominant tradition in rational thought throughout the ages. Scepticism did not deny justificationism: it only claimed that there was (and could be) no proven knowledge and *therefore* no knowledge whatsoever. For the sceptics 'knowledge' was nothing but animal belief. Thus justificationist scepticism ridiculed objective thought and opened the door to irrationalism, mysticism, superstition.

This situation explains the enormous effort invested by classical rationalists in trying to save the synthetical *a priori* principles of intellectualism and by classical empiricists in trying to save the certainty of an empirical basis and the validity of inductive inference. For all of them *scientific honesty demanded that one assert nothing that is unproven*. However, both were defeated: Kantians by non-Euclidean geometry and by non-Newtonian physics, and empiricists by the logical impossibility of establishing an empirical basis (as Kantians pointed out, facts cannot prove propositions) and of establishing an inductive logic (no logic can infallibly increase content). It turned out that *all theories are equally unprovable*.

Philosophers were slow to recognize this, for obvious reasons: classical justificationists feared that once they conceded that theoretical science is unprovable, they would have also to concede that it is sophistry and illusion, a dishonest fraud. The philosophical importance of *probabilism* (or '*neojustificationism*') lies in the denial that such a concession is necessary.

Probabilism was elaborated by a group of Cambridge philosophers who thought that although scientific theories are equally unprovable, they have different degrees of probability (in the sense of the calculus of probability) relative to the available empirical evidence.[7] *Scientific honesty then requires less than had been thought: it consists in uttering only highly probable theories: or even in merely specifying, for each scientific theory, the evidence, and the probability of the theory in the light of this evidence.*

Of course, replacing proof by probability was a major retreat for justificationist thought. But even this retreat turned out to be insufficient. It was soon shown, mainly by Popper's persistent efforts, that under very general conditions all theories have zero probability, whatever the evidence; *all theories are not only equally unprovable but also equally improbable*.[8]

Many philosophers still argue that the failure to obtain at least a probabilistic solution of the problem of induction means that we "throw over almost everything that is regarded as knowledge by science and common sense".[9] It is against this background that one must appreciate the dramatic change brought about by falsificationism in evaluating theories, and in general, in the standards of intellectual honesty. Falsificationism was, in a sense, a new and considerable retreat for rational thought. But since it was a retreat from utopian standards, it cleared away much hypocrisy and muddled thought, and thus, in fact, it represented an advance.

First I shall discuss a most important brand of falsificationism: dogmatic (or 'naturalistic') falsificationism. Dogmatic falsificationism admits the fallibility of *all* scientific theories without qualification, but it retains a sort of infallible empirical basis. It is strictly empiricist without being inductivist: it denies that the certainty of the empirical basis can be transmitted to theories. *Thus dogmatic falsificationism is the weakest brand of justificationism.*

It is extremely important to stress that admitting [fortified] empirical counterevidence as a final arbiter against a theory does not make one a dogmatic falsificationist. Any Kantian or inductivist will agree to such arbitration. But both the Kantian and the inductivist, while bowing to a negative crucial experiment, will also specify conditions of how to establish, entrench one unrefuted theory more than another. Kantians held that Euclidean geometry and Newtonian mechanics were established with certainty; inductivists held they had probability 1. For the dogmatic falsificationist, however, empirical *counter*evidence is the *one and only* arbiter which may judge a theory.

The hallmark of dogmatic falsificationism is then the recognition that all theories are equally conjectural. Science cannot *prove* any theory. But although science cannot *prove*, it can *disprove*: it "can perform with complete logical certainty [the act of] repudiation of what is false",[10] that is there is an absolutely firm empirical basis of facts which can be used to disprove theories. Falsificationists provide new – very modest – standards of scientific honesty: they are willing to regard a proposition as 'scientific' not only if it is a proven factual proposition, but even if it is nothing more than a falsifiable one, that is, if there are factual propositions available at the time with which it may clash, or, in other words, if it has potential falsifiers.[11]

Scientific honesty then consists of specifying, in advance, an experiment such that if the result contradicts the theory, the theory has to be given up.[12] The falsificationist demands that once a proposition is disproved, there must be no prevarication: the proposition must be unconditionally rejected. To (non-tautologous) unfalsifiable propositions the dogmatic falsificationist gives short shrift: he brands them 'metaphysical' and denies them scientific standing.

Dogmatic falsificationists draw a sharp demarcation between the theoretician and the experimenter: the theoretician proposes, the experimenter – in the name of Nature – disposes. As Weyl put it: "I wish to record my unbounded admiration for the work of the experimenter in his struggle to wrest interpretable facts from an unyielding Nature who knows so well how to meet our theories with a decisive *No* – or with an inaudible *Yes*".[13]Braithwaite gives a particularly lucid exposition of dogmatic falsificationism. He raises the problem of the objectivity of science: "To what extent, then, should an established scientific deductive system be regarded as a free creation of the human mind, and to what extent should it be regarded as giving an objective account of the facts of nature?" His answer is:

> The form of a statement of a scientific hypothesis and its use to express a general proposition, is a human device; what is due to Nature are the observable facts which refute or fail to refute the scientific hypothesis... [In science] we hand over to Nature the task of deciding whether any of the contingent lowest-level conclusions are false. This objective test of falsity it is which makes the deductive system, in whose construction we have very great freedom, a deductive system of scientific hypotheses. Man proposes a system of hypotheses: Nature disposes of its truth or falsity. Man invents a scientific system, and then discovers whether or not it accords with observed fact.[14]

According to the logic of dogmatic falsificationism, science grows by repeated overthrow of theories with the help of hard facts. For instance, according to this view, Descartes's vortex theory of gravity was refuted – and eliminated – by the *fact* that planets moved in ellipses rather than in Carteisan circles; Newton's theory, however, explained successfully the then available facts, both those which had been explained by Descartes's theory and those which refuted it. Therefore Newton's theory replaced Descartes's theory. Analogously, as seen by falsificationists, Newton's theory was, in turn, refuted – proved false – by the anomalous perihelion of Mercury, while Einstein's explained that too. Thus science proceeds by bold speculations, which are never proved or even made probable, but some of which are later eliminated by hard, conclusive refutations and then replaced by still bolder, new and, at least at the start, unrefuted speculations.

Dogmatic falsificationism, however, is untenable. It rests on two false assumptions and on a too narrow criterion of demarcation between scientific and non-scientific.

The *first assumption* is that there is a natural, *psychological* borderline between theoretical or speculative propositions on the one hand and factual or observational (or basic) propositions on the other. (I shall call this – following Popper – the *naturalistic doctrine of observation*.)

The *second assumption* is that if a proposition satisfies the psychological criterion of being factual or observational (or basic) then it is true; one may say that it was *proved* from facts. (I shall call this the *doctrine of observational (or experimental) proof.)*[15]

These two assumptions secure for the dogmatic falsificationist's deadly disproofs an empirical basis from which proven falsehood can be carried by deductive logic to the theory under test.

These assumptions are complemented by a *demarcation criterion*: only those theories are 'scientific' which forbid certain observable states of affairs and therefore are factually disprovable. *Or, a theory is 'scientific' if it has an empirical basis.*[16]

But both assumptions are false. Psychology testifies against the first, logic against the second, and, finally, methodological judgment testifies against the demarcation criterion. I shall discuss them in turn.

(1) A first glance at a few characteristic examples already undermines the *first assumption*. Galileo claimed that he could 'observe' mountains on the moon and spots on the sun and that these 'observations' refuted the time-honoured theory that celestial bodies are faultless crystal balls. But his 'observations' were not 'observational' in the sense or being observed by the – unaided – senses: their reliability depended on the reliability of his telescope – and of the optical theory of the telescope – which was violently questioned by his contemporaries. It was not Galileo's – pure, untheoretical – *observations* that confronted Aristotelian *theory* but rather Galileo's 'observations' in the light of his optical theory that confronted the Aristotelians' 'observations' in the light of their theory of the heavens.[17] This leaves us with two inconsistent theories, *prima facie* on a par. Some empiricists may concede this point and agree that Galileo's 'observations' were not genuine observations; but they still hold that there is a 'natural demarcation' between statements impressed on an empty and passive mind directly by the senses – only these constitute genuine 'immediate knowledge' – and between statements which are suggested by impure, theory-impregnated sensations. Indeed, *all* brands of justificationist theories of knowledge which acknowledge the senses as a source (whether as *one* source or as *the* source) of knowledge are bound to contain a *psychology of observation*. Such psychologies specify the 'right', 'normal', 'healthy', 'unbiased', 'careful' or 'scientific' state of the senses – or rather the state of mind as a whole – in which they observe truth as it is. For instance, Aristotle – and the Stoics – thought that the

right mind was the medically healthy mind. Modern thinkers recognized that there is more to the right mind than simple 'health'. Descartes's right mind is one steeled in the fire of sceptical doubt which leaves nothing but the final loneliness of the *cogito* in which the *ego* can then be re-established and God's guiding hand found to recognize truth. All schools of modern justificationism can be characterized by the particular *psycho-therapy* by which they propose to prepare the mind to receive the grace of proven truth in the course of a mystical communion. In particular, for classical empiricists the right mind is a *tabula rasa,* emptied of all original content, freed from all prejudice of theory. But it transpires from the work of Kant and Popper – and from the work of psychologists influenced by them – that such empiricist psychotherapy can never succeed. For there are and can be no sensations unimpregnated by expectations and therefore *there is no natural (i.e. psychological) demarcation between observational and theoretical propositions.*[18]

(2) But even if there was such a natural demarcation, logic would still destroy the *second assumption* of dogmatic falsificationism. For the truth-value of the 'observational' propositions cannot be indubitably decided: *no factual proposition can ever be proved from an experiment.* Propositions can only be derived from other propositions, they cannot be derived from facts: one cannot prove statements from experiences – "no more than by thumping the table".[19] This is one of the basic points of elementary logic, but one which is understood by relatively few people even today.[20]

If factual propositions are unprovable then they are fallible. If they are fallible then clashes between theories and factual propositions are not 'falsifications' but merely inconsistencies. Our imagination may play a greater role in the formulation of 'theories' than in the formulation of 'factual propositions',[21] but they are both fallible. Thus *we cannot prove theories and we cannot disprove them either.*[22] The demarcation between the soft, unproven 'theories' and the hard, proven 'empirical basis' is non-existent: *all* propositions of science are theoretical and, incurably, fallible.[23]

(3) Finally, even if there were a natural demarcation between observation statements and theories, and even if the truth-value of observation statements could be indubitably established, dogmatic falsificationism would still be useless for eliminating the most important class of what are commonly regarded as scientific theories. For even if experiments *could*

prove experimental reports, their disproving power would still be miserably restricted: *exactly the most admired scientific theories simply fail to forbid any observable state of affairs.*

To support this last contention, I shall first tell a characteristic story and then propose a general argument.

The story is about an imaginary case of planetary misbehaviour. A physicist of the pre-Einsteinian era takes Newton's mechanics and his law of gravitation (N), the accepted initial conditions, I, and calculates, with their help, the path of a newly discovered small planet, p. But the planet deviates from the calculated path. Does our Newtonian physicist consider that the deviation was forbidden by Newton's theory and therefore that, once established, it refutes the theory N? No. He suggests that there must be a hitherto unknown planet p' which perturbs the path of p. He calculates the mass, orbit, etc., of this hypothetical planet and then asks an experimental astronomer to test his hypothesis. The planet p' is so small that even the biggest available telescopes cannot possibly observe it: the experimental astronomer applies for a research grant to build yet a bigger one.[24] In three years' time the new telescope is ready. Were the unknown planet p' to be discovered, it would be hailed as a new victory of Newtonian science. But it is not. Does our scientist abandon Newton's theory and his idea of the perturbing planet? No. He suggests that a cloud of cosmic dust hides the planet from us. He calculates the location and properties of this cloud and asks for a research grant to send up a satellite to test his calculations. Were the satellite's instruments (possibly new ones, based on a little-tested theory) to record the existence of the conjectural cloud, the result would be hailed as an outstanding victory for Newtonian science. But the cloud is not found. Does our scientist abandon Newton's theory, together with the idea of the perturbing planet and the idea of the cloud which hides it? No. He suggests that there is some magnetic field in that region of the universe which disturbed the instruments of the satellite. A new satellite is sent up. Were the magnetic field to be found, Newtonians would celebrate a sensational victory. But it is not. Is this regarded as a refutation of Newtonian science? No. Either yet another ingenious auxiliary hypothesis is proposed or ... the whole story is buried in the dusty volumes of periodicals and the story never mentioned again.[25]

This story strongly suggests that even a most respected scientific theory, like Newton's dynamics and theory of gravitation, may fail to forbid any observable state of affairs.[26] Indeed, *some scientific theories forbid an event occurring in some specified finite spatio-temporal region (or briefly, a 'singular event')* only on the condition that no other factor (possibly hidden in some distant and unspecified spatio-temporal corner of the universe) *has any influence on it.* But then *such theories never alone contradict a 'basic' statement:* they contradict at most a conjunction of a basic statement describing a spatio-temporally singular event and of a universal non-existence statement saying that no other relevant cause is at work anywhere in the universe. And the dogmatic falsificationist cannot possibly claim that such universal non-existence statements belong to the empirical basis: that they can be observed and proved by experience.

Another way of putting this is to say that some scientific theories are normally interpreted as containing a *ceteris paribus* clause:[27] in such cases it is always a specific theory *together* with this clause which may be refuted. But such a refutation is inconsequential for the *specific* theory under test because by replacing the *ceteris paribus* clause by a different one the *specific* theory can always be retained whatever the tests say.

If so, the 'inexorable' disproof procedure of dogmatic falsificationism breaks down in these cases *even if* there were a firmly established empirical basis to serve as a launching pad for the arrow of the *modus tollens*: the prime target remains hopelessly elusive.[28] And as it happens, it is exactly the most important, 'mature' theories in the history of science which are *prima facie* undisprovable in this way. Moreover, by the standards of dogmatic falsificationism all probabilistic theories also come under this head: for no finite sample can ever *disprove* a universal probabilistic theory;[29] probabilistic theories, like theories with a *ceteris paribus* clause, have no empirical basis. But then the dogmatic falsificationist relegates the most important scientific theories *on his own admission* to metaphysics where rational discussion – consisting, by his standards, of proofs and disproofs – has no place, since a metaphysical theory is neither provable nor disprovable. The demarcation criterion of dogmatic falsificationism is thus still strongly antitheoretical.

(Moreover, *one can easily argue that ceteris paribus clauses are not exceptions, but the rule in science.* Science, after all, must be demarcated from a curiosity shop where funny local – or cosmic – oddities are collected and displayed. The assertion that 'all Britons died from lung cancer between 1950 and 1960 is logically possible, and might even have been true. But if it has been only an occurrence of an event with minute probability, it would have only curiosity value for the crankish fact-collector, it would have a macabre entertainment value, but no scientific value. A proposition might be said to be scientific only if it aims at expressing a causal connection: such connection between being a Briton and dying of lung cancer may not even be intended. Similarly, 'all swans are white', if true, would be a mere curiosity unless it asserted that swanness *causes* whiteness. But then a black swan would not refute this proposition, since it may only indicate *other causes* operating simultaneously. Thus 'all swans are white' is either an oddity and easily disprovable or a scientific proposition with a *ceteris paribus* clause and therefore undisprovable. *Tenacity of a theory against empirical evidence would then be an argument for rather than against regarding it as 'scientific'. 'Irrefutability' would become a hallmark of science.*[30])

To sum up: classical justificationists only admitted proven theories; neoclassical justificationists probable ones; dogmatic falsificationists realized that in either case no theories are admissible. They decided to admit theories if they are disprovable – disprovable by a finite number of observations. But even if there were such disprovable theories – those which can be contradicted by a finite number of observable facts – they are still logically too near to the empirical basis. For instance, on the terms of the dogmatic falsificationist, a theory like 'All planets move in ellipses' may be disproved by five observations; therefore the dogmatic falsificationist will regard it as scientific. A theory like 'All planets move in circles' may be disproved by four observations; therefore the dogmatic falsificationist will regard it as still more scientific. The acme of scientificness will be a theory like 'All swans are white' which is disprovable by one single observation. On the other hand, he will reject all probabilistic theories together with Newton's, Maxwell's, Einstein's theories, as unscientific, for no finite number of observations can ever disprove them.

If we accept the demarcation criterion of dogmatic falsificationism, *and* also the idea that facts can prove 'factual' propositions, we have to declare that the most important, if not all, theories ever proposed in the history of science are metaphysical, that most, if not all, of the accepted progress is pseudo-progress, that most, if not all, of the work done is irrational. If, however, still accepting the demarcation criterion of dogmatic falsificationism, we deny that facts can prove propositions, then we certainly end up in complete scepticism: then all science is undoubtedly irrational metaphysics and should be rejected. *Scientific theories are not only equally unprovable, and equally improbable, but they are also equally undisprovable.* But the recognition that not only the theoretical but *all* the propositions in science are fallible, means the total collapse of *all* forms of dogmatic justificationism as theories of scientific rationality.

(b) *Methodological Falsificationism. The 'Empirical Basis'*

The collapse of dogmatic falsificationism because of fallibilistic arguments seems to bring us back to square one. If *all* scientific statements are fallible theories, one can criticize them only for inconsistency. But then, in what sense, if any, is science empirical? If scientific theories are neither provable, nor probabilifiable, nor disprovable, then the sceptics seem to be finally right: science is no more than vain speculation and there is no such thing as progress in scientific knowledge. Can we still oppose scepticism? *Can we save scientific criticism from fallibilism?* Is it possible to have a fallibilistic theory of scientific progress? In particular, if scientific criticism is fallible, on what ground can we ever eliminate a theory?

A most intriguing answer is provided by *methodological falsificationism.* Methodological falsificationism is a brand of conventionalism; therefore in order to understand it, we must first discuss conventionalism in general.

There is an important demarcation between *'passivist' and 'activist' theories of knowledge.* 'Passivists' hold that true knowledge is Nature's imprint on a perfectly inert mind: mental *activity* can only result in bias and distortion. The most influential passivist school is classical empiricism. 'Activists' hold that we cannot read the book of Nature without mental activity, without interpreting them in the light of our expectations or theories.[31] Now *conservative 'activists'* hold that we are born with our basic expectations; with them we turn the world into 'our world' but must then live for ever in the prison of our world. The idea that we live and die in the prison of our 'conceptual frameworks' was developed primarily by Kant; pessimistic Kantians thought that the real world is for ever unknowable because of this prison, while optimistic Kantians thought that God created our conceptual framework to fit the world.[32] But *revolutionary activists* believe that conceptual frameworks can be developed and also replaced by new, *better* ones; it is *we* who create our 'prisons' and we can also, critically, demolish them.[33]

New steps from conservative to revolutionary activism were made by Whewell and then by Poincaré, Milhaud and Le Roy. Whewell held that theories are developed by trial and error – in the 'preludes to the inductive epochs'. The best ones among them are then 'proved' – during the 'inductive epochs' – by a long primarily *a priori* consideration which he called 'progressive intuition'. The 'inductive epochs' are followed by 'sequels to the inductive epochs'; cumulative developments of auxiliary theories.[34] Poincaré, Milhaud and Le Roy were averse to the idea of *proof* by progressive intuition and preferred to explain the continuing historical success of Newtonian mechanics by a *methodological decision* taken by scientists: after a considerable period of initial empirical success scientists may *decide* not to allow the theory to be refuted. Once they have taken this decision, they solve (or dissolve) the apparent anomalies by auxiliary hypotheses or other 'conventionalist stratagems'.[35] This *conservative conventionalism* has, however, the disadvantage of making us unable to get out of our self-imposed prisons, once the first period of trial-and-error is over and the great decision taken. It cannot solve the problem of the elimination of those theories which have been triumphant for a long period. According to conservative conventionalism, experiments may have sufficient power to refute young theories, but not to refute old, established theories: *as science grows, the power of empirical evidence diminishes.*[36]

Poincaré's critics refused to accept his idea, that, although the scientists build their conceptual frameworks, there comes a time when these frameworks turn into prisons which cannot be demolished. This criticism gave rise to two rival schools of *revolutionary conventionalism*: Duhem's simplicism and Popper's methodological falsificationism.[37]

Duhem accepts the conventionalists' position that no physical theory ever crumbles merely under the weight of 'refutations', but claims that it still may crumble under the weight of "continual repairs, and many tangled-up stays" when "the worm-eaten columns" cannot support "the tottering building" any longer[38]; then the theory loses its original simplicity and has to be replaced. But falsification is then left to subjective taste or, at best, to scientific fashion, and leaves too much leeway for dogmatic adherence to a favourite theory.

Popper set out to find a criterion which is both more objective and more hard-hitting. He could not accept the emasculation of empiricism, inherent even in Duhem's approach, and proposed a methodology which allows experiments to be powerful even in 'mature' science. Popper's methodological falsificationism is both conventionalist and falsificationist, but he "differs from the [conservative] conventionalists in holding that the statements decided by agreement are *not* [spatio-temporally] universal but [spatio-temporally] singular"[39]; and he differs from the dogmatic falsificationist in holding that the truth-value of such statements cannot be proved by facts but, in some cases, may be decided by agreement.[40]

The *conservative conventionalist* (or methodological justificationist, if you wish) makes unfalsifiable by *fiat* some (spatio-temporally) universal theories, which are distinguished by their explanatory power, simplicity or beauty. Our *revolutionary conventionalist* (or 'methodological falsificationist') makes unfalsifiable by *fiat* some (spatio-temporally) singular statements which are distinguishable by the fact that there exists at the time a 'relevant technique' such that 'anyone who has learned it' will be able to *decide* that the statement is 'acceptable'.[41] Such a statement may be called an 'observational' or 'basic' statement, but only in inverted commas.[42] Indeed, the very selection of all such statements is a matter of a decision, which is not based on exclusively psychological considerations. This decision is then followed by a second kind of decision concerning the separation of the set of *accepted* basic statements from the rest.

These *two decisions* correspond to the *two assumptions* of dogmatic falsificationism. But there are important differences. First, the methodological falsificationist is not a justificationist, he has no illusions about 'experimental proofs' and is fully aware of the fallibility of his decisions and the risks he is taking.

The methodological falsificationist realizes that in the 'experimental techniques' of the scientist fallible theories are involved,[43] 'in the light of which' he interprets the facts. In spite of this he 'applies' these theories, he regards them in the given context not as theories under test but as *unproblematic background knowledge* "which we accept (tentatively) as unproblematic while we are testing the theory".[44] He may call these theories – and the statements whose truth-value he decides in their light – 'observational': but this is only a manner of speech which he inherited from naturalistic falsificationism.[45] The methodological falsificationist *uses our most successful theories as extensions of our senses* and widens the range of theories which can be applied in testing far beyond the dogmatic falsificationist's range of strictly observational theories. For instance, let us imagine that a big radio-star is discovered with a system of radio-star satellites orbiting it. We should like to test some gravitational theory on this planetary system – a matter of considerable interest. Now let us imagine that Jodrell Bank succeeds in providing a set of space-time co-ordinates of the planets which is inconsistent with the theory. We shall take these statements as potential falsifiers. Of course, these basic state-ments are not 'observational' in the usual sense but only "'observational'". They describe planets that neither the human eye nor optical instruments can reach. Their truth-value is arrived at by an 'experimental technique'. This 'experimental technique' is based on the 'application' of a well-corroborated theory of radio-optics. Calling these statements 'observa-tional' is no more than a manner of saying that, in the context of his problem, that is, in testing our gravitational theory, the methodological falsificationist uses radio-optics uncritically, as 'background knowledge'. *The need for decisions to demarcate the theory under test from unproblem-atic background knowledge is a characteristic feature of this brand of methodological falsificationism.*[46] (This situation does not really differ from Galileo's 'observation' of Jupiter's satellites: moreover, as some of Gali-leo's contemporaries rightly pointed out, he relied on a virtually non-existent optical theory – which then was less corroborated, and even less articulated, than present-day radio-optics. On the other hand, calling the reports of our human eye 'observational' only indicates that we 'rely' on some vague physiological theory of human vision.[47])

This consideration shows the conventional element in granting – in a given context – the (methodologically) 'observational' status to a theory.[48] Similarly, there is a considerable conventional element in the decision concerning the actual truth-value of a basic statement which we take after we have decided which 'observational theory' to apply. One single obser-vation may be the stray result of some trivial error: in order to reduce such risks, methodological falsificationists prescribe some safety control. The simplest such control is to repeat the experiment (it is a matter of convention how many times); another is to 'fortify' the potential falsifier by a 'well-corroborated falsifying hypothesis'.[49]

The methodological falsificationist also points out that, as a matter of fact, these conventions are institutionalized and endorsed by the scientific community; the list of 'accepted' falsifiers is provided by the verdict of the experimental scientists.[50]

This is how the methodological falsificationist establishes his 'empirical basis'. (He uses inverted commas in order 'to give ironical emphasis' to the term.[51]) This 'basis' can be hardly called a 'basis' by justificationist standards: there is nothing proven about it – it denotes 'piles driven into a swamp'.[52] Indeed, if this 'empirical basis' clashes with a theory, the theory may be *called* 'falsified', but it is not falsified in the sense that it is disproved. Methodological 'falsification' is very different from dogmatic falsification. If a theory is falsified, it is proven false; if it is 'falsified', it may still be true. If we follow up this sort of 'falsification' by the actual 'elimination' of a theory, we may well end up by eliminating a true, and accepting a false, theory (a possibility which is thoroughly abhorrent to the old-fashioned justificationist).

Yet the methodological falsificationist advises that exactly this is to be done. The methodological falsificationist realizes that if we want to reconcile fallibilism with (non-justificationist) rationality, we *must* find a way to eliminate *some* theories. If we do not succeed, the growth of science will be nothing but growing chaos.

Therefore the methodological falsificationist maintains that "[if we want] to make the method of selection by elimination work, and to ensure that only the fittest theories survive, their struggle for life must be made severe".[53] Once a theory has been falsified, in spite of the risk involved, it must be eliminated: "[with theories we work only] as long as they stand up to tests".[54] The elimination must be methodologically conclusive: "In general we regard an inter-subjectively testable falsification as final... A corroborative appraisal made at a later date... can replace a positive degree of corroboration by a negative one, but not *vice versa*".[55] This is the methodological falsificationist's explanation of how we get out of a rut: "It is always the experiment which saves us from following a track that leads nowhere".[56]

The methodological falsificationist separates rejection and disproof, which the dogmatic falsificationist had conflated.[57] He is a fallibilist but his fallibilism does not weaken his critical stance: he turns fallible propositions into a 'basis' for a hard-line policy. On these grounds he proposes a *new demarcation criterion*: only those theories – that is, non-'observational' propositions – which forbid certain 'observable 'states of affairs, and therefore may be 'falsified' and rejected, are 'scientific': or, briefly, *a theory is 'scientific' (or 'acceptable') if it has an 'empirical basis'*. This criterion brings out sharply the difference between dogmatic and methodological falsificationism.[58]

This methodological demarcation criterion is much more liberal than the dogmatic one. Methodological falsificationism opens up new avenues of criticism: many more theories may qualify as 'scientific'. We have already seen that there are more 'observational' theories than observational theories, and therefore there are more 'basic' statements than basic statements. [59] Furthermore, probabilistic theories may qualify now as 'scientific': although they are not falsifiable they can be easily made 'falsifiable' by an *additional (third type) decision* which the scientist can make by specifying certain rejection rules which may make statistically interpreted evidence 'inconsistent' with the probabilistic theory.[60]

But even these three decisions are not sufficient to enable us to 'falsify' a theory which cannot explain anything 'observable' without a *ceteris paribus* clause. No finite number of 'observations' is enough to 'falsify' such a theory. However, if this is the case how can one reasonably defend a methodology which claims to "interpret natural laws or theories as... statements which are partially decidable, i.e. which are, for logical reasons, not verifiable but, in an asymmetrical way, falsifiable..."?[61] How can we interpret theories like Newton's theory of dynamics and gravitation as 'one-sidedly decidable'?[62] How can we make in such cases genuine "attempts to weed out false theories – to find the weak points of a theory in order to reject it if it is falsified by the test"?[63] How can we draw them into the realm of rational discussion? The methodological falsificationist solves the problem by making a further (*fourth type*) *decision*: when he tests a theory together with a *ceteris paribus* clause and finds that this conjunction has been refuted, he must decide whether to take the refutation also as a refutation of the specific theory. For instance, he may accept Mercury's 'anomalous' perihelion as a refutation of the treble conjunction N_3 of Newton's theory, the known initial conditions and the *ceteris paribus* clause. Then he tests the initial conditions 'severely'[64] and may decide to relegate them into the 'unproblematic background knowledge'. This decision implies the refutation of the double conjunction N_2 of Newton's theory and the *ceteris paribus* clause. Now he has to take the crucial decision: whether to relegate also the *ceteris paribus* clause into the pool of 'unproblematic background knowledge'. He will do so if he finds the *ceteris paribus* clause well corroborated.

How can one test a *ceteris paribus* clause severely? By assuming that there *are* other influencing factors, by specifying such factors, and by testing these specific assumptions. If many of them are refuted, the *ceteris paribus* clause will be regarded as well-corroborated.

Yet the decision to 'accept' a *ceteris paribus* clause is a very risky one because of the grave consequences it implies. If it is decided to accept it as part of such background knowledge, the statements describing Mercury's perihelion from the empirical basis of N_2 are turned into the empirical basis of Newton's specific theory N_1 and what was previously a mere 'anomaly' in relation to N_1, becomes now crucial evidence against it, its falsification. (We may call an event described by a statement A an '*anomaly* in relation to a theory T' if A is a potential falsifier of the conjunction of T and a *ceteris paribus* clause but it becomes a potential falsifier of T itself after having decided to relegate the *ceteris paribus* clause into "unproblematic background knowledge".) Since, for our savage falsificationist, falsifications are methodologically conclusive, the fateful decision amounts to the methodological elimination of Newton's theory, making further work on it irrational. If the scientist shrinks back from such bold decisions he will "never benefit from experience", "believing, perhaps, that it is his business to defend a successful system against criticism as long as it is not *conclusively disproved*".[65] He will degenerate into an apologist who may always claim that "the discrepancies which are asserted to exist between the experimental results and the theory are only apparent and they will disappear with the advance of our understanding."[66] But for the falsificationist this is "the very reverse of the critical attitude which is the proper one for the scientist",[67] and is impermissible. To use one of the methodological falsificationist's favourite expressions: the theory "must be made to stick its neck out".

The methodological falsificationist is in a serious plight when it comes to deciding where to draw the demarcation, even if only in a well-defined context, between the problematic and unproblematic. The plight is most dramatic when he has to make a decision about *ceteris paribus* clauses, when he has to promote one of the hundreds of 'anomalous phenomena' into a 'crucial experiment', and decide that in such a case the experiment was 'controlled'.[68]

Thus, with the help of this fourth type of decision,[69] our methodological falsificationist has finally succeeded in interpreting even theories like Newton's theory as 'scientific'.[70]

Indeed, there is no reason why he should not go yet another step. Why not decide that a theory – which even these four decisions cannot turn into an empirically falsifiable one – is falsified if it clashes with another theory which is scientific on some of the previously specified grounds and is also well-corroborated?[71] After all, if we reject one theory because one of its potential falsifiers is seen to be true in the light of an observational theory, why not reject another theory because it clashes *directly* with one that may be relegated into unproblematic background knowledge? This would allow us, by a *fifth type decision*, to eliminate even 'syntactically metaphysical' theories, that is, theories, which, like 'all-some' statements or purely existential statements,[72] because of their *logical form* cannot have spatio-temporally singular potential falsifiers.

To sum up: the methodological falsificationist offers an interesting solution to the problem of combining hard-hitting criticism with fallibilism. Not only does he offer a philosophical basis for falsification after fallibilism had pulled the carpet from under the feet of the dogmatic falsificationist, but he also widens the range of such criticism very considerably. By putting falsification in a new setting, he saves the attractive code of honour of the dogmatic falsificationist: that scientific honesty consists in specifying, in advance, an experiment such, that if the result contradicts the theory, the theory has to be given up.

Methodological falsificationism represents a considerable advance beyond both dogmatic falsificationism and conservative conventionalism. It recommends risky decisions. But the risks are daring to the point of recklessness and one wonders whether there is no way of lessening them.

Let us first have a closer look at the risks involved.

Decisions play a crucial role in this methodology – as in any brand of conventionalism. Decisions however may lead us disastrously astray. The methodological falsificationist is the first to admit this. But this, he argues, is the price which we have to pay for the possibility of progress.

One has to appreciate the dare-devil attitude of our methodological falsificationist. He feels himself to be a hero who, faced with two catastrophic alternatives, dared to reflect coolly on their relative merits and choose the lesser evil. One of the alternatives was sceptical fallibilism,

with its 'anything goes' attitude, the despairing abandonment of all intellectual standards, and hence of the idea of scientific progress. Nothing
can be established, nothing can be rejected, nothing even communicated:
the growth of science is a growth of chaos, a veritable Babel. For two
thousand years, scientists and scientifically-minded philosophers chose
justificationist illusions of some kind to escape this nightmare. Some of
them argued that *one has to choose between inductivist justificationism and
irrationalism*: "I do not see any way out of a dogmatic assertion that we
know the inductive principle or some equivalent; the only alternative is
to throw over almost everything that is regarded as knowledge by science
and common sense".[73] Our methodological falsificationist proudly rejects
such escapism: he dares to measure up to the full impact of fallibilism and
yet escape scepticism by a daring and risky conventionalist policy, with no
dogmas. He is fully aware of the risks but insists that *one has to choose between some sort of methodological falsificationism and irrationalism*. He
offers a game in which one has little hope of winning, but claims that it is
still better to play than to give up.[74]

Indeed, those critics of naive falsificationism who offer no alternative
method of criticism are inevitably driven to irrationalism. For instance,
Neurath's muddled argument, that the falsification and ensuing elimination of a hypothesis may turn out to have been "an obstacle in the progress of science",[75] carries no weight as long as the only alternative he
seems to offer is chaos. Hempel is, no doubt, right in stressing that "science
offers various examples [when] a conflict between a highly-confirmed
theory and an occasional recalcitrant experiential sentence may well be resolved by revoking the latter rather than by sacrificing the former"[76];
nevertheless he admits that he can offer no other 'fundamental standard'
than that of naive falsificationism.[77] Neurath – and, seemingly, Hempel –
reject falsificationism as 'pseudo-rationalism'[78]; but where is 'real
rationalism'? Popper warned already in 1934 that Neurath's permissive
methodology (or rather lack of methodology) would make science unempirical and therefore irrational:

We need a set of rules to limit the arbitrariness of 'deleting' (or else 'accepting') a protocol sentence. Neurath fails to give any such rules and thus unwittingly throws empiricism overboard... Every system becomes defensible if one is allowed (as everybody is,
in Neurath's view) simply to 'delete' a protocol sentence if it is inconvenient.[79]

Popper agrees with Neurath that all propositions are fallible; but he forcefully makes the crucial point that we cannot make progress unless we have a firm rational strategy or method to guide us when they clash.[80]

But is not the firm strategy of the brand of methodological falsification-ism hitherto discussed *too firm*? Are not the decisions it advocates bound to be *too arbitrary*? Some may even claim that all that distinguishes methodological from dogmatic falsificationism is that *it pays lip-service to fallibilism*!

To criticize a theory of criticism is usually very difficult. Naturalistic falsificationism was relatively easy to refute, since it rested on an empirical psychology of perception: one could show that it was simply *false*. But how can methodological falsificationism be falsified? No disaster can ever disprove a non-justificationist theory of rationality. Moreover, how can we ever recognize an epistemological disaster? We have no means to judge whether the verisimilitude of our successive theories increases or decreases.[81] At this stage we have not yet developed a general theory of criticism even for scientific theories, let alone for theories of rationality[82]; therefore if we want to falsify our methodological falsificationism, we have to do it before having a theory of how to do it.

If we look at history of science, if we try to see how some of the most celebrated falsifications happened, we have to come to the conclusion that either some of them are plainly irrational, or that they rest on rationality principles radically different from the ones we just discussed. First of all, our falsificationist must deplore the fact that stubborn theoreticians frequently challenge experimental verdicts and have them reversed. In the falsificationist conception of scientific 'law and order' we have described there is no place for such successful appeals. Further difficulties arise from the falsification of theories to which a *ceteris paribus* clause is appended.[83] Their falsification as it occurs in actual history is *prima facie* irrational by the standards of our falsificationist. By his standards, scientists frequently seem to be irrationally slow: for instance, eighty-five years elapsed be-tween the acceptance of the perihelion of Mercury as an anomaly and its acceptance as a falsification of Newton's theory, in spite of the fact that the *ceteris paribus* clause was reasonably well corroborated. On the other hand, scientists frequently seem to be irrationally rash: for instance, Galileo and his disciples accepted Copernican heliocentric celestial mechanics in spite of the abundant evidence against the rotation of the Earth; or Bohr and his disciples accepted a theory of light emission in spite of the fact that it ran counter to Maxwell's well-corroborated theory.

Indeed, it is not difficult to see at least two crucial characteristics common to both dogmatic and our methodological falsificationism which are clearly dissonant with the actual history of science: that (1) *a test is – or must be made – a two-cornered fight between theory and experiment so that in the final confrontation only these two face each other; and (2) the only interesting outcome of such confrontation is (conclusive) falsification: "[the only genuine] discoveries are refutations of scientific hypotheses"*.[84] However, history of science suggests that (1') tests are – at least – three-cornered fights between rival theories and experiment and (2') some of the most interesting experiments result, *prima facie*, in confirmation rather than falsification.

But if – as seems to be the case – the history of science does not bear out our theory of scientific rationality, we have two alternatives. One alternative is to abandon efforts to give a rational explanation of the success of science. Scientific method (or 'logic of discovery'), conceived as the discipline of rational appraisal of scientific theories – and of criteria of *progress* – vanishes. We, may, of course, still try to explain *changes* in 'paradigms' in terms of social psychology.[85] This is Polanyi's and Kuhn's way.[86] The other alternative is to try at least to *reduce* the conventional element in falsificationism (we cannot possibly eliminate it) and replace the *naive* versions of methodological falsificationism – characterized by the theses (1) and (2) above – by a *sophisticated* version which would give a new *rationale* of falsification and thereby rescue methodology and the idea of scientific *progress*. This is Popper's way, and the one I intend to follow.

(c) *Sophisticated Versus Naive Methodological Falsificationism. Progressive and Degenerating Problemshifts.*

Sophisticated falsificationism differs from naive falsificationism both in its rules of *acceptance* (or 'demarcation criterion') and its rules of *falsification* or elimination. For the naive falsificationist any theory which can be interpreted as experimentally falsifiable, is 'acceptable' or 'scientific'. For the sophisticated falsificationist a theory is 'acceptable' or 'scientific' only if it has corroborated excess empirical content over its predecessor (or rival), that is, only if it leads to the discovery of novel facts. This condition can be analysed into two clauses: that the new theory has excess empirical content (*'acceptability'*$_1$) and that some of this excess content is verified ('*acceptability*'$_2$). The first clause can be checked instantly by *a priori* logical analysis; the second can be checked only empirically and this may take an indefinite time.

Again, for the naive falsificationist a theory is *falsified* by a ('fortified') 'observational' statement which conflicts with it (or rather, which he decides to interpret as conflicting with it). The sophisticated falsificationist regards a scientific theory T as falsified if and only if another theory T' has been proposed with the following characteristics: (1) T' has excess empirical content over T: that is, it predicts *novel* facts, that is, facts improbable in the light of, or even forbidden, by T;[87] (2) T' explains the previous success of T, that is, all the unrefuted content of T is contained (within the limits of observational error) in the content of T'; and (3) some of the excess content of T' is corroborated.[88]

In order to be able to appraise these definitions we need to understand their problem background and their consequences. First, we have to remember the conventionalists' methodological discovery that no experimental result can ever kill a theory: any theory can be saved from counter-instances either by some auxiliary hypothesis or by a suitable reinterpretation of its terms. Naive falsificationists solved this problem by relegating – in crucial contexts – the auxiliary hypotheses to the realm of unproblematic background knowledge, eliminating them from the deductive model of the test-situation and thereby *forcing* the chosen theory into logical isolation, in which it becomes a sitting target for the attack of test-experiments. But since this procedure did not offer a suitable guide for a rational reconstruction of the history of science, we may just as well completely rethink our approach. Why aim at falsification at any price? Why not rather impose certain standards on the theoretical adjustments by which one is allowed to save a theory? Indeed, some such standards have been well-known for centuries, and we find them expressed in age-old wisecracks against *ad hoc* explanations, empty prevarications, face-saving, linguistic tricks.[89] We have already seen that Duhem adumbrated such standards in terms of 'simplicity' and 'good sense'. But *when* does lack of 'simplicity' in the protective belt of theoretical adjustments reach the point at which the theory *must* be abandoned?[90] In what sense was Copernican theory, for instance, 'simpler' than Ptolemaic?[91] The vague notion of Duhemian 'simplicity' leaves, as the naive falsificationist correctly argued, the decision very much to taste and fashion.

Can one improve on Duhem's approach? Popper did. His solution – a sophisticated version of methodological falsificationism – is more objective and more rigorous. Popper agrees with the conventionalists that theories and factual propositions can always be harmonized with the help of auxiliary hypotheses: he agrees that the problem is how to demarcate between scientific and pseudoscientific *adjustments*, between rational and irrational changes of theory. According to Popper, saving a theory with the help of auxiliary hypotheses which satisfy certain well-defined conditions represents scientific progress; but saving a theory with the help of auxiliary hypotheses which do not, represents degeneration. Popper calls such inadmissible auxiliary hypotheses *ad hoc* hypotheses, mere linguistic devices, 'conventionalist stratagems'.[92] But then any scientific theory has to be appraised together with its auxiliary hypotheses, initial conditions, etc., and, especially, together with its predecessors so that we may see by what sort of *change* it was brought about. Then, of course, what we appraise is a *series of theories* rather than isolated *theories*.

Now we can easily understand why we formulated the criteria of acceptance and rejection of sophisticated methodological falsificationism as we did. But it may be worth while to reformulate them slightly, couching them explicitly in terms of *series of theories*.

Let us take a series of theories, $T_1, T_2, T_3,\ldots$ where each subsequent theory results from adding auxiliary clauses to (or from semantical reinterpretations of) the previous theory in order to accommodate some anomaly, each theory having at least as much content as the unrefuted content of its predecessor. Let us say that such a series of theories is *theoretically progressive (or 'constitutes a theoretically progresssive problemshift')* if each new theory has some excess empirical content over its predecessor, that is, if it predicts some novel, hitherto unexpected fact. Let us say that a theoretically progressive series of theories is also *empirically progressive (or 'constitutes an empirically progressive problemshift')* if some of this excess empirical content is also corroborated, that is, if each new theory leads us to the actual discovery of some *new fact*.[93] Finally, let us call a problemshift *progressive* if it is both theoretically and empirically progressive, and *degenerating* if it is not.[94] We *'accept'* problemshifts as 'scientific 'only if they are at least theoretically progressive; if they are not, we *'reject'* them as 'pseudoscientific'. Progress is measured by the degree to which a problemshift is progressive, by the degree to which the series of theories leads us to the discovery of novel facts. We regard a theory in the series 'falsified' when it is superseded by a theory with higher corroborated content.

This demarcation between progressive and degenerating problemshifts sheds new light on the appraisal of *scientific – or, rather, progressive – explanations.* If we put forward a theory to resolve a contradiction between a previous theory and a counterexample in such a way that the new theory, instead of offering a content-increasing (scientific) *explanation,* only offers a content-decreasing (linguistic) *reinterpretation,* the contradiction is resolved in a merely semantical, unscientific way. *A given fact is explained scientifically only if a new fact is also explained with it.*[95]

Sophisticated falsificationism thus shifts the problem of how to appraise *theories* to the problem of how to appraise *series of theories.* Not an isolated *theory,* but only a series of theories can be said to be scientific or unscientific: to apply the term 'scientific' to one *single* theory is a category mistake.[96]

The time-honoured empirical criterion for a satisfactory theory was agreement with the observed facts. Our empirical criterion for a series of theories is that it should produce new facts. *The idea of growth and the concept of empirical character are soldered into one.*

This revised form of methodological falsificationism has many new features. First, it denies that "in the case of a scientific theory, our decision depends upon the results of experiments. If these confirm the theory, we may accept it until we find a better one. If they contradict the theory, we reject it".[97] It denies that "what ultimately decides the fate of a theory is the result of a test, i.e. an agreement about basic statements".[98] Contrary to naive falsificationism, *no experiment, experimental report, observation statement or well-corroborated low-level falsifying hypothesis alone can lead to falsification. There is no falsification before the emergence of a better theory.*[99] But then the distinctively negative character of naive falsificationism vanishes; criticism becomes more difficult, and also positive, constructive. But, of course, if falsification depends on the emergence of better theories, on the invention of theories which anticipate new facts, then falsification is *not* simply a relation between a theory and the empirical basis, but a multiple relation between competing theories, the original 'empirical basis', and the empirical growth resulting from the competition. Falsification can thus be said to have a *'historical character'.*[100]

Moreover, some of the theories which bring about falsification are frequently proposed *after* the 'counterevidence'. This may sound paradoxical for people indoctrinated with naive falsificationism. Indeed, this epistemological theory of the relation between theory and experiment differs sharply from the epistemological theory of naive falsificationism. The very term 'counterevidence' has to be abandoned in the sense that no experimental result must be interpreted directly as 'counterevidence'. If we still want to retain this time-honoured term, we have to redefine it like this: 'counterevidence to T_1' is a corroborating instance to T_2 which is either inconsistent with or independent of T_1 (with the *proviso* that T_2 is a theory which satisfactorily explains the empirical success of T_1). This shows that *'crucial counterevidence'* – or *'crucial experiments'* – can be recognized as such among the scores of anomalies only *with hindsight*, in the light of some superseding theory.[101]

Thus the crucial element in falsification is whether the *new theory* offers any novel, excess information compared with its predecessor and whether some of this excess information is corroborated. Justificationists valued 'confirming' instances of a theory; naive falsificationists stressed 'refuting' instances; for the methodological falsificationists it is the – rather rare – corroborating instances of the *excess* information which are the crucial ones; these receive all the attention. We are no longer interested in the thousands of trivial verifying instances nor in the hundreds of readily available anomalies: the few crucial *excess-verifying instances* are decisive.[102] This consideration rehabilitates – and reinterprets – the old proverb: *Exemplum docet, exempla obscurant.*

'Falsification' in the sense of naive falsificationism (corroborated counterevidence) is not a *sufficient* condition for eliminating a specific theory: in spite of hundreds of known anomalies we do not regard it as falsified (that is, eliminated) until we have a better one.[103] Nor is 'falsification' in the naive sense *necessary* for falsification in the sophisticated sense: a progressive problemshift does not have to be interspersed with 'refutations'. Science can grow without any 'refutations' leading the way. Naive falsificationists suggest a linear growth of science, in the sense that theories are followed by powerful refutations which eliminate them; these refutations in turn are followed by new theories.[104] It is perfectly *possible* that theories be put forward 'progressively' in such a rapid succession that the 'refutation' of the *n*-th appears only as the corroboration of the

$n+1$th. The problem fever of science is raised by proliferation of rival theories rather than counterexamples or anomalies.

This shows that the slogan of *proliferation of theories* is much more important for sophisticated than for naive falsificationism. For the naive falsificationist science grows through repeated experimental overthrow of theories; new rival theories proposed before such 'overthrows' may speed up growth but are not absolutely necessary[105], constant proliferation of theories is optional but not mandatory. For the sophisticated falsificationist proliferation of theories cannot wait until the accepted theories are 'refuted' (or until their protagonists get into a Kuhnian crisis of confidence).[106] While naive falsificationism stresses "the urgency of replacing a *falsified* hypothesis by a better one",[107] sophisticated falsificationism stresses the urgency of replacing *any* hypothesis by a better one. Falsification cannot "compel the theorist to search for a better theory",[108] simply because falsification cannot precede the better theory.

The problem-shift from naive to sophisticated falsificationism involves a semantic difficulty. For the naive falsificationist a 'refutation' is an experimental result which, by force of his decisions, is made to conflict with the theory under test. But according to sophisticated falsificationism one must not take such decisions before the alleged 'refuting instance' has become the confirming instance of a new, better theory. Therefore whenever we see terms like 'refutation', 'falsification', 'counterexample', we have to check in each case whether these terms are being applied in virtue of decisions by the naive or by the sophisticated falsificationist.[109]

Sophisticated methodological falsificationism offers new standards for intellectual honesty. Justificationist honesty demanded the acceptance of only what was proven and the rejection of everything unproven. Neo-justificationist honesty demanded the specification of the probability of any hypothesis in the light of the available empirical evidence. The honesty of naive falsificationism demanded the testing of the falsifiable and the rejection of the unfalsifiable and the falsified. Finally, the honesty of sophisticated falsificationism demanded that one should try to look at things from different points of view, to put forward new theories which anticipate novel facts, and to reject theories which have been superseded by more powerful ones.

Sophisticated methodological falsificationism blends several different traditions. From the empiricists it has inherited the determination to learn primarily from experience. From the Kantians it has taken the activist approach to the theory of knowledge. From the conventionalists it has learned the importance of decisions in methodology.

I should like to emphasize here a further distinctive feature of sophisticated methodological empiricism: the crucial role of excess corroboration. For the inductivist, learning about a new theory is learning how much confirming evidence supports it; about refuted theories one *learns* nothing (learning, after all, is to build up proven or probable *knowledge*). For the dogmatic falsificationist, learning about a theory is learning whether it is refuted or not; about confirmed theories one learns nothing (one cannot prove or probabilify anything), about refuted theories one learns that they are disproved.[110] For the sophisticated falsificationist, learning about a theory is primarily learning which new facts it anticipated: indeed, for the sort of Popperian empiricism I advocate, the only relevant evidence is the evidence anticipated by a theory, and *empiricalness (or scientific character) and theoretical progress are inseparably connected.*[111]

This idea is not entirely new. Leibnitz, for instance, in his famous letter to Conring in 1678, wrote: "It is the greatest commendation of an hypothesis (next to [proven] truth) if by its help predictions can be made even about phenomena or experiments not tried."[112] Leibnitz's view was widely accepted by scientists. But since appraisal of a scientific theory, before Popper, meant appraisal of its degree of justification, this position was regarded by some logicians as untenable. Mill, for instance, complains in 1843 in horror that "it seems to be thought that an hypothesis... is entitled to a more favourable reception, if besides accounting for all the facts previously known, it has led to the anticipation and prediction of others which experience afterwards verified".[113] Mill had a point: this appraisal was in conflict both with justificationism and with probabilism: why should an event *prove* more, if it was anticipated by the theory than if it was known already before? As long as *proof* was the only criterion of the scientific character of a theory, Leibnitz's criterion could only be regarded as irrelevant.[114] Also, the *probability* of a theory given evidence cannot possibly be influenced, as Keynes pointed out, by *when* the evidence was produced: the probability of a theory given evidence can depend only on the theory and the evidence,[115] and not upon whether the evidence was produced before or after the theory.

In spite of this convincing justificationist criticism, the criterion survived among some of the best scientists, since it formulated their strong dislike of merely *ad hoc* explanations, which "though [they] truly express the facts [they set out to explain, are] not born out by any other phenomena".[116]

But it was only Popper who recognized that the *prima facie* inconsistency between the few odd, casual remarks against *ad hoc* hypotheses on the one hand and the huge edifice of justificationist philosophy of knowledge must be solved by demolishing justificationism and by introducing new, nonjustificationist criteria for appraising scientific theories based on anti-adhocness.

Let us look at a few examples. Einstein's theory is not better than Newton's *because* Newton's theory was 'refuted' but Einstein's was not: there are many known 'anomalies' to Einsteinian theory. Einstein's theory is better than – that is, represents progress compared with – Newton's theory *anno 1916* (that is, Newton's laws of dynamics, law of gravitation, the known set of initial conditions; 'minus' the list of known anomalies such as Mercury's perihelion) *because* it explained everything that Newton's theory had successfully explained, and it explained also *to some extent* some known anomalies and, in addition, forbade events like transmission of light along straight lines near large masses about which Newton's theory had said nothing but which had been permitted by other well-corroborated scientific theories of the day; moreover, *at least some* of the unexpected excess Einsteinian content was in fact *corroborated* (for instance, by the eclipse experiments).

On the other hand, according to these sophisticated standards, Galileo's theory that the natural motion of terrestrial objects was circular, introduced no improvement since it did not forbid anything that had been not forbidden by the relevant theories he intended to improve upon (that is, by Aristotelian physics and by Copernican celestial kinematics). This theory was therefore *ad hoc* and therefore – from the heuristic point of view – valueless.[117]

A beautiful example of a theory which satisfied only the first part of Popper's criterion of progress (excess content) but not the second part (corroborated excess content) was given by Popper himself: the Bohr-Kramers–Slater theory of 1924. This theory was refuted in *all* its new predictions.[118]

Let us finally consider how much conventionalism remains in sophisticated falsificationism. Certainly *less* than in naive falsificationism. We need *fewer* methodological decisions. The *'fourth-type decision'* which was essential for the naive version has become completely redundant. To show this we only have to realize that if a scientific theory, consisting of some 'laws of nature', initial conditions, auxiliary theories (but without a *ceteris paribus* clause) conflicts with some factual propositions we do not have to decide which – explicit or 'hidden' – part to replace. We may try to replace *any* part and only when we have hit on an explanation of the anomaly with the help of some content-increasing change (or auxiliary hypothesis), and nature corroborates it, do we move on to eliminate the 'refuted' complex. Thus sophisticated falsification is a slower but possibly safer process than naive falsification.

Let us take an example. Let us assume that the course of a planet differs from the one predicted. Some conclude that this refutes the dynamics and gravitational theory applied: the initial conditions and the *ceteris paribus* clause have been ingeniously corroborated. Others conclude that this refutes the initial conditions used in the calculations: dynamics and gravitational theory have been superbly corroborated in the last two hundred years and all suggestions concerning further factors in play failed. Yet others conclude that this refutes the underlying assumption that there were no other factors in play except for those which were taken into account: these people may possibly be motivated by the metaphysical principle that any explanation is only approximative because of the infinite complexity of the factors involved in determining any single event. Should we praise the first type as *'critical'*, scold the second type as *'hack'*, and condemn the third as *'apologetic'*? No. We do not need to draw any conclusions about such 'refutation'. We never reject a specific theory simply by *fiat*. If we have an inconsistency like the one mentioned, we do not have to decide which ingredients of the theory we regard as problematic and which ones as unproblematic: we regard all ingredients as problematic in the light of the conflicting accepted basic statement and try to replace all of them. If we succeed in replacing some ingredient in a 'progressive' way (that is, the replacement has more corroborated empirical content than the original), we call it 'falsified'.

We do not need the *fifth type decision* of the naive falsificationist either. In order to show this let us have a new look at the problem of the appraisal of (syntactically) metaphysical theories – and the problem of their retention and elimination. The 'sophisticated' solution is obvious. We retain a syntactically metaphysical theory as long as the problematic instances can be explained by content-increasing changes in the auxiliary hypotheses appended to it. [119] Let us take, for instance, Cartesian metaphysics C: "in *all* natural processes *there is* a clockwork mechanism regulated by (*a priori*) animating principles". This is syntactically irrefutable: it can clash with no – spatiotemporally singular – 'basic statement'. It may, of course, clash with a refutable theory like N: "gravitation is a force equal to fm_1m_2/r^2 which *acts at a distance*". But N will only clash with C if 'action at a distance' is interpreted literally and possibly, in addition, as representing an *ultimate* truth, irreducible to any still deeper cause. (Popper would call this an 'essentialist' interpretation.) Alternatively we can regard 'action at a distance' as a mediate cause. Then we interpret 'action at a distance' figuratively, and regard it as a shorthand for some hidden mechanism of action by contact. (We may call this a 'nominalist' interpretation.) In this case we can attempt to explain N by C – Newton himself and several French physicists of the eighteenth century tried to do so. If an auxiliary theory which performs this explanation (or, if you wish, 'reduction') produces novel facts (that is, it is 'independently testable'), Cartesian metaphysics should be regarded as good, scientific, empirical metaphysics, generating a progressive problemshift. A progressive (syntactically) metaphysical theory produces a sustained progressive shift in its protective belt of auxiliary theories. If the reduction of the theory to the 'metaphysical' framework does not produce new empirical content, let alone novel facts, then the reduction represents a degenerating problemshift, it is a mere linguistic exercise. The Cartesian efforts to bolster up their 'metaphysics' in order to explain Newtonian gravitation is an outstanding example of such a merely linguistic reduction.[120]

Thus we do not eliminate a (syntactically) metaphysical theory if it clashes with a well-corroborated scientific theory, as naive falsificationism suggests. We eliminate it if it produces a degenerating shift in the long run and there is a better, rival, metaphysics to replace it. The methodology of a research programme with a 'metaphysical' core does not differ from the methodology of one with a 'refutable' core except perhaps for the logical level of the inconsistencies which are the driving force of the programme.

(It has to be stressed, however, that the very choice of the logical form in which to articulate a theory depends to a large extent on our method-ological decision. For instance, instead of formulating Cartesian meta-physics as an 'all-some' statement, we can formulate it as an 'all-state-ment': 'all natural processes are clockworks'. A 'basic statement' contra-dicting this would be: 'a is a natural process and it is not clockwork'. The question is whether according to the 'experimental techniques', or rather, to the interpretative theories of the day, 'x is not a clockwork' can be 'established' or not. Thus the rational choice of the logical form of a theory depends on the state of our knowledge; for instance, a meta-physical 'all-some' statement of today may become, with the change in the level of observational theories, a scientific 'all-statement' tomorrow. I have already argued that only series of theories and not theories should be classified as scientific or non-scientific; now I have indicated that even the logical form of a theory can only be rationally chosen on the basis of a critical appraisal of the state of the research programme in which it is embedded.)

The first, second, and third type decisions of naive falsificationism, however, cannot be avoided, but as we shall show, the conventional ele-ment in the second decision – and also in the third – can be slightly re-duced. We cannot avoid the decision which sort of propositions should be the 'observational' ones and which the 'theoretical' ones. We cannot avoid either the decision about the truth-value of some 'observational propositions'. These decisions are vital for the decision whether a prob-lemshift is empirically progressive or degenerating. But the sophisticated falsificationist may at least mitigate the arbitrariness of this second deci-sion by allowing for an *appeal procedure*.

Naive falsificationists do not lay down any such appeal procedure. They accept a basic statement if it is backed up by a well-corroborated falsi-fying hypothesis,[121] and let it overrule the theory under test – even though they are well aware of the risk.[122] But there is no reason why we should not regard a falsifying hypothesis – and the basic statement it supports – as being just as problematic as a falsified hypothesis. Now how exactly can we expose the problematicality of a basic statement? On what grounds can the protagonists of the 'falsified' theory appeal and win?

Some people may say that we might go on testing the basic statement (or the falsifying hypothesis) 'by their deductive consequences' until agreement is finally reached. In this testing we deduce – in the same deductive model – further consequences from the basic statement either with the help of the theory under test or some other theory which we regard as unproblematic. Although this procedure 'has no natural end', we always come to a point when there is no further disagreement.[123]

But when the theoretician appeals against the verdict of the experimentalist, the appeal court does not normally cross-question the basic statement directly but rather questions the *interpretative theory* in the light of which its truth-value had been established.

One typical example of a series of successful appeals is the Proutians' fight against unfavourable experimental evidence from 1815 to 1911. For decades Prout's theory T ("that all atoms are compounds of hydrogen atoms and thus 'atomic weights' of all chemical elements must be expressible as whole numbers") and falsifying 'observational' hypotheses, like Stas's 'refutation' R ("the atomic weight of chlorine is $35 \cdot 5$") confronted each other. As we know, in the end T prevailed over R.[124]

The first stage of any serious criticism of a scientific theory is to reconstruct, improve, its logical deductive articulation. Let us do this in the case of Prout's theory *vis à vis* Stas's refutation. First of all, we have to realize that in the formulation we just quoted, T and R were *not* inconsistent. (Physicists rarely articulate their theories sufficiently to be pinned down and caught by the critic.) In order to show them up as inconsistent we have to put them in the following form. T: "the atomic weight of all pure (homogeneous) chemical elements are multiples of the atomic weight of hydrogen", and R: "chlorine is a pure (homogeneous) chemical element and its atomic weight is $35 \cdot 5$". The last statement is in the form of a falsifying hypothesis which, if well corroborated, would allow us to use basic statements of the form B: "Chlorine X is a pure (homogeneous) chemical element and its atomic weight is $35 \cdot 5$" – where X is the proper name of a 'piece' of chlorine determined, say, by its space-time co-ordinates.

But how well-corroborated is R? Its first component depends on R_1: "Chlorine X is a pure chemical element". This was the verdict of the experimental chemist after a rigorous application of the 'experimental techniques' of the day.

Let us have a closer look at the fine-structure of R_1. In fact R_1 stands for a conjunction of two longer statements T_1 and T_2. The first statement, T_1, could be this: "If seventeen chemical purifying procedures $p_1, p_2 ... p_{17}$

are applied to a gas, what remains will be pure chlorine". T_2 is then: "X was subjected to the seventeen procedures $p_1, p_2 \ldots p_{17}$". The careful 'experimenter' carefully applied all seventeen procedures: T_2 is to be accepted. But the conclusion that therefore what remained *must* be pure chlorine is a 'hard fact' only in virtue of T_1. The experimentalist, while *testing T, applied T_1.* He *interpreted* what he saw in the light of T_1: the result was R^1. *Yet in the monotheoretical deductive model of the test situation this interpretative theory does not appear at all.*

But what if T_1, the interpretative theory, is false? Why not 'apply' T rather than T_1 and claim that atomic weights *must be* whole numbers? Then *this* will be a 'hard fact' in the light of T, and T_1 will be overthrown. Perhaps additional new purifying procedures must be invented and applied.

The problem is then *not* when we should stick to a '*theory*' in the face of '*known facts*' and when the other way round. The problem is *not* what to do when 'theories' clash with 'facts'. Such a 'clash' is only suggested by the '*monotheoretical deductive model*'. Whether a proposition is a '*fact*' or a '*theory*' in the context of a test-situation depends on our methodological decision. 'Empirical basis of a theory' is a mono-theoretical notion, it is *relative* to some mono-theoretical deductive structure. We may use it as first approximation; but in case of 'appeal' by the theoretician, we must use a *pluralistic model*. In the pluralistic model the clash is not 'between theories and facts' but between two high-level theories: between an *interpretative theory* to provide the facts and an *explanatory theory* to explain them; and the interpretative theory may be on quite as high a level as the explanatory theory. The clash is then not any more between a logically higher-level theory and a lower-level falsifying hypothesis. The problem should not be put in terms of whether a '*refutation*' is real or not. The problem is how to repair an *inconsistency* between the 'explanatory theory' under test and the – explicit or hidden – 'interpretative' theories; or, if you wish, *the problem is which theory to consider as the interpretative one which provides the 'hard' facts and which the explanatory one which 'tentatively' explains them.* In a mono-theoretical model we regard the higher-level theory as an *explanatory theory to be judged by the 'facts'* delivered from outside (by the authoritative experimentalist): in the case of a clash we reject the explanation.[125] In a pluralistic model we may decide, alternatively, to regard the higher level theory as an *interpretative theory to judge the 'facts'* delivered from outside: in case of a clash we may reject the 'facts' as 'monsters'. In a pluralistic model of testing, several theories – more or less deductively organized – are soldered together.

This argument alone would be enough to show the correctness of the conclusion, which we drew from a different earlier argument, that experiments do not simply overthrow theories, that no theory forbids a state of affairs specifiable in advance. It is not that we propose a theory and Nature may shout NO; rather we propose a maze of theories, and Nature may shout INCONSISTENT.[126]

The problem is then *shifted* from the old problem of replacing a theory refuted by 'facts' to the new problem of how to resolve inconsistencies between closely associated theories. Which of the mutually inconsistent theories should be eliminated? The sophisticated falsificationist can answer that question easily: one had to try to replace first one, then the other, then possibly both, and opt for that new set-up which provides the biggest increase in corroborated content, which provides the most progressive problemshift.[127]

Thus we have established an appeal procedure in case the theoretician wishes to question the negative verdict of the experimentalist. The theoretician may demand that the experimentalist specify his 'interpretative theory',[128] and he may then replace it – to the experimentalist's annoyance – by a better one in the light of which his originally 'refuted' theory may receive positive appriasal.[129]

But even this appeal procedure cannot do more than *postpone* the conventional decision. For the verdict of the appeal court is not infallible either. When we decide whether it is the replacement of the 'interpretative' or of the 'explanatory' theory that produces novel facts, we again must take a decision about the acceptance or rejection of basic statements. But then we have only *postponed* – and possibly *improved* – the decision, not avoided it.[130] The difficulties concerning the empirical basis which confronted 'naive' falsificationism cannot be avoided by 'sophisticated' falsificationism either. Even if we regard a theory as 'factual', that is, if our slow-moving and limited imagination cannot offer an alternative to it (as Feyerabend used to put it), we have to make, at least occasionally and temporarily, decisions about its truth-value. *Even then, experience still remains, in an important sense, the 'impartial arbiter'*[131] *of scientific controversy.* We cannot get rid of the problem of the 'empirical basis', if we want to

learn from experience[132]: but we can make our learning less dogmatic –
but also less fast and less dramatic. By regarding some observational
theories as problematic we may make our methodology more flexible: but
we cannot articulate and include *all* 'background knowledge' (or 'background ignorance'?) into our critical deductive model. This process is
bound to be piecemeal and some conventional line must be drawn at any
given time.

There is one objection even to the sophisticated version of methodological
falsificationism which cannot be answered without some concession to
Duhemian 'simplicism'. The objection is the so-called 'tacking paradox'.
According to our definitions, adding to a theory completely disconnected
low-level hypotheses may constitute a 'progressive shift'. It is difficult to
eliminate such makeshift shifts without demanding that the additional
assertions must be connected with the original assertion *more intimately*
than by mere conjunction. This, of course, is a sort of simplicity requirement which would assure the continuity in the series of theories which
can be said to constitute *one* problemshift.

This leads us to further problems. For one of the crucial features of
sophisticated falsificationism is that it replaces the concept of *theory* as the
basic concept of the logic of discovery by the concept of *series of theories*.
*It is a succession of theories and not one given theory which is appraised as
scientific or pseudo-scientific.* But the members of such series of theories
are usually connected by a remarkable *continuity* which welds them into
research programmes. This *continuity* – reminiscent of Kuhnian 'normal
science' – plays a vital role in the history of science; the main problems
of the logic of discovery cannot be satisfactorily discussed except in the
framework of a *methodology of research programmes*.

3. A METHODOLOGY OF SCIENTIFIC RESEARCH PROGRAMMES

I have discussed the problem of objective appraisal of scientific growth in
terms of progressive and degenerating problemshifts in series of scientific
theories. The most important such series in the growth of science are
characterized by a certain *continuity* which connects their members. This
continuity evolves from a genuine research programme adumbrated at
the start. The programme consists of methodological rules: some tell us

what paths of research to avoid (*negative heuristic*), and others what paths to pursue (*positive heuristic*).[133]

Even science as a whole can be regarded as a huge research programme with Popper's supreme heuristic rule: "devise conjectures which have more empirical content than their predecessors." Such methodological rules may be formulated, as Popper pointed out, as metaphysical principles.[134] For instance, the *universal* anti-conventionalist rule against exception-barring may be stated as the metaphysical principle: 'Nature does not allow exceptions'. This is why Watkins called such rules 'influential metaphysics'.[135]

But what I have primarily in mind is not science as a whole, but rather *particular* research programmes, such as the one known as 'Cartesian metaphysics'. Cartesian metaphysics, that is, the mechanistic theory of the universe – according to which the universe is a huge clockwork (and system of vortices) with push as the only cause of motion – functioned as a powerful heuristic principle. It discouraged work on scientific theories – like [the 'essentialist' version of] Newton's theory of action at a distance – which were inconsistent with it (*negative heuristic*). On the other hand, it encouraged work on auxiliary hypotheses which might have saved it from apparent counterevidence – like Keplerian ellipses (*positive heuristic*).[136]

(a) *Negative Heuristic: The 'Hard Core' of the Programme.*

All scientific research programmes may be characterized by their '*hard core*'. The negative heuristic of the programme forbids us to direct the *modus tollens* at this 'hard core'. Instead, we must use our ingenuity to articulate or even invent 'auxiliary hypotheses', which form a *protective belt* around this core, and we must redirect the *modus tollens* to *these*. It is this protective belt of auxiliary hypotheses which has to bear the brunt of tests and get adjusted and re-adjusted, or even completely replaced, to defend the thus-hardened core. A research programme is successful if all this leads to a progressive problemshift; unsuccessful if it leads to a degenerating problemshift.

The classical example of a successful research programme is Newton's gravitational theory; possibly the most successful research programme ever. When it was first produced, it was submerged in an ocean of 'anomalies' (or, if you wish, 'counterexamples'), and opposed by the observational theories supporting these anomalies. But Newtonians turned, with brilliant tenacity and ingenuity, one counter-instance after another into corroborating instances, primarily by overthrowing the original observational theories in the light of which this 'contrary evidence' was established. In the process they themselves produced new counter-examples which they again resolved. They "turned each new difficulty into a new victory of their programme".[137]

In Newton's programme the negative heuristic bids us to divert the *modus tollens* from Newton's three laws of dynamics and his law of gravitation. This 'core' is 'irrefutable' by the methodological decision of its protagonists: anomalies must lead to changes only in the 'protective' belt of auxiliary, 'observational' hypotheses and initial conditions.[138]

I have given a contrived micro-example of a progressive Newtonian problemshift.[139] If we analyse it, it turns out that each successive link in this exercise predicts some new fact; each step represents an increase in empirical content: the example constitutes a *consistently progressive theoretical shift*. Also, each prediction is in the end verified; although on three subsequent occasions they may have seemed momentarily to be 'refuted'.[140] While 'theoretical progress' (in the sense here described) may be verified immediately, 'empirical progress' cannot, and in a research programme we may be frustrated by a long series of 'refutations' before ingenious and lucky content-increasing auxiliary hypotheses turn a chain of defeats – *with hindsight* – into a resounding success story, either by revising some false 'facts' or by adding novel auxiliary hypotheses. We may then say that we must require that each step of a research programme be consistently content-increasing: that each step constitute a *consistently progressive theoretical problemshift*. All we need in addition to this is that at least every now and then the increase in content should be seen to be retrospectively corroborated: the programme as a whole should also display an *intermittently progressive empirical shift*. We do not demand that each step produce *immediately* an *observed* new fact. Our term '*intermittently*' gives sufficient *rational* scope for dogmatic adherence to a programme in face of *prima facie* 'refutations'.

The idea of 'negative heuristic' of a scientific research programme rationalizes classical conventionalism to a considerable extent. We may rationally decide not to allow 'refutations' to transmit falsity to the hard core as long as the corroborated empirical content of the protecting belt of auxiliary hypotheses increases. But our approach differs from Poincaré's justificationist conventionalism in the sense that, unlike Poincaré's, we maintain that if and when the programme ceases to anticipate novel facts, its hard core might have to be abandoned: that is, *our* hard core, unlike Poincaré's, may crumble under certain conditions. In this sense we side with Duhem who thought that such a possibility must be allowed for; but for Duhem the reason for such crumbling is purely *aesthetic*, while for us it is mainly *logical and empirical*.

(b) *Positive Heuristic: The Construction of the 'Protective Belt' and the Relative Autonomy of Theoretical Science.*

Research programmes, besides their negative heuristic, are also characterized by their positive heuristic.

Even the most rapidly and consistently progressive research programmes can digest their 'counter-evidence' only piecemeal: anomalies are never completely exhausted. But it should not be thought that yet unexplained anomalies – 'puzzles' as Kuhn might call them – are taken in random order, and the protective belt built up in an eclectic fashion, without any preconceived order. The order is usually decided in the theoretician's cabinet, independently of the *known* anomalies. Few theoretical scientists engaged in a research programme pay undue attention to 'refutations'. They have a long-term research policy which anticipates these refutations. This research policy, or order of research, is set out – in more or less detail – in the *positive heuristic* of the research programme. The negative heuristic specifies the 'hard core' of the programme which is 'irrefutable' by the methodological decision of its protagonists; the positive heuristic consists of a partially articulated set of suggestions or hints on how to change, develop the 'refutable variants' of the research-programme, how to modify, sophisticate, the 'refutable' protective belt.

The positive heuristic of the programme saves the scientist from becoming confused by the ocean of anomalies. The positive heuristic sets out a programme which lists a chain of ever more complicated *models* simulating reality: the scientist's attention is riveted on building his models following instructions which are laid down in the positive part of his programme. He ignores the *actual* counterexamples, the available '*data*'.[141] Newton first worked out his programme for a planetary system with a fixed point-like sun and one single point-like planet. It was in this model that he derived his inverse square law for Kepler's ellipse. But this model was forbidden by Newton's own third law of dynamics, therefore the model had to be replaced by one in which both sun and planet revolved round their common centre of gravity. This change was not motivated by any observation (the data did not suggest an 'anomaly' here) but by a theoretical difficulty in developing the programme. Then he worked out the programme for more planets as if there were only heliocentric but no interplanetary forces. Then he worked out the case where the sun and planets were not mass-points but mass-*balls*. Again, for this change he did not *need* the observation of an anomaly; infinite density was forbidden by an (inarticulated) touchstone theory, therefore planets *had* to be extended. This change involved considerable mathematical difficulties, held up Newton's work – and delayed the publication of the *Principia* by more than a decade. Having solved this 'puzzle', he started work on *spinning balls* and their wobbles. Then he admitted interplanetary forces and started work on *perturbations*. At this point he started to look more anxiously at the facts. Many of them were beautifully explained (qualitatively) by this model, many were not. It was then that he started to work on *bulging* planets, rather than round planets, etc.

Newton despised people who, like Hooke, stumbled on a first naive model but did not have the tenacity and ability to develop it into a research programme, and who thought that a first version, a mere aside, constituted a 'discovery'. He held up publication until his programme had achieved a remarkable progressive shift.[142]

Most, if not all, Newtonian 'puzzles', leading to a series of new variants superseding each other, were foreseeable at the time of Newton's first naive model and no doubt Newton and his colleagues *did* forsee them: Newton must have been fully aware of the blatant falsity of his first variants.[143] Nothing shows the existence of a positive heuristic of a research programme clearer than this fact: this is why one speaks of 'models' in research programmes. A '*model*' is a set of initial conditions (possibly together with some of the observational theories) which one knows is *bound* to be replaced during the further development of the programme, and one even knows, more or less, how. This shows once more how irrelevant 'refutations' of any specific variant are in a research programme: their existence is fully expected, the positive heuristic is there as the strategy both for predicting (producing) and digesting them. Indeed, if the positive heuristic is clearly spelt out, the difficulties of the programme are mathematical rather than empirical.[144]

One may formulate the 'positive heuristic' of a research programme as a 'metaphysical' principle. For instance one may formulate Newton's programme like this :"the planets are essentially gravitating spinning-tops of roughly spherical shape". This idea was never *rigidly* maintained: the planets are not *just* gravitational, they have also, for example, electromagnetic characteristics which may influence their motion. Positive heuristic is thus in general more flexible than negative heuristic. Moreover, it occasionally happens that when a research programme gets into a degenerating phase, a little revolution or a *creative shift* in its positive heuristic may push it forward again.[145] It is better therefore to separate the 'hard core' from the more flexible metaphysical principles expressing the positive heuristic.

Our considerations show that the positive heuristic forges ahead with almost complete disregard of 'refutations': it may seem that it is the '*verifications*'[146] rather than the refutations which provide the contact points with reality. Although one must point out that any 'verification' of the $n+1$th version of the programme is a refutation of the n'th version, we cannot deny that *some* defeats of the subsequent versions are always foreseen: it is the 'verifications' which keep the programme going, recalcitrant instances notwithstanding.

We may appraise research programmes, even after their 'elimination', for their *heuristic power*: how many new facts did they produce, how great was "their capacity to explain their refutations in the course of their growth"?[147]

(We may also appraise them for the stimulus they gave to mathematics. The real difficulties for the theoretical scientist arise rather from the *mathematical difficulties* of the programme than from anomalies. The greatness of the Newtonian programme comes partly from the development – by Newtonians – of classical infinitesimal analysis which was a crucial precondition of its success.)

Thus the methodology of scientific research programmes accounts for the *relative autonomy of theoretical science*: a historical fact whose rationality cannot be explained by the earlier falsificationists. Which problems scientists working in powerful research programmes rationally choose, is determined by the positive heuristic of the programme rather than by psychologically worrying (or technologically urgent) anomalies. The anomalies are listed but shoved aside in the hope that they will turn, in due course, into corroborations of the programme. Only those scientists have to rivet their attention on anomalies who are either engaged in trial-and-error exercises or who work in a degenerating phase of a research programme when the positive heuristic ran out of steam. (All this, of course, must sound repugnant to naive falsificationists who hold that once a theory is 'refuted' by experiment (by *their* rule book), it is irrational (and dishonest) to develop it further: one has to replace the old 're-futed' theory by a new, unrefuted one.)

NOTES

* From *Criticism and the Growth of Knowledge* (pp. 91–138), ed. by I. Lakatos and A. Musgrave. Copyright © 1970. Reprinted by permission of Cambridge University Press.

   This paper is a considerably improved version of my (1968b) and a crude version of my (1970). Some parts of the former are here reproduced without change with the permission of the Editor of the *Proceedings of the Aristotelian Society*. In the preparation of the new version I received much help from Tad Beckman, Colin Howson, Clive Kilmister, Larry Laudan, Eliot Leader, Alan Musgrave, Michael Sukale, John Watkins and John Worrall.

[1] The main contemporary protagonist of the ideal of 'probable truth' is Rudolf Carnap. For the historical background and a criticism of this position, cf. Lakatos (1968a).

[2] The main contemporary protagonists of the ideal of 'truth by consensus' are Polanyi and Kuhn. For the historical background and a criticism of this position, cf. Musgrave (1969a), Musgrave (1969b) and Lakatos (1970).

[3] Indeed he introduces his (1962) by arguing against the 'development-by-accumulation' idea of scientific growth. But his intellectual debt is to Koyré rather than to Popper. Koyré showed that positivism gives bad guidance to the historian of science, for the history of physics can only be understood in the context of a succession of 'metaphysical' research programmes. Thus scientific changes are connected with vast cataclysmic metaphysical revolutions. Kuhn develops this message of Burtt and Koyré and the vast success of his book was partly due to his hard-hitting, direct criticism of justificationist historiography – which created a sensation among ordinary scientists and historians of science whom Burtt's, Koyré's (or Popper's) message has not yet reached. But, unfortunately, his message had some authoritarian and irrationalist overtones.

[4] Cf. e.g. Watkins's and Feyerabend's contributions to *Criticism and the Growth of Knowledge*.

[5] Justificationists repeatedly stressed this asymmetry between singular factual statements and universal theories. Cf. e.g. Popkin's discussion of Pascal in Popkin (1968), p. 14 and Kant's statement to the same effect as quoted in the new *motto* of the third 1969 German edition of Popper's *Logik der Forschung*. (Popper's choice of this time-honoured cornerstone of elementary logic as a *motto* of the new edition of his classic shows his main concern: to fight *probabilism*, in which this asymmetry becomes irrelevant; for probabilists theories may become almost as well established as factual propositions.)

6 Indeed, even some of these few shifted, following Mill, the rather obviously insoluble problem of inductive proof (of universal from particular propositions) to the slightly less obviously insoluble problem of proving *particular* factual propositions from other *particular* factual propositions.

7 The founding fathers of probabilism were intellectualists; Carnap's later efforts to build up an empiricist brand of probabilism failed. Cf. my (1968a), p. 367 and also p. 361, footnote 2.

8 For a detailed discussion, cf. my (1968a), especially pp. 353ff.

9 Russell (1943), p. 683. For a discussion of Russell's justificationism, cf. my (1962), especially pp. 167 ff.

10 Medawar (1967), p. 144.

11 This discussion already indicates the vital importance of a demarcation between provable factual and unprovable theoretical propositions for the dogmatic falsificationist.

12 "*Criteria of refutation* have to be laid down beforehand: it must be agreed which observable situations, if actually observed, mean that the theory is refuted" (Popper [1963], p. 38, Footnote 3).

13 Quoted in Popper (1934), Section 85, with Popper's comment: "I fully agree".

14 Braithwaite (1953), pp. 367–8. For the 'incorrigibility' of Braithwaite's observed facts, cf. his (1938). While in the quoted passage Braithwaite gives a forceful answer to the problem of scientific objectivity, in another passage he points out that "except for the straightforward generalizations of observable facts... complete refutation is no more possible than is complete proof" ([1953], p. 19).

15 For these assumptions and their criticism, cf. Popper (1934), Sections 4 and 10. It is because of this assumption that – following Popper – I call this brand of falsificationism 'naturalistic'. Popper's 'basic propositions' should not be confused with the basic propositions discussed in this section; cf. *below*, note 42.

It is important to point out that these two assumptions are also shared by many justificationists who are not falsificationists: they may add to experimental proofs 'intuitive proofs' – as did Kant – or 'inductive proofs' – as did Mill. Our falsificationist accepts experimental proofs *only*.

16 The empirical basis of a theory is the set of its potential falsifiers: the set of those observational propositions which may disprove it.

17 Incidentally, Galileo also showed – with the help of his optics – that if the moon was a faultless crystal ball, it would be invisible (Galileo [1632]).

18 True, most psychologists who turned against the idea of justificationist sensationalism did so under the influence of pragmatist philosophers like William James who denied the possibility of any sort of objective knowledge. But, even so, Kant's influence through Oswald Külpe, Franz Brentano and Popper's influence through Egon Brunswick and Donald Campbell played a role in the shaping of modern psychology; and if psychology ever vanquishes psychologism, it will be due to an increased understanding of the Kant-Popper mainline of objectivist philosophy.

19 Cf. Popper (1934), Section 29.

20 It seems that the first philosopher to emphasize this might have been Fries in 1837 (cf. Popper [1934], Section 29, Footnote 3). This is of course a special case of the general thesis that logical relations, like probability or consistency, refer to *propositions*. Thus, for instance, the proposition 'nature is consistent' is false (or, if you wish, meaningless), for nature is not a proposition (or a conjunction of propositions).

21 Incidentally, even this is questionable.

[22] As Popper put it: "No conclusive disproof of a theory can ever be produced", those who wait for an infallible disproof before eliminating a theory will have to wait for ever and "will never benefit from experience" ([1934], Section 9).

[23] Kant and his English follower, Whewell, both realized that all scientific propositions, whether *a priori* or *a posteriori*, are equally theoretical; but both held that they are equally provable. Kantians saw clearly that the propositions of science are theoretical in the sense that they are not written by sensations on the *tabula rasa* of an empty mind, nor deduced or induced from such propositions. A factual proposition is only a special kind of theoretical proposition. In this Popper sided with Kant against the empiricist version of dogmatism. But Popper went a step further: in his view the propositions of science are not only theoretical but they are all also *fallible*, conjectural for ever.

[24] If the tiny conjectural planet were out of the reach even of the biggest *possible* optical telescopes, he might try some quite novel instrument (like a radiotelescope) in order to enable him to 'observe it', that is, to ask Nature about it, even if only indirectly. (The new 'observational' theory may itself not be properly articulated, let alone severely tested, but he would care no more than Galileo did.)

[25] At least not until a new research programme supersedes Newton's programme which happens to explain this previously recalcitrant phenomenon. In this case, the phenomenon will be unearthed and enthroned as a 'crucial experiment'.

[26] Popper asks: "What kind of clinical responses would refute to the satisfaction of the analyst not merely a particular diagnosis but psychoanalysis itself?" ([1963], p. 38, Footnote 3.) But what kind of observation would refute to the satisfaction of the Newtonian not merely a particular version but Newtonian theory itself?

[27] (*Added in press*:) This '*ceteris paribus*' clause must not normally be interpreted as a separate premise.

[28] Incidentally, we might persuade the dogmatic falsificationist that his demarcation criterion was a very naive mistake. If he gives it up but retains his two basic assumptions, he will have to ban theories from science and regard the growth of science as an accumulation of proven basic statements. This indeed is the final stage of classical empiricism after the evaporation of the hope that facts can prove or at least disprove theories.

[29] Cf. Popper (1934), Chapter VIII.

[30] For a *much* stronger case, cf. *below*, Sect. 3.

[31] This demarcation – and terminology – is due to Popper; cf. especially his (1934), Section 19 and his (1945), Chapter 23 and Footnote 3 to Chapter 25.

[32] No version of conservative activism explained why Newton's *gravitational* theory should be invulnerable; Kantians restricted themselves to the explanation of the tenacity of Euclidean geometry and Newtonian *mechanics*. About Newtonian *gravitation* and *optics* (or other branches of science) they had an ambiguous, and occasionally inductivist position.

[33] I do not include Hegel among 'revolutionary *activists*'. For Hegel and his followers change in conceptual frameworks is a predetermined, inevitable process, where individual creativity or rational criticism plays no essential role. Those who run ahead are equally at fault as those who stay behind in this 'dialectic'. The clever man is not he who creates a better 'prison' or who demolishes critically the old one, but the one who is always in step with history. Thus dialectic accounts for change without criticism.

[34] Cf. Whewell's (1837), (1840) and (1858).

[35] Cf. especially Poincaré (1891) and (1902); Milhaud (1896); Le Roy (1899) and (1901). It was one of the chief philosophical merits of conventionalists to direct the limelight to

the fact that any theory can be saved by 'conventionalist stratagems' from refutations. (The term 'conventionalist stratagem' is Popper's; cf. the critical discussion of Poincaré's conventionalism in his [1934], especially Sections 19 and 20.)

[36] Poincaré first elaborated his conventionalism only with regard to geometry (cf. his [1891]). Then Milhaud and Le Roy generalized Poincaré's idea to cover all branches of accepted physical theory. Poincaré's (1902) starts with a strong criticism of the Bergsonian Le Roy against whom he defends the empirical (falsifiable or 'inductive') character of all physics *except for* geometry and mechanics. Duhem, in turn, criticized Poincaré: in his view there was a possibility of overthrowing even Newtonian mechanics.

[37] The *loci classici* are Duhem's (1905) and Popper's (1934). Duhem was not a *consistent* revolutionary conventionalist. Very much like Whewell, he thought that conceptual changes are only *preliminaries* to the final – if perhaps distant – 'natural classification': "The more a theory is perfected, the more we apprehend that the logical order in which it arranges experimental laws is the reflection of an ontological order". In particular, he refused to see Newton's mechanics *actually* 'crumbling' and characterized Einstein's relativity theory as the manifestation of a "frantic and hectic race in pursuit of a novel idea" which "has turned physics into a real chaos where logic loses its way and commonsense runs away frightened" (Preface – of 1914 – to the second edition of his [1905]).

[38] Duhem (1905), Chapter VI, Section 10.

[39] Popper (1934), Section 30.

[40] *In this section I discuss the 'naive' variant of Popper's methodological falsificationism. Thus, throughout the section 'methodological falsificationism' stands for 'naive methodological falsificationism'.*

[41] Popper (1934), Section 27.

[42] *Op. cit.* Section 28. For the non-basicness of these methodologically 'basic' statements, cf. e.g. Popper (1934) *passim* and Popper (1959a), p. 35, Footnote *2.

[43] Cf. Popper (1934), end of Section 26 and also his (1968c), pp. 291–2.

[44] Cf. Popper (1963), p. 390.

[45] Indeed, Popper carefully puts 'observational' in quotes; cf. his (1934), Section 28.

[46] This demarcation plays a role both in the *first* and in the *fourth* type of decisions of the methodological falsificationist.

[47] For a fascinating discussion, cf. Feyerabend (1969).

[48] One wonders whether it would not be better to make a break with the terminology of naturalistic falsificationism and rechristen observational theories *'touchstone theories'*.

[49] Cf. Popper (1934), Section 22. Many philosophers overlooked Popper's important qualification that a basic-statement has no power to refute anything without the support of a well-corroborated falsifying hypothesis.

[50] Cf. Popper (1934), Section 30.

[51] Popper (1963), p. 387.

[52] Popper (1934), Section 30; also cf. Section 29: 'The Relativity of Basic Statements'.

[53] Popper (1957), p. 134. Popper, in other places, emphasizes that his method cannot 'ensure' the survival of the fittest. Natural selection may go wrong: the fittest may perish and monsters survive.

[54] Popper (1935).

[55] Popper (1934), Section 82.

[56] Popper (1934), Section 82.

[57] This kind of methodological 'falsification' is, unlike dogmatic falsification (disproof), a pragmatic, methodological idea. But then what exactly should we mean by it?

Popper's answer – which I am going to discard – is that methodological 'falsification' indicates an "urgent need of replacing a falsified hypothesis by a better one" (Popper [1959a], p. 87, Footnote *1). This shift is an excellent illustration of the process I described in my (1963–4) whereby critical discussion shifts the original *problem* without necessarily changing the old *terms*. The byproducts of such processes are *meaning-shifts*.

58 The demarcation criterion of the dogmatic falsificationist was: a theory is 'scientific' if it has an empirical basis.

59 Incidentally, Popper, in his (1934), does not seem to have seen this point clearly. He writes: "Admittedly, it is possible to interpret the concept of an *observable event* in a psychologistic sense. But I am using it in such a sense that it might just as well be replaced by 'an event involving position and movement of macroscopic physical bodies'". ([1934], Section 28.) In the light of our discussion, for instance, we may regard a positron passing through a Wilson chamber at time t_0 as an 'observable' event, in spite of the non-macroscopic character of the positron.

60 Popper (1934), Section 68. Indeed, this methodological falsificationism is the philosophical basis of some of the most interesting developments in modern statistics. The Neyman-Pearson approach rests completely on methodological falsificationism. Also cf. Braithwaite (1953), Chapter VI. (Unfortunately, Braithwaite reinterprets Popper's demarcation criterion as separating meaningful from meaningless rather than scientific from non-scientific propositions.)

61 Popper (1933).

62 Popper (1933).

63 Popper (1957), p. 133.

64 For a discussion of this important concept of Popperian methodology, cf. my (1968a), pp. 397 ff.

65 Popper (1934), Section 9.

66 *Ibid.*

67 *Ibid.*

68 The problem of '*controlled experiment*' may be said to be nothing else but the problem of arranging experimental conditions in such a way as to minimize the risk involved in such decisions.

69 This type of decision belongs, in an important sense, to the same category as the first decision: it demarcates, by decision, problematic from unproblematic knowledge.

70 Our exposition shows clearly the complexity of the decisions needed to define the 'empirical content' of a theory – that is, the set of its potential falsifiers. 'Empirical content' depends on our *decision* as to which are our 'observational theories' and which anomalies are to be promoted to counterexamples. If one attempts to compare the empirical content of different scientific theories in order to see which is 'more scientific', then one will get involved in an enormously complex and therefore hopelessly arbitrary system of decisions about their respective classes of 'relatively atomic statements' and their 'fields of application'. (For the meaning of these (very) technical terms, cf. Popper [1934], Section 38.) But such comparison is possible only when one theory supersedes another (cf. Popper, [1959a], p. 401, Footnote 7). And even then, there may be difficulties (which would not, however, add up to irremediable 'incommensurability').

71 This was suggested by J. O. Wisdom: cf. his (1963).

72 For instance: 'All metals have a solvent'; or 'There exists a substance which can turn all metals into gold'. For discussions of such theories, cf. especially Watkins (1957) and Watkins (1960).

73 Russell (1943), p. 683.

[74] I am sure that some will welcome methodological falsificationism as an 'existentialist' philosophy of science.

[75] Neurath (1935), p. 356.

[76] Hempel (1952), p. 621. Agassi, in his (1966), follows Neurath and Hempel, especially pp. 16ff. It is rather amusing that Agassi, in making this point, thinks that he is taking up arms against "the whole literature concerning the methods of science".

Indeed, many scientists were fully aware of the difficulties inherent in the "confrontation of theory and facts". (Cf. Einstein [1949], p. 27.) Several philosophers sympathetic to falsificationism emphasized that "the process of refuting a scientific hypothesis is more complicated than it appears to be at first sight" (Braithwaite [1953], p. 20). But only Popper offered a constructive, rational solution.

[77] Hempel (1952), p. 622. Hempel's crisp "theses on empirical certainty" do nothing but refurbish Neurath's – and some of Popper's – old arguments (against Carnap, I take it); but deplorably, he does not mention either his predecessors or his adversaries.

[78] Neurath (1935).

[79] Popper (1934), Section 26.

[80] Neurath's (1935) shows that he never grasped Popper's simple argument.

[81] I am using here 'verisimilitude' in Popper's sense: the difference between the truth content and falsity content of a theory. For the risks involved in estimating it, cf. my (1968a), especially pp. 395 ff.

[82] I tried to develop such a general theory of criticism in my (1970).

[83] The falsification of theories depends on the high degree of corroboration of the *ceteris paribus* clause. This however is not always the case. This is why the methodological falsificationist may advise us to rely on our 'scientific instinct' (Popper [1934], Section 18, Footnote 2) or 'hunch' (Braithwaite [1953], p. 20).

[84] Agassi (1959); he calls Popper's idea of science '*scientia negativa*' (Agassi [1968]).

[85] It should be mentioned here that the Kuhnian sceptic is still left with what I would call the '*scientific sceptic's dilemma*': any scientific sceptic will still try to explain changes in beliefs and will regard his own psychology as a theory which is more than simple belief, which, in some sense, is 'scientific'. Hume, while trying to show up science as a mere system of beliefs with the help of his stimulus-response theory of learning, never raised the problem of whether his theory of learning applies also to his own theory of learning. In contemporary terms, we might well ask, does the popularity of Kuhn's philosophy indicate that people recognize its *truth*? In this case it would be refuted. Or does this popularity indicate that people regarded it as an attractive new fashion? In this case, it would be 'verified'. But would Kuhn like *this* 'verification'?

[86] Feyerabend who contributed probably more than anybody else to the spread of Popper's ideas, seems now to have joined the enemy camp. Cf. his intriguing (1970).

[87] I use 'prediction' in a wide sense that includes 'postdiction'.

[88] *For a detailed discussion of these acceptance and rejection rules and for references to Popper's work,* cf. my (1968a), pp. 375–90.

[89] Molière, for instance, ridiculed the doctors of his *Malade Imaginaire*, who offered the *virtus dormitiva* of opium as the answer to the question as to why opium produced sleep. One might even argue that Newton's famous dictum *hypotheses non fingo* was really directed against *ad hoc* explanations – like his own explanation of gravitational forces by an aether-model in order to meet Cartesian objections.

[90] Incidentally, Duhem agreed with Bernard that experiments alone – without simplicity considerations – can decide the fate of theories in physiology. But in physics, he argued, they cannot ([1905], Chapter VI, Section 1).

[91] Koestler correctly points out that only Galileo created the myth that the Copernican theory was simple (Koestler [1959], p. 476); in fact, "the motion of the earth [had not] done much to simplify the old theories, for though the objectionable equants had disappeared, the system was still bristling with auxiliary circles" (Dreyer [1906], Chapter XIII).

[92] Popper (1934), Sections 19 and 20. I have discussed in some detail – under the heads 'monster-barring', 'exception-barring', 'monster-adjustment' – such stratagems as they appear in informal, quasi-empirical mathematics; cf. my (1963–4).

[93] If I already know P_1: 'Swan A is white', P_ω: 'All swans are white' represents no progress, because it may only lead to the discovery of such further similar facts as P_2: 'Swan B is white'. So-called 'empirical generalizations' constitute no progress. A *new* fact must be improbable or even impossible in the light of previous knowledge.

[94] The appropriateness of the term 'problemshift' for a series of theories rather than of problems may be questioned. I chose it partly because I have not found a more appropriate alternative – 'theoryshift' sounds dreadful – partly because theories are always problematical, they never solve all the problems they have set out to solve. Anyway, in the second half of the paper, the more natural term 'research programme' will replace 'problemshifts' in the most relevant contexts.

[95] Indeed, in the original manuscript of my (1968a) I wrote : "A theory without excess corroboration has no excess explanatory power; *therefore, according to Popper, it does not represent growth and therefore it is not 'scientific'; therefore, we should say, it has no explanatory power*" (p. 386). I cut out the italicized half of the sentence under pressure from my colleagues who thought it sounded too eccentric. I regret it now.

[96] Popper's conflation of 'theories' and 'series of theories' prevented him from getting the basic ideas of sophisticated falsificationism across more successfully. His ambiguous usage led to such confusing formulations as "Marxism [as the core of a series of theories or of a 'research programme'] is irrefutable" and, at the same time, "Marxism [as a particular conjunction of this core and some specified auxiliary hypotheses, initial conditions and a *ceteris paribus* clause] has been refuted." (Cf. Popper [1963].)

Of course, there is nothing wrong in saying that an isolated, single theory is 'scientific' if it represents an advance on its predecessor, as long as one clearly realizes that in this formulation we appraise the theory as the outcome of – and in the context of – a certain historical development.

[97] Popper (1945), Vol. II, p. 233. Popper's more sophisticated attitude surfaces in the remark that "concrete and practical consequences can be *more* directly tested by experiment" (*ibid.*, my italics).

[98] Popper (1934), Section 30.

[99] "In most cases we have, before falsifying a hypothesis, another one up our sleeves" (Popper [1959a], p. 87, Footnote *1). But, as our argument shows, we *must* have one. Or, as Feyerabend put it: "The best criticism is provided by those theories which can replace the rivals they have removed" ([1965], p. 227). He notes that in *some* cases "alternatives will be quite indispensable for the purpose of refutation" (*ibid.* p. 254). But according to our argument *refutation without an alternative shows nothing but the poverty of our imagination in providing a rescue hypothesis.*

[100] Cf. my (1968a), pp. 387ff.

[101] In the distorting mirror of naive falsificationism, new theories which replace old refuted ones, are themselves born unrefuted. Therefore they do not believe that there is a relevant difference between anomalies and crucial counterevidence. For them, anoma-

ly is a dishonest euphemism for counterevidence. But in actual history new theories are born refuted: they inherit many anomalies of the old theory. Moreover, frequently it is *only* the new theory which dramatically predicts that fact which will function as crucial counterevidence against its predecessor, while the 'old' anomalies may well stay on as 'new' anomalies.

All this will be still clearer when we introduce the idea of 'research programme'.

[102] *Sophisticated falsificationism adumbrates a new theory of learning.*

[103] It is clear that the theory T' may have excess corroborated empirical content over another theory T even if both T and T' are refuted. Empirical content has nothing to do with truth or falsity. Corroborated contents can also be compared irrespective of the refuted content. Thus we may see the rationality of the elimination of Newton's theory in favour of Einstein's, even though Einstein's theory may be said to have been born – like Newton's – 'refuted'. We have only to remember that 'qualitative confirmation' is a euphemism for 'quantitative disconfirmation'. (Cf. my [1968a], pp. 384–6.)

[104] Cf. Popper (1934), Section 85, p. 279 of the 1959 English translation.

[105] It is true that a certain type of *proliferation of rival theories* is allowed to play an accidental heuristic role in falsification. In many cases falsification heuristically "depends on [the condition] that sufficiently many and sufficiently different theories are offered" (Popper [1940]). For instance, we may have a theory T which is apparently unrefuted. But it may happen that a new theory T', inconsistent with T, is proposed which equally fits the available facts: the differences are smaller than the range of observational error. In such cases the inconsistency prods us into improving our 'experimental techniques', and thus refining the 'empirical basis' so that either T or T' (or, incidentally, both) can be falsified: "We need [a] new theory in order to find out where the old theory was deficient" (Popper [1963], p. 246). But the role of this proliferation is *accidental* in the sense that, once the empirical basis is refined, the fight is between this refined empirical basis and the theory T under test; the rival theory T' acted only as a *catalyst*.

[106] Also cf. Feyerabend (1965), pp. 254–5.

[107] Popper (1959a), p. 87, Footnote *1.

[108] Popper (1934), Section 30.

[109] [*Added in press:*] Possibly it would be better in future to abandon these terms altogether, just as we have abandoned terms like 'inductive (or experimental) proof'. Then we may call (naive) 'refutations' anomalies, and (sophisticatedly) 'falsified' theories 'superseded' ones. Our 'ordinary' language is impregnated not only by 'inductivist' but also by falsificationist dogmatism. A reform is overdue.

[110] For a defence of this theory of 'learning from experience', cf. Agassi (1969).

[111] *These remarks show that 'learning from experience' is a normative idea; therefore all purely 'empirical' learning theories miss the heart of the problem.*

[112] Cf. Leibnitz (1678). The expression in brackets shows that Leibnitz regarded this criterion as second best and thought that the best theories are those which are proved. Thus Leibnitz's position – like Whewell's – is a far cry from fully fledged sophisticated falsificationism.

[113] Mill (1843), Vol. II, p. 23.

[114] This was J. S. Mill's argument (*ibid.*). He directed it against Whewell, who thought that 'consilience of inductions' or successful prediction of improbable events *verifies* (that is, *proves*) a theory. (Whewell [1858], pp. 95–6.) No doubt, *the basic mistake both in Whewell's and in Duhem's philosophy of science is their conflation of predictive power and proven truth. Popper separated the two.*

[115] Keynes (1921), p. 305. But cf. my (1968a), p. 394.

[116] This is Whewell's critical comment on an *ad hoc* auxiliary hypothesis in Newton's theory of light (Whewell [1857], Vol. II, p. 317).

[117] In the terminology of my (1968a), this theory was '*ad hoc$_1$*' (cf. my [1968a], p. 389, Footnote 1); the example was originally suggested to me by Paul Feyerabend as a paradigm of a *valuable ad hoc theory*.

[118] In the terminology of my (1968a), this theory was not '*ad hoc$_1$*', but it was '*ad hoc$_2$*' (cf. my [1968a], p. 389, Footnote 1). For a simple but artificial illustration, see *ibid.* p. 387, Footnote 2.

[119] *We can formulate this condition with striking clarity only in terms of the methodology of research programmes to be explained in §3: we retain a syntactically metaphysical theory as the 'hard core' of a research programme as long as its associated positive heuristic produces a progressive problemshift in the 'protective belt' of auxiliary hypotheses.*

[120] This phenomenon was described in a beautiful paper by Whewell (1851); but he could not explain it methodologically. Instead of recognizing the victory of the *progressive* Newtonian programme over the *degenerating* Cartesian programme, he thought this was the victory of proven truth over falsity. For details cf. my (1973): for a general discussion of the demarcation between progressive and degenerating reduction cf. Popper (1969).

[121] Popper (1934), Section 22.

[122] Cf. e.g. Popper (1959a), p. 107, Footnote *2.

[123] This is argued in Popper (1934), Section 29.

[124] Agassi claims that this example shows that we may "stick to the hypothesis in the face of known facts in the hope that the facts will adjust themselves to theory rather than the other way round" ([1966]. p. 18) .But *how* can facts 'adjust themselves'? Under which *particular* conditions should the theory win? Agassi gives no answer.

[125] The decision to use some monotheoretical model is clearly vital for the naive falsificationist to enable him to reject a theory on the *sole* ground of experimental evidence. *It is in line with the necessity for him to divide sharply, at least in a test-situation, the body of science into two: the problematic and the unproblematic. It is only the theory he decides to regard as problematic which he articulates in his deductive model of criticism.*

[126] Let me here answer a possible objection: "Surely we do not need Nature to tell us that a set of theories is *inconsistent*. Inconsistency – unlike falsehood – can be ascertained without Nature's help". But Nature's actual 'NO' in a monotheoretical methodology takes the form of a fortified 'potential falsifier', that is a sentence which, in this way of speech, we claim Nature had uttered and which is the *negation of our theory*. Nature's actual 'INCONSISTENCY' in a pluralistic methodology takes the form of a 'factual' statement couched in the light of one of the theories involved, which we claim Nature had uttered and which, if added to our proposed theories, yields an *inconsistent system.*

[127] For instance, in our earlier example some may try to replace the gravitational theory with a new one and others may try to replace the radio-optics by a new one: we choose the way which offers the more spectacular growth, the more progressive problemshift.

[128] Criticism does not *assume* a fully articulated deductive structure: it creates it. (Incidentally, this is the main message of my [1963–4].)

[129] A classical example of this pattern is Newton's relation to Flamsteed, the first Astronomer Royal. For instance, Newton visited Flamsteed on 1 September 1694, when working full time on his lunar theory; told him to reinterpret some of his data since they contradicted his own theory; and he explained to him exactly how to do it. Flamsteed

obeyed Newton and wrote to him on 7 October: "Since you went home, I examined the observations I employed for determining the greatest equations of the earth's orbit, and considering the moon's places at the times... I find that (*if, as you intimate, the earth inclines on that side the moon then is*) you may abate abt 20″ from it..." Thus Newton constantly criticized and corrected Flamsteed's observational theories. Newton taught Flamsteed, for instance, a better theory of the refractive power of the atmosphere; Flamsteed accepted this and corrected his original 'data'. One can understand the constant humiliation and slowly increasing fury of this great observer, having his data criticized and improved by a man who, on his own confession, made no observations himself: it was this feeling – I suspect – which led finally to a vicious personal controversy.

130 The same applies to the third type of decision. If we reject a stochastic hypothesis only for one which, in our sense, supersedes it, the exact form of the 'rejection rules' becomes *less* important.

131 Popper (1945), Vol. II, Chapter 23, p. 218.

132 Agassi is then wrong in his thesis that "observation reports may be accepted as false and hence the problem of the empirical basis is thereby disposed of" (Agassi [1966], p. 20).

133 One may point out that the negative and positive heuristic gives a rough (implicit) definition of the 'conceptual framework' (and consequently of the language). The recognition that the history of science is the history of research programmes rather than of theories may therefore be seen as a partial vindication of the view that the history of science is the history of conceptual frameworks or of scientific languages.

134 Popper (1934), Sections 11 and 70. I use 'metaphysical' as a technical term of naive falsificationism: a contingent proposition is 'metaphysical' if it has no 'potential falsifiers'.

135 Watkins (1958). Watkins cautions that "the logical gap between statements and prescriptions in the metaphysical-methodological field is illustrated by the fact that a person may reject a [metaphysical] doctrine in its fact-stating form while subscribing to the prescriptive version of it" (*Ibid.* pp. 356–7).

136 For this Cartesian research programme, cf. Popper (1958) and Watkins (1958), pp. 350–1.

137 Laplace (1796), Livre IV, Chapter ii.

138 The actual hard core of a programme does not actually emerge fully armed like Athene from the head of Zeus. It develops slowly, by a long, preliminary process of trial and error. In this paper this process is not discussed.

139 For *real* examples, cf. my (1973).

140 The 'refutation' was each time successfully diverted to 'hidden lemmas'; that is, to lemmas emerging, as it were, from the *ceteris paribus* clause.

141 If a scientist (or mathematician) has a positive heuristic, he refuses to be drawn into observation. He will "lie down on his couch, shut his eyes and forget about the data". (Cf. my [1963–64], especially pp. 300ff., where there is a detailed case study of such a programme.) Occasionally, of course, he will ask Nature a shrewd question: he will then be encouraged by Nature's *YES*, but not discouraged by its *NO*.

142 Reichenbach, following Cajori, gives a different explanation of what delayed Newton in the publication of his *Principia*: "To his disappointment he found that the observational results disagreed with his calculations. Rather than set any theory, however beautiful, before the facts, Newton put the manuscript of his theory into his drawer. Some twenty years later, after new measurements of the circumference of the

earth had been made by a French expedition, Newton saw that the figures on which he had based his test were false and that the improved figures agreed with his theoretical calculation. It was only after this test that he published his law... The story of Newton is one of the most striking illustrations of the method of modern science" (Reichenbach [1951], pp. 101–2). Feyerabend criticizes Reichenbach's account (Feyerabend [1965], p. 229), but does not give an alternative *rationale*.

[143] For a further discussion of Newton's research programme, cf. my (1973).

[144] For this point cf. Truesdell (1960).

[145] Soddy's contribution to Prout's programme or Pauli's to Bohr's (old quantum theory) programme are typical examples of such creative shifts.

[146] A 'verification' is a corroboration of excess content in the expanding programme. But, of course, a 'verification' does not *verify* a programme: it shows only its heuristic power.

[147] Cf. my (1963–4), pp. 324–30. Unfortunately in 1963–4 I had not yet made a clear terminological distinction between theories and research programmes, and this impaired my exposition of a research programme in informal, quasi-empirical mathematics. There are fewer such shortcomings in my (1974).

BIBLIOGRAPHY

Agassi, J.: 1959, 'How are Facts Discovered?', *Impulse* 3, 2–4.

Agassi, J.: 1962, 'The Confusion between Physics and Metaphysics in the Standard Histories of Sciences', in the *Proceedings of the Tenth International Congress of the History of Science*, 1964, Vol. 1, pp. 231–8.

Agassi, J.: 1964, 'Scientific Problems and Their Roots in Metaphysics', in Bunge (ed.), *The Critical Approach to Science and Philosophy*, 1964, pp. 189–211.

Agassi, J.: 1966, 'Sensationalism', *Mind* 75, 1–24.

Agassi, J.: 1968, 'The Novelty of Popper's Philosophy of Science', *International Philosophical Quarterly* 8, 442–63.

Agassi, J.: 1969, 'Popper on Learning from Experience', in Rescher (ed.), *Studies in the Philosophy of Science*, 1969.

Ayer, A. J.: 1936, *Language, Truth and Logic*, 1936; second edition 1946.

Bartley, W. W.: 1968, 'Theories of Demarcation between Science and Metaphysics', in Lakatos and Musgrave (eds.), *Problems in the Philosophy of Science*, 1968, pp. 40–64.

Braithwaite, R.: 1938, 'The Relevance of Psychology to Logic', *Aristotelian Society Supplementary Volumes* 17, 19–41.

Braithwaite, R.: 1953, *Scientific Explanation*, 1953.

Carnap, R.: 1932–3, 'Über Protokollsätze', *Erkenntnis* 3, 215–28.

Carnap, R.: 1935, Review of Popper's (1934), *Erkenntnis* 5, 290–4.

Dreyer, J. L. E.: 1906, *History of the Planetary Systems from Thales to Kepler*, 1906.

Duhem, P.: 1906, *La Théorie Physique, Son Objet et Sa Structure*, 1905. English translation of the second (1914) edition: *The Aim and Structure of Physical Theory*, 1954.

Einstein, A.: 1949, 'Autobiographical Notes', in Schilpp (ed.), *Albert Einstein, Philosopher-Scientist*, Vol. 1, pp. 2–95.

Feyerabend, P. K.: 1959, 'Comments on Grünbaum's "Law and Convention in Physical Theory"', in Feigl and Maxwell (eds.), *Current Issues in the Philosophy of Science*, 1961, pp. 155–61.

Feyerabend, P. K.: 1965, 'Reply to Criticism', in Cohen and Wartofsky (eds.), *Boston Studies in the Philosophy of Science*, Vol. II, pp. 223–61.

Feyerabend, P. K.: 1968–9, 'On a Recent Critique of Complementarity', *Philosophy of Science* **35**, 309–31 and 36, 82–105.
Feyerabend, P. K.: 1969, 'Problems of Empiricism II', in Colodny (ed.), *The Nature and Function of Scientific Theory*, 1969.
Feyerabend, P. K.: 1970, 'Against Method', *Minnesota Studies for the Philosophy of Science* **4**, 1970.
Galileo, G.: 1632, *Dialogo dei Massimi Sistemi*, 1632.
Grünbaum, A.: 1959a, 'The Falsifiability of the Lorentz-Fitzgerald Contraction Hypothesis', *British Journal for the Philosophy of Science* **10**, 48–50.
Grünbaum, A.: 1959b, 'Law and Convention in Physical Theory', in Feigl and Maxwell (eds.), *Current Issues in the Philosophy of Science*, 1961, pp. 40–155.
Grünbaum, A.: 1960, 'The Duhemian Argument', *Philosophy of Science* **2**, 75–87.
Grünbaum, A.: 1966, 'The Falsifiability of a Component of a Theoretical System', in Feyerabend and Maxwell (eds.), *Mind, Matter and Method: Essays in Philosophy and Science in Honor of Herbert Feigl*, 1966, pp. 273–305.
Grünbaum, A.: 1969, 'Can We Ascertain the Falsity of a Scientific Hypothesis?', *Studium Generale* **22**, 1061–93.
Hempel, C. G.: 1937, Review of Popper's (1934), *Deutsche Literaturzeitung*, 1937, pp. 309–14.
Hempel, C. G.: 1952, 'Some Theses on Empirical Certainty', *The Review of Metaphysics* **5**, 620–1.
Keynes, J. M.: 1921, *A Treatise on Probability*, 1921.
Koestler, A.: 1959, *The Sleepwalkers*, 1959.
Kuhn, T. S.: 1962, *The Structure of Scientific Revolutions*, 1962.
Kuhn, T. S.: 1965, 'Logic of Discovery or Psychology of Research', pp. 1–23.
Lakatos, I.: 1962, 'Infinite Regress and the Foundations of Mathematics', *Aristotelian Society Supplementary Volume* **36**, 155–84.
Lakatos, I.: 1963–4, 'Proofs and Refutations', *The British Journal for the Philosophy of Science* **14**, 1–25, 120–39, 221–43, 296–342.
Lakatos, I.: 1968a, 'Changes in the Problem of Inductive Logic', in Lakatos (ed.), *The Problem of Inductive Logic*, 1968, pp. 315–417.
Lakatos, I.: 1968b, 'Criticism and the Methodology of Scientific Research Programmes', in *Proceedings of the Aristotelian Society* **69**, 149–86.
Lakatos, I.: 1971, 'Popper zum Abgrenzungs- und Induktionsproblem', in H. Lenk (ed.), *Neue Aspekte der Wissenschaftstheorie*, 1971; the English version to appear under the title 'Popper on Demarcation and Induction' in Schilpp (ed.), *The Philosophy of Sir Karl Popper*.
Lakatos, I.: 1972a, 'History of Science and its Rational Reconstructions', in R. C. Buck and R. S. Cohen (eds.), *Boston Studies in the Philosophy of Science*, Vol. 8. Reidel Publishing House, 1972, pp. 91–135.
Lakatos, I.: 1972b, 'Replies to Critics', in R. C. Buck and R. S. Cohen (eds.), *Boston Studies in the Philosophy of Science*, Vol. 8. Reidel Publishing House, 1972, pp. 174–82.
Lakatos, I.: 1973, *The Changing Logic of Scientific Discovery*, 1973.
Lakatos, I.: 1974, *Proofs and Refutations and Other Essays in the Philosophy of Mathematics*, 1974.
Laplace, P.: 1796, *Exposition du Système du Monde*, 1796.
Laudan, L.: 1965, 'Grünbaum on "The Duhemian Argument"', *Philosophy of Science* **32**, 295–9.
Leibnitz, G. W.: 1678, Letter to Conring, 19.3.1678.

Le Roy, E.: 1899, 'Science et Philosophie', *Revue de Métaphysique et de Morale* **7**, 375–425, 503–62, 706–31.
Le Roy, E.: 1901, 'Un Positivisme Nouveau', *Revue de Métaphysique et de Morale* **9**, 138–53.
Medawar, P. B.: 1967, *The Art of the Soluble*, 1967.
Medawar, P. B.: 1969, *Induction and Intuition in Scientific Thought*, 1969.
Milhaud, G.: 1896, 'La Science Rationnelle', *Revue de Métaphysique et de Morale* **4**, 280–302.
Mill, J. S.: 1843, *A System of Logic, Ratiocinative and Inductive, Being a Connected View of the Principles of Evidence, and the Methods of Scientific Investigation*, 1843.
Musgrave, A.: 1968, 'On a Demarcation Dispute', in Lakatos and Musgrave (eds.), *Problems in the Philosophy of Science*, 1968, pp. 78–88.
Musgrave, A.: 1969a, *Impersonal Knowledge*, Ph. D. Thesis, University of London, 1969.
Musgrave, A.: 1969b, Review of Ziman's 'Public Knowledge: An Essay Concerning the Social Dimensions of Science', in *The British Journal for the Philosophy of Science* **20**, 92–4.
Musgrave, A.: 1973, 'The Objectivism of Popper's Epistemology', in Schilpp (ed.), *The Philosophy of Sir Karl Popper*, 1973.
Nagel, E.: 1967, 'What is True and False in Science: Medawar and the Anatomy of Research', *Encounter* **29**, 68–70.
Neurath, O.: 1935, 'Pseudorationalismus der Falsifikation', *Erkenntnis* **5**, pp. 353–65.
Poincaré, J. H.: 1891, 'Les géométries non euclidiennes', *Revue Générale des Sciences Pures et Appliquées* **2**, 769–74.
Poincaré, J. H.: 1902, *La Science et l'Hypothèse*, 1902.
Polanyi, M.: 1958, *Personal Knowledge, Towards a Post-critical Philosophy*, 1958.
Popkin, R.: 1968, 'Scepticism, Theology and the Scientific Revolution in the Seventeenth Century', in Lakatos and Musgrave (eds.), *Problems in the Philosophy of Science*, 1968, pp. 1–28.
Popper, K. R.: 1933, 'Ein Kriterium des empirischen Charakters theoretischer Systeme', *Erkenntnis* **3**, 426–7.
Popper, K. R.: 1934, *Logik der Forschung*, 1935 (expanded English edition, Popper [1959a]).
Popper, K. R.: 1935, 'Induktionslogik und Hypothesenwahrscheinlichkeit', *Erkenntnis* **5**, 170–2; published in English in his (1959a), pp. 315–17.
Popper, K. R.: 1940, 'What is Dialectic?', *Mind*, N. S. **49**, 403–26; reprinted in Popper 1963, pp. 312–35.
Popper, K. R.: 1945, *The Open Society and Its Enemies*, I-II, 1945.
Popper, K. R.: 1957a, 'The Aim of Science', *Ratio* **I**, 24–35.
Popper, K. R.: 1957b, *The Poverty of Historicism*, 1957.
Popper, K. R.: 1958, 'Philosophy and Physics'; published in *Atti del XII Congresso Internazionale di Filosofia*, Vol. 2, 1960, pp. 363–74.
Popper, K. R.: 1959a, *The Logic of Scientific Discovery*, 1959.
Popper, K. R.: 1959b, 'Testability and "ad-Hocness" of the Contraction Hypothesis', *British Journal for the Philosophy of Science* **10**, p. 50.
Popper, K. R.: 1963, *Conjectures and Refutations*, 1963.
Popper, K. R.: 1965, 'Normal Science and its Dangers', pp. 51–8.
Popper, K. R.: 1968a, 'Epistemology without a Knowing Subject', in Rootselaar and Staal (eds.), *Proceedings of the Third International Congress for Logic, Methodology and Philosophy of Science*, Amsterdam, 1968, pp. 333–73.

Popper, K. R.: 1968b, 'On the Theory of the Objective Mind', in *Proceedings of the XIV International Congress of Philosophy* **1** (1968), 25–53.

Popper, K. R.: 1968c, 'Remarks on the Problems of Demarcation and Rationality', in Lakatos and Musgrave (eds.), *Problems in the Philosophy of Science*, 1968, pp. 88–102.

Popper, K. R.: 1969, 'A Realist View of Logic, Physics and History', in Yourgrau and Breck (eds.), *Physics, Logic and History*, 1969.

Quine, W. V.: 1953, *From a Logical Point of View*, 1953.

Reichenbach, H.: 1951, *The Rise of Scientific Philosophy*, 1951.

Russell, B.: 1914, *The Philosophy of Bergson*, 1914.

Russell, B.: 1943, 'Reply to Critics', in Schilpp (ed.), *The Philosophy of Bertrand Russell*, 1943, pp. 681–741.

Russell, B.: 1946, *History of Western Philosophy*, 1946.

Truesdell, C.: 1960, 'The Program Toward Rediscovering the Rational Mechanics in the Age of Reason', *Archive of the History of Exact Sciences* **I**, 3–36.

Watkins, J.: 1957, 'Between Analytic and Empirical', *Philosophy* **32**, 112–31.

Watkins, J.: 1958, 'Influential and Confirmable Metaphysics', *Mind*, N.S. **67**, 344–65.

Watkins, J.: 1960, 'When are Statements Empirical?', *British Journal for the Philosophy of Science* **10**, 287–308.

Watkins, J.: 1968, 'Hume, Carnap and Popper', in Lakatos (ed.), *The Problem of Inductive Logic*, 1968, pp. 271–82.

Whewell, W.: 1837, *History of the Inductive Sciences, from the Earliest to the Present Time*, Three volumes, 1837.

Whewell, W.: 1840, *Philosophy of the Inductive Sciences, Founded upon their History*, Two volumes, 1840.

Whewell, W.: 1851, 'On the Transformation of Hypotheses in the History of Science', *Cambridge Philosophical Transactions* **9**, 139–47.

Whewell, W.: 1858, *Novum Organon Renovatum*. Being the second part of the philosophy of the inductive sciences, Third edition, 1858.

Whewell, W.: 1860, *On the Philosophy of Discovery, Chapters Historical and Critical*, 1860.

Wisdom, J. T.: 1963, 'The Refutability of "Irrefutable" Laws', *The British Journal for the Philosophy of Science* **13**, 303–6.

ADOLF GRÜNBAUM

IS IT *NEVER* POSSIBLE TO FALSIFY A
HYPOTHESIS IRREVOCABLY?*

In his book *The Aim and Structure of Physical Theory*, Duhem denied the feasibility of crucial experiments in physics. Said he:

> ... the physicist can never subject an isolated hypothesis to experimental test but only a whole group of hypotheses; when the experiment is in disagreement with his predictions, what he learns is that at least one of the hypotheses constituting this group is unacceptable and ought to be modified; *but the experiment does not designate which one should be changed* (my italics)[1].

Duhem illustrates and elaborates this contention by means of examples from the history of optics. And in each of these cases, he maintains that "If physicists had attached some value to this task",[2] any one component hypothesis of optical theory such as the corpuscular hypothesis (or so-called emission hypothesis) could have been preserved in the face of seemingly refuting experimental results such as those yielded by Foucault's experiment. According to Duhem, this continued espousal of the component hypothesis could be justified by "shifting the weight of the experimental contradiction to some other proposition of the commonly accepted optics".[3] Here Duhem is maintaining that the refutation of a component hypothesis H is at least usually no more certain than its verification could be.

In terms of the notation H and A which we have been using, Duhem is telling us that we could blame an experimentally false consequence C of the total optical theory T on the falsity of A while upholding H. In making this claim, Duhem is quite clear that the falsity of *A* no more *follows* from the experimental falsity of C than does the falsity of H. But his point is that this fact does not logically prevent us from postulating that A is false while H is true. And he is telling us that altogether the pertinent empirical facts *allow* us to reject A as false. Hence we are under no deductive logical constraint to infer the falsity of H. In questioning Dicke's purported refutation of the GTR, I gave sanction to this Duhemian contention in this instance precisely on the grounds that Dicke's auxiliary A is *not* known to be true with certainty.

In contemporary philosophy of science, a *generalized* version of

Duhem's thesis with its ramifications has been attributed to Duhem and has been highly influential. I shall refer to this elaboration of Duhem's philosophical legacy in present-day philosophy of science as 'the D-thesis'. But in doing so, my concern is with the philosophical credentials of this legacy, not with whether this attribution to Duhem himself can be uniquely sustained exegetically as against rival interpretations given by Duhem scholars. But I should remark that at least one such scholar, L. Laudan, has cited textual evidence which casts some doubt on this attribution.[4] The present philosophical appraisal is intended to supersede some parts of my earlier published critique of the D-thesis. In his Pittsburgh Doctoral Dissertation, Philip Quinn has pointed out that the version of the D-thesis with which I am concerned can be usefully stated in the form of two subtheses D1 and D2, and has argued that Laudan's attributional doubts are warranted only with respect to D2 but not with respect to D1.

The two subtheses are:

D1. No constituent hypothesis H of a wider theory can *ever* be sufficiently isolated from some set or other of auxiliary assumptions so as to be separately falsifiable observationally. H is here understood to be a constituent of a wider theory in the sense that no observational consequence can be deduced from H alone.

It is a corollary of this subthesis that *no* such hypothesis H *ever* lends itself to a crucially falsifying experiment any more than it does to a crucially verifying test.

D2. In order to state the second subthesis D2, we let T be a theory of *any* domain of empirical knowledge, and we let H be *any* of its component subhypotheses, while A is the collection of the remainder of its subhypotheses. Also, we assume that the observationally testable consequence O entailed by the conjunction H·A is taken to be empirically false, because the observed findings are taken to have yielded a result O' *incompatible* with O. Then D2 asserts the following: For all potential empirical findings O' of this kind, there exists at least one suitably revised set of auxiliary assumptions A' such that the conjunction of H with A' *can be held to be true and explains O'*. Thus D2 claims that H can be held to be true *and* can be used to explain O' no matter what O' turns out to be, i.e., *come what may.*

Note that if D2 did not assert that A' can be held to be true in the face of the evidence no less than H, then H could not be claimed to explain O' via A'. For premises which are already *known* to be false are not scientifically acceptable as bases for explanation.[5] Hence the part of D2 which asserts that H and A' can each be held to be *true* presupposes that either they could not be separately falsified or that neither of them has been separately falsified.

In my prior writings on the D-thesis, I made three main claims concerning it:

(1)   There are quite trivial senses in which D1 and D2 are uninterestingly true and in which no one would wish to contest them.[6]

(2)   In its non-trivial form, D2 has not been demonstrated.[7]

(3)   D1 is false, as shown by counterexamples from physical geometry.[8]

Since then, Gerald Massey has called my attention to yet another defect of D2 which any proponent of that thesis would presumably endeavor to remedy. Massey has pointed out that, as it stands, D2 attributes *universal* explanatory relevance and power to any one component hypothesis H. For let O* be *any* observationally testable statement *whatever* which is compatible with $\sim$O, while O' is the conjunction $\sim$O·O*. Assume that $(H \cdot A) \to O$. Then D2 asserts the existence of an auxiliary A' such that the theoretical conjunction H·A' explains the putative observational finding $\sim$O·O*. As Joseph Camp has suggested, the proponent of D2 might reply that A' itself may potentially explain O* *without* H, even though H is essential for explaining $\sim$O via A'. But the advocate of D2 has no guarantee that he can circumvent the difficulty in this way.

Instead, he might perhaps wish to require that O' must pertain to *the same kind of phenomena* as O, thereby ruling out 'extraneous' findings O*. Yet even if he can articulate such a restriction or provide a viable alternative to it, there is the following further difficulty noted by Massey: D2 gratuitously asserts the existence of a *deductive* explanation for any event whatever. This existential claim is gratuitous. For there *may* be individual occurrences (in the domain of quantum phenomena or elsewhere) which cannot be explained deductively, because of the irreducibly statistical character of its pertinent laws.

Our governing concern here is the question: 'Is there *any* component hypothesis H whatever whose falsity we can ascertain?' I shall try to answer this question by giving reasons for now *qualifying* my erstwhile charge that D1 is false. Hence I shall modify the *third* of my earlier contentions about the D-thesis. But before doing so, I must give my reasons for *not* also retracting either of the first two of these contentions in response to the critical literature which they have elicited. These reasons will occupy a number of the pages that follow. The first of my earlier claims was that D1 and D2 are each true in trivial senses which are respectively exemplified by the following two examples, which I had given.[9]

i. Suppose that someone were to assert the presumably false empirical hypothesis H that 'Ordinary buttermilk is highly toxic to humans'. Then the English sentence expressing this hypothesis could be 'saved from refutation' in the face of the observed wholesomeness of ordinary buttermilk by making the following change in the theoretical system constituted by the hypothesis: changing the rules of English usage so that the intension of the term 'ordinary buttermilk' is that of the term 'arsenic' as customarily understood. In this Pickwickian sense, D1 could be sustained in the case of this particular H.

ii. In an endeavor to justify D2, let someone propose the use of an A′ which is itself of the form

$$\sim H \vee O'.$$

In that case, it is certainly true in standard systems of logic that

$$(H \cdot A') \to O'.$$

Let us now see by reference to these two examples why I regard them as exemplifications of trivially true versions of D1 and D2 respectively.

i. *D1*

Here H was the hypothesis that 'ordinary buttermilk is highly toxic to humans'. When the proponent of D1 was challenged to save this H from refutation, what he did 'save' was *not* the proposition H but the *sentence* H expressing it, as reinterpreted in the following respect: Only the term 'ordinary buttermilk' was given a new semantical usage, and *no constraint was imposed on its new usage other than that the ensuing reinterpretation turn the sentence H into a true proposition.*

If one does countenance such *unbridled* semantical instability of some of the theoretical language in which H is stated, then one can indeed thereby uphold D1 in the form of Quine's epigram: "Any statement can be held true come what may, if we make drastic enough adjustments elsewhere in the system."[10] But, in that case, D1 turns into a thoroughly unenlightening truism.

I took pains to point out, however, that the commitments of D1 can also be trivially or uninterestingly fulfilled by semantical devices far more sophisticated and restrictive than the unbridled reinterpretation of some of the vocabulary in H. As an example of such a trivial fulfillment of D1 by devices whose feasibility is itself not at all trivial in other respects, I cited the following:

... suppose we had two particular substances I_1 and I_2 which are isomeric with each other. That is to say, these substances are composed of the same elements in the same proportions and with the same molecular weight but the arrangement of the atoms within the molecule is different. Suppose further that I_1 is not at all toxic while I_2 is highly toxic, as in the case of two isomers of trinitrobenzene. [At this point, I gave a footnote citation of 1,3,5-trinitrobenzene and of 1,2,4-trinitrobenzene respectively.] Then if we were to call I_1 'aposteriorine' and asserted that 'aposteriorine is highly toxic', this statement H could also be trivially saved from refutation in the face of the evidence of the wholesomeness of I_1 by the following device: only *partially* changing the meaning of 'aposteriorine' so that its intension is the second, highly toxic isomer I_2, thereby leaving the chemical 'core meaning' of 'aposteriorine' intact. To avoid misunderstanding of my charge of triviality, let me point out precisely what I regard as trivial here. The preservation of H from refutation in the face of the evidence by a *partial* change in the meaning of 'aposteriorine' is trivial in the sense of being only a *trivial* fulfillment of *the expectations raised by the D-thesis* [D 1]. But, in my view, *the possibility as such of preserving H by this particular kind of change in meaning is* not at all trivial. For this possibility as such reflects a fact about the world: the existence of isomeric substances of radically different degrees of toxicity (allergenicity)![11]

Mindful of this latter kind of example, I emphasized that a construal of D1 which allows itself to be sustained by this kind of alteration of the intension of 'trinitrobenzene' is no less trivial *in the context of the expectations raised by the D-thesis* than one which rests its case on calling arsenic 'buttermilk'.[12] And hence I was prompted to conclude that "a necessary condition for the nontriviality of Duhem's thesis is that the theoretical language be semantically stable in the relevant respects [i.e., with respect to the vocabulary used in H]"[13].

Mary Hesse took issue with this conclusion in her review of the essay in which I made these claims. There she wrote:

... it is not clear... that 'semantic stability' *is* always required when a hypothesis is nontrivially saved in face of undermining evidence. The law of conservation of momentum is in a sense saved in relativistic mechanics, and yet the usage of 'mass' is changed – it becomes a function of velocity instead of a constant property. But further argument along these lines is idle without more detailed analysis of what it is for a hypothesis to be the 'same', and what is involved in 'semantic stability'.[14]

This criticism calls for several comments.

(a) Hesse calls for a "more detailed analysis of what it is for a hypothesis to be the 'same', and what is involved in 'semantic stability'." To this I say that the *primary* onus for providing that more detailed analysis falls on the shoulders of the Duhemian. For he wishes to claim that his thesis D1 is interestingly true. And we saw that if he is to make such a claim, he must surely *not* be satisfied with the mere retention of the *sentence* H in some interpretation or other.

Fortunately, both the proponent and the critic of D1 can avail themselves of an apparatus of distinctions proposed by Peter Achinstein to confer specificity on what it is for the vocabulary of H to remain semantically stable.

Achinstein introduces semantical categories for describing the various possible relationships between properties of X and the term 'X'. And he appeals to the normal, standard or typical scientific *use* of a term 'X' at a given time as distinct from special uses.[15] He writes:

... I must introduce the concept of relevance and speak of a property as relevant for being an X. By this I mean that if an item is known to possess certain properties and lack others, the fact that the item possesses (or lacks) the property in question normally will count, at least to some extent, in favor of (or against) concluding that it is an X; and if it is known to possess or lack sufficiently many properties of certain sorts, the fact that the item possesses or lacks the property in question may justifiably be held to settle whether it is an X.[16] ...

Two distinctions are now possible. The first is between positive and negative relevance. If the fact that an item has P tends to count more in favor of concluding that it is an X than the fact that it lacks P tends to count against it, P can be said to have more positive than negative relevance for X. The second distinction is between semantical and nonsemantical relevance and is applicable only to certain cases of relevance.

Suppose one is asked to justify the claim that the reddish metallic element of atomic number 29, which is a good conductor and melts at 1083°C, is copper. One reply is that such properties tend to count in and of themselves, to some extent, toward classifying something as copper. By this I mean that an item is correctly classifiable as copper solely in virtue of having such properties; they are among the properties which constitute a final court of appeal when considering matters of classification; such properties are, one might say, intrinsically copper-making ones. Suppose, on the other hand, one is asked to justify the claim that the substance constituting about 10^{-4} percent of the igneous rocks in the earth's crust, that is mined in Michigan, and that was used by the ancient Greeks, is copper. Among the possible replies is *not* that such properties tend to count in and of themselves, to some extent, toward something's being classifiable as copper – that is, it is not true that something is classifiable as copper solely in virtue of having such properties. These properties do not constitute a final court of appeal when considering matters of classification. They are not intrinsically copper-making ones. Rather, the possession of such properties (among others) counts in favor of classifying something as copper solely because it allows one to infer that the item possesses other properties such as being metallic and having the atomic number 29, properties that are intrinsically copper-making ones, in virtue of which it is classifiable as copper.[17]...

Suppose $P^1, ..., P_n$ constitutes some set of relevant properties of X. If the properties in this set tend to count in and of themselves, to some extent, toward an item's being classifiable as an X, I shall speak of them as *semantically relevant* for X. If the possession of properties by an item tends to count toward an X-classification solely because it allows one to infer that the item possesses properties of the former sort, I shall speak of such properties as *nonsemantically relevant* for X. This distinction is not meant to apply to all relevant properties of X, for there will be cases on or near the borderline not clearly classifiable in either way.[18]

... in the latter part of the eighteenth century, with the systematic chemical nomenclature of Bergman and Lavoisier, the chemical composition of compounds began to be treated as semantically relevant; and ... chemical composition ... provided the basis for classification of compounds.

I have used the labels semantical and nonsemantical relevance because X's semantically relevant properties have something to do with the meaning or use of the term 'X' in a way that X's nonsemantically relevant properties do not.[19]...

Suppose you learn the semantically relevant properties of items denoted by the term 'X.' Then you will know those properties a possession of which by actual and hypothetical substances in and of itself tends to count in favor of classifying them as ones to which the term 'X' is applicable.[20]... Properties semantically relevant for X... include those that are logically necessary or sufficient....

Consider a term 'X' and the properties or conditions semantically relevant for X. It is perfectly possible that there be two different theories in which the term 'X' is used, where the same set of semantically relevant properties of X (or conditions for X) are presupposed in each theory (even though other properties attributed to X by these theories, properties not semantically relevant for X, might be different). If so, the term 'X' would not mean something different in each theory.[21]

Let us now employ Achinstein's distinctions to characterize a semantically stable use of the *sentence* H. I would say that one is engaged in a *semantically stable use* of the term 'X', if and only if no changes are made in the membership of the set of properties which are semantically relevant to being an item denoted by the term 'X'. And similarly for the semantically stable use of the various terms 'X_1' ($i = 1, 2, 3... n$) constituting the vocabulary of the sentence H in which a particular hypothesis (proposition) is expressed, even though these terms will, of course, not be confined to substance words or to the three major types of terms treated by Achinstein.[22] By the same token, if the entire sentence H is used in a semantically stable manner, then the hypothesis H has remained the same in the face of other changes in the total theory. Moreover, employing different terminology, I dealt with the concrete case of geometry and optics in earlier publications to point out the following: when the rejection of a certain presumed law of optics leads to a change in the membership of the properties *non*semantically relevant to a geometrical term 'X', this change does *not* itself make for a semantic instability in the use of 'X'.[23] I shall develop the latter point further below after discussing Hesse's objection.

In the case of my example of the two isomers of trinitrobenzene, it is clear from the very names of the two isomers that the molecular structure is semantically relevant or even logically necessary to being 1, 2, 4-trinitrobenzene as distinct from 1, 3, 5-trinitrobenzene or conversely. It will be recalled that presumably only the *former* of these two isomers is toxic. The Duhemian should be concerned to be able to uphold the hypothesis '1, 3, 5-trinitrobenzene is toxic' (H) in a semantically stable manner. Hence, if he is to succeed in doing so, his use of the word '1, 3, 5-trinitrobenzene' must leave all of the semantically relevant properties of 1, 3, 5-trinitrobenzene unchanged. And thus it would constitute only a trivial fulfillment of D1 in this case to adopt a *new use* of '1, 3, 5-trinitrobenzene' which suddenly confers *positive* semantic relevance on the particular properties constituting the molecular structure of 1, 2, 4-trinitrobenzene. I do say that upholding H by such a 'meaning-switch' is trivial *for the purposes of D1*. But I do *not* thereby affirm at all the biochemical triviality of the fact which makes it possible to uphold H in this particular fashion. For I would be the first to grant that the existence of isomers of trinitrobenzene differing greatly in toxicity is biochemically significant.

(b) Hesse believes that Einstein's replacement of the Newtonian law of conservation of momentum by the relativistic one is a case in which "a hypothesis is non-trivially saved in face of undermining evidence" amid a violation of semantic stability. And she views this case as a counter-example to my claim that semantic stability is a necessary condition for the non-trivial *fulfillment of D1*. The extension of my remarks about the trinitrobenzene example to this case will now serve to show why I do not consider this objection cogent: it rests on a conflation of being non-trivial in some respect or other with being non-trivial vis-à-vis D1.

One postulational base of the special relativistic dynamics of particles combines *formal homologues* of the two principles of the conservation of mass and momentum with the kinematical Lorentz transformations.[24] It is then shown that it is possible to satisfy the two formal conservation principles such that they are Lorentz-covariant, only on the assumption that the mass of a particle depends on its velocity. And the exact form of that velocity-dependence is derived via the requirement that the conservation laws go over into the classical laws for moderate velocities, i.e., that the relativistic mass m assume the value of the Newton mass m_0 for vanishing velocity.[25] This latter fact certainly makes it non-trivial and useful *for mechanics* to use the word 'mass' in the case of both Newtonian and relativistic mass. And hence Einstein's *formal* retention of the conservation principles is certainly *not* an instance of an unbridled reinterpretation of them. Yet despite the interesting common feature of the term 'mass', and the formal homology of the conservation principles, the two theories disagree here.

Using Achinstein's concept of mere *relevance*, we can say that the Newtonian and relativity theories disagree here as to the membership of the set of properties *relevant* to 'mass'. For it is clear that in Newton's theory, velocity-*in*dependence is positively *relevant*, in Achinstein's sense, to the term 'mass'. And it is likewise clear that velocity-*dependence* is similarly positively relevant in special relativistic dynamics. What is unclear is whether velocity *in*dependence and dependence respectively are *semantically* or *non*semantically relevant. Thus, it is unclear whether relativistic dynamics modified the Newtonian conservation principles in semantically relevant ways or preserved them semantically while modifying only the properties *non*semantically relevant to 'mass' and 'velocity'.

Let us determine the bearing of each of these two possibilities on the force of Hesse's purported counterexample to my claim that semantic stability is a necessary condition for the non-triviality of the D-thesis.

In the event of semantic relevance, semantic stability has been violated in a manner already illustrated by my trinitrobenzene example. For in that event, Newton's theory can be said to assert the conservation principles in an interpretation which assigns to the abstract word 'mass' the value m_0 as its denotatum, whereas relativity theory rejects these principles as generally false in that interpretation. But a *non*-trivial fulfillment of D1 here would have required the retention of the hypothesis of conservation of momentum in an interpretation which preserves *all* of the properties semantically relevant to 'mass', and to 'velocity' for that matter. Since this requirement is not met on this construal, the *formal* relativistic retention of this conservation principle cannot qualify as a nontrivial fulfillment of D1. But the success of Einstein's particular semantical reinterpretation is, of course, highly illuminating in other respects.

On the other hand, suppose that relativistic dynamics has modified only the properties which are *non*-semantically relevant to 'mass' and 'velocity'. In that event, its retention of the conservation hypothesis is indeed non-trivial vis-à-vis D1. But on that alternative construal, the relativistic introduction of a velocity-dependence of mass would *not* involve a violation of semantic stability. And then this retention cannot furnish Hesse with a viable counterexample to my claim that semantic stability is necessary for non-triviality vis-à-vis D1.

Finally, suppose that here we are confronted with one of Achinstein's borderline cases such that the distinction between semantic and non-semantic relevance does not apply in the case of mass to the relevant property of velocity-dependence. In that case, we can characterize the transition from the Newtonian to the relativistic momentum conservation law with respect to 'mass' as merely a repudiation of the Newtonian claim that velocity-*in*dependence is positively relevant in favor of asserting that velocity-dependence is thus relevant. And then this transition cannot be adduced as a retention of a hypothesis H which violates semantic stability while being non-trivial vis-à-vis D1.

Since there are borderline cases between semantic and non-semantic relevance, it is interesting that Duhem has recently been interpreted as denying that the distinction between them is ever scientifically pertinent. C. Giannoni reads Duhem as maintaining that actual scientific practice always accords the same semantic role to *all* relevant properties.[26] Thus, C. Giannoni writes:

Duhem is concerned primarily with quantities which are the subject of derivative measurement rather than fundamental measurement.... Quantities which are measured by other means than the means originally used in introducing the concept are derivative relative to this method of measurement. For example, length can be fundamentally measured by using meter sticks, but it also can be measured by sending a light beam along the length and back and measuring the time which it takes to complete the round trip. We can then calculate the length by finding the product of the velocity of light (c) and the time. Such a method of measurement is dependent not only on the measurement of time as is the first type of derivative quantity, but also on the law of nature that the velocity of light is a constant equal to c.[27]

Duhem further substantiates his view... by noting that when several methods are available for measuring a certain property, no one method is taken as the absolute criterion relative to which the other methods are derivative in the second sense of derivative measurement noted above [i.e., in Achinstein's sense of being nonsemantically relevant]. Each method is used as a check against the others.[28]

If this thesis of *semantic parity* among *all* of the relevant properties were correct, then one might argue that a semantically stable use of the sentence H would simply not be possible in conjunction with each of two different auxiliary hypotheses A and A′ of the following kind: A and A′ contain *incompatible* law-like statements each of which pertains to one or more entities that are *also* designated by terms in the sentence H. In earlier publications, I used physical geometry and optics as a test case for these claims. There I considered the actual use of geometrical and optical language in standard relativity physics[29] as well as a hypothetical Duhemian modification of the optics of special relativity.[30] Let me now develop the import of these considerations. We shall see that it runs counter to the thesis of universal semantic *parity* and sustains the feasibility of semantic stability.

The Michelson-Morley experiment involves a comparison of the round-trip times of light along equal closed spatial paths in different directions of an inertial system. As is made explicit in the account of this experiment in standard treatises, the spatial distances along the arms of the interferometer *are measured by rigid rods*.[31] The issue which arose from this experiment was conceived to be the following: Are the round-trip times *along the equal spatial paths* in an inertial frame generally *unequal*, as claimed by the aether theory, or equal as asserted by special relativity? Hence both parties to the dispute agreed that, to within a certain accuracy, equal numerical verdicts furnished by rigid rods were positively semantically relevant to the geometrical relation term 'spatially congruent' (as applied to line segments).[32] But the statement of the dispute shows that the *positive* relevance of the equality of the round-trip times of light to *spatial* congruence was made *contingent* on the particular *law of optics* which would be borne out by the Michelson-Morley experiment. Let X be the relational property of being congruent as obtaining among intervals of space. And let Y be the relational property of being traversed by light in equal round-trip times, as applied to spatial paths as well. Then we can say that the usage of 'X' at the turn of the current century was such that the relation Y was only *non*semantically relevant to the relation term 'X', whereas identical numerical findings of *rigid rods* – to be called 'RR' – were held to be positively semantically relevant to it.

Relativity physics then added the following results pertinent to the kind of relevance which can be claimed for Y: (1) Contrary to the situation in inertial frames, in the socalled 'time-orthogonal' *non*-inertial frames the round-trip times of light in different directions along spatially *equal* paths are generally unequal.[33] (2) On a disk rotating uniformly in an inertial system I, there is generally an *inequality* among the transit times required by light departing from the same point to traverse a given closed polygonal path in opposite senses. Thus, suppose that two light signals are *jointly* emitted from a point P on the periphery of a rotating disk and are each made to traverse that periphery in opposite directions so as to return to P. Then the two oppositely directed light pulses are indeed said to traverse *equal spatial distances*, i.e., the relation X is asserted for their paths on the strength of the obtaining of the relation RR. But the one light pulse which travels in the direction *opposite* to that of the disk's rotation with respect to frame I will travel a shorter path in I, and hence will return to the disk point P earlier than the light pulse traveling in the direction in which the disk rotates. Hence the round-trip times of light for these *equal* spatial paths will be *unequal*.[34] (3) Whereas the spatial geometry yielded by a rigid rod on a stationary disk is Euclidean, such a rod yields a hyperbolic non-Euclidean geometry on a rotating disk.[35]

What is the significance of these several results? We see that space intervals which are RR are called 'congruent' ('X') in reference frames in which Y is relevant to their being *non*-X, no less than RR intervals are called 'X' in those frames in which Y is relevant to their being X! Contrary to Giannoni's general contention, this important case from geometry and optics exhibits the pertinence of the distinction between semantic and non-semantic relevance. Here RR is positively semantically relevant to being X. But Y has only *non*semantical relevance to X, as shown by the fact that while Y is relevant to X in inertial systems, it is relevant to *non*-X in at least two kinds of non-inertial systems. And this refutes the generalization which Giannoni attributes to Duhem, viz., that there is parity of positive *semantic* relevance among properties such as RR and Y, so long as there are instances in which they are all merely positively *relevant* to X.

This lack of *semantic* parity between Y and RR with respect to 'X' allows a semantically *stable* geometrical use of 'X' on both the stationary and rotating disks, unencumbered by the *incompatibility* of the optical laws which relate the property Y to X in these two different reference frames. Indeed, by furnishing a *tertium comparationis*, this semantic stability confers *physical* interest on the contrast between the incompatible laws of optics in the two disks, and on the contrast between their two incompatible spatial geometries! For in *both* frames, RR is alike centrally semantically relevant to being X, i.e., in each of the two frames, rigid rod coincidences serve alike to determine what space intervals are assigned equal measures $ds = \sqrt{(g_{ik}dx^i dx^k)}$. This sameness of RR, which makes for the semantic stability of 'X', need not, however, comprise a sameness of the correctional physical laws that enable us to allow computationally for the thermal and other deviations from rigidity.

I used certain formulations of relativity theory as a basis for the preceding account of the properties semantically and nonsemantically relevant to 'X' ('spatially congruent'). But this account does not, of course, gainsay the legitimacy of alternative uses of 'X' that would issue in alternative, albeit physically equivalent, formulations. For example, the relativistic interpretation of the upshot of the Michelson-Morley experiment could alternatively have been stated as follows: In any inertial system, distances which are equal in the metric based on light-propagation are equal also in the metric which is based on rigid measuring rods.[36] I would say that this particular alternative formulation places Y on a par with RR as a *candidate* for being semantically relevant to 'X'. Moreover, I have strongly emphasized elsewhere that we are indeed free to formulate certain physical theories alternatively so as to make Y rather than RR semantically relevant to 'X'.[37]

So much for matters pertaining to trivial fulfillments of D1.

ii. *D2*

It will be recalled that my example of a trivial fulfillment of D2 involved the use of an A' which is itself of the form $\sim H \vee O'$. In that example, the triviality did *not* arise from a violation of semantic stability. Instead, the fulfillment of D2 by the formal validity of the statement $[H \cdot (\sim H \vee O')] \rightarrow O'$ is only trivial because the entailment holds independently of the specific assertive content of H, unless a restrictive kind of entailment is used as that of the system E of Anderson and Belnap. No matter what H happens to be about, and no matter whether H is substantively relevant to the specific content of O' or not, O' will be entailed by H via the particular A' used here. Hence H does *not* serve to explain via A' the facts asserted by O', thereby satisfying D2 here only trivially, *if at all*.

As Philip Quinn has shown,[38] one of the fallacies which vitiates J. W. Swanson's purported syntactical proof that D2 holds interestingly[39] is the failure to take cognizance of this particular kind of semantic trivialization or even outright falsity of D2.

This brings us to the second of my previously published claims. The latter was that in its non-trivial form, D2 has not been demonstrated by means of D1. Having dealt with several *necessary* conditions for the non-triviality of A', I therefore went on to comment on a possible *sufficient* condition by writing:

I am unable to give a formal and completely general *sufficient* condition for the *non*-triviality of A'. And, so far as I know, neither the originator nor any of the advocates of the D-thesis [D2] have ever shown any awareness of the need to circumscribe the class of *non*trivial revised auxiliary hypotheses A' so as to render the D-thesis [D2] interesting. I shall therefore assume that the proponents of the D-thesis intend it to stand or fall on the kind of A' which we would all recognize as *non*trivial *in any given case*, a kind of A' which I shall symbolize by A'_{nt}.[40]

And I added that D2 was *undemonstrated*, since it does *not* follow from D1 that

$$(\exists A'_{nt}) \, [(H \cdot A'_{nt}) \rightarrow O'].$$

Mary Hesse discusses my charge that D2 is a non-sequitur, as well as my erstwhile objections to D1, which I shall qualify below. And she addresses herself to a particular example of mine which was of the following kind: the *original* auxiliary hypothesis A ingredient in the conjunction H·A is highly confirmed quite separately from H, but the conjunction H·A does entail a presumably incorrect observational consequence. Apropos of this example, she writes:

... Grünbaum admits that... "A is only more or less highly confirmed" (p. 289) – a significant retreat from the claim of truth. He thinks nevertheless that it is crucial that confirmation of A can be *separated* from that of H. But surely this is not sufficient for his purpose. For as long as it is not empirically demonstrable, but only likely, that A is true, it will always be possible to reject A in order to save H.... Which is where the Duhem thesis came in. It must be concluded that D [the D-thesis] still withstands Grünbaum's assaults.... [41]

It is not clear how Hesse wants us to construe her phrase "in order to save H", when she tells us that "it will always be possible to reject A in order to save H". If she intends that phrase to convey merely the same as "with a view to attempting to save H", then her objection pertains only to my critique of D1, which is not now at issue. And in that case, she cannot claim to have shown that the D-thesis "still withstands Grünbaum's assaults" with respect to my charge of non-sequitur against D2. But if she interprets "in order to save H" in the sense of D2 as meaning the same as "with the assurance that there exists an A'_{nt} that permits H to explain O'", then I claim that she is fallaciously deducing D2 from the assumed truth of D1. I quite agree that if D1 is assumed to be true *and* IF THERE EXISTS AN A'_{nt} such that H does explain O' via A'_{nt}, then D1 would guarantee the feasibility of upholding A'_{nt} or H (but not necessarily both) as true. But this fact does not suffice at all to establish that there *exists* such an A'_{nt}! Thus, my charge of non-sequitur against the purported deduction of D2 from D1 does *not* rest on the complaint that the Duhemian cannot assure *a priori* being able to marshal supporting evidence *for* A'_{nt}.

Thus, if Hesse's criticisms were intended, in part, to invalidate my charge of non-sequitur against D2, then I cannot see that they have been successful. And since Philip Quinn has shown [42] that D2, in turn, does *not* entail D1, I maintain that D1 and D2 are logically independent of one another.

This concludes the statement of my reasons for not retracting either of the first two of my three erstwhile claims, which were restated early in this section. Hence we are now ready to consider whether there are any known counterexamples to D1.

Thus, our question is: Is there any component hypothesis H of some theory T or other such that H can be sufficiently isolated so as to lend itself to separate observational falsification? It is understood here that H *itself* does not have any observational consequences. The example which I had adduced to answer this question affirmatively in earlier publications was drawn from physical geometry. I now wish to re-examine my claim that the example in question is a counterexample to D1. In order to do so, let us consider an arbitrary surface S. And suppose that the Duhemian wants to uphold the following component hypothesis H about S: *if* lengths $ds = \sqrt{(g_{ik}dx^i dx^k)}$ are assigned to space intervals by means of *rigid* unit rods, then Euclidean geometry is the metric geometry which prevails physically on the given surface S. Before investigating whether and how this hypothesis H might be falsified, we must be mindful of an important assumption which is ingredient in the antecedent of its if-clause.

That antecedent tells us that the numbers furnished by *rigid* unit rods are to be centrally semantically relevant to the interval measures ds of the theory. Thus, it is a necessary condition for the *consistent* use of rigid rods in assigning lengths to space intervals that any collection of two or more initially coinciding unit solid rods of whatever chemical constitution can thereafter be used *interchangeably* everywhere in the region S, *unless* they are subjected to independently designatable perturbing influences. Thus, the assumption is made here that there is a concordance in the coincidence behavior of solid rods such that no inconsistency would result from the subsequent interchangeable use of initially coinciding unit rods which remain *unperturbed* or 'rigid' in the specified sense. In short, there is concordance among rigid rods such that all rigid rods *alike* yield the same metric ds and thereby the same geometry. Einstein stated the relevant empirical law of concordance as follows:

All practical geometry is based upon a principle which is accessible to experience, and which we will now try to realize. We will call that which is enclosed between two boundaries, marked upon a practically-rigid body, a tract. We imagine two practically-rigid bodies, each with a tract marked out on it. These two tracts are said to be 'equal to one another' if the boundaries of the one tract can be brought to coincide permanently with the boundaries of the other. We now assume that:

If two tracts are found to be equal once and anywhere, they are equal always and everywhere.

Not only the practical geometry of Euclid, but also its nearest generalisation, the practical geometry of Riemann and therewith the general theory of relativity, rest upon this assumption.[43]

I shall refer to the empirical assumption just formulated by Einstein as 'Riemann's concordance assumption', or, briefly, as 'R'. Note that the assumption R predicates the concordance among rods of different chemical constitution on their being rigid or free from perturbing influences. Perturbing influences are exemplified by inhomogeneities of temperature and by the presence of electric, magnetic and gravitational fields. Thus, two unit rods of different chemical constitution which initially coincide when at the same standard temperature will generally experience a destruction of their initial coincidence if brought to a place at

which the temperature is different, and the amount of their thermal elon-
gation will depend on their chemical constitution. By the same token, a
wooden rod will shrink or sag more in a gravitational field than a steel
one. Since such perturbing forces produce effects of different magnitude
on different kinds of solid rods, Reichenbach has called them 'differential forces'. The perturbing influences which qualify as differential are
linked to designatable sources, and their presence is certifiable in ways
other than the fact that they issue in the destruction of the initial coincidence of chemically different rods. But the set of perturbing influences is
open-ended, since physics cannot be presumed to have discovered *all* such
influences in nature by now. This open-endedness of the class of perturbing forces must be borne in mind if one asserts that the measuring rods in
a certain region of space are free from perturbing influences or 'rigid'.

It is clear that if our surface S is indeed free from perturbing influences,
then it follows from the assumption R stated by Einstein that *any* two
rods of different chemical constitution which initially coincide in S will
coincide everywhere else in S, independently of their respective paths of
transport. This result has an important bearing on whether it might be
possible to falsify the putative Duhemian hypothesis H that the geometry
G of our surface S is Euclidean.

For suppose now that S does satisfy the somewhat idealized condition
of actually being macroscopically free from perturbing influences. Let us
call the auxiliary assumption that S is thus free from perturbing influences 'A'. Then the conjunction H·A entails that measurements carried
out on S should yield the findings required by Euclidean geometry.
Among other things, this conjunction entails, for example, that the ratio
of the periphery of a circle to its diameter should be π on S. But suppose
that the surface S is actually a sphere rather than a Euclidean plane and
that the measurements carried out on S with presumedly rigid rods yield
various values significantly *less* than π for this ratio, depending on the
size of the circle. How then is the Duhemian going to uphold his hypothesis H that if lengths ds on S are measured with rigid rods, the geometry
G of S will be Euclidean so that all circles on S will exhibit the ratio π independently of their size?

Clearly, he will endeavor to do so by denying the initial auxiliary hypothesis A, and asserting instead the following: the rods in S are not rigid
after all but are being subjected to perturbing influences constituted by

differential forces. Obviously, just as in the case of Dicke's refutation of Einstein's theory, the falsification of H itself requires that the truth of A be established, so that Duhem would not be able to deny A. Suppose that we seek to establish A on the strength of the following further experimental findings: However different in chemical appearance, all rods which are found to coincide initially in S preserve their coincidences everywhere in S to within experimental accuracy independently of their paths of transport. How well does this finding establish the truth of A, i.e., that S is *free* from all perturbing influences which would act differentially on rods of diverse chemical constitution?

Let us use the letter 'C' to denote the statement that whatever their chemical constitution, any and all rods invariably preserve their initial coincidences under transport in S. Then we can say that Riemann's concordance assumption R states the following: If A is true of *any* region S, then C is true of that region S. Clearly, therefore, C follows from the conjunction A·R, but *not* from A alone. Hence if the observed preservation of coincidence or concordance in S establishes anything here, what it does establish at best is the conjunction A·R rather than A alone, since I would *not* wish to argue against the Duhemian holist that any finding C which serves inductively to establish a conjunction A·R must be held to establish likewise each of the conjuncts separately.[44] This fact does not, however, give the Duhemian a basis for an objection to my attempt to establish A. To see this, recall that R must be assumed by the Duhemian if his claim that S is Euclidean is to have the physical significance which is asserted by his hypothesis H. Since the Duhemian wants to uphold H, he could not contest R but only A. Hence, in challenging my attempt to establish A, the Duhemian will *not* be able to object that the observed concordance of the rods in S could establish only the conjunction A·R rather than establishing A *itself*. The issue is therefore one of establishing A. And for the reasons given in Einstein's statement of R above, R is assumed both by the Duhemian, who claims that our surface S is a Euclidean plane, and by the anti-Duhemian, who maintains that S is a sphere.

The Duhemian can argue that the observed concordance cannot conclusively establish A for the following two reasons: (1) He can question whether C can establish A, even if the *apparent* concordance is taken to establish the *universal* statement C indubitably, and (2) he can challenge the initial inference from the seeming concordance to the physical truth of C. Let me first articulate each of these two Duhemian objections in turn. Thereafter, I shall comment critically on them.

Objection 1

The Duhemian grants that the mere presence of a *single* perturbing influence of significant magnitude would have produced differential effects on chemically different rods. Thus, he admits that the presence of a *single* such influence would be incompatible with C. But he goes on to point out that C *is* compatible with the joint presence of several perturbing influences of possibly unknown physical origin which just *happen* to act as follows: different effects produced on the rods by each one of these perturbing influences are of just the right magnitude to combine into the same *total* deformation of each of the rods. And all being deformed *alike* by the collective action of differential forces *emanating from physical sources*, the rods behave in accord with C. In short, the Duhemian claims that such a *superposition* of differential effects is compatible with C and that it therefore cannot be ruled out by C. Hence he concludes that C cannot be held to establish the truth of A beyond question. Since this Duhemian objection is based on the possibility of conjecturing the specified kind of superposition, I shall refer to it as "the superposition objection".[45]

Objection 2

This second Duhemian objection calls attention to the fact that in inferring C from the *apparent* concordance, we have made the following several inductive inferences: (i) We have taken differences in chemical *appearance* to betoken actual differences in chemical constitution, (ii) we have taken apparent coincidence to betoken actual coincidence, at least to within a certain accuracy, and (iii) we have inductively inferred the *invariable* preservation of coincidence by *all* kinds of rods everywhere in S from only a limited number of trials. The first two of these three inferences invoke collateral hypotheses. For they rest on presumed laws of psychophysics and neurophysiology to the following effect: certain appearances, which are contents of the awareness of the human organism, are lawfully correlated with the objective physical presence of the respective states of affairs which they are held to betoken. Hence let us say that the first two of these three inferences infer a conclusion of the form 'Physical item P *has* the property Q' from the premise 'P *appears* to be Q', or that they infer *being* Q from *appearing* to be Q.

These objections prompt two corresponding sets of comments.

1. It is certainly true that A is *not* entailed by the conjunction R·C. For R is the conditional statement 'If A, then C'. Since A does not follow deductively from C·R, the premise C·R *cannot* be held to rule out the rival superposition conjecture DEDUCTIVELY. Thus, given R, C *deductively allows* the superposition conjecture as a *conceivable* alternative to A. Nevertheless, we shall now see that in the context of R, the inductive confirmation of A by C is so enormously high that A can be reregarded as *well-nigh established* by R·C. Since my impending inductive argument will make use of Bayes' theorem, it will be unconvincing to those who question the applicability of that theorem to the probability of a *hypothesis*. For example, Imre Lakatos registered an objection in this vein.[46]

I shall adopt the non-standard probability notation employed by H. Reichenbach[47] and W. C. Salmon[48] to let 'P(L, M)' mean 'the probability *from* L to M' or 'the probability of M, given L'. In the usual notations, the two arguments L and M are reversed.

Recalling that R says 'If A, then C' (for any region S), we wish to evaluate the probability $P(R \cdot C, A)$ of A, given that R·C. To do so, we use a special form of Bayes' theorem[49] and write

$$P(R \cdot C, A) = \frac{P(R, A)}{P(R, C)} \times P(R \cdot A, C).$$

C follows *deductively* from R·A. Hence C is guaranteed by R·A, so that

$$P(R \cdot A, C) = 1.$$

Thus, our formula shows that our desired posterior probability $P(R \cdot C, A)$ will be very high, if the ratio $P(R, A)/P(R, C)$ is itself close to 1, as indeed it will now turn out to be. Needless to say, the *ratio* of these probabilities can be close to 1, even though they are individually quite small.

We can compare the two probabilities in this ratio with respect to magnitude without knowing their individual magnitudes. To effect this comparison, we first consider the conditional probability of C, if we are given that R is true while A is *false*. We can now see that this probability $P(R \cdot \sim A, C)$ is very low, whereas we recall that $P(R \cdot A, C) = 1$. For suppose that A is false, i.e., that S *is* subject to differential forces. Then C can hold only in the *very rare* event that there *happens* to be just the right superposition. Incidentally in the case of our surface S, there is no evidence at all for the existence of the *physical sources* to which the Duhemian needs to attribute his superposed differential effects. Hence we are confronted here with a situation in which the Duhemian wants C to hold under conditions in which superposed differential forces are actually operative, although there is no independent evidence whatever for them.[50]

Since $P(R \cdot \sim A, C)$ is very low, we can say that the probability $P(R,C)$ of concordance among the chemically different rods is *only slightly greater, if at all*, than the probability $P(R, A)$ of the occurrence of a region S which is free from differential forces. And this comparison among these probabilities holds here, notwithstanding the fact that there may be as yet undiscovered kinds of differential forces in nature, as I stressed above. It follows that $P(R \cdot C, A)$ is close to 1. And since

$$P(R \cdot C, \sim A) = 1 - P(R \cdot C, A),$$

the conditional probability of $\sim A$ is close to zero.[51]

Thus, the Duhemian's first objection does not detract, and indeed may not be intended to detract from the fact that A is *well-nigh* established by $R \cdot C$.

2. The second objection noted that being Q is inferred inductively from appearing to be Q in the case of chemical difference and of coincidence. Furthermore, it called attention to the inductive inference of the *universal* statement C from only a limited number of test cases. What is the force of these remarks?

Let me point out that the claim of mere chemical *difference* does *not* require the definite identification of the particular chemical constitution of each of the rods as, say, iron or wood.[52] Now suppose that the apparent chemically relevant differences among most of the tested rods are striking. And note that they can be further enriched by bringing additional rods to S. Would it then be helpful to the Duhemian to postulate that all of these prima facie differences in chemical appearance mask an underlying chemical identity? If he wishes to avoid resorting to the fantastic superposition conjecture, the Duhemian might be tempted to postulate that the great differences in chemical appearance belie a true crypto-identity of chemical constitution. For by postulating his crypto-identity, he could hope to render the observed concordance among the rods un-problematic for $\sim A$ *without* the superposition conjecture. Crypto-identity might make superposition dispensable because the Duhemian might then be able to assert that S is subject to only *one* kind of differ-ential force.[53] But there is no reason to think that the crypto-identity con-jecture is any more probable than the incredible superposition conjecture. And either conjecture may try a working scientist's patience with philosophers.[54]

As for the inductive inferences of physical coincidence from apparent coincidence, the degree of accuracy to which it can be certified inductively is, of course, limited and macroscopic. But what would it avail the Duhe-mian to assume in lieu of superposition the imperceptibly slight differen-tial deformations which are compatible with the limited accuracy of the observations of coincidence? Surely the imprecision involved here does not provide adequate scope to modify A sufficiently to be able to reconcile the substantial observed deviations of the circular ratios from π with the Euclidean H that he needs to save.

What of the inductive inference of the universal statement C from only a limited number of test cases? Note that C need not be established in its full universality in order to create inductive difficulties for the superposi-tion conjecture and to provide strong support for A. For as long as all the chemically different rods actually tested satisfy C, the Duhemian must make his superposition conjecture or crypto-identity assumption credible with respect to *these* rods. Otherwise, he could not reconcile the measure-ments of circular ratios furnished by them on S with his Euclidean H.

The Duhemian might point out that I have inferred A inductively in a two-step inductive inference, which first infers at least instances of the physical claim C from apparent coincidences of prima facie different kinds of rods, and then proceeds to infer A inductively from these in-stances of C, coupled with R. And he might object that the relation of inductive support is not transitive. It is indeed the case that the relation of inductive support fails to be transitive.[55] But my inductive inference of A does not require an intermediate step via C, for cases of *apparent* coin-cidence, no less than cases of *physical* coincidence, are correlated with $A \cdot R$, and R is no more in question here than before. Thus, given R, the inference can proceed to A directly via the assumption that most cases of apparent coincidence of prima facie chemically different rods are correlates of cases of A.[56] And my inductive inference of A is thus predicated on this correlation.

In conclusion, we can now try to answer two questions: (i) To what extent has our verification of A been separate from the assumption of H?, and (ii) to what extent has H *itself* been falsified? When I speak here of a hypothesis as having been 'falsified' in asking the latter question, I mean that the *presumption* of its falsity has been established, *not* that its falsity has been established with certainty or irrevocably. This construal of 'falsify', is of course, implicit in the presupposition of my question that there are different degrees of falsification or differences in the extent to which a hypothesis can be falsified.

(i) Our verification of A did proceed in the context of the assumption of R. And while we saw that R is ingredient in the Euclidean H of our example, as specified, R is similarly ingredient in the rival hypothesis that our surface S is not a Euclidean plane but a spherical surface. Thus, our verification of A was separate from the assumption of the *distinctive* physical content of the particular Duhemian H.

(ii) Duhem attributed the inconclusiveness of the falsification of a component hypothesis H to the legitimacy of denying instead any of the collateral hypotheses A which enter into any test of H. Our analysis has shown that *the denial of A is legitimate precisely to the extent that its* VERIFICATION *suffers from inductive uncertainty*. Moreover, in each of our examples of attempted falsification, the inconclusiveness is attributable *entirely* to the inductive uncertainty besetting the following two *verifications*: the verification of A, and the verification of the so-called observation statement which entails the *falsity* of the conjunction H·A. In short, the inconclusiveness of the falsification of a component H derives wholly from the inconclusiveness of verification. And the falsification of H itself is inconclusive or revocable in the sense that the falsity of H is *not* a *deductive* consequence of premises *all* of which can be known to be true with certainty.

Hence if the falsification of H denied by Duhem's D1 is construed as *irrevocable*, then I agree with Mary Hesse[57] that my geometrical example does not qualify as a counterexample to D1. But I continue to claim that it does so qualify, if one requires only the *very strong presumption* of falsity. And to the extent that my geometrical example does falsify the A'_{nt} which D2 invokes in conjunction with the H of that example, the example also refutes D2's claim that A'_{nt} can justifiably be held to be true.

Subject to an important *caveat* to be issued presently, I maintain, therefore, that there are cases in which we can establish a strong presumption of the falsity of a component hypothesis, although we cannot falsify H in these cases beyond any and all possibility of subsequent rehabilitation. Thus, I emphatically do allow for the following possibility, though not likelihood, in the case of an H which has been falsified in my merely presumptive sense: A daring and innovative scientist who continues to entertain H, albeit as part of a new research program which he envisions as capable of vindicating H, *may* succeed in incorporating H in a theory so subsequently fruitful and well-confirmed as to retroactively alter our assessment of the initial falsification of H. And my *caveat* is that my conception of the falsification of H as establishing the strong presumption of falsity is certainly *not* tantamount to a stultifying injunction to any and all imaginative scientists to cease entertaining H forthwith, whenever such falsification obtains at a given time! Nor is this conception of falsification to be construed as being committed to the historically false assertion that no inductively *un*warranted and daring continued espousal of H has ever been crowned with success in the form of subsequent vindication.

APPENDIX

One might ask: Are there no *other* cases at all in which H can be justifiably rejected as *irrevocably* falsified by observation? On the basis of a schema given by Philip Quinn, it may *seem* that *if* we can ignore such uncertainty as attaches to our observations when testing H, then this question can be answered affirmatively for reasons to be stated presently. When asserting the prima facie existence of this kind of irrevocable falsification as a fact of logic, Quinn recognized that its relevance to actual concrete cases in the pursuit of empirical science is at best very limited. Specifically, Quinn invites consideration of the following kind of logical situation. Let the observation statement O_3 be pairwise *incompatible* with each of the two observation statements O_1 and O_2, so that $O_3 \rightarrow \sim O_1$, and $O_3 \rightarrow \sim O_2$. And suppose further that

$$[(H \cdot A) \rightarrow O_1] \cdot [(H \cdot \sim A) \rightarrow O_2].$$

Assume also that observations made to test H are taken to establish the truth of O_3. Then we can deduce via *modus tollens* that

$$\sim (H \cdot A) \cdot \sim (H \cdot \sim A).$$

Using the law of excluded middle to assert $A \vee \sim A$, we can write

$$(A \vee \sim A) \cdot [\sim (H \cdot A) \cdot \sim (H \cdot \sim A)].$$

The application of the distributive law and the omission of one of the two conjuncts in the brackets from each of the resulting disjuncts yields

$$[A \cdot \sim (H \cdot A)] \vee [\sim A \cdot \sim (H \cdot \sim A)].$$

Using the principle of the complex constructive dilemma, we can deduce $\sim H \vee \sim H$ and hence $\sim H$.

Although the assumed truth of the observation statement O_3 allows us to deduce the falsity of H in this kind of case, its relevance to empirical science is at best very limited for the following reasons.

In most, if not all, cases of actual science, the conjunction $H \cdot \sim A$ will not be rich enough to yield an observational consequence O_2, if the conjunction $H \cdot A$ does yield an observational consequence O_1. The reason is that the mere denial of A is not likely to be sufficiently specific in content. What, for example, is the observational import of conjoining the *denial* of Darwin's theory of evolution to a hypothesis H concerning the age of the earth? Thus this feature at least severely limits the relevance of this second group of logically possible cases to actual science.

Furthermore, the certainty of the conclusion that H is false in this kind of case rests on certainty that the observational statement O_3 is true. Such certainty is open to question. But *if* we can ignore our doubts on this score, then it would seem that H can be held to have been falsified beyond any possibility of subsequent rehabilitation. Yet, alas, John Winnie has pointed out that despite appearances to the contrary, the H in Quinn's schema does not qualify as a component hypothesis after all.

For Winnie has noted that Quinn's basic premise $[(H \cdot A) \rightarrow O_1] \cdot [(H \cdot \sim A) \rightarrow O_2]$ permits the deduction of

$$H \rightarrow (O_1 \vee O_2).$$

Thus, if $O_1 \vee O_2$ can be held to qualify as an observationally testable statement, then we find that H entails it *without* the aid of the auxiliary A or of $\sim A$. But in that case, Winnie notes that H does not qualify as a constituent of a wider theory in the sense specified in our statement of D1 above.[58]

NOTES

* From Chapter 17 of *Philosophical Problems of Space and Time*, second, enlarged ed., pp. 585–629. Published by D. Reidel, Boston and Dordrecht, 1974 as Vol. XII in the *Boston Studies in the Philosophy of Science* (ed. by R. S. Cohen and M. Wartofsky). In this form the essay also appeared in M. Mandelbaum (ed.), *Observation and Theory in Science*, Johns Hopkins Press, Baltimore, 1971. Reprinted by permission.

[1] Pierre Duhem, *The Aim and Structure of Physical Theory*, Princeton University Press, Princeton, 1954, p. 187.

[2] *Ibid.*

[3] *Ibid.*, p. 186.

[4] Laurens Laudan, 'On the Impossibility of Crucial Falsifying Experiments: Grünbaum on "The Duhemian Argument"', *Philosophy of Science* 32 (1965), 295–99.

[5] For a discussion of the epistemic requirement that explanatory premises must *not* be known to be false, see Ernest Nagel, *The Structure of Science*, Harcourt, Brace and World, New York, 1961, pp. 42–43.

[6] A. Grünbaum, 'The Falsifiability of a Component of a Theoretical System', in *Mind, Matter, and Method: Essays in Philosophy and Science in Honor of Herbert Feigl*, P. K. Feyerabend and G. Maxwell, eds., University of Minnesota Press, Minneapolis, 1966, pp. 276–80.

[7] *Ibid.*, pp. 280–81.

[8] *Ibid.*, pp. 283–95; and A. Grünbaum, *Geometry and Chronometry in Philosophical Perspective*, University of Minnesota Press, Minneapolis, 1968, Chap. III, pp. 341–51. In Chap. III, Section 9.2, pp. 351–69 of the latter book, I present a counterexample to H. Putnam's *particular* geometrical version of D2. For a brief summary of Putnam's version, see Note 55.

[9] Grünbaum, 'The Falsifiability of a Component of a Theoretical System', pp. 277–78.

[10] W. V. O. Quine, *From a Logical Point of View* (revised ed.), Harvard University Press, Cambridge, Mass., 1961, pp. 43 and 41n.

[11] Grünbaum, 'The Falsifiability of a Component of a Theoretical System', pp. 279–80.

[12] *Ibid.*, p. 279.

[13] *Ibid.*, p. 278.

[14] Mary Hesse, *The British Journal for the Philosophy of Science* 18 (1968), 334.

[15] Peter Achinstein, *Concepts of Science*, Johns Hopkins Press, Baltimore, 1968, pp. 3ff.

[16] *Ibid.*, p. 6.

[17] *Ibid.*, pp. 7–8.

[18] *Ibid.*, pp. 8–9.

[19] *Ibid.*, p. 9.

[20] *Ibid.*, p. 35.

[21] *Ibid.*, p. 101.

[22] *Ibid.*, p. 2.

[23] See pp. 143–44 of A. Grünbaum, *Philosophical Problems of Space and Time*, second, enlarged ed., D. Reidel Publ. Co., Boston and Dordrecht, 1974; and A. Grünbaum, *Geometry and Chronometry...*, pp. 314–17.

[24] See, for example, Tolman, *Relativity, Thermodynamics and Cosmology*, pp. 42–45.

[25] See P. G. Bergmann, *Introduction to the Theory of Relativity*, Prentice-Hall, New York, 1946, pp. 86–88.

[26] C. Giannoni, 'Quine, Grünbaum, and The Duhemian Thesis', *Nous* **1** (1967), 288.

[27] *Ibid.*, pp. 286–87.

[28] *Ibid.*, p. 288. The example which Giannoni then goes on to cite from Duhem is one in which a very *weak* electric current was running through a battery and where, therefore, one or more indicators may fail to register its presence. In that case, the current will still be said to flow if one or another indicator yields a positive response. This particular case may well be a borderline one as between semantic and nonsemantic relevance.

[29] Grünbaum, *Geometry and Chronometry...*, pp. 314–17.

[30] See pp. 143–44 of Grünbaum, *Philosophical Problems of Space and Time*.

[31] E.g., in Bergmann, *Introduction to the Theory of Relativity*, pp. 23–26.

[32] Since the aether-theoretically expected time difference in the second order terms is only of the order of 10^{-15} second, allowance had to be made in practice for the absence of a corresponding accuracy in the measurement of the equality of the two arms. This is made feasible by the fact that, on the aether theory, the effect of any discrepancy in the lengths of the two arms should *vary*, on account of the earth's motion, as the apparatus is rotated. For details, see Bergmann, *Introduction to the Theory of Relativity*, pp. 24–26, and J. Aharoni, *The Special Theory of Relativity*, Oxford University Press, Oxford, 1959, pp. 270–73. Indeed, slightly unequal arms are needed to produce neat interference fringes.

[33] Grünbaum, *Geometry and Chronometry in Philosophical Perspective*, pp. 316–17.

[34] *Ibid.*, pp. 314–15.

[35] See, for example, C. Møller, *Theory of Relativity*, Oxford University Press, Oxford, 1952, Ch. VIII, Section 84.

For a rebuttal to John Earman's objections to ascribing a spatial geometry to the rotating disk, see the detailed account of the status of metrics in A. Grünbaum, 'Space, Time and Falsifiability', Part I, *Philosophy of Science* **37** (1970), 562–65 (Chapter 16 of Grünbaum, *Philosophical Problems of Space and Time*, pp. 542–545).

[36] This formulation is given in E. T. Whittaker, *From Euclid to Eddington*, Cambridge University Press, London, 1949, p. 63.

[37] Grünbaum, *Geometry and Chronometry...*, *passim*.

[38] Philip Quinn, 'The Status of the D-Thesis', Section III, *Philosophy of Science* **36**, (1969), No. 4.

[39] J. W. Swanson, 'On the D-Thesis', *Philosophy of Science* **34** (1967), 59–68.

[40] Grünbaum, 'The Falsifiability of a Component of a Theoretical System', p. 278.

[41] Hesse, *The British Journal for the Philosophy of Science* **18**, 334–35.

[42] University of Pittsburgh Doctoral Dissertation (1970).

[43] A. Einstein, 'Geometry and Experience', in *Readings in the Philosophy of Science*, in H. Feigl and M. Brodbeck (eds.), Appleton-Century-Crofts, New York, 1953, p. 192.

[44] By the same token, if *alternatively* the initial coincidence of the rods had been observed to *cease conspicuously* in the course of transport, the latter observations could *not* be held to falsify A in isolation from R. This inconclusiveness with respect to the falsity of A itself would obtain in the face of the *apparent* discordances, even if the latter are taken as *indubitably* falsifying the claim C of universal physical concordance. For C is entailed by A·R rather than by A alone.

[45] This kind of objection is raised by L. Sklar, 'The Falsifiability of Geometric

Theories', *The Journal of Philosophy* **64** (1967), 247–53. Swanson, 'On the D-Thesis', presents a model universe for which he conjectures superposition of differential effects *in name only*. For, as Philip Quinn has noted (in 'The Status of the D-Thesis'), Swanson effectively adopts the convention that even in the *absence* of perturbing influences, all initially coinciding rods will be held to be *non*-rigid by being assigned lengths that *vary alike* with their positions and/or orientations. Thus, Swanson renders the superposition conjecture physically empty by ignoring the crucial requirement that the alleged differential effects are held to emanate individually from physical sources. Swanson's model universe will be discussed further in Note 54.

46 See Imre Lakatos and Alan Musgrave (eds.), *Criticism and Growth of Knowledge*, Cambridge University Press, New York, 1970, p. 187. But if he countenances entertaining the superposition conjecture $\sim$A on the grounds that A cannot be said to be highly confirmed, how will this $\sim$A escape being a *metaphysical* proposition?

47 H. Reichenbach, *The Theory of Probability*, University of California Press, Berkeley, 1949.

48 W. C. Salmon, *The Foundations of Scientific Inference*, University of Pittsburgh Press, Pittsburgh, 1967, p. 58.

49 For this form of the theorem, see Reichenbach, *Theory of Probability*, p. 91, Equation (6).

50 This lack of evidence is here construed as obtaining at the given time. An alternative construal in which there avowedly would *never* be any such evidence at all is tantamount to an admission that A is true after all, or that $\sim$A is physically empty. See Philip Quinn ['The Status of the D-Thesis',] for a discussion of the import of this point for L. Sklar's *particular* superposition objection.

51 H. Putnam claims to have shown in 'An Examination of Grünbaum's Philosophy of Geometry', in *Philosophy of Science, The Delaware Seminar*, Vol. 2, B. Baumrin (ed.), Interscience Publishers, New York, 1963, pp. 247–55, that a superposition of so-called gravitational, electromagnetic, and interactional differential forces can always be invoked here to assert $\sim$A. In particular, he is committed to the following: Suppose A is assumed for a region S (say, on the basis of C) and that a particular metric tensor g_{ik} then results from measurements conducted in S with rods that are presumed to be rigid in virtue of A. Then, says Putnam, we can *always* assert $\sim$A of S by reference to the specified three kinds of forces. And we can do so in such a way that rods corrected for the effects of these forces would yield any desired different metric tensor g'_{ik}. The latter tensor would enable the Duhemian to assert his chosen geometric H of S.

But I have demonstrated elsewhere in detail [Grünbaum, *Geometry and Chronometry in Philosophical Perspective*, Ch. III, Section 9] that the so-called gravitational, electromagnetic and interactional forces which Putnam wishes to invoke in order to assert $\sim$A do *not* qualify *singly* as differential. Hence Putnam's proposed scheme cannot help the Duhemian.

52 For some pertinent details, see Grünbaum, 'The Falsifiability of a Component of a Theoretical System', pp. 286–87.

53 It is not obvious that the Duhemian could make this assertion in this context with impunity: even in the presence of only one kind of differential force, there may be rods of one and the same chemical constitution which exhibit the hysteresis behavior of ferromagnetic materials in the sense that their coincidences at a given place will depend on their paths of transport. Cf. Grünbaum, *Geometry and Chronometry...*, p. 359.

54 Swanson, 'On the D-Thesis', pp. 64–65, maintains that when the Duhemian contests A in my geometric example, he would *not* invoke a crypto *identity* conjecture, as I have

him do, but rather a conjecture of a crypto-*difference* among prima facie chemically *identical* rods. And that crypto-difference might arise from a kind of undiscovered isomerism. He says:

It is not the assertion of the chemical *difference* of two rods that the D-theorist would claim to be theory-laden. Grünbaum is certainly right in showing that the D-theorist could not argue for *that*. Rather, it is the assertion of their *identity* that would be theory-laden. For suppose that according to a given chemical theory T_1, two rods a and b were found to be chemically identical to one another.... We can now construe T_1 as some sort of pre-'isomeric' chemistry... and then go on to imagine a post-'isomeric' T_2 such that upon the fulfillment of certain tests... it is found that $a \neq b$.... But if identity of a and b is thus theory-laden, it does little good to argue that non-identical rods can be clearly discriminated.

Swanson then considers a model universe of three rods, two of which are prima facie chemically identical though crypto-different, while being pairwise objectively and perceptibly different from the third. And he then introduces two (ghostlike) perturbing influences whose *superposition* imparts the same total deformation to each of the three rods. Thus, in the context of his superposition model, Swanson attaches significance to the fact that the first two of his three rods, a_1 and a_2, are crypto-different.

But nothing in his argument from superposition depends on the hypothesis that a_1 and a_2 are crypto-different instead of identical, nor does his superposition model gain anything from that added hypothesis. Surely the differential character of his two perturbing influences would not be gainsaid by the mere supposition that a_1 and a_2 are chemically identical and that either of the two perturbing influences therefore affects them alike. If there is superposition issuing in the same total deformation, as contrasted with the absence of differential perturbations, then chemically identical rods can exhibit the same total deformation no less than conspicuously different or crypto-different ones! Far from lending added plausibility to the superposition conjecture, Swanson's introduction of the hypothesis of crypto-difference merely compounds the inductive felonies of the two.

In fact, Swanson overlooked the non-sequitur which I did commit apropos of crypto-identity in my 1966 paper. As is clear from the present essay, the Duhemian could couple his denial of A with the superposition conjecture *rather than* with the hypothesis of crypto-identity. It was therefore incorrect on my part to assert in 1966 that, when denying A, the Duhemian 'must' assert crypto-identity (see 'The Falsifiability of a Component of a Theoretical System', p. 287, *item (3)*).

[55] For a lucid explanation of the *non*-transitivity (as distinct from intransitivity) of the relation of inductive support and of its ramifications, see W. C. Salmon, 'Consistency, Transitivity, and Inductive Support', *Ratio* 7 (1965), 164–68.

[56] This claim of direct inductive inferability is made here only with respect to the specifics of this example. And it is certainly not intended to deny the existence of other situations, such as those discussed by R. C. Jeffrey, *The Logic of Decision*, McGraw-Hill, New York, 1965, pp. 155–56, in which a perceptual experience can reasonably prompt us to entertain a variety of relevant propositions.

[57] Hesse, *The British Journal for the Philosophy of Science* **18**, 334.

[58] I am indebted to Philip Quinn and Laurens Laudan for reading a draft of this paper and suggesting some improvements in it. I also wish to thank Wesley Salmon and Allen Janis for some helpful comments and references.

PAUL K. FEYERABEND

THE RATIONALITY OF SCIENCE

(From 'Against Method')

1. DISCOVERY AND JUSTIFICATION; OBSERVATION AND THEORY*

Let us now use the material of the [earlier sections of 'Against Method'] to throw light on the following features of contemporary empiricism: first, the distinction between a context of discovery and a context of justification; second, the distinction between observational terms and theoretical terms; third, the problem of incommensurability.

One of the objections which may be raised against the preceding discussion is that it has confounded two contexts which are essentially separate, viz. a context of discovery and a context of justification. *Discovery* may be irrational and need not follow any recognized method. *Justification*, on the other hand, or, to use the Holy Word of a different school, *criticism*, starts only after the discoveries have been made and proceeds in an orderly way. Now, if the example given here and the examples I have used in earlier papers show anything, then they show that the distinction refers to a situation that does not arise in practice at all. And, if it does arise, it reflects a temporary stasis of the process of research. Therefore, it should be eliminated as quickly as possible.

Research at its best is an *interaction* between new theories which are stated in an explicit manner and older views which have crept into the observation language. It is not a one-sided *action* of the one upon the other. Reasoning within the context of justification, however, presupposes that one side of this pair, viz. observation, has frozen, and that the principles which constitute the observation concepts are preferred to the principles of a newly invented point of view. The former feature indicates that the discussion of principles is not carried out as vigorously as is desirable; the latter feature reveals that this lack of vigor may be due to some unreasonable and perhaps not even explicit preference. But is it wise to be dominated by an inarticulate preference of this kind? Is it wise to make it the raison d'être of a distinction that separates two entirely different modes of research? Or should we not rather demand that our methodology treat explicit and implicit assertions, doubtful and intuitively evident theories, known and unconsciously held principles, in exactly the same way, and that it provide means for the discovery and the criticism of the latter? Abandoning the distinction between a context of discovery and a context of justification is the first step toward satisfying this demand.

Another distinction which is clearly related to the distinction between discovery and justification is the distinction between *observational terms and theoretical terms*. It is now generally admitted that the distinction is not as sharp as it was thought to be only a few decades ago. It is also admitted, in complete agreement with Neurath's original views, that *both* theories *and* observation statements are open to criticism. Yet the distinction is still held to be a useful one and is defended by almost all philosophers of science. But what is its point? Nobody will deny that the sentences of science can be classified into long sentences and short sentences, or that its statements can be classified into those which are intuitively obvious and others which are not. But nobody will put particular weight on these distinctions, or will even mention them, *for they do not now play any role in the business of science*. (This was not always so. Intuitive plausibility, for example, was once thought to be a most important guide to the truth; but it disappeared from methodology the very moment intuition was replaced by experience.) Does experience play such a role in the business of science? Is it as essential to refer to experience as it was once thought essential to refer to intuition? Considering what has been said earlier, I think that these questions must be answered in the negative. True – much of our thinking *arises* from experience, but there are large portions which do not arise from experience at all but are firmly grounded on intuition, or on even deeper lying reactions. True – we often test our theories by experience, but we equally often *invert* the process; we *analyze* experience with the help of more recent views and we *change* it in accordance with these views (see the preceding discussion of Galileo's procedure). Again, it is true that we often rely on experience in a way that suggests that we have here a solid foundation of knowledge, but such reliance turns out to be just a psychological quirk, as is shown whenever the testimony of an eyewitness or of an expert crumbles under cross-examination. Moreover, we equally firmly rely on general principles so that even our most solid *perceptions* (and not only our *assumptions*) become indistinct and ambiguous when they clash with these principles. The symmetry between observation and theory that emerges from such remarks is perfectly reasonable. Experience, just as our theories, contains natural interpretations which are abstract and even metaphysical ideas. For example, it contains the idea of an observer-independent existence. It is incontestable that these abstractions, these speculative ideas, are connected with sensations and perceptions. But, first of all, this does not give them a privileged position, unless we want to assert that perception is an infallible authority. And, secondly, it is quite possible to altogether *eliminate* perception from all the essential activities of science.... All that remains is that some of our ideas are *accompanied* by strong and vivid psychological processes, 'sensations', while others are not. This, however, is just a peculiarity of human existence which is as much in need of examination as is anything else.

Now, if we want to be 'truly scientific' (dreaded words!), should we then not regard the theses "experience is the foundation of our knowledge" and "experience helps us to discover the properties of the external world" as (very general) hypotheses? And must these hypotheses not be examined just like any other hypothesis, *and perhaps even more vigorously*, as so much depends on their truth? Furthermore, will not such an examination be rendered impossible by a method that either justifies or criticizes "on the basis of experience"? These are some of the questions which arise in connection with the customary distinctions between observation and theory, discovery and justification. None of them is really new. They are known to philosophers of science, and are discussed by them at length. But the inference that the distinction between theory and observation has now ceased to be relevant either is not drawn or is explicitly rejected.[1] Let us take a step forward, and let us abandon this last remainder of dogmatism in science!

1. RATIONALITY AGAIN

Incommensurability, which I shall discuss next, is closely connected with the question of the rationality of science. Indeed, one of the most general objections, either against the *use* of incommensurable theories or even against the idea that *there are* such theories to be found in the history of science, is the fear that they would severely restrict the efficacy of traditional, nondialectical *argument*. Let us, therefore, look a little more closely at the critical *standards* which, according to some people, constitute the content of a 'rational' argument. More especially, let us look at the standards of the Popperian school with whose ratiomania we are here mainly concerned.

Critical rationalism is either a meaningful idea or a collection of slogans (such as 'truth'; 'professional integrity'; 'intellectual honesty') designed to intimidate yellow-bellied opponents (who has the fortitude, or even the insight, to declare that Truth might be unimportant, and perhaps even undesirable?).

In the former case it must be possible to produce rules, standards, restrictions which permit us to separate critical behavior (thinking, singing, writing of plays) from other types of behavior so that we can *discover* irrational actions and *correct* them with the help of concrete suggestions. It is not difficult to produce the standards of rationality defended by the Popperian school.

These standards are standards of *criticism*: rational discussion consists in the attempt to criticize, and not in the attempt to prove, or to make probable. Every step that protects a view from criticism, that makes it safe, or 'well founded', is a step away from rationality. Every step that makes it more vulnerable is welcome. In addition it is recommended that ideas which have been found wanting be abandoned, and it is forbidden to retain them in the face of strong and successful criticism unless one can present a suitable counterargument. Develop your ideas so that they can be criticized; attack them relentlessly; do not try to protect them, but exhibit their weak spots; and eliminate them as soon as such weak spots have become manifest – these are some of the rules put forth by our critical rationalists.

These rules become more definite and more detailed when we turn to the philosophy of science, and especially to the philosophy of the natural sciences.

Within the natural sciences criticism is connected with experiment and observation. The content of a theory consists in the sum total of those basic statements which contradict it; it is the class of its potential falsifiers. Increased content means increased vulnerability; hence theories of large content are to be preferred to theories of small content. Increase of content is welcome; decrease of content is to be avoided. A theory that contradicts an accepted basic statement must be given up. Ad hoc hypotheses are forbidden – and so on and so forth. A science, however, that accepts the rules of a critical empiricism of this kind will develop in the following manner.

We start with a *problem* such as the problem of the planets at the time of Plato. This problem is not merely the result of *curiosity*, it is a *theoretical result*, it is due to the fact that certain *expectations* have been disappointed: On the one hand it seemed to be clear that the stars must be divine; hence one expects them to behave in an orderly and lawful manner. On the other hand one cannot find any easily discernible regularity. The planets, to all intents and purposes, move in a quite chaotic fashion. How can this fact be reconciled with the expectation and with the principles that underlie the expectation? Does it show that the expectation is mistaken? Or have we failed in our analysis of the facts? This is the problem.

It is important to see that the elements of the problem are not simply *given*. The 'fact' of irregularity, for example, is not accessible without further ado. It cannot be discovered by just anyone who has healthy eyes and a good mind. It is only through a certain expectation that it becomes an object of our attention. Or, to be more accurate: this fact of irregularity *exists* because there is an expectation of regularity. After all, the term 'irregularity' makes sense only if we have a rule. In our case the rule (which is a more specific part of the expectation that has not yet been mentioned) asserts circular motion with constant angular velocity. The fixed stars agree with this rule and so does the sun if we trace its path relative to the fixed stars. The planets do not obey the rule, neither directly, with respect to the earth, nor indirectly, with respect to the fixed stars.

(In the case just discussed the rule is formulated explicitly, and it can be discussed. This need not be the case. Recognizing a color as red is made possible by deep-lying assumptions concerning the structure of our surroundings and recognition does not occur when these assumptions cease to be available.)

To sum up this part of the Popperian doctrine: Research starts with a problem. The problem is the result of a conflict between an expectation and an observation which in turn is constituted by the expectation. It is clear that this doctrine differs from the doctrine of inductivism where objective facts mysteriously enter a passive mind and leave their traces there. It was prepared by Kant, by Dingler, and, in a very different manner, by Hume.

Having formulated a problem one tries to *solve* it. Solving a problem means inventing a theory that is relevant, falsifiable (to a larger degree than any alternative solution), but not yet falsified. In the case mentioned above (planets at the time of Plato) the problem was to find circular motions of constant angular velocity for the purpose of saving the planetary phenomena. It was solved by Eudoxos.

Next comes the *criticism* of the theory that has been put forth in the attempt to solve the problem. Successful criticism removes the theory *once and for all* and creates a new problem, viz. to explain (a) why the theory has been successful so far; (b) why it failed. Trying to solve *this* problem we need a new theory that produces the successful consequences of the older theory, denies its mistakes, and makes additional predictions

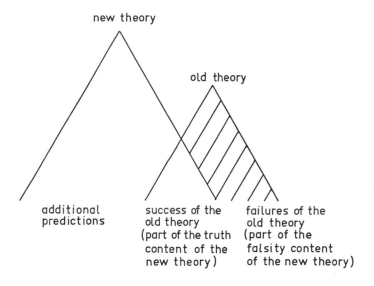

new theory

old theory

additional predictions

success of the old theory (part of the truth content of the new theory)

failures of the old theory (part of the falsity content of the new theory)

not made before. These are some of the *formal conditions* which *a suitable successor of a refuted theory* must satisfy. Adopting the conditions one proceeds, by conjectures and refutations, from less general theories to more general theories and expands the content of human knowledge. More and more facts are *discovered* (or constructed with the help of expectations) and are then *connected* in a reasonable manner. There is no guarantee that man will solve every problem and replace every theory that has been refuted with a successor satisfying the formal conditions.

The invention of theories depends on our talents and other fortuitous circumstances, such as a satisfactory sex life. But as long as these talents hold out the accompanying scheme is a correct account of the growth of a knowledge that satisfies the rules of critical rationalism.

Now, at this point we may raise two questions:

1. Is it *desirable* to live in accordance with the rules of a critical rationalism?

2. Is it *possible* to have both a science as we know it and these rules?

As far as I am concerned the first question is far more important than the second. True – science and other depressing and narrow-minded institutions play an important part in our culture and they occupy the center of interest of most philosophers. Thus the ideas of the Popperian school were obtained by generalizing solutions for methodological and epistemological problems. Critical rationalism arose from the attempt to solve Hume's problem and to understand the Einsteinian revolution, and it was then extended to politics, and even to the conduct of one's private life (Habermas and others therefore seem to be justified in calling Popper a positivist). Such a procedure may satisfy a *school philosopher* who looks at life through the spectacles of his own specific problems and recognizes hatred, love, happiness only to the extent to which they occur in these problems. But if we consider the interests of *man* and, above all, the question of his freedom (freedom from hunger, despair, from the tyranny of constipated systems of thought, *not* the academic 'freedom of the will'), then we are proceeding in the worst possible fashion.

For is it not possible that science as we know it today (the science of critical rationalism that has been freed from all inductive elements) or a 'search for the truth' in the style of traditional philosophy will create a monster? Is it not possible that it will harm man, turn him into a miserable, unfriendly, self-righteous mechanism without charm and without humor? "Is it not possible", asks Kierkegaard, "that my activity as an objective [or a critico-rational] observer of nature will weaken my strength as a human being?" [2] I suspect the answer to all these questions must be affirmative and I believe that a reform of the sciences that makes it more anarchistic and more subjective (in Kierkegaard's sense) is therefore urgently needed. But this is not what I want to discuss in the present essay. Here I shall restrict myself to the second question and I shall ask: is it possible to have both a science as we know it and the rules of a critical rationalism as just described? And to *this* question the answer seems to be a resounding *no*.

To start with we have seen, though rather briefly, that the actual development of institutions, ideas, practices, and so on often does not start from a problem but rather from some irrelevant activity, such as playing, which, as a side effect, leads to developments which later on can be interpreted as solutions to unrealized problems. Are such developments to be excluded? And if we *do* exclude them, will this not considerably reduce the number of our adaptive reactions and the quality of our learning process?

Secondly, we have seen that a strict principle of falsification, or a 'naive falsificationism' as Imre Lakatos calls it, combined with the demand for maximum testability and non-adhocness would wipe out science as we know it, and would never have permitted it to start. This has been realized by Lakatos who has set out to remedy the situation.[3] His remedy is not mine, it is not anarchism. His remedy consists in slight modification of the 'critical standards' he adores. (He also tries to show, with the help of amusing numerological considerations, that it is already foreshadowed in Popper.)

According to naive falsificationism, a theory is judged, i.e., either accepted or condemned, as soon as it is introduced into the discussion. Lakatos gives a theory time, he permits it to develop, to show its hidden strength, and he judges it only 'in the long run'. The 'critical standards' *he* employs provide for an interval of hesitation. They are applied 'with hindsight'. If the theory gives rise to interesting new developments, if it engenders 'progressive problem shifts', then it may be retained despite its initial vices. If on the other hand the theory leads nowhere, if the ad hoc hypotheses it employs are not the starting point but the end of all research, if the theory seems to kill the imagination and to dry up every resource of speculation, if it creates 'degenerating problem shifts', i.e., changes which terminate in a dead end, then it is time to give it up and to look for something better.

Now it is easily seen that standards of this kind have practical force only if they are combined with a *time limit*. What looks like a degenerating problem shift may be the beginning of a much longer period of advance, so – how long are we supposed to wait? But if a time limit *is* introduced, then the argument against the more conservative point of view, against 'naive falsificationism', reappears with only a minor modification. For if you can wait, then why not wait a little longer? Besides there are theories which for centuries were accompanied by degenerating problem shifts until they found the right defenders and returned to the stage in full bloom. The heliocentric theory is one example. The atomic theory is another. We see that the new standards which Lakatos wants to defend either are *vacuous* – one does not know when and how to apply them – or else can be *criticized* on grounds very similar to those which led to them in the first place.

In these circumstances one can do one of the following two things. One can stop appealing to permanent standards which remain in force throughout history, and govern every single period of scientific development and every transition from one period to another. Or one can retain such standards as a *verbal ornament*, as a memorial to happier times when it was still thought possible to run a complex and catastrophic business like science by a few simple and 'rational' rules. It seems that Lakatos wants to choose the second alternative.

Choosing the second alternative means abandoning permanent standards *in fact*, though retaining them *in words*. *In fact* Lakatos's position now is identical with the position of Popper as summarized in the marvelous (because self-destructive) Appendix i/15 of the fifth edition of the *Open Society*.[4] According to Popper, we do not "need any... definite frame of reference for our criticism", we may revise even the most fundamental rules and drop the most fundamental demands if the need for a different measure of excellence should arise.[5] Is such a position irrational? Yes and no. Yes, because there no longer exists a single set of rules that will guide us through all the twists and turns of the history of thought (science), either as participants or as historians who want to reconstruct its course. One can of course *force* history into a pattern, but the results will always be poorer and less interesting than were the actual events. No, because each particular episode is rational in the sense that some if its features can be explained in terms of reasons which were either accepted at the time of its occurrence or invented in the course of its development. Yes, because even these local reasons which change from age to age are never sufficient to explain *all* the important features of a particular episode. One must add accidents, prejudices, material conditions, e.g., the existence of a particular type of glass in one country and not in another for the explanation of the history of optics, the vicissitudes of married life (Ohm!), superficiality, pride, oversight, and many other things, in order to get a complete picture. No, because, transported into the climate of the period under consideration and endowed with a lively and curious intelligence, we might have had still more to say; we might have tried to overcome accidents, and to 'rationalize' even the most whimsical sequence of events. But, and now I come to a decisive point for the discussion of incommensurability, how is the transition from certain standards to other standards to be achieved? More especially, what happens to our standards, as opposed to our theories, during a period of revolution? Are they changed in the manner suggested by Mill by a critical discussion of alternatives, or are there processes which defy a rational analysis? Well, let us see!

That standards are not always adopted on the basis of argument has been emphasized by Popper himself. Children, he says, "learn to imitate others... and so learn to look upon standards of behavior as if they consisted of fixed, 'given' rules... and such things as sympathy and imagination may play an important role in this development".[6] Similar considerations apply to those grownups who want to continue learning, and who are intent on expanding both their knowledge and their sensibility. This we have already discussed in Section 1. Popper also admits that new standards may be discovered, invented, accepted, imparted to others in a very irrational manner. But, he points out, one can criticize them *after* they have been adopted, and it is *this* possibility which keeps our knowledge rational. "What, then, are we to trust?" he asks after a survey of possible sources for standards.[7]

What are we to accept? The answer is: whatever we accept we should trust only tentatively, always remembering that we are in possession, at best, of partial truth (or rightness), and that we are bound to make at least some mistake or misjudgement somewhere – not only with respect to facts but also with respect to the adopted standards; secondly, we should trust (even tentatively) our intuition only if it has been arrived at as the result of many attempts to use our imagination; of many mistakes, of many tests, of many doubts, *and of searching criticism.*

Now this reference to tests and to criticism, which is supposed to guarantee the rationality of science, and, perhaps, of our entire life, may be either to *well-defined procedures* without which a criticism or test cannot be said to have taken place, or to a purely *abstract* notion, so that it is left *to us* to fill it now with this, and now with that concrete content. The first case has just been discussed. In the second case we have again but a verbal ornament. The questions asked in the last paragraph but one remain unanswered in either case.

In a way even this situation has been described by Popper, wo says that "rationalism is necessarily far from comprehensive or self-contained."[8] But our present inquiry is not whether *there are* limits to our reason; the question is *where* these limits are *situated*. Are they outside the sciences so that science itself remains entirely rational; or are irrational changes an essential part even of the most rational enterprise that has been invented by man? Does the historical phenomenon 'science' contain ingredients which defy a rational analysis, although they may be described with complete clarity in psychological or sociological terms? Can the abstract aim to come closer to the truth be reached in an entirely rational manner, or is it perhaps inaccessible to those who decide to rely on argument only? These are the problems which were raised, first by Hegel and then, in quite different terms, by Kuhn. They are the problems I wish to discuss.

In discussing these further problems, Popper and Lakatos reject considerations of sociology and psychology, or as Lakatos expresses himself, 'mob psychology', and assert the rational character of *all* science. According to Popper, it is possible to arrive at a judgment as to which of two theories is closer to the truth, even if the theories should be separated by a catastrophic upheaval such as a scientific or other revolution. (A theory is closer to the truth than another theory if the class of its true consequences, its truth content, exceeds the truth content of the latter without an increase of falsity content.) According to Lakatos, the apparently unreasonable features of science occur only in the material world and in the world of (psychological) thought; they are absent from the "world of ideas, from Plato's and Popper's 'third world'". It is in this third world that the growth of knowledge takes place, and that a rational judgment of all aspects of science becomes possible.

Now in regard to this convenient flight into higher regions, it must be pointed out that the scientist is, unfortunately, dealing with the world of matter and of psychological (i.e., subjective) thought also. It is *mainly* this material world he wants to change and to influence. And the rules which create order in the third world will most likely be entirely inappropriate for creating order in the brains of living human beings (unless these brains and their structural features are put in the third world also, a point that does not become clear from Popper's account[9]). The numerous deviations from the straight and rather boring path of rationality which one can observe in actual science may well be *necessary* if we want to achieve progress with the brittle and unreliable material (instruments; brains; assistants; etc.) at our disposal.

However, there is no need to pursue this objection further. There is no need to argue that science as we know it may differ from its third-world shadow *in precisely those respects* which make progress possible.[10] For the Popperian model of an approach to the truth breaks down even if we confine ourselves to ideas entirely. It breaks down because there are incommensurable theories.

2. INCOMMENSURABILITY

Scientific investigation, says Popper, *starts* with a problem, and it proceeds by *solving* it.

This characterization does not take into account that problems may be wrongly formulated, that one may inquire about properties of things or processes which later research declares to be nonexistent. Problems of this kind are not *solved*, they are *dissolved* and removed from the domain of legitimate inquiry. Examples are the problem of the absolute velocity of the earth, the problem of the trajectory of an electron in an interference pattern, or the important problem whether incubi are capable of producing offspring or whether they are forced to use the seeds of men for that purpose.[11]

The first problem was dissolved by the theory of relativity which denies the existence of absolute velocities. The second problem was dissolved by the quantum theory which denies the existence of trajectories in interference patterns. The third problem was dissolved, though much less decisively so, by modern (i.e., post-sixteenth century) psychology and physiology as well as by the mechanistic cosmology of Descartes.

Now changes of ontology such as those just described are often accompanied by conceptual changes.

The discovery that certain entities do not exist may force the scientist to redescribe the events, processes, observations which were thought to be manifestations of them and were therefore described in terms assuming their existence. Or, rather, it may force him to use new *concepts* as the older *words* will remain in use for a considerable time. Thus the term 'possessed' which was once used for giving a causal description of the behavioral peculiarities connected with epilepsy was retained, but it was voided of its devilish connotations.

An interesting development occurs when the faulty ontology is *comprehensive*, that is, when its elements are thought to be present in every process in a certain domain. In *this* case *every* description inside the domain must be changed and must be replaced by a different statement (or by no statement at all). Classical physics is a case in point. It has developed a comprehensive terminology for describing the most fundamental mechanical properties of our universe, such as shapes, speeds, and masses. The conceptual system connected with this terminology assumes that the properties *inhere* in objects and that they change only if one interferes with the objects, not otherwise. The theory of relativity teaches us, at least in one of its interpretations, that there are no such inherent properties in the world, neither observable, nor unobservable, and it produces an entirely new conceptual system for description inside the domain of mechanics. This new conceptual system does not just *deny* the existence of classical states of affairs, it does not even permit us to *formulate statements* expressing such states of affairs (there is no arrangement in the Minkowski diagram that corresponds to a classical situation). It does not, and cannot, share a single statement with its predecessor. As a result the formal conditions for a suitable successor of a refuted theory (it has to repeat the successful consequences of the older theory, deny its false consequences, and make additional predictions) cannot be satisfied in the case of relativity versus classical physics and the Popperian scheme of progress breaks down. It is not even possible to connect classical statements and relativistic statements by an *empirical hypothesis*.[12] Formulating such a connection would mean formulating statements of the type "whenever there is possession by a demon there is discharge in the brain" which perpetuate rather than eliminate the older ontology. Comprehensive theories of the kind just mentioned are therefore completely disjointed, or *incommensurable*. The existence of incommensurable theories provides another difficulty for critical rationalism (and, a fortiori, for its more positivistic predecessors). We shall discuss this difficulty by discussing and refuting objections against it.

It was pointed out that progress may lead to a complete replacement of statements (and perhaps even of descriptions) in a certain domain. More especially, it may replace certain natural interpretations by others.... Galileo replaces the idea of the operative character of all motion by his relativity principle in order to accommodate the new views of Copernicus. It is entirely natural to proceed in this way. A cosmological theory such as the heliocentric theory, or the theory of relativity, or the quantum theory (though the last one only with certain restrictions) makes assertions about the world as a whole. It applies to observed and to unobserved (unobservable, 'theoretical') processes. It can therefore demand to be used always, and not only on the theoretical level. Now such an adaption of observation to theory, and this is the gist of the *first objection*, removes conflicting observation reports and saves the theory in an ad hoc manner. Moreover, there arises the *suspicion* that observations which are interpreted in terms of a new theory can no longer be used to refute that theory. It is not difficult to reply to these points.

As regards the objection we point out, in agreement with what has been said before..., that an inconsistency between theory and observation may reveal a fault of our *observational terminology* (and even of our sensations) so that it is quite natural to change this terminology, to adapt it to the new theory, and to see what happens. Such a change gives rise, and should give rise, to new auxiliary subjects (hydrodynamics, theory of solid objects, optics in the case of Galileo) which may more than compensate for the empirical content lost by the adaptation. And as regards the suspicion we must remember that the predictions of a theory depend on its postulates, the associated grammatical rules, *as well as* on initial conditions while the meaning of the 'primitive' notions depends on the postulates (and the associated grammatical rules) only.[13] In those rare cases, however, where a theory *entails* assertions about possible initial conditions,[14] we can refute it with the help of *self-inconsistent observation reports* such as "object A does not move on a geodesic" which, if analyzed in accordance with the Einstein-Infeld-Hoffmann account reads "singularity α which moves on a geodesic does not move on a geodesic".

The *second objection* criticizes the interpretation of science that brings about incommensurability. To deal with it we must realize that the question "are two particular comprehensive theories, such as classical celestial mechanics (CM) and the special theory of relativity (SR) incommensurable?" is not a complete question. Theories can be interpreted in different ways. They will be commensurable in some interpretations, incommensurable in others. Instrumentalism, for example, makes commensurable all those theories which are related to the same observation language and are interpreted on its basis. A realist, on the other hand, wants to give a unified account, both of observable and of unobservable matters, and he will use the most abstract terms of whatever theory he is contemplating for that purpose.[15] This is an entirely natural procedure. SR, so one would be inclined to say, does not just invite us to rethink *unobserved* length, mass, duration; it would seem to entail the relational character of all lengths, masses, durations, whether observed or unobserved, observable or unobservable.

Now, and here we only repeat what was said not so long ago, extending the concepts of a new theory, T, to all its consequences, observational reports included, may change the interpretation of these consequences to such an extent that they disappear from the consequence classes either of earlier theories or of the available alternatives. These earlier theories and alternatives will then all become incommensurable with T. The relation between SR and CM is a case in point. The concept of length as used in SR and the concept of length as presupposed in CM are different concepts. Both are *relational* concepts, and very complex relational concepts at that (just consider determination of length in terms of the wave length of a specified spectral line). But relativistic length, or relativistic shape, involves an element that is absent from the classical concept and is in principle excluded from it. It involves the relative velocity of the object concerned in some reference system. It is of course true that the relativistic scheme very often yields numbers which are practically identical with the *numbers* obtained from CM, but this does not make the *concepts* more similar. Even the case $c \to \infty$ (or $v \to 0$) which yields identical predictions cannot be used as an argument for showing that the concepts must coincide, at least in this special case. Different magnitudes based on different concepts may give identical values on their respective scales without ceasing to be different magnitudes. The same remark applies to the attempt to identify classical mass with relativistic *rest* mass.[16] This conceptual disparity, if taken seriously, infects even the most 'ordinary' situations. The relativistic concept of a certain *shape*, such as the shape of a table, or of a certain temporal sequence, such as my saying 'Yes', will differ from the corresponding classical concept also. It is therefore futile to expect that sufficiently long derivations may eventually return us to the older ideas.[17] The consequence classes of SR and CM are not related in any way. A comparison of content and a judgment of verisimilitude cannot be made.[18]

The situation becomes even clearer when we use the Marzke-Wheeler interpretation of SR. For it can be easily shown that the methods of measurement provided by these authors, while perfectly adequate in a relativistic universe, either collapse or give nonsensical results in a classical world (length, for example, is no longer transitive, and in some coordinate systems it may be impossible to assign a definite length to *any* object).

We are now ready to discuss the *second* and most popular objection against incommensurability. This objection proceeds from the version of realism described above. "A realist", we said, "will want to give a unified account, both of observable and of unobservable matters, and he will use the most abstract terms of whatever theory he is contemplating for his purpose". He will use such terms in order either to *give* meaning to observation sentences or else to *replace* their customary interpretation. (For example, he will use the ideas of SR in order to replace the customary CM-interpretation of everyday statements about shapes, temporal sequences, and so on.) Against this, it is pointed out that theoretical terms receive their interpretation by being connected with a preexisting observation language, or with another theory that has already been connected with such an observation language, and that they are devoid of content without such connection. Thus Carnap asserts[19] that

[t]here is no independent interpretation for L_T [the language in terms of which a certain theory, or a certain world view, is formulated]. The system T [the axioms of the theory and the rules of derivation] is in itself an uninterpreted postulate system. [Its] terms ... obtain only an indirect and incomplete interpretation by the fact that some of them are connected by the [correspondence] rules C with observation terms....

Now, if theoretical terms have no "independent interpretation", then surely they cannot be used for correcting the interpretation of the observation statements, which is the one and only source of their meaning. It follows that realism as described here is an impossible doctrine.

The guiding idea behind this very popular objection is that new and abstract languages cannot be introduced in a direct way, but must be first connected with an already existing, and presumably stable, observational idiom.[20]

This guiding idea is refuted at once by noting the way in which children learn to speak and in which anthropologists and linguists learn the unknown language of a newly discovered tribe.

The first example is instructive for other reasons also, for incommensurability plays an important role in the early months of human development. As has been suggested by Piaget and his school[21] the child's perception develops through various stages before it reaches its relatively stable adult form. In one stage objects seem to behave very much like afterimages,[22] and they are treated as such. In this stage the child follows the object with his eyes until it disappears, and he does not make the slightest attempt to recover it, even if this would require but a minimal physical (or intellectual) effort, an effort, moreover, that is already within the child's reach. There is not even a tendency to search; and this is quite appropriate, 'conceptually' speaking. For it would indeed be nonsensical to 'look for' an afterimage. Its 'concept' does not provide for such an operation.

The arrival of the concept and of the perceptual image of material objects changes the situation quite dramatically. There occurs a drastic reorientation of behavioral patterns, and, so one may conjecture, of thought. Afterimages, or things somewhat like them, still exist, but they are now difficult to find and must be discovered by special methods. (The earlier visual world therefore *literally disappears*.) Such special methods proceed from a new conceptual scheme (afterimages occur in *humans*, not in the outer physical world, and are tied to them) and cannot lead back to the exact phenomena of the previous stage (these phenomena should therefore be called by a different name, such as 'pseudo-afterimages'). Neither afterimages nor pseudo-afterimages are given a special position in the new world. For example, they are not treated as 'evidence' on which the new notion of a material object is supposed to rest. Nor can they be used to *explain* this notion: afterimages arise *togeher with it*, and are absent from the minds of those who do not yet recognize material objects. And pseudo-afterimages *disappear* as soon as such recognition takes place. It is to be admitted that every stage possesses a kind of observational 'basis' to which one pays special attention and from which one receives a multitude of suggestions. However, this basis (i) *changes* from stage to stage;

and (ii) is *part* of the conceptual apparatus of a given stage; it is *not* its one and only source of interpretation.

Considering developments such as these, one may suspect that the family of concepts centering upon 'material object' and the family of concepts centering upon 'pseudo-afterimage' are incommensurable in precisely the sense that is at issue here. Is it reasonable to expect that conceptual and perceptual changes of this kind occur in childhood only? Should, we welcome the fact, if it is a fact, that an adult is stuck with a stable perceptual world and an accompanying stable conceptual system which he can modify in many ways, but whose general outlines have forever become immobilized? Or is it not more realistic to assume that fundamental changes, entailing incommensurability, are still possible, and that they should be encouraged lest we remain forever excluded from what might be a higher stage of knowledge and of consciousness?... Besides, the question of the mobility of the adult stage is at any rate an empirical question, which must be attacked by *research* and which cannot be settled by methodological *fiat*. The attempt to break through the boundaries of a given conceptual system and to escape the reach of 'Popperian spectacles' (Lakatos) is an essential part of such research (and should be an essential part of any interesting life).[23]

Looking now at the second element of the refutation, anthropological field work, we see that what is anathema here (and for very good reasons) is still a fundamental principle for the contemporary representatives of the philosophy of the Vienna Circle. According to Carnap, Feigl, Nagel, and others, the terms of a theory receive their interpretation in an indirect fashion, by being related to a different conceptual system which is either an older theory or an observation language.[24] This older theory, this observation language, is not adopted because of its theoretical excellence. It cannot possibly be: the older theories are usually refuted. It is adopted because it is "used by a certain language community as a means of communication".[25] According to this method, the phrase 'having much larger relativistic mass than...' is partially *interpreted* by first connecting it with some *prerelativistic* terms (classical terms, common-sense terms), which are 'commonly understood' (presumably, as the result of previous teaching in connection with crude weighing methods), and it is *used* only after such connection has given it a well-defined meaning.

This is even worse than the once quite popular demand to clarify doubtful points by translating them into Latin. For while Latin was chosen because of its precision and clarity, and also because it was conceptually richer than the slowly evolving vulgar idioms,[26] the choice of an observation language or of an older theory as a basis for interpretation is justified by saying that they are 'antecedently understood': the choice is based on sheer *popularity*. Besides, if *prerelativistic* terms which are pretty far removed from reality (especially in view of the fact that they come from an incorrect theory implying a nonexistent ontology) can be taught ostensively, for example, with the help of crude weighing methods (and one must assume that they can be so taught, or the whole scheme collapses), then why should one not introduce the *relativistic* terms *directly*, and *without* assistance from the terms of some other idiom? Finally, it is but plain common sense that the teaching or the learning of new and unknown languages must not be contaminated by external material. Linguists remind us that a perfect translation is never possible, even if one is prepared to use complex contextual definitions. This is one of the reasons for the importance of *field work* where new languages are learned *from scratch*, and for the rejection, as inadequate, of any account that relies on 'complete' or 'partial' translation. *Yet just what is anathema in linguistics is taken for granted by logical empiricism*, a mythical 'observation language' replacing the English of the translators. Let us commence field work in this domain also, and let us study the language of new theories not in the definition factories of the double language model, but in the company of those metaphysicians, theoreticians, playwrights, courtesans who have constructed new world views! This finishes my discussion of the guiding principle behind the second objection against realism and the possibility of incommensurable theories.

Another point that is often made is that there exist *crucial experiments* which refute one of two allegedly incommensurable theories and confirm the other (example: the Michelson-Morley experiment, the variation of the mass of elementary particles, the transverse Doppler effect, are said to refute CM and confirm SR). The answer to this problem is not difficult either: adopting the point of view of relativity, we find that the experiments, *which of course will now be described in relativistic terms*, using the relativistic notions of length, duration, speed, and so on, are relevant to the theory. And we also find that they support the theory. Adopting CM (with, or without an ether), we again find that the experiments, which are now described in the very different terms of classical physics, i.e., roughly in the manner in which Lorentz described them, are relevant. But we also find that they *undermine* CM, i.e., the conjunction of classical electrodynamics and of CM. Why should it be necessary to possess terminology that allows one to say that it is the *same* experiment which confirms one theory and refutes the other? But did we not ourselves use such terminology? Well, for one thing it should be easy though somewhat laborious to express what was just said *without* asserting identity. Secondly, the identification is of course not contrary to our thesis, for we are now not *using* the terms of either relativity or classical physics, as is done in a test, but are *referring* to them and their relation to the physical world. The language in which *this* discourse is carried out can be classical, or relativistic, or ordinary. It is no good insisting that scientists act as if the situation were much less complicated. If they act that way, then they are either instrumentalists (see above) or mistaken (many scientists are nowadays interested in *formulas*, while the subject here is *interpretations*). It is also possible that being well acquainted with both CM and SR, they change back and forth between these theories with such speed that they seem to remain within a single domain of discourse.

It is also said that by admitting incommensurability into science we can no longer decide whether a new view explains what it is supposed to explain, or whether it does not wander off into different fields.[27] For example, we would not know whether a newly invented physical theory is still dealing with problems of space and time or whether its author has not by mistake made a biological assertion. But there is no need to possess such knowledge. For once the fact of incommensurability has been admitted, the question which underlies the objection does not arise. Conceptual progress often makes it impossible to ask certain questions and to explain certain things; thus we can no longer ask for the absolute velocity of an object, at least as long as we take relativity seriously. Is this a serious loss for science? Not at all! Progress was made by the very same 'wandering off into different fields' whose undecidability now so greatly exercises the critic: Aristotle saw the world as a super *organism*, as a *biological* entity, while one essential element of the new science of Descartes, Galileo, and their followers in medicine and in biology is its exclusively *mechanistic* outlook. Are such developments to be forbidden? And if they are not, what, then, is left of the complaint?

A closely connected objection starts from the notion of *explanation* or *reduction* and emphasizes that this notion presupposes continuity of concepts; other notions could be used for starting exactly the same kind of argument. (Relativity is supposed to explain the valid parts of classical physics; hence it cannot be incommensurable with it!) The reply is again obvious. As a matter of fact it is a triviality for anyone who has only the slightest acquaintance with the Hegelian philosophy: why should the relativist be concerned with the fate of classical mechanics except as part of a historical exercise? There is only *one* task we can legitimately demand of a theory, and it is that it should give us a correct account of the world, i.e., of the totality of facts *as seen through its own concepts*. What have the principles of explanation got to do with this demand? Is it not reasonable to assume that a point of view such as the point of view of classical mechanics that has been found wanting in various respects, that gets in difficulty *with its own facts* (see above, on crucial experiments), and must therefore be regarded as self-inconsistent (another application of Hegelian principles!), cannot have entirely adequate concepts? Is it not equally reasonable to try replacing its concepts with those of a more promising cosmology? Besides, why should the notion of explanation be burdened by the demand for conceptual continuity? This notion has been found to be too narrow before (demand of derivability), and it had to be widened so as to include partial and statistical connections. Nothing prevents us from widening it still further and admitting, say, 'explanations by equivocation'.

Incommensurable theories, then, can be *refuted* by reference to their own respective kinds of experience, i.e., by discovering the *internal contradictions* from which they are suffering (in the absence of commensurable alternatives these refutations are quite weak, however [28]). Their *content* cannot be compared, nor is it possible to make a judgment of *verisimilitude* except within the confines of a particular theory. None of the methods which Popper (or Carnap, or Hempel, or Nagel) want to use for rationalizing science can be applied, and the one that *can* be applied, refutation, is greatly reduced in strength. What remains are esthetic judgments, judgments of taste, and our own subjective wishes.[29] Does this mean that we are ending up in subjectivism? Does this mean that science has become arbitrary, that it has become an element of the general relativism which so much exercises the conscience of some philosophers? Well, let us see.

3. THE CHOICE BETWEEN COMPREHENSIVE IDEOLOGIES

To start with, it seems to me that an enterprise whose human character can be seen by all is preferable to one that looks 'objective' and impervious to human actions and wishes.[30] The sciences, after all, are our own creation, including all the severe standards they seem to impose on us. It is good to be constantly reminded of this fact. It is good to be constantly reminded of the fact that science as we know it today is not inescapable, and that we can construct a world in which it plays no role whatever. (Such a world, I venture to suggest, would be more pleasant to behold than the world we live in today, both materially and intellectually.) What better reminder is there than the realization that the choice between theories which are sufficiently general to yield a comprehensive world view and which are empirically disconnected may become a matter of taste? *That the choice of a basic cosmology may become a matter of taste?*

Secondly, matters of taste are not completely beyond the reach of argument. Poems, for example, can be compared in grammar, sound structure, imagery, rhythm, and can be evaluated on such a basis (cf. Ezra Pound on progress in poetry[31]). Even the most elusive mood can be analyzed *and should be analyzed* if the purpose is to present it in a manner that either can be enjoyed or increases the emotional, cognitive, perceptual, etc., inventory of the reader. Every poet who is worth his salt compares, improves, argues until he finds the correct formulation of what he wants to say.[32] Would it not be marvelous if this free and entertaining[33] process played a role in the sciences also?

Finally, there are more pedestrian ways of explaining the same matter which may be somewhat less repulsive to the tender ears of a professional philosopher of science. One may consider the *length* of derivations leading from the principles of a theory to its observation language, and one may also draw attention to the number of *approximations* made in the course of the derivation. All derivations must be standardized for this purpose so that unambiguous judgments of length can be made. (This standardization concerns the form of the derivation, it does not concern the *content*.) Smaller length and smaller number of approximations would seem to be preferable. It is not easy to see how this requirement can be made compatible with the demand for simplicity and generality which, so it seems, would tend to increase both parameters. However that may be, there are many ways open to us once the fact of incommensurability is understood, and taken seriously.

4. CONCLUSION

The idea that science can and should be run according to some fixed rules, and that its rationality consists in agreement with such rules, is both unrealistic and vicious. It is *unrealistic*, since it takes too simple a view of the talents of men and of the circumstances which encourage, or cause, their development. And it is *vicious*, since the attempt to enforce the rules will undoubtedly erect barriers to what men might have been, and will reduce our humanity by increasing our professional qualifications. We can *free* ourselves from the idea and from the power it may possess over us (i) by a detailed study of the work of revolutionaries such as Galileo, Luther, Marx, or Lenin; (ii) by some acquaintance with the Hegelian philosophy and with the alternative provided by Kierkegaard; (iii) by remembering that the existing separation between the sciences and the arts is artificial, that it is a side effect of an idea of professionalism one should eliminate, that a poem or a play can be intelligent as well as informative (Aristophanes, Hochhuth, Brecht), and a scientific theory pleasant to behold (Galileo, Dirac), and that we can change science and make it agree with our wishes. We can turn science from a stern and demanding mistress into an attractive and yielding courtesan who tries to anticipate every wish of her lover. Of course, it is up to us to choose either a dragon or a pussycat as our companion. So far mankind seems to have preferred the latter alternative: "The more solid, well defined, and splendid the edifice erected by the understanding, the more restless the urge of life... to escape from it into freedom". We must take care that we do not lose our ability to make such a choice.

NOTES

* Sections 11 through 15 of 'Against Method: Outline of an Anarchistic Theory of Knowledge', by Paul K. Feyerabend. From Volume IV, *Minnesota Studies in the Philosophy of Science*, University of Minnesota Press, Minneapolis. © Copyright 1970 by the University of Minnesota. Reprinted by permission.
1 "Neurath fails to give... rules [which distinguish empirical statements from others] and thus unwittingly throws empiricism overboard." K. R. Popper, *The Logic of Scientific Discovery*, Basic Books, New York, 1959, p. 97.

[2] *Papirer*, ed. by P. A. Heiberg (Copenhagen, 1909), VII, Part I, see A, No. 182. Cf. also Sections 7ff of my forthcoming paper 'Abriss einer anarchistischen Erkenntnislehre'.

[3] 'Criticism and the Methodology of Scientific Research Programs', in *Criticism and the Growth of Knowledge*, ed. by I. Lakatos and A. Musgrave, North-Holland, Amsterdam, 1969. Quotations are from the typescript of the paper which Lakatos distributed liberally before its publication. In this typescript the reference is mostly to Popper. Had Lakatos been as careful with acknowledgments as he is when the Spiritual Property of the Popperian Church is concerned, he would have pointed out that his liberalization which sees knowledge as a *process* is indebted to Hegel.

[4] Popper, *The Open Society and Its Enemies*, pp. 388ff.

[5] *Ibid.*, p. 390. Cf. also Note 28.

[6] *Ibid.* Cf. Note 22 and the corresponding text.

[7] *Ibid.*, p. 391.

[8] *Ibid.*, p. 231.

[9] I am referring here to the following two papers: 'Epistemology Without a Knowing Subject', in Bob Van Rootselaar and J. F. Staal (eds.), *Logic, Methodology and a Knowledge of Science*, Vol. III, North-Holland, Amsterdam, 1968, as well as 'On the Theory of the Objective Mind'. In the first paper, *birdnests* are assigned to the 'third world' (p. 341) and an interaction is assumed between them and the remaining worlds. They are assigned to the third world *because of their function*. But then stones and rivers can be found in this third world too, for a bird may sit on a stone, or take a bath in a river. As a matter of fact, everything that is noticed by some organism will be found in the third world, which will therefore contain the whole material world and all the mistakes mankind had made. It will also contain 'mob psychology'.

[10] Cf. again 'Problems of Empiricism, Part II'.

[11] Cf. *Malleus Maleficarum*, trans. by Montague Summers (Pushkin Press, London, 1928), Part II, question I, Chapter IV: "Here follows the way whereby witches copulate with those Devils known as Incubi", second item, as to the acts, "whether it is always accompanied with the injection of semen received from some other man". The theory goes back to St. Thomas Aquinas.

[12] It is of course possible to establish correlations between the *sentences* of the two theories, but one must realize that the elements of the correlation, when interpreted, cannot be both meaningful, or both true: if relativity is true, then classical descriptions are either always false or are always nonsensical. Continued use of classical sentences must therefore be regarded as an abbreviation for sentences of the following kind: "Given conditions C, the classical sentence S was uttered by a classical physicist whose sense organs are in order, and who understands his physics" – and sentences of this kind, if taken together with certain psychological assumptions, can be used for a test of relativity. However, the *statements* which are expressed by these sentences are part of the *relativistic* framework, for they use relativistic terms. This situation is overlooked by Lakatos who argues as if classical terms and relativistic terms can be combined at will and who infers from this assumption the nonexistence of incommensurability.

[13] This became clear to me in a discussion with Mr. L. Briskman, in Professor Watkins's seminar at the London School of Economics.

[14] This seems to occur in certain versions of the general theory of relativity. Cf. A. Einstein, L. Infeld, and B. Hoffmann, 'The Gravitational Equations and the Problem of Motion', *Annals of Mathematics* **39** (1938), 65, and Sen, *Fields and/or Particles*, pp. 19ff.

[15] This consideration has been raised into a principle by Bohr and Rosenfeld, *Kgl.*

THE RATIONALITY OF SCIENCE

Danske Videnskab. Sekskab, Mat. Fys. Medd., 12, No. 8 (1933), and, more recently, by Robert F. Marzke and John A. Wheeler, 'Gravitation as Geometry I', in Chiu and Hoffmann (eds.), *Gravitation and Relativity*, p. 48: "every proper theory should provide in and by itself its own means for defining the quantities with which it deals. According to this principle, classical general relativity should admit to calibrations of space and time that are altogether free of any reference to [objects which are external] to it such as rigid rods, inertial clocks, or atomic clocks [which involve] the quantum of action."

[16] For this point and further arguments see A. S. Eddington, *The Mathematical Theory of Relativity*, Cambridge University Press, Cambridge, 1963, p. 33. The more general problem of concepts and numbers has been treated by Hegel, *Logik*, I, Das Mass.

[17] This takes care of an objection which Professor J. W. N. Watkins has raised on various occasions.

[18] For further details, especially concerning the concept of mass, the function of 'bridge laws' or 'correspondence rules', and the two-language model, see Section iv of 'Problems of Empiricism'. It is clear that, given the situation described in the text, we cannot derive classical mechanics from relativity, not even approximately. For example, we cannot derive the classical law of mass conservation from relativistic laws. The possibility of connecting the *formulas* of the two disciplines in a manner that might satisfy a pure mathematician, or an instrumentalist, is, however, not excluded. For an analogous situation in the case of quantum mechanics see Section 3 of my paper 'On a Recent Critique of Complementarity'. See also Section 2 of the same article for more general considerations.

[19] R. Carnap, 'The Methodological Character of Theoretical Concepts', *Minnesota Studies in the Philosophy of Science*, Vol. I, ed. by H. Feigl and M. Scriven (University of Minnesota Press, Minneapolis, 1956), p. 47.

[20] An even more conservative principle is sometimes used when discussing the possibility of languages with a logic different from our own: "Any allegedly new possibility must be capable of being fitted into, or understood in terms of, our present conceptual or linguistic apparatus". B. Stroud, 'Conventionalism and the Indeterminacy of Translations', *Synthese*, 1968, p. 173.

[21] As an example the reader is invited to consult J. Piaget, *The Construction of Reality in the Child*, Basic Books, New York, 1954.

[22] *Ibid.*, pp. 5ff.

[23] For the condition of research formulated in the last sentence see Section 8 of 'Reply to Criticism', *Boston Studies in the Philosophy of Science*, Vol. II, ed. by Cohen and Wartofsky. For the role of observation see Section 7 of the same article. For the application of Piaget's work to physics and, more especially, to the theory of relativity see the appendix of Bohm, *The Special Theory of Relativity*. Bohm and Schumacher have also carried out an analysis of the various informal structures which underlie our theories. One of the main results of their work is that Bohr and Einstein argued from incommensurable points of view. Seen in this way the case of Einstein, Podolsky, and Rosen cannot refute the Copenhagen Interpretation and it cannot be refuted by it either. The situation is, rather, that we have two theories, one permitting us to formulate EPR, the other not providing the machinery necessary for such a formulation. We must find independent means for deciding which one to adopt. For further comments on this problem see Section 9 of my 'On a Recent Critique of Complementarity'.

[24] For what follows cf. also my review of Nagel's *Structure of Science* on pp. 237–249 of the *British Journal for the Philosophy of Science* 6 (1966), 237–249.

[25] Carnap, 'The Methodological Character of Theoretical Concepts', p. 40. Cf. also

C. G. Hempel, *Philosophy of Natural Science*, Prentice-Hall, Englewood Cliffs, N.J., 1966, pp. 74ff.

[26] lt was for this reason that Leibniz regarded the German of his time and especially the German of the artisans as a perfect observation language, while Latin, for him, was already too much contaminated by theoretical notions. See his 'Unvorgreifliche Gedancken, betreffend die Ausübung und Verbesserung der Teutschen Sprache', published in *Wissenschaftliche Beihefte zur Zeitschrift des allgemeinen deutschen Sprachvereins* **IV**, 29 (F. Berggold, Berlin, 1907), pp. 292ff.

[27] This objection was raised at a conference by Prof. Roger Buck.

[28] For this point see section I of 'Reply to Criticism', as well as the corresponding sections in 'Problems of Empiricism'.

[29] That the choice between comprehensive theories rests on one's interests entirely and reveals the innermost character of the one who chooses has been emphasized by Fichte in his 'Erste Einleitung in die Wissenschaftslehre'. Fichte discusses the opposition between idealism and materialism which he calls dogmatism. He points out that there are no facts and no considerations of logic which can force us to adopt either the one or the other position. "... we are here faced", he says (*Erste und Zweite Einleitung in die Wissenschaftslehre*, Felix Meiner, Hamburg, 1961, p. 19), "with an absolutely first act that depends on the freedom of thought entirely. It is therefore determined in an arbitrary manner [durch *Willkür*] and, as an arbitrary decision must have a reason nevertheless, by our *inclination* and our *interest*. The final reason for the difference between the idealist and the dogmatist is therefore the difference in their interests".

[30] Here once more the familiar problem of *alienation* arises: what is the result of our own activity becomes separated from it, and assumes an existence of its own. The connection with our intentions and our wishes becomes more and more opaque so that in the end we, instead of leading, follow slavishly the dim outlines of our shadow whether this shadow manifests itself *objectively*, in certain institutions, or *subjectively*, in what some people are pleased to call their 'intellectual honesty', or their 'scientific integrity'. ("... Luther eliminates *external* religiousness and turns religiousness into the *inner* essence of man... he negates the raving parish-priest outside the layman because he puts him into the very heart of the layman". Marx, *Nationaloekonomie und Philosophie*; quoted from Marx, *die Frühschriften*, ed. by Landshut, p. 228.)

In the *economic field* the development is very clear: "In antiquity and in the Middle Ages exploitation was regarded as an obvious, indisputable, and unchangeable fact by both sides, by the free as well as by the slaves, by the feudal lords as well as by their bondsmen. It was precisely because of this knowledge on the part of both parties that the class structure was so transparent; and it was precisely because of the dominance of agriculture that the exploitation of the lower classes *could be seen in the strict sense of the word*. In the Middle Ages the serf worked, say, four days and a half per week on his own plot of land and one day and a half on the land of his master. The place of work for himself was distinctly separated from the place of serfdom... Even the language was clear, it spoke of 'bondsmen' ['Leibeigene', i.e., those whose bodies are owned by someone else]... of 'compulsory service' ['Fronarbeit'] and so on. Thus the class distinctions could not only be *seen*, they could also be *heard*. Language did not conceal the class structure, it expressed it in all desirable clarity. That was true in Egypt, Greece, the European Middle Ages, in Asiatic as well as in European languages. It is no longer true in our present epoch... Workers in early capitalism spent their whole time in the factory. There was neither a spatial nor a temporal separation between the period they worked for their own livelihood and the period they slaved for the capitalist. This led to the phenomenon I have called ... the 'sociology of repression'. The fact of exploitation was no longer admitted and the repression was facilitated because exploitation could no longer be *seen*". Fritz Sternberg. *Der Dichter und die Ratio; Erinnerungen an Bertolt Brecht* (Sachse und Pohl, Göttingen, 1963), pp. 47ff. *Exactly the same development occurred between Galileo and, say, Laplace.* Science ceased to be a variable human instrument for exploring and changing the world and became a solid block of 'knowledge', impervious to human dreams, wishes, expectations. At the same time the scientists themselves became more and more remote, 'serious', greedy for recognition, and incapable and unwilling to express themselves in a way that could be understood and enjoyed by all. Einstein and Bohr, and Boltzmann before them, were notable exceptions. But they did not change the general trend. There are only a few physicists now who share the humor, the modesty, the sense of perspective, and the philosophical interests of these extraordinary people. All of them have taken over their physics, but they have thoroughly ruined it.

It is even worse in the philosophy of science. For some details, see my papers 'Classical Empiricism' and 'On the Improvement of the Sciences and the Arts, and the Possible Identity of the Two', in *Boston Studies in the Philosophy of Science*, Vol. III, ed. by R. S. Cohen and M. W. Wartofsky (Reidel, Dordrecht, 1968).

[31] Popper has repeatedly asserted, both in his lectures and in his writings, that while there is progress in the sciences there is no progress in the arts. He bases his assertion on the belief that the content of succeeding theories can be compared and that a judgment of verisimilitude can be made. The refutation of this belief eliminates an important difference, and perhaps the *only* important difference, between science and the arts, and makes it possible to speak of styles and preferences in the first, and of progress in the second.

[32] Cf. B. Brecht, 'Ueber das Zerpfluecken von Gedichten', *Über Lyrik* (Suhrkamp, Frankfurt, 1964). In my lectures on the theory of knowledge I usually present and discuss the thesis that finding a new theory for given facts is exactly like finding a new production for a well-known play. For painting see also E. Gombrich, *Art and Illusion* (Pantheon, New York, 1960).

[33] "The picture of society which we construct for the river-engineers, for the gardeners ... and for the revolutionaries. All of them we invite into our theater, and we ask them not to forget their interest in *entertainment* when they are with us, for we want to turn over the world to their brains and hearts so that they may change it according to their wishes." Brecht, 'Kleines Organon für das Theater', *Schriften zum Theater* (Suhrkamp, Frankfurt, 1964), p. 20; my italics.

INDEX OF NAMES

SYNTHESE LIBRARY

Monographs on Epistemology, Logic, Methodology,
Philosophy of Science, Sociology of Science and of Knowledge, and on the
Mathematical Methods of Social and Behavioral Sciences

Managing Editor:

JAAKKO HINTIKKA (Academy of Finland and Stanford University)

Editors:

ROBERT S. COHEN (Boston University)
DONALD DAVIDSON (The Rockefeller University and Princeton University)
GABRIËL NUCHELMANS (University of Leyden)
WESLEY C. SALMON (University of Arizona)

1. J. M. BOCHEŃSKI, *A Precis of Mathematical Logic.* 1959, X + 100 pp.
2. P. L. GUIRAUD, *Problèmes et méthodes de la statistique linguistique.* 1960, VI + 146 pp.
3. HANS FREUDENTHAL (ed.), *The Concept and the Role of the Model in Mathematics and Natural and Social Sciences, Proceedings of a Colloquium held at Utrecht, The Netherlands, January 1960.* 1961, VI + 194 pp.
4. EVERT W. BETH, *Formal Methods. An Introduction to Symbolic Logic and the Study of Effective Operations in Arithmetic and Logic:* 1962, XIV + 170 pp.
5. B. H. KAZEMIER and D. VUYSJE (eds.), *Logic and Language. Studies dedicated to Professor Rudolf Carnap on the Occasion of his Seventieth Birthday.* 1962, VI + 256 pp.
6. MARX W. WARTOFSKY (ed.), *Proceedings of the Boston Colloquium for the Philosophy of Science, 1961–1962, Boston Studies in the Philosophy of Science* (ed. by Robert S. Cohen and Marx W. Wartofsky), Volume I. 1973, VIII + 212 pp.
7. A. A. ZINOV'EV, *Philosophical Problems of Many-Valued Logic.* 1963, XIV + 155 pp.
8. GEORGES GURVITCH, *The Spectrum of Social Time.* 1964, XXVI + 152 pp.
9. PAUL LORENZEN, *Formal Logic.* 1965, VIII + 123 pp.
10. ROBERT S. COHEN and MARX W. WARTOFSKY (eds.), *In Honor of Philipp Frank, Boston Studies in the Philosophy of Science* (ed. by Robert S. Cohen and Marx W. Wartofsky), Volume II. 1965, XXXIV + 475 pp.
11. EVERT W. BETH, *Mathematical Thought. An Introduction to the Philosophy of Mathematics.* 1965, XII + 208 pp.
12. EVERT W. BETH and JEAN PIAGET, *Mathematical Epistemology and Psychology.* 1966, XII + 326 pp.
13. GUIDO KÜNG, *Ontology and the Logistic Analysis of Language. An Enquiry into the Contemporary Views on Universals.* 1967, XI + 210 pp.
14. ROBERT S. COHEN and MARX W. WARTOFSKY (eds.), *Proceedings of the Boston Colloquium for the Philosophy of Science 1964–1966, in Memory of Norwood Russell Hanson, Boston Studies in the Philosophy of Science* (ed. by Robert S. Cohen and Marx W. Wartofsky), Volume III. 1967, XLIX + 489 pp.

15. C. D. Broad, *Induction, Probability, and Causation. Selected Papers*. 1968, XI + 296 pp.
16. Günther Patzig, *Aristotle's Theory of the Syllogism. A Logical-Philosophical Study of Book A of the Prior Analytics*. 1968, XVII + 215 pp.
17. Nicholas Rescher, *Topics in Philosophical Logic*. 1968, XIV + 347 pp.
18. Robert S. Cohen and Marx W. Wartofsky (eds.), *Proceedings of the Boston Colloquium for the Philosophy of Science 1966–1968*, Boston Studies in the Philosophy of Science (ed. by Robert S. Cohen and Marx W. Wartofsky), Volume IV. 1969, VIII + 537 pp.
19. Robert S. Cohen and Marx W. Wartofsky (eds.), *Proceedings of the Boston Colloquium for the Philosophy of Science 1966–1968*, Boston Studies in the Philosophy of Science (ed. by Robert S. Cohen and Marx W. Wartofsky), Volume V. 1969, VIII + 482 pp.
20. J. W. Davis, D. J. Hockney, and W. K. Wilson (eds.), *Philosophical Logic*. 1969, VIII + 277 pp.
21. D. Davidson and J. Hintikka (eds.), *Words and Objections: Essays on the Work of W. V. Quine*. 1969, VIII + 366 pp.
22. Patrick Suppes, *Studies in the Methodology and Foundations of Science. Selected Papers from 1911 to 1969*, XII + 473 pp.
23. Jaakko Hintikka, *Models for Modalities. Selected Essays*. 1969, IX + 220 pp.
24. Nicholas Rescher et al. (eds.). *Essay in Honor of Carl G. Hempel. A Tribute on the Occasion of his Sixty-Fifth Birthday*. 1969, VII + 272 pp.
25. P. V. Tavanec (ed.), *Problems of the Logic of Scientific Knowledge*. 1969, XII + 429 pp.
26. Marshall Swain (ed.), *Induction, Acceptance, and Rational Belief*. 1970. VII + 232 pp.
27. Robert S. Cohen and Raymond J. Seeger (eds.), *Ernst Mach; Physicist and Philosopher*, Boston Studies in the Philosophy of Science (ed. by Robert S. Cohen and Marx W. Wartofsky), Volume VI. 1970, VIII + 295 pp.
28. Jaakko Hintikka and Patrick Suppes, *Information and Inference*. 1970, X + 366 pp.
29. Karel Lambert, *Philosophical Problems in Logic. Some Recent Developments*. 1970, VII + 176 pp.
30. Rolf A. Eberle, *Nominalistic Systems*. 1970, IX + 217 pp.
31. Paul Weingartner and Gerhard Zecha (eds.), *Induction, Physics, and Ethics, Proceedings and Discussions of the 1968 Salzburg Colloquium in the Philosophy of Science*. 1970, X + 382 pp.
32. Evert W. Beth, *Aspects of Modern Logic*. 1970, XI + 176 pp.
33. Risto Hilpinen (ed.), *Deontic Logic: Introductory and Systematic Readings*. 1971, VII + 182 pp.
34. Jean-Louis Krivine, *Introduction to Axiomatic Set Theory*. 1971, VII + 98 pp.
35. Joseph D. Sneed, *The Logical Stricture of Mathematical Physics*. 1971, XV + 311 pp.
36. Carl R. Kordig, *The Justification of Scientific Change*. 1971, XIV + 119 pp.
37. Milič Čapek, *Bergson and Modern Physics*, Boston Studies in the Philosophy of Science (ed. by Robert S. Cohen and Marx W. Wartofsky), Volume VII, 1971, XV + 414 pp.
38. Norwood Russell Hanson, *What I do not Believe, and other Essays*, ed. by Stephen Toulmin and Harry Woolf, 1971, XII + 390 pp.

39. ROGER C. BUCK and ROBERT S. COHEN (eds.), *PSA 1970. In Memory of Rudolf Carnap*, Boston Studies in the Philosophy of Science (ed. by Robert S. Cohen and Marx W. Wartofsky, Volume VIII. 1971, LXVI + 615 pp. Also available as a paperback.
40. DONALD DAVIDSON and GILBERT HARMAN (eds.), *Semantics of Natural Language*. 1972, X + 769 pp. Also available as a paperback.
41. YEHOSHUA BAR-HILLEL (ed)., *Pragmatics of Natural Languages*. 1971, VII + 231 pp.
42. SÖREN STENLUND, *Combinators, λ-Terms and Proof Theory*. 1972, 184 pp.
43. MARTIN STRAUSS, *Modern Physics and Its Philosophy. Selected Papers in the I ogic, History, and Philosophy of Science*. 1972, X + 297 pp.
44. MARIO BUNGE, *Method, Model and Matter*. 1973, VII + 196 pp.
45. MARIO BUNGE, *Philosophy of Physics*. 1973, IX + 248 pp.
46. A. A. ZINOV'EV, *Foundations of the Logical Theory of Scientific Knowledge (Complex Logic)*, Boston Studies in the Philosophy of Science (ed. by Robert S. Cohen and Marx W. Wartofsky), Volume IX. Revised and enlarged English edition with an appendix, by G. A. Smirnov, E. A. Sidorenko, A. M. Fedina, and L. A. Bobrova 1973, XXII + 301 pp. Also available as a paperback.
47. LADISLOV TONDL, *Scientific Procedures*, Boston Studies in the Philosophy of Science (ed. by Robert S. Cohen and Marx W. Wartofsky), Volume X. 1973, XII + 268 pp. Also available as a paperback.
48. NORWOOD RUSSELL HANSON, *Constellations and Conjectures*, ed. by Willard C. Humphreys, Jr. 1973, X + 282 pp.
49. K. J. J. HINTIKKA, J. M. E. MORAVCSIK, and P. SUPPES (eds.), *Approaches to Natural Language. Proceedings of the 1970 Stanford Workshop on Grammar and Semantics*. 1973, VIII + 526 pp. Also available as a paperback.
50. MARIO BUNGE (ed.), *Exact Philosophy – Problems, Tools, and Goals*. 1973, X + 214 pp.
51. RADU J. BOGDAN and ILKKA NIINILUOTO (eds.), *Logic, Language, and Probability*. A selection of papers contributed to Sections IV, VI, and XI of the Fourth International Congress for Logic, Methodology, and Philosophy of Science, Bucharest, September 1971. 1973, X + 323 pp.
52. GLENN PEARCE and PATRICK MAYNARD (eds.), *Conceptual Chance*. 1973, XII + 282 pp.
53. ILKKA NIINILUOTO and RAIMO TUOMELA, *Theoretical Concepts and Hypothetico-Inductive Inference*. 1973, VII + 264 pp.
54. ROLAND FRAÏSSÉ, *Course of Mathematical Logic* – Volume I: *Relation and Logical Formula*. 1973, XVI + 186 pp. Also available as a paperback.
55. ADOLF GRÜNBAUM, *Philosophical Problems of Space and Time*. Second, enlarged edition, Boston Studies in the Philosophy of Science (ed. by Robert S. Cohen and Marx W. Wartofsky), Volume XII. 1973, XXIII + 884 pp. Also available as a paperback.
56. PATRICK SUPPES (ed.), *Space, Time, and Geometry*. 1973, XI + 424 pp.
57. HANS KELSEN, *Essays in Legal and Moral Philosophy*, selected and introduced by Ota Weinberger. 1973, XXVIII + 300 pp.
58. R. J. SEEGER and ROBERT S. COHEN (eds.), *Philosophical Foundations of Science. Proceedings of an AAAS Program, 1969*. Boston Studies in the Philosophy of Science (ed. by Robert S. Cohen and Marx W. Wartofsky), Volume XI. 1974, X + 545 pp. Also available as paperback.
59. ROBERT S. COHEN and MARX W. WARTOFSKY (eds.), *Logical and Epistemological*

Studies in Contemporary Physics, Boston Studies in the Philosophy of Science (ed. by Robert S. Cohen and Marx W. Wartofsky), Volume XIII. 1973, VIII + 462 pp. Also available as paperback.

60. ROBERT S. COHEN and MARX W. WARTOFSKY (eds.), *Methodological and Historical Essays in the Natural and Social Sciences. Proceedings of the Boston Colloquium for the Philosophy of Science, 1969–1972*, Boston Studies in the Philosophy of Science (ed. by Robert S. Cohen and Marx W. Wartofsky), Volume XIV. 1974, VIII + 405 pp. Also available as paperback.

61. ROBERT S. COHEN, J. J. STACHEL, and MARX W. WARTOFSKY (eds.), *For Dirk Struik. Scientific, Historical and Political Essays in Honor of Dirk J. Struik*, Boston Studies in the Philosophy of Science (ed. by Robert S. Cohen and Marx W. Wartofsky), Volume XV. 1974, XXVII + 652 pp. Also available as paperback.

62. KAZIMIERZ AJDUKIEWICZ, *Pragmatic Logic*, transl. from the Polish by Olgierd Wojtasiewicz. 1974, XV + 460 pp.

63. SÖREN STENLUND (ed.), *Logical Theory and Semantic Analysis. Essays Dedicated to Stig Kanger on His Fiftieth Birthday*. 1974, V + 217 pp.

64. KENNETH F. SCHAFFNER and ROBERT S. COHEN (eds.), *Proceedings of the 1972 Biennial Meeting, Philosophy of Science Association*, Boston Studies in the Philosophy of Science (ed. by Robert S. Cohen and Marx W. Wartofsky), Volume XX. 1974, IX + 444 pp. Also available as paperback.

65. HENRY E. KYBURG, JR., *The Logical Foundations of Statistical Inference*. 1974, IX + 421 pp.

66. MARJORIE GRENE, *The Understanding of Nature: Essays in the Philosophy of Biology*, Boston Studies in the Philosophy of Science (ed. by Robert S. Cohen and Marx W. Wartofsky), Volume XXIII. 1974, XII + 360 pp. Also available as paperback.

67. JAN M. BROEKMAN, *Structuralism: Moscow, Prague, Paris*. 1974, IX + 117 pp.

68. NORMAN GESCHWIND, *Selected Papers on Language and the Brain*, Boston Studies in the Philosophy of Science (ed. by Robert S. Cohen and Marx W. Wartofsky), Volume XVI. 1974, XII + 549 pp. Also available as paperback.

69. ROLAND FRAÏSSÉ. *Course of Mathematical Logic* – Volume II: *Model Theory*. 1974, XIX + 192 pp.

70. ANDRZEJ GRZEGORCZYK, *An Outline of Mathematical Logic. Fundamental Results and Notions Explained with all Details*. 1974, X + 596 pp.

71. FRANZ VON KUTSCHERA, *Philosophy of Language*. 1975, VII + 305 pp.

75. JAAKKO HINTIKKA and UNTO REMES, *The Method of Analysis. Its Geometrical Origin and Its General Significance*. 1974, XVIII + 144 pp.

76. JOHN EMERY MURDOCH and EDITH DUDLEY SYLLA, *The Cultural Context of Medieval Learning. Proceedings of the First International Colloquium on Philosophy, Science, and Theology in the Middle Ages – September 1973*. Boston Studies in the Philosophy of Science (ed. by Robert S. Cohen and Marx. W. Wartofsky), Volume XXVI. 1975, X + 566 pp. Also available as paperback.

77. STEFAN AMSTERDAMSKI, *Between Experience and Metaphysics. Philosophical Problems of the Evolution of Science*. Boston Studies in the Philosophy of Science (ed. by Robert S. Cohen and Marx W. Wartofsky), Volume XXXV. 1975, XVIII + 193 pp. Also available as paperback.

SYNTHESE HISTORICAL LIBRARY

Texts and Studies
in the History of Logic and Philosophy

Editors:

N. KRETZMANN (Cornell University)
G. NUCHELMANS (University of Leyden)
L. M. DE RIJK (University of Leyden)

1. M. T. BEONIO-BROCCHIERI FUMAGALLI, *The Logic of Abelard.* Translated from the Italian. 1969, IX + 101 pp.

2. GOTTFRIED WILHELM LEIBNITZ, *Philosophical Papers and Letters.* A selection translated and edited with an introduction, by Leroy E. Loemker. 1969, XII + 736 pp.

3. ERNST MALLY, *Logische Schriften,* ed. by Karl Wolf and Paul Weingartner. 1971, X + 340 pp.

4. LEWIS WHITE BECK (ed.), *Proceedings of the Third International Kant Congress.* 1972, XI + 718 pp.

5. BERNARD BOLZANO, *Theory of Science,* ed. by Jan Berg. 1973, XV + 398 pp.

6. J. M. E. MORAVCSIK (ed.), *Patterns in Plato's Thought. Papers arising out of the 1971 West Coast Greek Philosophy Conference.* 1973, VIII + 212 pp.

7. NABIL SHEHABY, *The Propositional Logic of Avicenna: A Translation from al-Shifā': al-Qiyās,* with Introduction, Commentary and Glossary. 1973, XIII + 296 pp.

8. DESMOND PAUL HENRY, *Commentary on De Grammatico: The Historical-Logical Dimensions of a Dialogue of St. Anselm's.* 1974, IX + 345 pp.

9. JOHN CORCORAN, *Ancient Logic and Its Modern Interpretations.* 1974. X + 208 pp.

10. E. M. BARTH, *The Logic of the Articles in Traditional Philosophy.* 1974, XXVII + 533 pp.

11. JAAKKO HINTIKKA, *Knowledge and the Known. Historical Perspectives in Epistemology.* 1974, XII + 243 pp.

12. E. J. ASHWORTH, *Language and Logic in the Post-Medieval Period.* 1974, XIII + 304 pp.

13. ARISTOTLE, *The Nicomachean Ethics.* Translated with Commentaries and Glossary by Hyppocrates G. Apostle. 1975, XXI + 372 pp.

14. R. M. DANCY, *Sense and Contradiction: A Study in Aristotle.* 1975, XII + 184 pp.

15. WILBUR RICHARD KNORR, *The Evolution of the Euclidean Elements. A Study of the Theory of Incommensurable Magnitudes and Its Significance for Early Greek Geometry.* 1975, IX + 374 pp.

16. AUGUSTINE, *De Dialectica.* Translated with the Introduction and Notes by B. Darrell Jackson. 1975, XI + 151 pp.

Edit: Thurston Davis

(co

Th
or
fr
the
the ecum
meaning of H
untried Peace

Old subscrib
glad to peruse fa
former years. Tho:
AMERICA for the fir:
will find that they
vastly new and cha
the mind."

Augustine's *City o*
fascinating monument
and erudition—one of th
sessions of all mankind.

The editors of *Betu*
make a more modest cl
of selections. They ho
struct as well as entert
who dwell perforce i
sider themselves b
another commonw
all earthly dignit
shifting scene."

Between Two Cities
God and Man in AMERICA

A COMMENTARY ON OUR TIMES

Cover design by Robert Borja

Library of Congress Catalog Card Number: 62-10907

Between Two Cities

God and Man in America

Edited by

Fathers of The Society of Jesus

Thurston N. Davis

Donald R. Campion

L. C. McHugh

LOYOLA UNIVERSITY PRESS

Chicago 13, Illinois

1962

IMPRIMI POTEST: John R. Connery, S.J., *Provincial of the Chicago Province,* December 11, 1961. NIHIL OBSTAT: John B. Amberg, S.J., *Censor deputatus,* December 11, 1961. IMPRIMATUR: ✠ Albert Cardinal Meyer, *Archbishop of Chicago,* December 12, 1961.

Preface

The title imprinted on the spine of this book
suggests that between these two covers a reader
might expect to find some sort of "tale of two
cities." It is in fact such a tale, or a collection of
such tales, but not of just any two cities.

An armchair traveler can readily locate many
sets of paired cities around the globe: Paris and
Rome, Berlin and Munich, Moscow and Leningrad, London and Birmingham. Here in the United
States we could name San Francisco and Los Angeles on our west coast, St. Paul and Minneapolis
in Minnesota, Baltimore and Washington on our
east coast.

However, to those of us who have shared the
experience of putting this book together, the twin
cities suggested by the title of this volume might
well be Chicago and New York. Chicago—because
it is the home of the book's publisher, Loyola University Press; New York City—because on the western rim of New York's colorful midtown area
stands Campion House, residence of the editors of
AMERICA, the National Catholic Weekly Review.
In a real though secondary sense the title of this
volume acknowledges both these points of origin.

The fraternal bonds between AMERICA on one
hand and Loyola University Press on the other
have for years been tight and strong. Our two publishing establishments operate in distinct fields
and according to different editorial tempos, but on
each side we share the concern of brothers and

I

friends for the persons and the work of the other.

The late Father Austin G. Schmidt, S.J., former director of the Loyola University Press, was, all through his dedicated and scholarly life as a publisher, an object of admiration to the editors of AMERICA. Moreover, throughout the years we have come to know and esteem his able assistant, now Father Schmidt's successor, Father John B. Amberg, S.J. And it was Robert B. McCoy, a recent addition to the lively staff of Loyola University Press, who first suggested and has enthusiastically taken on the task of publishing this book. We are grateful to him.

But what are the two cities between which this volume is really poised? They are, of course, the City of God and the City of Man. AMERICA's editors, following the path marked out for all mortal *viatores*, are on pilgrimage from one to the other. In our professional lives as priests, we move back and forth, over and over again, from one of these cities to the other. Moreover, in our peculiar vocation as editors and journalists, we make this journey, back and forth every seven days, in obedience to the rigorous demands of an unending and inevitable series of press deadlines. From a study of the latest scarifying communique out of the Kremlin; from a discussion of the plight of the *braceros* in California and the Southwest; from a volume on demography and world agricultural statistics; from the halls of the United Nations or the Congress on Capitol Hill, we pack up and go each week—galleys and page proofs in hand—to a point beyond time and out of space, into a realm of unchanging truths, fixed principles, and stable values. Then back we come with another issue of AMERICA. In other words, we humbly attempt a weekly raid on the City of God in order to publish, in the City of Man, a journal that talks common Christian sense about the world of human events.

This may sound a bit pretentious. It is not so intended. We have no hidden or special revelations. We know no secret back door into the mind of God. But we do have certain advantages: an excellent library, a sound and long tradition as a journalistic team, and the circumstances of life that allow us the leisure (a very busy leisure, I can assure you) to explore and think and argue and write. Moreover, we are blessed in that we do all these things in the congenial company of able and hard-working colleagues.

In other words, AMERICA and its contributors are concerned with both God and man, and with the cities of both of them. Within the pages of this book a reader will find a hundred paths that crisscross the complicated world of contemporary affairs. But through it all, and written between every line, is the conviction that in all its diversity and change the world of man is God's world and that he who does not labor to return it to God redeemed in some small measure by his tears and worry and dedication has missed the meaning of man's job on earth.

This little preface gives me the occasion publicly to acknowledge my personal esteem for my colleagues, past and present, on the staff of AMERICA. To each and every one of this band of brothers who, for more than fifty years, have striven to publish a Review worthy of its name, I am deeply in debt for the inspiration and example of their lives and work. Finally, it should be noted that this book would not have seen the light of day without the collaboration of my two associates, Donald R. Campion, S.J. and L. C. McHugh, S.J., and of a valued colleague, William Holub, general manager of the AMERICA Press.

THURSTON N. DAVIS, s.j.
Editor in Chief
AMERICA

To
Austin G. Schmidt
and
Vincent A. McCormick
of the Society of Jesus

Table of Contents

Poems

IX

What would *you* talk about, if you were one of an army of pilgrims hiking to the cathedral of Chartres? Perhaps you would talk of shoes and ships and sealing wax, but it is more likely that you would talk of God and man; and your conversation might turn with regret to the scandal of Christian disunity. Why has the constructive dialogue on ecumenism, so far advanced in Europe, made so little progress in America, the land chosen to exercise leadership in freedom's darkest hour? Is there any way in which priest and layman can help make the Church overwhelmingly evident as the way to God? Or are we doomed to go on bickering over the proper highway to heaven, even as humanity begins to explore its destiny on the endless frontiers of space?

God and Man

The Story of a Roman Catholic

John LaFarge, S.J.

IF YOU WRITE a book that people like, they are apt to say nice things. In itself, this is not so bad, since it may help you to discover what is really worth your effort. Nice words also humble you, because the response you get, even if most kindly, is not quite what you expected. Favorite passages remain unnoticed, while people seem to delight in matters on which you set little store.

They were reading my autobiography aloud at meals, a couple of years ago, in a monastery refectory. After four chapters had been completed, word came that they were to cease and desist. Certain items in these first four chapters, matters that did not seem to me particularly humorous, had caused chuckles among the brethren. The Very Reverend Prior Visitor, who was of another nationality and language, reminded the brethren that reading at table should excite compunction, not hilarity. Other monasteries and convents, so it is reported, have read the book through without remorse. Even Père Jean Minéry's recent French translation (which, incidentally, is a swell one) is being read in refectories abroad. All this was a surprise to me, since

1

I had aimed at a merely factual narrative. Life itself, however, is full of anomalies.

More to be expected are honest corrections and disagreements. They are considerably helpful, even if they inflict minor wounds. I was much distressed when I received a strong protest from a very dear relative—now no longer with us—because on one occasion I had referred to the (temporary) color of her mother's hair. On consideration, I saw that it was possibly unfair to select a single item when so very much could be rightly said about her mother, a lovely, gifted and generous soul. So I made amends in the paper-back edition. Yet I stood by the accuracy of my original narrative. I cannot *prove* the occurrence, yet I distinctly recall it. So I invoked a principle, the guide of autobiographers. What you set down on paper is simply what you personally remember, as you remember it. You check the facts to the best of your ability, but certain of them rest in the realm of merely personal memories.

This is why I stood by my narrative when I expressed the disagreeable impression made upon me, as an undergraduate, by the late George Santayana, or the favorable one I received concerning the heroic Cardinal Stepinac, when I took supper with him in Zagreb just before the war. Stepinac was—and still is—a national hero of the Croatians, and Serbians and Croats have many bitter griefs against one another. But the best you can do, if you are not attempting to write history, is to record impressions. And on the other slope of that national divide, what finer impression could anyone receive than that afforded by AMERICA's valiant Serbian-Irish Balkan correspondent over the many years, the gifted Miss Annie Christich?

As I leaf through the apparently endless series of messages and comments that have continued from the publication of the book to this very day, or recall recent conversations along the same line, some do specially stand out. They give you heart, though they are of interest only to yourself. The confirmation of my own memory that perhaps I have most prized came last summer as I talked in Boston with Mrs. John Potter (Ellen Hooper) and her sister Mrs. Ward Thoron,

2

nieces of Henry Adams, and friends of my college days. "You did a perfect picture of your father," they said. "Just as we knew him."

Why, after all, would such a story attract people? "What I notice in your story," writes one critic, "is that you were attracted by the personality of Jesus." Quite so. If that attraction has seized a person, especially one who speaks for Jesus, others may be attracted to him, and wish for his story. People after all are drawn to the priest, not by any tricks of his own personality, but by the very fact that he is the bearer and the messenger of the Church's imperishable hope. And people today are desperately in need of hope.

Words to that effect come from many sources. Dr. Karl Menninger, that famous oracle of psychic health, said recently that in his opinion people today are most of all concerned over the loss of hope. The crisis, as many think, is more of hope that it is even of faith. As Fr. Vincent McCorry puts it (AM. 1/16/60, p. 484):

> Some ages in Christian history have needed more religious faith; such, in general, was the 19th century. Other eras needed more religious love; such as the 18th century, insofar as it was the age of Jansenism. Our time almost certainly needs more religious hope.

The priest speaks for hope. He is not just a foreteller of some future glorious event. By his sacramental action as well as by his words and activities, he conveys to the world Him who *is* our hope. If the priest himself is accessible in the world of time, it is easier for us to communicate with the world of the Eternal, who has made His home—pitched His tent, as St. John says—in time.

I remember watching the dial of my fat old silver-cased watch the night of December 31, 1899, before turning out the gas light over my bed, and asking myself whether the new 20th century could ever equal the marvels of the one just expired. (Oh, I know, I was a year too soon, and all that.) The 19th-century tradition did everything possible to persuade people they could scrap their faith and yet—or thereby—be completely and comfortably cheerful. (Every Englishman might enjoy

3

his cup of hot cocoa on a fast electric train.) So life's rhythm could still proceed in secular comfort even though deprived of the logical foundation of any assurance. Today, the sabotaging of faith has developed its equally logical conclusion. Many of those who preserve the faith are not unaffected by the prevailing atmosphere.

STRESS ON CATHOLICITY

Yet, I must ask myself, has not the story something more personal to tell, something quite distinctive? If I review in my mind this chain of events and experiences, I note a lifelong refrain. This was a deep and abiding sense of the universal mission—to all classes and peoples —of the Church, that spiritual Kingdom of which I became a citizen on a cold day back in February some 80 years ago. (The pastor, incidentally, didn't enter my name in the parish register, since I was not expected to live. It was only years later, through the affidavit of the two witnesses to the baptism, that it was made historically certain that I did really enter the Kingdom. My mother habitually lived on hope. Indeed, I had only emerged into this world at all, because, as Ed Lahey remarked recently in the Boston *Globe*, Mrs. LaFarge was not over-concerned about the population explosion. But that is another story.)

Some may legitimately ask: what difference does your abiding view make to the rest of us? Perhaps none. Nevertheless, if the Church is to make its deserved impact upon the confused and divided contemporary world, its universal mission, and thereby its catholicity, must be made overwhelmingly evident. Unless that is done, and done with great imagination and comprehension, the Church's other famous "marks" of unity, holiness and apostolic origin are spoken to deaf ears and closed minds. I do not venture this remark just of myself. I find this same stress on catholicity running through the utterances of the Holy See in our times— as for instance in the great Christmas discourses of Pope Pius XII. Yet my own experience has forced it upon me.

Catholicity, to mean anything, must be in force here

and now: in my city, in my block or suburban section, in my own parish. In line with just such a viewpoint, for instance, the Holy See declared on February 22, 1951, in response to an inquiry from a Catholic diocese in the United States, that no more parishes could be set up on racial lines. By catholicity I mean not only that the mission of the Church is to all peoples and all mankind alike, but also—in a vertical sense—to all that is contained in man himself. It is certainly a vital part of the Church's present-day message that it loves the created world, as God's free gift, as it has evolved through countless ages; the natural world in its intricacy and multiplicity. This world remains, and will apparently remain until the end of time as a battleground between the Spirit of God and the Enemy of the human race. Yet this is the world to which God has said Amen, and waits for us to respond with our own Amen, having given us the awesome power of accepting or refusing His grace.

Or of ignoring the moral law, which has a catholicity of its own.

There is a catholicity of love, which is the forward-moving wave of the Church. But there is also a catholicity of justice, of the integral teaching and exemplifying of the moral law, without which love cannot freely operate. This I believe is the issue especially confronting us Catholics during the year 1960.

The moral issue is being put with sharpness and poignancy from various and contradictory angles, about matters of social or political conduct that most people are not used to consider as falling under the Ten Commandments. The use of nuclear weapons and the testing of hydrogen bombs have put a startling new face upon the Fifth (Protestant and Hebrew Sixth) Commandment: Thou shalt not kill. This commandment, as Herman Wouk notices in his remarkable book, *This Is My God,* isn't even expressed by that much rhetoric in its original form. The Hebrew simply says: Don't kill, *lô tirtzah;* don't steal; don't covet, etc. The searching test case of true catholicity at Deerfield Park, Illinois, and the defeat last summer of the Fair Housing Bill in my native Rhode Island put sharp moral queries to Catholic

5

consciences. Other queries face young married couples planning for future families, or candidates for high public office, not to speak of the moral questions that afflict the quiz-hero selling his presumed talents in the isolation booth, or the practitioners of economic power in management or labor.

As I see it, Christian moral teaching, if it is to be convincing to the contemporary mind, must also be truly catholic or universal. It can admit of no preferences or exceptions, for the simple reason that man is an integral whole, not a collection of propositions. Our teaching must speak the entire moral message in all its bearings: state the theory, but also apply it remorselessly to tangible and well-known situations. Prudence—that important but frequently distorted virtue—may suggest that we speak only a partial version and soft-pedal the rest. I suppose it would be highly imprudent today, at least in Governor-elect Jimmy Davis's Louisiana, to recall how, on the eve of World War II, a filibustering senator of that same sovereign State (where my paternal grandfather lived until he decided to sell his plantation, manumit his slaves, and leave *"pour les Etats-Unis"*), filled the halls of the U. S. Senate for days with pagan appeals to race prejudice and shreds of Hitler's "Aryan" lore, yet sent, over his own signature, along with Senators Connally and Pepper, a fervent message of congratulation to the Reds of anti-Franco Spain. But should we conceal these apparent inconsistencies?

CLIMATE FOR DIALOGUE

Indeed the same prudence, considering the present international scene, can be in fact the highest imprudence. So much so, that I would consider any evasive course today as suicidal for us Catholics, or indeed for believing people in general. The slight advantages we gain by avoiding one or the other painful issue are entirely out of proportion to the respect that is earned by frankness upon all the great moral issues.

I believe there is no better or sounder climate for fruitful dialogue with all persons of sincere religious belief, especially here in the United States, than one created by a mutual confrontation of the entire gamut of moral problems, as they affect the individual, the family, society and community, or the nation. This would mean considering them, not—as is so frequently done—in splendid isolation, but as related to one another, and related to the great central idea of the full realization of man's destiny under God.

Such advice, however, must take into account that on many of these specific points our Catholic doctrine as yet is much subject to discussion. The obstacles we encounter frequently reveal connections and implications that otherwise would be passed over. The challenge of the birth-control issue, for instance, is making intelligent people on both sides of the dispute look more sharply into the allied moral questions of the nature of the family, of our country's international obligations, of the aims and purposes of American agriculture, as well as into the nature of moral law and its relation to revealed religion. Enthusiastic Planned Parenthood proponents are becoming somewhat more guarded in their utterances, while Catholics recognize they must present the issue not just as a one-two precept, but as framed in the wider context of all human experience.

Yet today how much general encouragement is given to long and patient conference and investigation, such— to take but a single instance—as the year-round program of the Catholic Association for International Peace? This is a work based upon the best knowledge obtainable from every source, and upon the scrupulously faithful study of the Church's official documents and the advice of competent scholars from every direction. There is a time, King Solomon says, to speak and a time to keep silent. So there are occasions that call for stern warning, the sensational exposé, even the harangue. But a continued diet of this emergency language is self-defeating. High school children write in to me and ask when the Pope will open "that letter." A widely read popular columnist in Catholic papers recently exhorted our clergy to drop every other subject and concentrate all their preaching upon the sole issue of U. S. Communist infiltration. That way madness lies.

Certainly the preacher or writer who adroitly avoids any positive moral commitments enjoys many tactical advantages. He can always create the impression of saying something, without suffering any inconvenience by what is said. He can't be caught off base, because there is nothing definite to catch on to. Today's whirlwind means of publicity, too, favor the dedicated partisan, the professional liberal or anti-liberal. In both courses there are evident tactical advantages. Yet—and I offer this as the result of cumulative experience—the odds in the long run rest with the relatively small number who climb the long, uphill road and try to speak the many-sided moral message as an integrated whole. They will be reproached for mumbling where others speak with a clarion voice. But in the long run they succeed in being heard, and what they say sticks permanently.

A few weeks ago I spoke at St. Mary's City, Maryland, at a meeting to commemorate the 325th Anniversary of the founding, on that spot, of the original Maryland Colony. The event had for me a special significance, since I had spent many years as pastor of that territory, and had often tried to create anew in my own mind the happenings of those remote historical days.

LESSON OF MARYLAND

The gentlemen or burgesses of early Maryland—mostly, even if not all, Catholic—were not as free as they would have liked to make clear, public statements on certain delicate issues. Jealous royal eyes across the sea were watching them. Yet they stoutly held their position on civil liberties, as Englishmen and heirs to the tradition of Magna Carta. Respecting these same liberties, they proclaimed in guarded language the freedom of religious worship and personal freedom as well. For the first half-century of their existence under the tutelage of the Lords Baltimore—that is to say, until the proprietary or personal colony was forced to become a royal colony—these burgesses sedulously practiced the tolerance that they preached.

Today, we in the United States suffer no restraints upon our freedom of speech, and there is nothing but human respect to hinder our conveying to the public the total integrity of our faith, or of our moral message. Our spiritual strength lies not in the dialectical discomfiture of those who happen to disagree with us, and still remain unconvinced. Our strength lies rather in the partnership of all good men, of whatever origin or condition, in seeking agreement by the best means available on these capital issues, at least as far as is possible under the very different set of presuppositions by which we approach them.

"I find no difficulty," said a prominent Protestant clergyman to me of late, as we were discussing certain recent utterances of an ethical nature, "I find no difficulty about the term 'catholic.' Indeed, I like to consider myself a catholic in my own way. But the term 'Roman' causes me trouble. It seems to limit the term catholic, and associates the Church with a particular discipline. 'Catholic' would seem to imply the wide and comprehensive view, on the great issues of the day. 'Roman,' those of a sect."

It was not then the occasion to discuss the position of Rome, guardian of the tomb of St. Peter, and custodian of the sacred deposit of Christian faith. But I did call his attention to the "catholic" spirit in which Rome, through its official utterances, particularly the discourses and letters of the Holy See, approaches the intricate web of great moral problems. One cannot long read these documents, without being impressed by their agreeable lack of one-sidedness: not an owlish affectation of impartiality, such as little minds easily assume, but rather a concern so deep, a general perspective so broad, that each single question is seen in a broader setting. Questions are studied not only in a further context of kindred moral issues, but in the still wider perspective of the Church's entire view of man, the world, history and time. Attempts therefore to give a partisan twist to papal utterances are usually doomed to failure.

There is of course the other side to the picture. Papal discourses are necessarily so broadly pitched to a world audience, so traditional in style and so solicitous about

9

how future generations may interpret them, that they usually are not as specific and concrete as we would like. The specifying is our job, and the job of the best minds in each country. In the meanwhile we continue to do everything in our power to clear up the strictly theological and historical misconceptions that hinder our progress on the rocky road to religious unity.

As I have said, the more truly catholic is our catholicity, the more easily all misunderstandings will be cleared away. Certainly, the more thoroughly Roman, that is, the more thoroughly integral and devoid of one-sidedness is our affirmation of every item in the whole entire gamut of moral rights and duties, the sooner will such a constructive dialogue be possible, and the better prepared we shall be for whatever the coming ecumenical council may bring forth. This is one reason that I am glad I wrote the life story of a Catholic, and of a Roman Catholic as well.

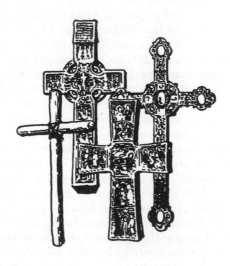

American Protestants

Erik von Kuehnelt-Leddihn

As a lecturer I frequently have occasion to tell American audiences about the increasingly fruit- ful relations which are developing in most of the free European countries between the Catholic Church and the denominations that stem from the Reformation. Almost invariably, when I speak on this subject, I am asked why this movement finds no counterpart in the United States. The answer is not simple. Responsibility for the situation in America rests not only on the shoulders of some Catholics and, in my opinion, in even larger measure on the "other side," but must also be attributed to certain theological, historical and social facts. Yet these, at least in part, could be overcome if only a real determination to bridge the abyss actually existed.

Catholics, needless to say, should be the very last to remain uncritical of themselves. Our Church continues to be "God's strength in the guise of human weakness"; our first Pope was *piscator et peccator*—fisherman and sinner; the smitten ear of Malchus and the mocking crow of the rooster haunt our memories down the cen- turies.

11

Though we never surrender the essentials, it is still true that our very catholicity, our innate tendency to annex, baptize and incorporate anything that is true, good or beautiful, as well as our inner urge to expand in every imaginable direction (what Karl Barth critically eyes as *das katholische Und*—"The Catholic AND") sometimes prompts us to get lost in what is purely marginal and unessential. Popular devotions have a way of ballooning to the point where they distract from the actual deposit of faith. The result is that we find ourselves exposed to perfectly legitimate criticism from the heirs of the Reformation.

Another subject calling for self-examination on our part is the not infrequent phenomenon of the "Catholicist" (as opposed to the Catholic) in the United States. He looks for all the world like a partisan of a sect rather than a member of the Church Universal. His "Catholicism"—a term coined in the Counter-Reformation and once criticized by Pius XII—is a genuine "ism." He knows practically nothing about the theological content and the spiritual yearnings of the Reformation and tends to read the attitudes of latter-day, relativistic Protestantism—now largely vanished from the European continent—into the history of the 16th century. This accounts, for instance, for the credence he lends to the fairy-tale version of Luther as an early liberal who believed in "private interpretation." Worst of all, the "Catholicist" (admittedly not a breed found exclusively in the American hemisphere) remains profoundly suspicious of the label "Christian."

ATTITUDE OF EUROPEAN PROTESTANTS

On the European continent, the Evangelical (Lutheran) and even the Reformed (Calvinist) Christians always had some sort of conscious `and subconscious realization that they had sprung from the Catholic Church. It is her renewed vigor which baffles them, over and over again. Thus, Harnack never concealed his surprise that the Catholic Church was still producing saints.

The members of the Reformation faiths in the United States, on the other hand, are frequently exposed to

12

the illusion of having some sort of historical primacy. This stems, understandably, from the fact that, aside from Maryland and the Southwest, they actually had settled this country before Catholics came on the scene in any numbers.

A German Lutheran, however, worshipping in Nuremberg's St. Sebaldus or in Marburg's St. Elizabeth, where medieval art is so much better preserved than in many a Catholic church and where images of our Lady abound, cannot but be aware of the old unity. The *Simultankirchen*—churches owned in common since the Reformation—vividly bring home the fact of an even earlier common history through fifteen hundred years. But to the American Protestant, his colonial-style meeting house, tabernacle or church has no predecessor other than the wilderness which harbored the Indians. Little wonder that he resents the present Romanist challenge to his ancient birthright.

Even the continental Calvinist is aware of the common past. In Geneva's St. Peter's, the mother church of Presbyterianism, one still can see a papal tiara in the stained glass not far from where the altar once stood. Traces of a Catholic "prehistory" are also to be seen in the buildings of the Reformed Church in the Netherlands. These and other such Catholic memories and survivals lend a psychological assist to theological encounters in Holland, France and Switzerland.

Still, the fact should not be overlooked that the bulk of the followers of the Reformation on the continent of Europe are Lutherans and not Calvinists. One has only to compare the *Confessio Augustana* with either the spirit of Calvin's *Institutions* or the text of the Anglican *Thirty-Nine Articles* to discover that the split issuing from Geneva was of a far more radical nature than the schism begun by Luther. The dry, sober French lawyer did a far more thorough job than the German Augustinian—a "wrestler with Christ"—who, had his lot been other, might have become a great mystic. And whatever the contribution of Lutherans to the spirit of American Protestantism, the fact remains that the nation's cultural roots have been watered far more by the springs of Calvinism, Congregationalism, Anglicanism, Unitarianism, and even of the Quaker persuasion.

Thus, in spite of the many external affinities of Episcopalianism to the Catholic Church, the position of Catholics in the German Lutheran world was always more relaxed and better protected legally than in the British Isles or in colonial America. The Catholic monarch of Saxony was able to rule a 92-per-cent Lutheran population, while the Catholic subjects of the Hohenzollerns did not suffer the indignities of third-class citizenship imposed on their co-religionists in an otherwise more liberal country. All this is not so surprising, however, if we keep in mind that the German princes until 1806 were, at least in name, the vassals of the Holy Roman Emperor—whose very title could not have been more Catholic.

There is a German nationalistic spirit in Lutheranism, but it was always kept within limits. It could not have been otherwise, because the political center of the Germanies until the dawn of the 19th century was Catholic Vienna; Catholics inhabited the most ancient historical sectors of Germany (Westphalia, the Rhineland, Bavaria, Alsace and Austria); Catholic thought and concepts had influenced some of the most important non-Catholic German artists and thinkers—Goethe, Schiller, Novalis, Bach and Leibniz.

THE AMERICAN EXPERIENCE

In the United States, on the other hand, as in Britain, the various denominations which date their rise from the Reformation and its aftermath entered into the spirit of a nascent nationalism which, in America far more than in Britain, was linked with the rising tide of democracy. Indeed, as Prof. Ray A. Billington and others have pointed out, the religious tolerance practiced by the London Government in its overseas domains (in the Quebec Act, for instance) kindled the flames of American revolutionary fervor. Antimonarchical sentiment swelled in close harmony with hatred of the Church. This was the sense of the popular ditty dating from that day: "If Gallic papists have the right / To worship their own way / Then farewell to the

14

liberties / Of poor Amerikay!"

It is evident that the prejudices imported from Britain played a significant role in the fermenting of this potent brew of religious and nationalistic drives. When liberal philosophers of 18th-century England denied full citizenship to Catholics, they acted entirely in harmony with prevailing sentiments. To the average Englishman, even today, "dark men begin at Calais." The shadow of the Armada, the fear of Jesuits in disguise, the threat of a foreign conquest by tyrannical potentates or of a new Inquisition by clerics lusting after power and hostile to sincere doubt or free research, the twin specter of ignorance and superstition emanating from the continent—all these partly real, but mostly imaginary, dangers have for centuries weighed on Britain's soul. And almost all these traumatic apprehensions have been transposed to the American scene.

The anti-Catholic biases of American Protestantism, however, are of a dual nature. In the United States, it must be remembered, two diametrically opposed attitudes of *protest* exist side by side. One of these is critical of the Catholic Church and faith after the manner of the Reformers. The other, representing a Protestantism influenced by the spirit of the Enlightenment and of relativism, passes a very different judgment on us. We have only to ask a preacher in the mountains of Kentucky what sort of people we Catholics really are. He will tell us that we appear to be half-pagans given to dancing, smoking, drinking, merrymaking and to sinful games of chance such as bingo. But a minister in some progressive, suburban church might insist that Catholics are medievalists ground under the heels of an infallible Pope and a greedy clergy, a melancholy and backward tribe deprived of birth control, divorces, salty books and breezy movies.

ROLE OF THE REFORMATION

The common bond between these two radically different views continues to be the emphatic rejection of the Catholic faith. But whereas the puritanical and, even yet, often fundamentalist outlook of an earlier

15

orthodoxy is on the wane in America, the symbolists, relativists and crypto-Unitarians still flourish. It is true that the clarion call of a return to the Reformation and its ideals and tenets has also been heard in the United States. But it sounds much louder on the European continent. There the Reformation remains part of the historical past. Its memory—with the help of monumental "visual aids"—can be more easily revived. And this in spite of the fact that certain theologians, like Rudolf Bultmann, are still fighting a rear-guard action.

We Catholics can only applaud this conscious return to the ideas and ideals of the Reformation because it implies a going back to the point where our separated brethren branched off from the common road. The discussion between Luther and Dr. Eck, which had been carried on in a murmur all through the centuries, has once again become articulate. In turn, our confrontation with the momentous problems and questions addressed to the Church four hundred years ago, and only superficially answered in many instances by our apologists, is being renewed.

Thus, we again stand face to face with the Reformation. Such an earnest book as Pastor Max Lackmann's *Katholische Einheit und Augsburger Konfession* (Graz, 1959), as well as the seriously questioning essays of Prebendary Hans Asmussen, merit well-pondered replies. Berlin's Bishop (now Cardinal) Julius Doepfner, last year during the Unity Octave, rightly said that we do not expect that our Evangelical brethren ought to come back into the Church on their knees and with empty hands. "Already [there are] so many institutions and insights deriving from Evangelical theology which we have incorporated into ours and which we would not like to miss."

It is true that such language is rarely heard among high Catholic ecclesiastics in America. Is this the case solely because America's theologians are more narrow, more "partisan" in their outlook? Or because America's Protestant theology has a less enriching effect on our theological thinking? Or is this due to other reasons which have an intrinsic link with the survival of Protestantism as opposed to an *evangelisches Christentum?*

Indeed, it is difficult to overlook the fact that the label "Protestant," originally attached to the Lutherans, today survives largely in the more or less non-Lutheran United States. There it continues to be proudly used by denominations which have nothing at all to do with the famous Protestation of the Lutheran estates in 1529.

On the other hand, we see how precisely a recent Lutheran meeting in the United States affirmed Luther's old dictum that papalism *is* anti-Christ. All of this shows that the old orthodoxy, in spite of the great watering down by the Enlightenment and relativism, can still be preserved in spots. The old orthodoxy, however, is by no means the new orthodoxy, which is the result of a return to initial positions after long and fruitless peregrinations in the desert of near-unbelief.

European neo-orthodoxy can look at the mysterious Church of Rome without fear or hatred. But it finds no *full-fledged* American counterpart. Hence the curious alliance of a "professional" U. S. Protestantism with nationalist sentiments and the consequent outcry that the Catholic faith is an alien religion. (It is this accusation which incites many U. S. Catholics to that 200-per-cent Americanism which, in turn, lies at the bottom of much of their anti-intellectualism.) Hence also the unpardonable ganging up with non-Christian forces—witness the POAU—in outbursts of singular pettiness and narrow patriotism so foreign to truly Christian groups.

The history of our Church is full of very dark pages, and intelligent Catholics are thoroughly aware of that grim fact. But I really cannot imagine a group calling itself "Catholics and Other Austrians United" trying to curtail Evangelical demands for Lutheran (or Presbyterian) instruction in our Austrian public schools. As a matter of fact, in Austria we have issued a series of stamps with a surcharge in order to raise money for the reconstruction of our oldest Evangelical school destroyed by bombs in World War II. Some of these stamps bear the portraits of outstanding Evangelical Austrians from Kepler to Hansen. Moreover, our Evangelical brethren have an established right to a holiday on October 31, the anniversary of a landmark in Ref-

17

ormation history. And all this takes place in a country which, together with Spain, was the driving force in the Counter-Reformation and in which Evangelicals today constitute only six per cent of the total population.

Much of the animosity of American Protestants towards the Catholic Church seems to be due to a spirit of uneasy competitiveness as well as to an absence of that broad historical outlook which recognizes—at least up to a point—the providential role the Catholic Church has played in the story of Christianity. It is precisely in this respect that they differ radically from Europe's Evangelical Christians. On the other hand, American Catholics should be more mindful of St. Paul's words *oportet et haereses esse* ("for there must be factions" I Cor. 11:19). They may rightly regret the dissent, but must rejoice in the survival of authentic Christianity wherever they find it. In their midst there is still too much naive denigration of the Reformation, too much of the neurotic-monk-wanting-to-marry-a-nun formula used in Catholic instruction about the Reformation.

Again, a great deal of the latent as well as open anti-Catholicism among Protestants in the United States may be traced to another aspect of their fundamentally parochial attitude. This is a tendency to insist childishly that without Calvin's *Institutions* there would be no dental care in the common schools, no modern plumbing, jet planes or voting machines. On top of it all there is the fear that without the Reformation America would have been just another glorified Quebec or Bolivia. According to this view, if the United States ever becomes Catholic, it will also be magically transformed into a Latin nation. To this must be added the nightmare of Rum, Romanism and Rebellion—plus bingo and Roman Catholic kids riding public school buses. Of course, the specter of wild demonstrations in Colombia or of a confiscated novel in Ireland helps to reinforce this curious pattern of thinking.

Yet, in the heart of the continent where once the Thirty Years War was fought in all its bitterness, the situation is radically different. It is true that Pastor Niemoeller still sticks to his formula, "Rather Moscow than Rome," but the learned institutes studying the opposite faiths are flourishing, Evangelical and Catholic Christians are writing for each other's periodicals, and mixed theological conferences are taking place. For now the 400-year-old division of the Church in the West clearly appears to both sides to be what it really was: a catastrophe of the first order for which the guilt, as Pope John has hinted, must be shared by both sides. And while it may be the role of Europe's torn and bleeding heartland to close the wound where it first was opened, it seems especially tragic that the healing process is slowest in the country which history has appointed to be the leader in freedom's most crucial struggle—a country where the mutual understanding of all Christians is therefore a command of the hour.

Royal Road to Chartres

Robert J. O'Connell

A BRIGHT, HOT sun shone down on the pilgrims around me as we rounded the Gare Montparnasse. My first glimpse was of a mob of students lining the Rue Vaugirard. The last stragglers of the preceding group, called Route Topaz, were trooping into the station. That meant the thousand still waiting belonged to our group of the pilgrimage, Route Amethyst. Almost all in hiking attire, they clustered round their chapter emblems beguiling the time with talk. Each chapter was theoretically broken down into ten teams of five, and just looking down the long street at that buzzing crowd, it was hard to imagine that six such "routes" were going that day, and that the following weekend would see a second wave start out in comparable numbers. Indeed, each year (depending on weather and other factors) some 15,000 students make the pilgrimage to Chartres.

No doubt it was a bit *à la mode,* and therefore all complexions and mentalities were represented—the fervent, the Catholic Actioneers, the tepid, the doubters, even some out-and-out unbelievers. Péguy himself, that strange figure whose ambiguous history is at the root of the Chartres pilgrimage, casts a shadow large enough to envelop them all. And, after all, a convivial hike of some 40 or 50 kilometers along those flat roads of the Beauce all aflower with early May—what more could a young student ask?

This was my first meeting with our chapter. Circumstances and malevolent chance had made it impossible for me to be present at their preliminary briefings, which were aimed at preparing the discussion of this year's theme, "The Priest and Us." Well, that topic should cause some minor explosions, given the position of the clergy on the French scene. Is the priest just a member of a caste, separated by privilege and taboo from ordinary humanity? And if Christ was admittedly the only priest in the fullest sense of the term, of what use is this cassocked creature of breviary and celibacy and all the other apparatus of clericalism? And if the faithful themselves are a priestly people, how does that fit in with this apparent relic of medieval and pre-medieval society? It was almost disconcerting to see how much the literature aimed at nourishing the discussion seemed calculated to sharpen every arm the critical intelligence of the French *universitaire* could lay hold of. Just two or three clever unbelievers in the group, I thought, and this would turn into quite a circus indeed!

At last came the departure signal, and off we wound into the trains waiting to take us out to the edge of the metropolis, where the columns could snake along side roads without tying up the main auto routes. The train ride allowed for some introductions, too, but, in spite of that, the atmosphere seemed somehow reserved, almost distant. Then we arrived at Epernon; there was a blessing of the group and their emblems; and the route to Chartres stretched out before us.

The first portion of each stage was devoted to discussion. The group chaplain flitted from team to team, inciting, encouraging, sharpening a point here and there, trying above all to assure that openness was the rule— the minimum of orthodoxy could be assured later. An hour or more of this and we arrived at our first brief halt. Out came the canteens, the oranges and lemons, and all at once it was apparent that the original reserve was beginning to evaporate. The first marks of that strange fellowship among the Chartres *routiers* started to show. Someone had brought too many oranges, too much water or tea, and wanted (naturally!) to lighten his pack—anybody have some? Chocolate, Père? These almonds you'll find good against thirst. Do have some! The budding intellectualism of the discussion gave way momentarily to banter and smiles and an easy camaraderie. The magic was beginning to operate.

The second stage was devoted to meditation. A few themes for reflection, and off we went. This time the chaplain hung back in case any of the lads wanted to confess en route. Some forty minutes of this, and the chapter leader led us in the Our Father. Then we sang the rosary—*Je vous salue Marie . . .*—to a marching air, and soon we had come to our halting place for dinner.

Once we had settled in the shade, the work of blazing sun, hot roads and heavy packs made itself felt. The first blisters began to sting, the news that water was available was hailed with parched enthusiasm, and, in spite of fatigue, a friendly rivalry set in to see who would go and fetch it for the group. Dinner over, the time came for "chapter"; a report from each team on the issues raised in their discussion, the questions they proposed to the chaplain, then the latter's replies. (Where was this religious eggheadedness? These youngsters were talking about *life!* And the doubters, why something was bothering them, there was something they wanted to see!)

The sun was kinder now, the evening cool and melting into night as we trudged off for Esclimont and the spacious grounds of the chateau where the vigil service was to be held. Here, especially, the organization of the pilgrimage made itself felt; it was not over-organized, but remarkably adequate to an almost impossible situation. Places were marked for each chapter's packs and for the hearing of confessions for both girls and boys. In the flat center of a natural amphitheater, a huge canvas, floodlit and gracefully stretched on guyed poles, hovered like a great white bird over the altar. Hymns and chants and readings, a polyphonic choir excellently trained, light and dark, candle and torch and vestment all blended into a symphony for ear and eye and—the lines for confession and Communion seemed to say—for the heart as well. We were at one of the high moments of the pilgrimage, a moment when its religious core was laid suddenly bare. The last doubts began to wither; these youngsters were not just off on a weekend hike.

Hours later, as I tossed to find a comfortable position in a sleeping bag that was laid almost directly onto concrete, any residual doubts were gone. One didn't march some fifty or so kilometers only to spend an all but sleepless night, unless there was something deeper to this thing than mere fashion or fellowship. The next day, Sunday, confirmed that impression again and again.

The early morning sun soon turned from warm to hot. As the march wore on, with discussion leading into meditation, a certain gingerly quality in everyone's walk became more and more manifest; the blisters, so methodically punctured and taped the night before, had taken up their work again. The halt for lunch was welcomed. Then our road wound into the blazing eye of the afternoon sun. Again and again a boy would be seen carrying two packs instead of one, while a protesting girl hobbled alongside remonstrating for the right to carry her own burden. We had stopped at noon in one of the few bits of forest shade discoverable in that portion of the Beauce, but for the afternoon stop there was little such comfort.

Dinner, chapter and a few final thoughts for the last stretch all took place practically full under the cruel sun. Still there was no complaining; even the joking references to blisters lacked the slightest tinge of the dramatic. And suddenly it dawned that in a single day's time the group had insensibly become a unit, welded at a depth of unspoken understanding that few could understand and that no one alluded to. The doubters still doubted, yes. Their contribution was almost invariably an objection, a note of skepticism, a tired "Oh, Père, those are fine words, but. . . ." The others listened, patiently and without resentment, with no desire for cheap dialectical triumphs. This was a pilgrimage, a

23

time for another kind of action, another kind of thought.

It was appropriate that the last lap came after discussion and chapter on the priesthood of the laity, for that last lap was the pure stuff of sacrifice. Now the moment for reflecting on that sacrifice was past. Dead ahead into the declining sun the column moved with sweat and fatigue and discomfort and sometimes naked pain. Minds had lost their edge; voices more sighed than sang the Aves; packs changed hands; arms wound round some slogging figure that refused to drop out so near to the goal. And yet, how often, when two glances accidentally met, there was that sudden, spontaneous smile!

Thank you, students of Paris, for that image, and for that résumé of the twilight journey we call the Christian life! From the sanctuary on which our hearts were set streamed a love that drew us all together at the same time it drew us on. When the spires of Chartres rose slowly over the purpling fields, and the greeting went up from a thousand throats—*Salve, Regina!*— which of us would have guessed there was that much spirit left in him?

Other Worlds – for Man

Joseph A. Breig and L. C. McHugh, S.J.

I: MAN STANDS ALONE

THE MORE one appreciates the marvel called man and the infinite wonder of the creation, Incarnation and Redemption, the less is one impressed by the thought that there may be, on some heavenly planet other than earth, creatures like us—that is, composed of material bodies and immortal spiritual souls, able to know and love and serve God consciously, and destined for an everlasting sharing in His divine life.

So, at any rate, it strongly seems to me.

Every consideration higher than the barest logic—every instinct and feeling for poetry and artistry, and every religious sense of the divine fitness of things, argues, I think, that mankind is utterly unique and will forever remain so; and that if the cosmos is to be populated by thinking beings, the populating will be done by us in breathtaking migrations through space, much like man's migrations across oceans and continents.

My case, I believe, can best be made not by negative debating against the idea that God may have created other beings like us, but by a positive presentation of the electrifying extraordinariness of what we know about ourselves, both by our own observation and by divine revelation.

THE NUB OF THE ARGUMENT

It strikes me that for those who share my religious beliefs, I can point up the whole matter by inquiring whether they can imagine more than one Virgin Mary— more than one Mother of God. This query, for me, goes to the heart of the question.

What confronts us is the following body of knowledge about the human race—that is, specifically, about you and me.

We are made in God's image and likeness. We have God's word on that—the word of God speaking to us concerning His will in creation and concerning creation's crowning splendor.

Such is the stunning unity—the oneness in diversity— of mankind that the fall from God's grace of the first, the representative man, was the fall of all.

Yet so loved—and so somehow valuable in God's eyes —were all of us that God Himself, in the Person of His Son became one of us for our redemption. He assumed a human nature in order to be born among us, flesh of our flesh and bone of our bone, of a woman who for that reason was preserved from the spiritual effect of our otherwise universal fall from the privilege of full companionship with God.

God, thus become incarnate, lived among us, shared our miseries and our joys, comforted and counseled and taught us, and by dying in unutterable anguish of soul and body and affections, reopened the way for us to enter into participation in His own eternal bliss through the door of our own deaths, incorporated into His.

What is it about human beings that makes them so priceless in the divine plan?

The answer that I see is this: God, in bringing man out of nothingness, created a being so marvelous that in him all the material creation, unto the farthest spiral nebula and the most distant boundaries of space, is gathered up and united with an immortal spirit destined for knowledge and love and service of the infinite Creator.

So it is that if I, by the grace of God—God grant it!—lift my mind and heart to adore and love and thank and petition my Maker and Saviour, in me the cosmos is lifted up in prayer. Not merely for myself do I pray, but for all my fellow men past and present and future, and for the stars known and unknown; for the planets and moons and meteors and meteorites; for the mineral and plant and animal kingdoms; for all that materially exists anywhere.

Adam was uniquely the Representative Man; but you and I too are representatives to God of God's material creation. In our bodies we are made of matter in its manifold forms: even the cosmic rays enter into us as messengers to us from outer space. And in our souls we are from God, breathed directly into existence. Principally in soul, but also mysteriously in body, we are His images and likenesses. Moreover, we are brothers of Christ through the Incarnation. We are caught up in Him and live by His life; we are members of Him; in us and through us He continues the work on earth which He began in His own body as one of us.

Here, it seems to me, we come to the crux of the matter. In the Incarnation, by taking flesh and blood and bone from us through the Virgin, God in a divinely wonderful way united to Himself not only mankind, but through mankind the whole of the visible creation. Thus we have the amazingly beautiful, and the divinely magnificent, ascent of creation from its lowest and most elementary forms into union with spirit, first in man, then in the God-man, and through Him with God, the infinite pure Spirit. We have the coming together of the finite and the infinite, of the lowest and the highest. The perfection and splendor of this divine work seem to me to leave no place for supposing that it is not unique.

To me, there is a divine rightness in the concept of this singular unity of mankind, of the cosmos and of the Creator, which cannot be present in any theory that there may be one or more other races of thinking beings composed of matter and spirit.

More than one Incarnation of God? I hold that the

answer is No.

More than one Mother of God? No.

More than one race into which God has poured His image and likeness? No.

But all this is not to say that the cosmos, save for our own planet, is to remain forever uninhabited by beings able to speak for the visible creation in adoration of its Maker. It seems to me rather likely that God's commandment to men and women to love each other, and in their love to multiply and fill the earth and subdue it, may include the conquest of space and the colonizing of frontiers more astounding than the frontiers already conquered by human migrations through the thousands of years—the hundreds of thousands of years—that lie behind us in history. Indeed, I think that our times call upon us, not to neglect history or to cast it away from us, but to look forward and upward and outward to staggeringly glorious visions of the future and of our amazing destiny in it.

Perhaps, in the technological achievements of the past few years, God is calling upon us to rise above the dreadful quarrels we have lately endured, and to lift our eyes to the countless worlds beckoning to us. Possibly He is reminding us of—and chiding us with—the fact that He made us more wonderful than we realize, with a dignity and splendor and power we have half-forgotten; and that there are immense works at hand to which we ought to be addressing ourselves.

In fact, it may be that by turning the minds of the people of Russia away from foolish Communist grubbing and toward the limitless reaches of space, God is preparing them for a return to Him, for a rejection of empty atheism and for a future of accomplishment beyond their wildest dreams. It is not to the conquest of other nations that the peoples of the Soviet Union ought to be addressing themselves, but to the crossing of the oceans of space—and to magnificent new insights into the majesty and goodness of God and the immeasurable riches with which He endowed mankind when He gave us worlds innumerable to explore and to represent in noble adoration.

JOSEPH A. BREIG

28

II: OTHERS OUT YONDER

I FEEL that intelligent life is common in those planetary oases which dot the "deserts of vast eternity" that stretch through space. The family of Adam, or *homo terrenus,* as I shall dub him, is not a lonely wayfarer in a wilderness of glowing cinders and icy cosmic dust.

At this moment, there is not one scrap of "hard" evidence to support my case. I have only suggestive analogies, a priori and statistical probabilities, and an individualized sense of the fitness of things. Frankly, therefore, I admit in advance that my position must be stated in arguments that show the forms of logic but have the probative strength of straws.

Already, serious astronomers are striving to find meaningful patterns amid all the random static that pours into the dish of a big radio telescope in West Virginia. That story was told by Ray Bradbury in the October 24th issue of *Life.* Project Ozma's outlay of tax money is not a boondoggle. It stems from the statistical implications and the intuitional thrust of modern cosmogony and biochemistry.

Increasingly, biochemistry favors the view that life will arise as a normal result of chemical evolution whenever conditions are right. If favorable circumstances prevail over sufficient eons of time, the evolutive quality of life verges naturally toward higher and eventually rational forms. As for astronomy, current theories of cosmic development generally suggest that planet formation is a common event in the evolution of the quintillions of stars which lie within our view. It is exceedingly probable that billions of planets occupy the "golden zones" of distant suns where temperature and other energy factors favor the emergence of life. Perhaps then the highest forms of organic existence are a widespread climax of cosmic history, rather than an isolated discontinuity that has appeared just once as a sort of "contamination" of the earth's mantle.

I have put this summary in materialistic terms that prescind from finality. But where science talks of the

29

statistical probability that life may arise by chance and develop through random variations, the Catholic philosopher holds that there can be no real chance in a world created by God. There are only lines of causality, unintelligible to us but unified in the divine Mind and somehow expressive of His purposes. On the other hand, since we increasingly regard the origin of life as an event of the natural order rather than as a metaphysical problem demanding the miraculous intervention of creative power, there is no insurmountable difficulty in granting that life may arise whenever and wherever the apt conditions are present.

Moreover, in life as we know it, there are remarkable powers of adaptability, differentiation and movement toward rational organization. Organic life on earth evolved toward a specialized animal form that in God's design was apt material for the infusion of a spiritual soul, while at the same time the lower forms were ordained to serve as a substratum for rational existence and its needs. Why should these things not be generally true in a physical universe characterized by uniformity of law and process? Hence, where suitable conditions exist on a planetary mantle, we may expect life to emerge, perhaps as a normal aspect of development. Where the conditions persist for billions of years, may we not expect that God is preparing for a divine incursion of the biosphere which will manifest His glory in the phenomenon of man?

God made the universe for His glory, not as something to be gained but as a benefit to be bestowed; and above all it consists in sharing His happiness with rational creatures. Does it not seem strange to say that His power, immensity, beauty and eternity are displayed with lavish generosity through unimaginable reaches of space and time, but that the knowledge and love which alone give meaning to all this splendor are confined to this tiny globe where self-conscious life began to flourish a few millennia ago?

Let me put that differently. We teach that irrational creatures are God's "footprints," but that men are His images—radically through the possession of mind and will, accidentally through the doing of virtuous deeds.

Does it not seem odd to say that God has left His footprints everywhere, but has established His image only here? Faceless indeed seems the vast cosmos, if there is heard nowhere but on earth an echo of His knowledge, freedom and, above all, His holiness.

When God set Adam in Eden, He gave him a mandate to "subdue the earth." In the context of *Genesis,* it may be argued that man's image-quality is chiefly manifest in his lordship over created things, a lordship that does not consist merely in superiority of status but in proprietary dominion. But the note of genuine dominion is effective occupation. It would seem, then, that if the physical world is meant to serve genuinely utilitarian purposes and not be just an object of contemplation, it must lie open to exploitation by other  rational beings than ourselves. For undoubtedly, most of the universe is forever closed to penetration by *homo terrenus.*

In his *Contra Gentes,* while discussing the providence of God, Thomas Aquinas notes that "the first thing aimed at in creatures is their multiplication . . . and to the gaining and securing of this end all things else seem to be subordinated." Aquinas held this position within the narrow framework of Aristotelian physics and astronomy. If he had considered the hierarchy of being within the framework of an expanding universe, how would he have expressed himself? Would he perchance have suggested that the irrational creation, where numbers proliferate, cannot meaningfully establish and conserve that to which it is ordained, unless rational life is also common in the physical universe?

Aquinas further argued that the good of the species transcends that of the individual, and that therefore the multiplication of species is a greater addition to the good of the universe than the multiplication of individuals of one species. He applied this principle with consistency even to the upper end of the spectrum of creation—the angelic realm; his stand was that every angel is of a different species and that the multitude of the angels exceeds every material number. If he had possessed our knowledge of the cosmos, would he perhaps have held that a few billions of our human stock, inhabiting an isolated corner of creation, inadequately manifest the goodness of God as it can be shown in rational animals? "Multiplication and variety was needful in the creation, to the end that the perfect likeness of God might be found in creatures according to their measure."

Perhaps then there are races that were never elevated to grace. We sons of Adam are a race elevated, fallen and redeemed—a testimony to His mercy. Who will say there are no races elevated but not fallen, as testimony to His holiness? Who can say that there are no races elevated, fallen but forever without redemption, to show forth the mystery of His justice?

I am not aware of any grave theological objections to the probabilities I have described. The purpose of revelation was to teach us what we needed for salvation, not to instruct us in science or philosophy.

I am not perturbed, for example, at the imagination of a second Mother of God, for I do not even find any difficulty in conceiving a thousand incarnations of each or all the Persons of the Trinity. Aquinas discussed several such possibilities centuries ago and found none of them repugnant to sound theology, though of course he did not suggest that they were actually realized.

Why do we normally resist the thought that man may not be alone in the universe? I think that our perennial jealousy of our status largely springs from the erroneous science which long ago placed us at the physical center of the visible creation. It was inevitable that we regarded ourselves as unique, when all the obvious signs indicated that the heavens revolved about us day and night.

During the course of the years, as we know, the supposed center of the cosmos moved from the earth to the sun, from the sun to the galaxy, until today man resides he knows not where in a metagalaxy of undisclosed structure and extent. Yet now that we have lost our physical centrality, we still cling to the prejudice that we are unique, solely on the grounds of our intellectual superiority.

Where does the evidence for that uniqueness lie? Man as such surely has no centrality in the total creation. The hub of that complex lies in the angelic world, not in any aspect of material creation. We on earth cannot argue uniqueness from the fact that we are images of God in a material mold: for the dignity of being such an image is endlessly shareable. And neither does our worth give any ground for complacency; the redemptive love which the Father showed us in sending His Only-Begotten Son was a proof of His generosity in the face of our desperate need, not an indication that God found some vein of gold in the corruption of our fallen nature.

Despite our lowliness, I feel that even our race has a splendid destiny in space. We may not be able to assign a farthest limit to our lordship over nature, but it will not be limited to the earth alone. Earthly man will go as far as his ambition and inventiveness can carry him. And since he will take the Mystical Body with him, wherever he goes, his colonization of space may be looked upon as a providential extension of the Incarnation in space and time. The Church, perhaps, has a physical dimension beyond our ken.

But, as I already observed, our destiny among the stars will be severely prescribed by the frontiers of impassable distances. C. S. Lewis, speculating on the formidable distances about us, conjectured that they might be God's quarantine precautions. "They prevent the spiritual infection of a fallen species from spreading." Perhaps God has restricted the ambit of our wayfaring, precisely lest we play the role of the primeval Serpent in Edens beyond our reach.

So I rest my slender case. My arguments were no

more than bumbling thrusts into the unknown. But while we wait for real evidence to be acquired, I like to think that the physical universe is a flourishing commonwealth, not a boundless waste where a lonely pilgrim sojourns in the silent dark. And when the evidence accumulates, I should not be surprised to find that Alice Meynell intuitively grasped the full truth in her poem "Christ in the Universe":

O, be prepared, my soul!
To read the inconceivable, to scan
The million forms of God those stars unroll
When, in our turn, we show to them a Man.

L. C. McHugh, s.j.

Moral Rearmament

Robert A. Graham, S.J.

W ITH THE INSTINCT of true public-relations men,
the leaders of Moral Rearmament found
Premier Khrushchev's visit to the United
States the ideal peg on which to hang their message.
From the Swiss headquarters of MRA at Caux, an ex-
Comintern official, Eudocio Ravines of Peru, found
space in the U. S. newspapers of September 29 to declare that the Cold War was now entering a new and
more perilous phase—and to add the operative part of
the story: "Moral Rearmament is the total and categorical answer to the Communist danger." Thus the promoters of one of the most original moral movements of
our times used the front-page news to draw attention
to their work and to 'their claims.

Moral Rearmament, formerly known as Buchmanism
and later as the Oxford Group, is four decades old. Its
founder, Dr. Frank N. D. Buchman, now in his eighties,
remains its guiding spirit, even though his activities
have been necessarily reduced with his advancing years.
Originating as a sort of revivalist movement in the
YMCA tradition, it grew with the years and now operates on a wholesale scale in the international field. To-

day there appears no abatement of its progress. With abundant resources, a dedicated and experienced staff, and a lofty, appealing message pitched to the tastes of the top echelons of national and international society, the school of Dr. Buchman moves on with an obvious assurance. It has encountered ridicule, scoffing, suspicion; it has been condemned for its philosophy, its morals and its motivation. In 1940 it appeared totally discredited, but it survived this crisis as it had survived earlier setbacks. Though its full-time workers are still estimated at a bare two thousand at most, its program reaches many thousands of what MRA likes to term "key people."

If, as alleged, Moral Rearmament is *the* answer to communism, then it surely deserves attention. And, indeed, there are distinguished Americans who say it is just that. Thus Admiral William H. Standley, former U. S. envoy in Moscow, is quoted as saying only last March: "The choice for America is Moral Rearmament or communism." This is quite an alternative to put before the American public. The sweeping terms with which this claim is put forward suggest, on the part of its authors, more than an ordinary confidence in the validity of their mission and their message. It must be admitted that the recent record of MRA activities demonstrates that, for good or ill, the movement has made itself felt on the international scene and that the end is not yet in sight.

This past summer, at its two major world centers— one at Caux and the other at Cedar Point, Mackinac Island, Mich.—MRA was going full blast. Successive waves of invitees underwent the "courses" that MRA has designed to implant its message. The visitors consisted of men and women, young and old, from a wide variety of occupations, but all leaders in their own circles. They were flown in from all points of the world and maintained in that first-class-resort comfort —tobacco and liquor aside—to which most of them were accustomed. As the Mackinac Island delegates approached the scene of their coming indoctrination, their eyes met a sign at the gate of Cedar Point: "People of every nation, race and color come here from every

continent for training in an idea to unite men and remake the world." For the following week, or rather
weeks, carefully segregated from the ordinary vaca-
tioners on the island, they took part in what is entitled
the "Summit Strategy Conference." In a spacious hall
designed to resemble a wigwam—presumably to suggest the idea of a council of the sachems of all nations
—they met to hear ex-Communists, ex-agnostics, authors,
diplomats and other persons of note, or at least of interest, tell of the triumph in their lives of the four
Absolutes of Honesty, Purity, Unselfishness and Love.
In the evenings they attended morality plays and movies
created by MRA to bring out the central point of personal change. Here, in short, they were exposed to the
"only answer" to communism. One must admit that
only the Communists seem to carry on, at anywhere
near the same scale, such a systematic formation of
the elite of all nations and races. And there is more
to come. Next year, if rumors stirring island folk are
true, MRA will take over additional property. Mackinac
Island bids fair to become one big MRA citadel.

AN IDEOLOGY? OR A RELIGION?

Any way one looks at it, Moral Rearmament cannot
be ignored. It claims that it is not a religion; it claims
also not to be interested in political action. Yet it operates in a zone avowedly related to religion and politics. By its own description MRA is an ideological force.
An ideology, says the movement's literature, is "an idea
that dominates the whole of the person—his motives,
his thinking, his living—and fights with a strategy to
get everybody to live the same way." But this definition,
in its first part, necessarily involves a religious or quasi-
religious commitment, while the second part points to
inevitable involvement in the political or quasi-politi-
cal field. In this respect MRA compares itself to the
ideology of communism. It feels that only MRA has
the elements to cope successfully with the ideology
that emanates from Moscow.

The followers of Dr. Buchman first conceived themselves as the answer to communism in the aftermath

37

of the second World War. With the typically generous self-praise it habitually indulges in, MRA now claims to have had a major share in saving Western Europe from communism. According to a booklet issued this past July, three things saved Western Europe in 1947. One was the announcement of the Marshall Plan, the second the ascent of Robert Schuman to power in France—and the third "the ideological initiative of Moral Rearmament in creating, through changed men, totally committed to a superior ideology, the conditions under which the Marshall Plan and the Schuman Plan could work." The movement's literature also claims that through its work in the Ruhr, and in the British coal fields as well, it prevented a Communist takeover. The four Absolutes had dissolved the conflicts by changing men.

These claims are not taken seriously by those in a position to know the details of the developments in question. It remains, however, that MRA did succeed in attracting to itself a number of former Communists for whom MRA has obviously become an alternative to Marxism. One of these was Señor Ravines, already mentioned. Another, currently featured in MRA events, is Angelo Passetto of Italy, once a propagandist for Red chief Palmiro Togliatti. He is now engaged in creating songs and dramatizations of the "idea" behind Moral Rearmament. Other ex-Communists appear frequently on the movement's programs.

On the whole, the technical efficiency and professional polish of MRA's activities are in marked contrast to the naïveté of its efforts in the social and political field. For a while MRA directed its attention to the labor field. It succeeded only in earning for itself the indignation of labor organizations for its uninvited and unwelcome interference and its disregard for the legitimate procedures long established in the solution of labor disputes. In its well-intentioned fight against communism, MRA has given cause for great doubt as to its ability to carry through its program without serious mistakes.

The case history of *The Vanishing Island* is an instance in point and one that will not soon be forgotten by the United States Government. This is the title of an MRA "ideological musical play," one of the morality productions used (quite effectively) in transmitting the MRA message. Like most of the plays in the MRA repertory, this one was written by Peter Howard, former British newspaperman and for many years one of Dr. Buchman's closest aides. The play was presented in mid-1955 in Washington before a distinguished audience from all branches of the Government and from diplomatic society. It concerned the fortunes of the citizens of two countries, "I Love Me" and "We Hate You." In the former (the capitalist world, the United States in particular) there reign corruption, selfishness, materialism, licentiousness and hypocrisy. This nation is threatened by the other country, which carries on a vigorous anti-"I Love Me" propaganda. The leaders of the West are confused and weak, because they have no ideas. On the Eastern side, however, the rulers know precisely what they want and how to get it. As the play develops, war seems inevitable until an MRA team arrives on the scene and reconciles the postulates of capitalism and communism (somehow) through the four Absolutes. All admit their respective faults and peace is assured.

The play's over-all effect was anything but complimentary to the West. The Washington audience reacted with understandable coolness but seems to have taken it as they would take a sermon preached against well-known local abuses. What the invitees did not know was that plans were already in readiness to take this indictment of the West on a grand tour of the Far East, the Middle East and Africa, where at that precise moment neutralism was causing U. S. foreign policy the most acute concern. Still less did they know that the Secretary of the Air Force, urged by influential congressional friends of MRA, had agreed to put at MRA's disposal three Air Force planes to carry the troupe on what was described as a "Statesmen's Mission" for world peace. And so it came about that in June, 1955 this compromising spectacular, with its nearly two hundred people, left Washington on a mission that was bound to confirm the neutralism of the uncommitted peoples and make the United States, to boot, look like an idiot for appearing to sponsor this self-condemnatory exhibition.

Belatedly informed, the State Department issued instructions to U. S. representatives abroad that *The Vanishing Island* had no official backing whatsoever. When public opinion reacted to news of the junket at the expense of the U. S. Government, the tour's directors arranged to make compensation. But the damage was already done. In 16 countries the United States and all Western civilization had been given a black eye, and communism only the mildest of rebuffs. The tour ended at Geneva in September, but it soon took off again, this time labeled the MRA Ideological Mission, to present *The Vanishing Island* in West Germany and the capitals of Scandinavia. Altogether, according to an MRA statement, the musical play was seen by 243,000 people in 24 nations during a 40,000-mile journey. This was MRA's "answer" to communism. No one really can measure the ultimate harm done by this naive meddling in international politics.

The claim to be a bulwark against communism is being made by Moral Rearmament not only to Governments but also to the Catholic Church. MRA believes it is the natural vehicle which can bind together those world forces which are equally threatened by the onrushing tide of Marxist materialism. One of the relatively few Catholics active in the movement, a Filipino newspaperman, Vicente Villamin, reported early this year that he had presented to authorities at the Vatican a scheme according to which Moral Rearmament was to become a sort of common ground for all efforts to face and resist communism, particularly in Asia. If such a plan was ever seriously considered in Rome, there is no evidence of it.

THREE CATHOLIC DOCUMENTS

While the Holy See no doubt welcomes the stress on personal reform and spiritual idealism which is the essence of the MRA program, it cannot overlook the fact that in many respects this "ideology" is tantamount to a religion. Hence the extreme reserve manifested in

40

the past toward the Caux-Mackinac axis. It is not necessary to review here the Catholic stand on Moral Rearmament. Two Catholics studies in English are particularly useful. One is by the Auxiliary Bishop of Malines, Most Rev. Léon-Joseph Suenens, and is entitled *The Right View of Moral Rearmament* (Burns and Oates, 1953). The other is a full-length article by Edward Duff, S.J., "Verdict on MRA," which appeared in the June, 1956 *Social Order*. In a class by itself, because of its official character, is the pastoral instruction on faith which the Bishop of Marquette, Mich., Most Rev. Thomas L. Noa, issued to his faithful under date of August 15, 1958. (Mackinac Island is located in that diocese.) The instruction terminated by stating that Catholics living in the diocese "may not attend the meetings of MRA, or participate in or promote its activities." The same prohibition the bishop extended to all other Catholics "whenever they may be within the limits of the jurisdiction of the Diocese of Marquette."

The conclusions of these three sources, not to speak of earlier condemnations and warnings emanating from various bishops in Europe and elsewhere, are negative. Nevertheless, no general ban on cooperation with MRA has yet been issued by the Holy See, at least as applicable to lay persons. Consequently there are still a certain number of Catholics who are full-time workers. And there are many more who annually accept invitations to attend MRA assemblies or "Summit Conferences," at least at Caux. For these, obviously, the argument that MRA presents the "answer" to communism still has some validity.

It would be unfair at this point to fail to acknowledge that MRA has been, for many people, a sort of halfway house to the Catholic Church. The number of indifferent or fallen-away Catholics who attribute their return to the Church to their first contacts with Moral Rearmament is too large to be just a coincidence. The same is true for a fair number of converts. For this reason and for perhaps others, many Catholics can be cited as loyal defenders of the movement. These range, to mention only names appearing in recent MRA literature, from the late Joseph Scott of Los Angeles to

41

Gabriel Marcel, the French existentialist. Even Richard Cardinal Cushing is cited as paying tribute to what MRA has done for souls, although just what the Archbishop of Boston really said is not a matter of record outside of MRA literature.

Don Luigi Sturzo, who died last August, is also quoted in Moral Rearmament literature. In this connection, however, it seems that MRA has overdone its zeal, at the expense of accuracy, if not of veracity. The founder of Italian Christian Democracy is quoted as follows: "In no other place do I feel the presence of God as at MRA Assemblies. Here one learns with the fullness of divine teaching how men can live together. These ideas must become the policy of nations." This is a startling encomium that, on the face of it, Don Sturzo could not possibly have uttered. The Don Luigi Sturzo Foundation in Rome, to which this writer referred the quote for verification, states that Don Sturzo never made any such statement and, in fact, that he never attended an MRA assembly in his life.

This loose and casual liberty taken with words is characteristic of the unguided course of MRA's relations with the Church. A few years ago, it appeared that some safeguards to eliminate potential dangers in theology and ascetical practice had been set up by an agreement between MRA and diocesan authorities in Switzerland. However, this "agreement" was later annulled by an MRA spokesman on the grounds that, after all, Moral Rearmament is not an "organization" but an organism, and that consequently there could be no question of an "agreement" with anyone. The effect of this repudiation was to restore the former ambiguities in MRA that have long troubled Catholic theologians. These indications are not calculated to bolster the Holy See's confidence in Moral Rearmament as an ideological alternative for communism, whether for Catholics or anyone else.

To the extent that Moral Rearmament with its message of spiritual rebirth awakens dormant souls and leads them to a serious reconsideration of the life they have been leading, the efforts of Dr. Buchman deserve the well-wishing of mankind. Who cannot but be sympathetic toward a cause whose object is "to unite men and remake the world"? But this is fundamentally a religious task, and Moral Rearmament has hovered perilously close to the religious field at all times. In view of the astounding progress of MRA since its beginnings and its undoubted influence on human lives, "Frank," as Dr. Buchman is known to the workers in the movement, should go down in history as one of the greatest Protestant missionaries in these modern times.

It is understandable, on the other hand, that the Catholic Church shies away when MRA presents itself as a guide in the spiritual redemption of souls, as it does in its "ideological" campaign. On the nonreligious plane, MRA can also be welcomed to the extent that it can unveil to confused minds the vision of a world where personal change can transform the atmosphere of society. The evidence thus far, however, is that MRA has neither the political sophistication nor the doctrinal cohesion that could enable it, at this moment, to cope with communism as one ideology against another. Men should be glad if MRA accomplishes good for individuals and for society; they will be vastly relieved, and even surprised, if it does not do harm to Governments or to souls.

This is an age of explosions: nuclear, technological, demographic—and cultural. Yes, art and literature are "busting out all over," at least if you rely on statistics. But what is the worth of the cultural explosion when people question whether art and letters are a true imitation of nature and life? Some critics look on modern painting and sculpture as a sort of private joke. Some think literature has so debased shock values that it runs the risk of becoming a mere commentary on sex and sadism. Where is the cultural explosion likely to end, when even an artist like Hemingway confesses that he does not know the meaning of Man?

Arts and Letters

The Culture Explosion

C. J. McNaspy

IT WAS REASSURING during the recent Presidential campaign to find both candidates stoutly in favor of "deeper and wider education, and the intellectual curiosity in which culture flourishes." These are Mr. Kennedy's words, but Mr. Nixon said much the same. A Secretary of Culture? No, replied our next President. Such a department "might even stultify the arts, if wrongly administered. We have more than enough conformity now." It was clear that America's new frontiers had room for the free, vigorous expansion of culture.

What seems significant here is not precisely what Mr. Kennedy stated, nor even his courage in quoting T. S. Eliot's poetry during a campaign speech, nor his eggheadish preference for Moussorgsky and Berlioz—a riskier choice than Mr. Nixon's choice of "Oklahoma!" and "Swan Lake." It is rather that both candidates should have had to submit to a cultural inquisition at all. Did General de Gaulle? Fearlessly, Mr. Kennedy's advisers did not stop there. A "Committee of Arts, Letters & Science for Kennedy for President" unabashedly brandished the most formidable list of intellectuals (from Aaron Copland to Thornton Wilder) in an open plea "to those who have not decided." Clearly, the eggheads now have it. After exhausting the "religious issue," political analysts should peer into the "cultural issue."

I have just read or reread some dozen recent books on culture in America. Most of them are unflattering, self-deflation being a national pastime. Now made acutely aware of our deficiencies as a society, we are, it seems, hideously and variously wasteful; alienated, over-organized, lonely, other-directed; victimized by power elites and all manner of pacifiers and hidden persuaders; in a word, a lot of ugly, status-seeking, vulgarian operators. This iconoclastic list could be endlessly prolonged, but with the title of one book at least we can hardly take exception: Eric Larrabee is surely safe when calling us *The Self-Conscious Society*.

This self-consciousness takes many shapes. In the campaign, however, it became increasingly alembicated into one simple essence—how was our *prestige* coming along? What was our *image* abroad? In a word, the question was *culture*—culture taken in the broad anthropological sense, but not excluding the popular meaning of education, fine arts, and what are generally called the "higher" things.

It would come as a shock to many of Mr. Casey Stengel's devotees to learn that in 1959 more than twice as many people viewed the art shown in New York's Metropolitan Museum as that performed in Yankee Stadium. The score, to be precise, was 3,947,365 to 1,552,030 in favor of the Metropolitan. In defense of our "national sport" one may point out that baseball has a shorter season than painting, that baseball isn't the only professional sport, that statistics can prove anything (almost). In rebuttal, the protagonists of indoor culture may add to their score the 568,744 who paid to enter the Museum of Modern Art, the 2,356,221 who visited the Museum of Natural History, the 435,943 viewers of the Brooklyn Art Museum (obviously after the Dodgers' departure), and some ten million visitors to other museums in New York. True, not all are "art" museums, but all are "cultural."

Dropping idle and odious comparisons (besides,

athletics and the fine arts are not incompatible), we note that the enormous number of museum visitors in New York is not a local phenomenon. Chicago's Art Institute reports 1,093,958 viewers in 1959, with 4,306,-528 visitors to other municipal museums. The Smithsonian Institution in Washington had 6,662,126 visitors that year, and the National Gallery 961,883. In the West, Los Angeles reports 916,996 viewers in its County Museum, and the California Academy of Sciences, in San Francisco, 2,405,270.

Philadelphia's Franklin Institute numbered 410,443 visitors; Detroit's Art Museum 963,391; Boston's Museum of Fine Arts (not including the Isabella Stewart Gardner Museum or the Fogg Museum) 485,961 (with a paying, supporting membership of 9,153). The South now boasts several impressive museums. Among them, Houston's Museum of Fine Arts in the same year numbered 172,809 viewers. In these and other museums that I have visited in various parts of the country, the average is always roughly proportionate to population.

It is interesting to find that viewers in American museums are not outnumbered abroad, where American tourists do more than their part to swell attendance. The venerable Louvre reports 1,671,000 (Americans included) visitors in one year, less than half the number attending the Metropolitan in New York. The British Museum reports 752,826, less than Detroit. Amsterdam's celebrated Rijksmuseum had 490,426, far less than the Museum of Modern Art. In 1958 the world-famed museum of The Hague entertained 148,145 visitors, less than Houston.

Another noteworthy fact about our American museums is the steady, often spectacular increase in popular attendance. With few exceptions (such as during the temporary closing of the Museum of Modern Art, caused by fire in 1958), each of the museums studied shows continuous growth. In 1956, for example, the museums of New York registered 12,281,602, and the following year 13,190,033. The Metropolitan attendance grew from 1,326,955 in 1939 to 3,947,365 last year—an increase of almost 300 per cent. In 1924 the Houston museum had 33,959 viewers; last year, 172,809—a truly

47

Texan boom of more than 500 per cent. (I am grateful to all the curators who helped me assemble this information.)

The quality of our museums is almost as striking as their quantity. Eloise Spaeth's new book on *American Art Museums and Galleries* studies 84 important museums from Ogunquit, Maine, to Phoenix, Arizona, and 125 galleries (she had to choose, since there are over 300 in New York alone). To those who know their painting from books it may be surprising to discover that many world masterpieces are not in Europe but here. The Gainsborough *Blue Boy* is in California, the Brueghel *Wedding Dance* (the best-known version) in Detroit, El Greco's *Assumption* in Chicago. The richest collection of French impressionists is in Baltimore, not Paris; that of Vermeers in Washington and New York's Frick Gallery, not Amsterdam.

C ONCERT music's growth in popularity has been so spectacular in the last generation that it is hard to give a fair sketch in a few words. While in 1920 there were not a hundred symphony orchestras in the United States, today there are 1,142, more than half the world's total, and 35 million concert-goers. Today just about everyone has a collection of "classical" records or has a friend who does. Such music is now available on radio almost all day and night in and near large cities. In fact, it has become almost as hard to escape being engulfed by Wagner or Brahms as by "rockabilly." The hi-fi industry is a multi billion-dollar one, and always growing.

The theatre, despite heralds of doom, shows signs of a vitality hardly matched in our history—and seldom perhaps in anyone's history. The American Theatre Society's subscription list alone has grown from 84,000 ten years ago to 116,332 this year. Alice Griffin shows that there are some five thousand theatres in the country today, with vigorous regional performing groups at a high professional level. "Gone," she says, "are the days of the militantly amateur group which performed to entertain its members rather than its audience."

"Nowadays nobody reads any more," is a common lament. Facts taken from the 1961 edition of the *American Library and Book Trade Annual* suggest the opposite. Not counting high school libraries, we find presently in the United States 12,852 libraries (7,257 public, with 3,566 branches, and 1,948 college libraries). The public library circulation for 1956 (the latest date available) was 489,520,000. Presumably, some of these books were read after being withdrawn. The number of new books printed in 1920 was 5,101; in 1958 it had doubled to 11,012. *Publishers' Weekly* for June, 1959, showed that the average per capita American expenditure for books was $7.00 for that year. *U. S. Income and Output* showed the following growth in annual expenditure for 1950 and 1957: books, $677 million to $1.026 billion; magazines and newspapers, $1.47 billion to $2.71 billion. If nobody reads today, a great deal of money is being spent on nothing.

These are a few of the facts regarding America's "culture consumption." While they need to be interpreted and appraised, they cannot be gainsaid or simply bypassed. Why then are our critics of culture so unhappy? Why does almost every week bring again the hue and cry of alarm? What is there to be said against the indisputable data of this cultural explosion?

The general objection is a real or imaginary application of Gresham's Law to culture. This law, it will be remembered, states that bad money will drive out good money. It is the law of debasement through dispersion. The renowned historian Rostovtzeff puts it well and has often been quoted: "Is it possible to extend a higher civilization to the lower classes without debasing its standard and diluting its quality to the vanishing point?" Stanley Edgar Hyman calls it the Law of Raspberry Jam: the wider you spread it the thinner it gets.

Doubtless there has been an increase in America of public contact with the higher culture, but has this brought a decrease of depth and meaningfulness? Is the whole cultural explosion real or just a superficial puff? Is "instant culture," if I may so call it, a genuine or a spurious brew? After Russell Lynes coined his triple cultural criterion—high-brow, low-brow, middle-brow—

the statusmongers began providing taste signals. Unsure of oneself, one was now shown just what were the proper reactions and even how to achieve them, not merely to pretend. Visual aids were provided, of course, by *Life* magazine: illustrated charts specifying the proper drinks, hobbies, likes and dislikes to fit the level of "brow" one aspired to.

The problem is not altogether new. In his *Phaedrus* Plato naturally wrestled with it, and so did Aristotle. More recently, De Tocqueville, with the hauteur befitting his class, feared the "tyranny of the majority," while Lord Bryce felt that democracy would lead us to a "self-distrust, a despondency, a disposition to fall into line, to acquiesce in the dominant opinion, to submit thought as well as action to the encompassing power of numbers." Americans, too, have always been concerned, and the booming H. L. Mencken leveled murderous attacks against the "booboisie."

A MONG THE LEADING critics of the "pessimistic school" are Dwight Macdonald, Hannah Arendt, Oscar Handlin and Ernest van den Haag. These and other respected scholars are deeply disquieted at the hollowness discerned in much enthusiasm for culture. Even

more, the "mass media," they feel, are a serious danger, promoting standardization rather than standards, homogenization rather than humanism, passivity not creativity. The power, money and prestige of these media seduce and beguile; men of authentic gifts are lured away and lost to true culture. Mass culture has debased and prostituted class culture, lowering it to the level of entertainment. The mass media do not foster art; they replace it.

It is manifestly impossible to discuss fairly, in a brief summary, this serious position. Let me urge my readers to study the debate on mass culture in the special Spring issue of *Daedalus* (1960), in the symposium edited by Bernard Rosenberg and David Manning White, *Mass Culture* (Free Press, 1957), and issues 3, 4 and 5 of *Partisan Review* (1952).

The main source of malaise and the principal critical target today is, of course, television. Now that well over 90 per cent of our American homes possess (or are possessed by) these 50-some million sets, the average instrument is in operation for more than 38½ hours per week (see *Fortune*, December, 1958). It would be hard not to blame TV for our ills.

During the past thirty years, Mack Hanan points out in *The Pacifiers*, the average workweek has been shortened by the number of hours in a whole day. We are now a society of the leisure-stricken. Despite this, if Vance Packard's estimates are accurate, only one American adult in three hundred reads serious books with any regularity, and, again on his word, a recent poll showed that most Americans could not recall reading any kind of book in the past year. Moreover, despite our large investment in book publication, we were greatly surpassed in 1958 by the USSR, Japan, Great Britain and West Germany. It is suggested, if not stated, that TV is both cause and effect. Ennui is never-ending, and we now find people desperately escaping from leisure to new employment, causing the current problem of "moonlighting." Perhaps explicitizing T. S. Eliot's famed bang-and-whimper, Arthur Morgan has predicted: "America will not perish from a bomb. It will perish from boredom."

No one should underestimate the usual vacuity of TV. However, its positive danger, as well as the positive influence of most mass media, seems to have been considerably exaggerated. Joseph T. Klapper's new book, *The Effects of Mass Communication,* torpedoes many gruesome generalizations. Apathy, passivity and a playpen atmosphere seem, on the whole, the real threat. Having personally spent many hundreds of hours in television studios, as well as my normal quota in vigil before the "Happy Screen," I believe myself not insensitive to the problems and limitations of the medium.

The critics do seem oblivious to so much, especially the growing movement of Education Television throughout the nation. In New Orleans, for example, station WYES-TV provides regular programs of instruction, in cooperation with local universities and with a national exchange of select videotapes, in many areas of the liberal arts. In Boston and throughout New England, WGBH-TV and related stations have for several years carried out an elaborate series of courses in cultural subjects. In addition, they have provided a unique public service in cooperation with the Boston Museum of Fine Arts. The museum is completely wired for television and has produced some six hundred live programs, many of which are available on tape for other stations. More than fifty non-commercial educational stations now cover most of our country's highly populated centers.

This may be only one swallow and not yet a true spring, but it does urge us to return to even keel and reflect on some of our blessings. In one of the wisest books I have read on American culture, *God's Country and Mine,* Jacques Barzun reminds us that "what we have undertaken, no other society has tried: we do not suppress half of mankind to refine part of the other half." Without indulging in naive chauvinism of time or space, we can be humbly grateful for unprecedented opportunities. Had we been born at another time, how many of us who criticize mass culture would enjoy sufficient culture to be able to criticize it? In any but our own favored society, could we pass judgment from the Olympian heights?

Athens, it is true, created a high culture and gave us much that is best; but at the time it was for the few, and the few killed Socrates, the critic of culture. Florence could boast its proud moments and monuments; yet how many Medici enjoyed fully the golden fruits of a golden age? What percentage of the masses of Vienna or Salzburg had leisure and freedom to relish Mozart?

And how many contemporaries ever heard Beethoven's late quartets? "Class culture" is fine for the classes, but most of us are the masses.

Granted, the day of the finger-bowl has passed, save for chaplains in a few other-worldly convents. Lordly manners, too, that once characterized (did they really?) a few lords have yielded to unaffected considerateness. While delinquency, adult and juvenile, is always with us, life is notably safer (except for the possibility of complete annihilation) and the human dignity of more people more respected now than ever. Numberless millions of us share patterns of culture formerly enjoyed by a tiny enclave of the privileged. When millions of workers prolong their hours in adult education programs, it would be ungenerous to suggest that this is only conformism and a doltish quest for cultural status symbols. And how can a society that faithfully reads "Peanuts" by altogether philistine?

If we find the TV ratings of Welk and Westerns too high and their quality too low, it may be because we all have been given chances to know better things. Besides, TV sets are still equipped with a button that brings instant, blessed release. And the paperbacks (1,912 new ones last year alone) make the best—from Homer to Hemingway—comfortably available to everyone who still reads. For all its blemishes, the culture explosion can surely serve the good of man.

The Lid Life of
Herbert Matthews

Thurston N. Davis

C AN THERE POSSIBLY be a sort of Matthews' Law—a
rule-of-thumb by which to tell whether Herbert
L. Matthews did or didn't write a given editorial
in the New York *Times?* The question is worth asking
because, within a wide circle of *Times* readers, no name
in recent newspaperdom creates more controversial
reactions. Rightly or wrongly, many believe, and have
for years believed, that certain of the self-confessed
biases and strongly personal evaluations of Mr. Matthews' signed stories represent a point of view several
degrees at variance with the general editorial positions
of his newspaper. This viewpoint, they argue, must
inevitably be reflected in the unsigned editorials he contributes. Hence, if such a thing as Matthews' Law exists,
thousands of readers will find it a useful daily tool.

Herbert Matthews, for almost forty years a news-
gatherer and news-analyst for the *Times*, has since 1949
been a member of the editorial staff of that esteemed
newspaper. For a generation, under the Matthews by-
line, the public has been accustomed to read his interpretations of Franco Spain. In fact, at the *Times*, all
Iberian and Latin American affairs come under Herbert

Matthews' eye in one way or another. Moreover, until recently, Mr. Matthews was the *Times'* "man in Havana"—the specialist assigned to watch the progress of the Cuban revolution of Fidel Castro. He no longer writes on this topic under his own by-line.

Editorials in the New York *Times* carry great weight in many quarters. Because they are unsigned, the full force and authority of the newspaper lie behind them. Since Mr. Matthews is a member of the *Times* editorial staff, it may reasonably be assumed that at least certain editorials on Spain, Portugal, Cuba and other Latin American countries come from his typewriter.

One's interest in trying to detect which are, and which are not, the editorials of Herbert Matthews is whetted by his own analysis of what it means to be a journalist. He expounds his theory in his recent enlarged edition of *The Yoke and the Arrows* (Braziller, 1961):

> I would never dream of hiding my own bias or denying it. I did not do so during the Spanish Civil War and I do not do so now. In my credo, as I said before, the journalist is not one who must be free of bias or opinions or feelings. Such a newspaperman would be a pitiful specimen, to be despised rather than admired. (p, 225)

Moreover, in a foreword to the same edition of that book, dated February, 1961, Mr. Matthews discusses his relation as editorialist to the newspaper he works for:

> The editorial policy of the New York *Times* toward the Franco regime is blamed on me. (As a matter of fact, editorials in the *Times* are expressions of the newspaper's opinions and when it comes to Franco Spain it does not matter who writes the editorials.)

One wonders whether the same two sentences could have been written with "Castro Cuba" substituted for the references to Spain and Franco. But this point need not be belabored here.

The modest purpose of this article is to suggest how certain of Mr. Matthews' editorials in the New York *Times* can be identified.

One discovers, without a taxing amount of research,

a sort of metaphorical thumb-print on the unsigned editorials contributed by Herbert Matthews.

Examine this recent specimen. It is from the July 18, 1961 editorial page of the *Times*—a piece of indubitably genuine Matthews copy on "The Spanish Civil War." Commemorating the 25th anniversary of the start of that war on July 18, 1936, the *Times* said:

> It was a bitter, bloody, heroic struggle whose greatest tragedy was that it left Spain unchanged. In fact, it turned Spain backward to its autocratic, hierarchical, feudal system of army, Church, land-owning, banking and industrial elite. The same Gen. Franco Bahamonde *sits on the lid*, waiting for the end which must come to him, but which will be a beginning for Spain. . . . (*Emphasis added here and below*)

Earlier this year, when Premier Salazar's Portugal began to make front-page headlines, a *Times* editorial ("Portugal's 'Wind of Change'") said in part on May 4:

> The result is that a growing opposition is *pressing against the lid* with which he [Salazar] would keep it down.

Again, an excerpt from another unsigned editorial bears the tell-tale trace. Dated March 14, 1958, it discussed the state of affairs in the Cuba of General Batista:

> Now he [Batista] has *clamped the lid down again* and reimposed censorship of the press and radio, but he cannot turn the clock back.

The thumb-print on a Matthews editorial checks perfectly with that found in Mr. Matthews' signed articles. Here, from the Sunday *Times* of May 6, 1956, is a snippet of a lengthy discussion of Batista's regime in Cuba:

> The uprising last week was symptomatic of the ferment *under the lid* that General Batista sits upon. One of these days *the lid may blow off*—but not now.

Again, in the first installment of an extended signed report on Spain, in the *Times* for Sept. 17, 1956, we read:

> On top all is tranquil because there is nothing to be seen but *the Generalissimo sitting on a lid*.

Five "lids," however, do not make a law. Can we get this project securely out of the realm of hypothesis and assure ourselves that there really is such a thing as Matthews' Law? I believe we can.

Turn once again to *The Yoke and the Arrows*. The passages cited below can all be found, identically the same, in both the first (1957) and revised (1961) editions, although page references are supplied only for the revised version.

Early in this book, discussing the centuries of Spanish history that lie behind the regime of General Franco, Mr. Matthews writes:

> Each time the people rose or freedom asserted itself, one or other of them, or all together ["the kings, the aristocrats, the generals and the priests"] would *clamp the lid back again*. . . . That is what happened in 1939 when a *caudillo*—Generalissimo Francisco Franco—*slammed another lid down and sat on it*. (p. 25)

Then, within six pages:

> Franco *slammed a lid down and sat on it*. (p. 31)

Again:

> The irony of it! Here we all are in the year 1961 with Italian fascism a sordid memory and with Bolshevism triumphant, while *a Caudillo still sits uneasily on a lid* in Spain. (p. 54)

Some pages later:

> What Franco had to do was *to clamp a lid down—or a number of lids*—and *sit on them*. (pp. 64-65)

Perhaps the finest specimen of all is:

> In those years after the Civil War, Franco had far more to do than just to *sit on the lid of a Spain* which, in any event, was prostrate and licking its wounds. (p. 67)

Then, in rapid succession, we run across two more:

> If one had to seek a single reason why, twenty-two years after the Spanish Civil War, *the same Caudillo is sitting on the same lid* in Spain, this is it. (pp. 98-99)

and

> On top, all is tranquil because, as stated before,

there is nothing to be seen but *the Generalissimo sitting on a lid.* (p. 99)

Farther along:

There is *another lid on which Franco sits.* (p. 105)

Then, too:

A student of Spain can only be amazed and heartsick at seeing generation after generation of the Spanish clergy repeating the same mistakes and building up the same forces of hatred that take such a terrible toll when *the lid blows off.* As of today, it is *Franco who sits on that lid.* ... (p. 169)

Finally, two more:

Meanwhile *Generalissimo Francisco Franco sits on the lid*—this one, too. (p. 190)

and

Is it any wonder that we who write about Spain refer again and again to *the "lid" on which the Caudillo sits?* (p. 200)

A student for the doctorate in journalism might some day—by wearisome research through thirty or more years of the *Times* on microfilm—push Matthews' Law on to its fully scientific conclusions. If he chooses this topic for his dissertation, he will doubtless find his best material in Mr. Matthews' pre-1961 period.

58

Christ or Credit Card?

Moira Walsh

HOLLYWOOD'S NEWEST multimillion-dollar, color-and-wide-screen, stereophonic-sound, three-hour-long biblical epic, *King of Kings*, which is being released by MGM, is the logical culmination of a gigantic fraud perpetrated by the film industry on the movie-going public. The fraud consists in persuading people that these films, as a group, have substantial religious or at least edifying qualities. It is closer to the truth to say that, with some notable exceptions such as *Ben Hur*, they are disedifying and even antireligious.

These are harsh words. They should be tempered by admitting that the public continues with unabated enthusiasm to support this fraud. And frequently enough it has had the well-intentioned backing of religious and community leaders. In defense of the religious and community leaders of today it should be said that their attitude is based on assumptions arrived at many years ago when the implications of the motion picture medium were very imperfectly understood by the public, by serious film observers and by the movie industry itself. If they now have doubts about the wisdom of this policy, as many of them do, they are caught up in an institutionalized tradition for which they feel a certain personal responsibility and from which it is extremely difficult to break away.

In defense of the film industry it might be said that their motives in making these pictures may be quite sincere. They have always had a stultifying tendency to believe in their own publicity and in box-office receipts as a sure test of cinematic excellence. When their own high estimate of their screen treatments of Holy Writ are confirmed, not only by the financial support of the mass audience but also by the unrestrained praise of the "opinion-making" segment of the community, which is generally freer with a denunciation than an endorsement, the movie makers can hardly be blamed for basking in a self-righteous glow.

In any case the sincerity of the members of this "society for the admiration of biblical movies" is not the point at issue. The question is: Are they, wittingly or not, caught up in an ultimately self-destructive vicious circle?

An increasing number of lucid and informed voices are saying so. The first group to express skepticism about the worth of these cinematic forays into Scripture were the serious secular critics. They were generally dismissed by the film industry as "out of touch with the tastes of the mass public," a remark which also served to indicate how far out of touch the industry was with the purpose of serious criticism. And film executives were not the only ones who were outraged by the critics' lack of enthusiasm for the Bible à la Hollywood. Their remarks were often, and usually wrongly, interpreted by religious spokesmen as contemptuous and even subversive of Christian values.

At the same time Catholics have always harbored reservations about some of these films, beginning with their objections to Cecil B. de Mille's *Sign of the Cross*, a "Christians-to-the-lions" epic enthusiastically spiked with pagan decadence, which was one of the contributing factors in the formation of the Legion of Decency. And the Legion itself has frequently taken note of one suspicious trend in these films—the tendency to give sex and/or sadism top billing over religion—by marking a number of them "objectionable in part"—for example, *Salome, David and Bathsheba, The Prodigal, Solomon and Sheba* and *Esther and the King*.

Another undoubted shortcoming of the gospel according to Hollywood is the habit of watering down the Bible in our pluralistic society so that it gives the least possible offense to the religious sensibilities of all shades of believers and unbelievers in the audience. Occasionally the Legion of Decency will make a notation to this effect, while at the same time classifying the picture in question as unobjectionable—as in the case of *The Robe*. More often they will let the omissions or distortions pass without comment.

Justice and charity sometimes dictate this course of action. In a fictional story which deals only tangentially with biblical incidents (and most of the films under discussion fall into this category), inaccuracies and elisions can in most cases be presumed to result from the clouded vision of the fictional characters rather than the shilly shallying of the producer. At other times the Legion's "hands-off" policy may be pragmatic in a perfectly legitimate sense.

Contrary to popular belief both inside and outside the Church, the Legion's power to influence the content of films has always been extremely limited. And it has maintained what influence it possesses by the skillful practice of the art of the possible. In most cases dogmatic or scriptural accuracy has been too much to expect of Hollywood, so the Legion has reluctantly accepted what it could not change.

King of Kings, however, presents an entirely different kind of a problem. This is not a well-meaning bit of pious fiction in which Christ appears briefly as an off-screen voice or a faceless white-robed figure. Rather, it purports to portray the life of Christ with a popular actor, Jeffrey Hunter, playing the role front and center; besides, it bears a title which is one of the traditional titles under which Christ is invoked. Some regard for facts and some spiritual comprehension are required if the film is not to be literally blasphemous despite its impeccable air of surface reverence. The Legion found it so wanting in this regard that it placed the picture in its "Separate Classification" with the observation:

> While acknowledging the inspirational intent of this motion picture, the poetic license taken in the development of the life of Christ renders the film theologically, historically and scripturally inaccurate.

No doubt this verdict elicited cries of anguish, astonishment and outraged innocence from the movie's sponsors. If so, I must admit to having a certain sympathy for them. The reactions of religious spokesmen to biblical films over the years have certainly given the producers some grounds for concluding that there was absolutely nothing they could not get away with except possibly too much sex and violence. Yet if the Legion had not taken a stand on *King of Kings,* it would have found it almost impossible to take one in the future on any film that violated scriptural or religious truth. And with our expanding knowledge of the screen medium it is becoming increasingly clear that violations of other Commandments than the Sixth and Ninth must be reckoned with, and that films sinning against the intellect and the human personality are, in the long run, the greater threat to faith and morals.

My sympathy for the film's sponsors, faced with the bad news that their sure-fire, time-tested formula for appeasing everyone has failed in this case to appease one important segment of the public, was short-lived, however. It has been replaced by a certain grudging awe for the virtuosity with which the company is using the "gamesmanship" approach in promoting the film among religious groups. The purpose of this maneuver is to divide and conquer, to confuse, to throw off balance, to blunt potential criticism by tricking some segments of the various groups into giving a seeming endorsement. Its chief weapon is apparent altruism—for example, the gratis distribution in schools of handsomely gotten-up film strips equipped with impeccably Catholic or impeccably Protestant captions.

An even grander gesture consists of the donation of preview theatre performances of the picture in key cities all over the world to religious leaders with no strings attached, except the concealed one that the company is hardly going to hide its candle under a bushel. Its publicity releases about these newsworthy events are designed to make it sound as much as possible as though the religious leaders were endorsing the film.

The company's sleight-of-hand technique with a press release was further illustrated a few days before the film's New York opening when they announced that "*King of Kings* has received the highest rating from the Catholic Cinematographic Center, the Vatican's film rating and reviewing organization." All this means is that the Italian equivalent of the Legion of Decency classified the film as their equivalent of "unobjectionable for general patronage." To grasp this, however, one must know that in context the words "highest rating" have no positive connotation, that the C.C.C. is an Italian Catholic national group and not "the Vatican's film rating and reviewing organization," and that Catholic film-classifying groups in various countries often come to different conclusions on the same film.

What the uninitiated can be counted on to infer from the release is 1) that Pope John himself has personally pinned a medal on the picture and 2) that, in expressing reservations about it, the American Legion of Decency is just not showing proper respect for the central authority of the Church. As long as film companies persist in these tactics, the chance of a fruitful dialogue between the churches and the film industry in this country seems fairly remote.

The task of tracking down all the liberties the picture takes with Scripture rightly belongs to a biblical scholar, which I am not. I mention one crucial falsification with reluctance. This is the trait common to most modern biblical films of rearranging facts to absolve the Jews of all blame for Christ's death. I mention it with reluctance because I acknowledge that the Jews have suffered for nineteen centuries from a terrible misunderstanding of this event by Christians. Or perhaps they have suffered rather because individual Christians were unwilling to accept the awful and essential truth about the Incarnation—that Christ was crucified for *my* sins—and have sought a scapegoat instead. But however legitimate is the grievance of the Jews on this point and however understandable their desire to rid themselves of an unjustly applied stigma, no good can be accomplished by letting Hollywood rewrite history according to wishful thinking. From a purely practical point of view it produces an effect diametrically opposed to the one sought after.

The application of this revisionist principle to *King of Kings* makes for an impossible situation dramatically as well as historically. The picture cuts from Christ being examined privately by Pilate (Hurd Hatfield) to Lucius (Ron Randell) releasing the imprisoned Barabbas (Harry Guardino) and saying in effect: "You are free. Christ will be crucified. Your followers yelled louder than the others." (Lucius is the good pagan centurion who figures, without aging, throughout the film's 33-year span, and Barabbas is pictured as the prototype of an Israeli freedom fighter.)

Thus, with one line of dialogue, the crucial matter of Christ's condemnation, how it came about and who was present at it, is casually glossed over and misrepresented. In the film the unspecified "others" are not present at the Crucifixion either. Nobody is there, in fact, except Christ's faithful few and some Roman soldiers.

How Christ came to be crucified in the first place is still more inexplicable. In the gospel according to scenarist Philip Yordan, His betrayal resulted from a not-ignoble miscalculation. Judas (Rip Torn) was a friend of Barabbas and, like him, an advocate of armed rebellion. When a premature rising of Barabbas' followers on Palm Sunday was mowed down by the Roman phalanx (in one of the film's few dramatically and pictorially exciting sequences), Judas, close to despair, had only one last hope. If the Master were delivered into the hands of His enemies, He would be forced to use His supernatural powers to strike down the oppressors and establish His kingdom on earth. To whom Judas betrayed Him is not clear, since the film introduces us to no one who had anything against Him. The thirty pieces of silver are nowhere in evidence.

Other stratagems employed by the film to keep it from having what might euphemistically be described as "sectarian orientation" might be cited. For example, any miracles that Christ is seen to perform are ones that could have a psychological explanation. The primacy of Peter among the apostles is not mentioned, but for that matter neither is the sending of the Twelve on their apostolic mission. And when the centurion (Lucius again) utters his great act of faith, it is not the ringing biblical phrase, "Truly this was the Son of God," but comes out instead: "Truly this was the Christ."

But criticism of the film on the grounds of scriptural inaccuracy only operates to discredit individual parts. It has to my mind a more fundamental failing which vitiates the whole. It is bad art, and bad art of a particular and precise sort designed to convey a surface impression to the unwary that is totally at odds with its real content and impact.

To understand the significance of this indictment it is necessary to understand the essential dynamics of films, how they actually work and how they affect audiences. The above-mentioned secular critics, by and large, were the first to perceive this all-important point. All too often the policies of film-betterment groups were formulated without sufficient comprehension of the screen medium, which has led them, at one time or another, to endorse some very strange films indeed.

A few months ago in AMERICA I attempted haltingly to explain how the film form operates:

> The essential impact of a work of art flows, not from what is said, but from the intangible synthesis or climate that is created by the collision of the artist's vision of life and his chosen material. Furthermore, this climate, this implicit outlook on life, is the basic quality that a film communicates

65

to the public and the communication is subliminal, on the level of emotion rather than intelligence.

These remarks were made as a preliminary explanation for taking the position that a superficially disedifying film, such as *La Dolce Vita,* can be in reality both moral and worth-while. They serve equally well to demonstrate how a superficially inspirational film such as *King of Kings* can prove in actuality to be nothing of the sort.

If a film about our Lord is going to communicate to the audience "an intangible synthesis or climate cre-

ated by the collision of the artist's vision of life and his chosen material," then producer Samuel Bronston, director Nicholas Ray and scenarist Yordan have to start off by answering the biblical question: "What think ye of Christ?" But from the movie it is obvious that these gentlemen have no opinion on the subject except that at the moment He is a "hot" box-office property if properly exploited. This attitude, or lack of it, poisons the wellsprings of the movie.

Christ is there as a physical presence, but His spirit is absent. No serious attempt is made to establish who He is, what His motives and purpose are or what His relationship is to the various social forces around Him. Consequently there is no "dramatic engagement" between Him and the audience, and not the slightest possibility that anyone will derive from the film any meaningful insight into what Christ's life and sufferings signify for us. On the contrary, the picture bends all its efforts in the other direction—to keep Christ as neutral and undynamic as possible while at the same time lulling the audience into a pleasurable state of pietistic euphoria by parading the familiar words and images of the New Testament in pageant style before them.

The Sermon on the Mount sequence, staged with mobs of people and an effective eye to cinematic movement, is a case in point. Having spoken the Beatitudes with great earnestness and some skill, Hunter falls to answering some of the famous questions proposed to Christ in the Gospels. The words sound fine but they have no integral connection with any of the film's dramatic action. Therefore in the context of the picture they have no more impact than audience participation on a Jack Paar show.

This avoidance of the fundamental issues is not due to ineptitude or even to the muddleheaded but understandable wish to avoid offending anyone's religious sensibilities. Rather it is based on the same shrewd commercial instinct that poisons so much of the entertainment shown in the mass media today. An important function of art after all is to make us see ourselves more clearly. This can be a very painful experience, so painful in fact that vast numbers of the mass audience have demonstrated that they will not support it with their ticket or grocery purchases. They prefer the kind of entertainment with the opposite but ultimately de-Christianizing message: "You are fine as you are."

"One of the weaknesses of some popular religion today is that it aims to make people feel good rather than be good," said Methodist minister Ralph W. Sockman recently. That is certainly the weakness of *King of Kings*. A life of Christ should be an irresistible challenge to man's conscience. Instead, this one is a tranquilizing drug.

Bishop Paul J. Hallinan of Charleston, S.C., has pinpointed this dichotomy even more pungently:

The greatest challenge to Christianity today is a popular, bland, respectable faith termed secular humanism. It is often called "the American Way of Life." It is not godless but it keeps God in His place—the pulpit. It equates the Christian moral code with such terms as decency, brotherhood, the Golden Rule. It is the orthodoxy of non-believers, but it is a ready refuge for the half-believer too. . . . It simply says: "Take up your credit card and follow me."

King of Kings is perfectly at home in a credit-card society. It is at opposite poles from Christ's exhortation: "Take up your cross and follow Me."

What Do the Scrolls Tell Us?

Raymond E. Brown, S.S.

IT IS NOW close to fifteen years since the first scrolls were discovered in a cave near Qumran. Today unbroken silence cloaks the ruins of the quasi-monastic settlement that has been excavated at Qumran in the intervening years. Eleven caves that yielded up documentary treasures once more open their gaping mouths only to bats and flies, and some passing Bedouin. If the dust seems finally to be settling at Qumran, so too in the scientific world of biblical studies a certain peace seems at last to hold sway in the matter of the scrolls.

As Joseph A. Fitzmyer, S.J., pointed out in AMERICA earlier this year (3/18), the mavericks with their cries of hoax are heard in ever fewer numbers. Also, calm weighing of the evidence has silenced or, at least, exposed in an embarrassing light the professional agnostics and the opportunistic scholars who saw a chance to make headlines by claiming that the new scrolls challenged the uniqueness of Jesus Christ. The years of investigation have resulted in a generally accepted view on the dating, nature and significance of the scrolls.

It is commonly agreed today that the scrolls come from the library of a group of Essenes who inhabited

the Qumran area from shortly after 150 B.C. until
68 A.D. The Essenes, according to the Jewish historian
Josephus, were one of the three major Jewish sects in
the first century of the Christian era. But the Essenes
were more truly a sect than their rivals, the Phari-
sees and Sadducees of Gospel fame, who were actually
parties of political and theological opinion. As
reconstructed from the scrolls and from sources like
Josephus, Pliny and Philo, Essene history began with
a breaking away from the main body of Judaism shortly after the Maccabean revolt of 166 B.C.

Most pious Jews cheered the attempts of the Macca-
bee brothers, Judas, Jonathan and Simon, to free Judea
from the political and religious tyranny of the Syrians.
Not all were equally enthusiastic about the usurpation
of the high priesthood by Jonathan (152 B.C.), a usurpation that was more or less legalized for Simon and his
descendants in 140 B.C. (I Mac. 14:41). The Essenes
seem to have sprung from a revolt of the real high-
priestly family (descended from Sadoc) and of their
followers against that usurpation. Withdrawing into
the wilderness of Judea, these priests and laymen
formed a community whose purpose, as they wrote, was
"to clear the way of the Lord in the wilderness." They
were to be a penitent nucleus which, by strict observ-
ance of the ancient law and adherence to the sacred
solar calendar, and by absolute purity in matters sexual
and ritual, would be prepared for the imminent coming
of God.

The Qumranians were persecuted by the Maccabean
high priests and by their successors, the Hasmoneans,
who earn in the Qumran writings such epithets as
"Wicked Priest," "Man of Lies" and "Vessels of Violence."

Shortly after the Essene movement began, there
arose at Qumran a man who bore the title of "the Righteous Teacher." This nameless priest was a man of great
spiritual attainments, for, according to the scrolls, God
revealed to him secrets which He had held back even
from the prophets of old. Although the personality of
the Righteous Teacher left its mark on the community's
ideals, we know little of his career, save that he was

persecuted. Of his claim to be a messiah, of his crucifixion and resurrection, the writings of the community say nothing. For these items one must consult the vivid and creative imagination of certain modern writers on the scrolls.

During the century before Christ, the community at Qumran grew in numbers. The buildings were enlarged about 110 B.C. They soon included an elaborate system for conducting and preserving water, a bakery, a pottery shop, storehouses and a community dining room. Most of the sectarians must have lived in tents and caves near this settlement. The swelling numbers may have included fugitives from the increasingly irreligious activities of the Hasmonean high priests like John Hyrcanus (d. 104) and Alexander Jannaeus (103-76). Jannaeus' wife, Salome Alexandra, who ruled from 76-67, is mentioned by name in the Qumran writings, as is the first Roman governor of Syria, Aemilius Scaurus (62).

Even the troubled times that saw the rise to power of Herod the Great (c. 40 B.C.) seem to have left their mark at Qumran, for fire and earthquake desolated the building. For about forty years, roughly contemporary with Herod's reign, the settlement was virtually deserted. Shortly after Herod's death in 4 B.C., however, the sect returned to settle at Qumran and remained there until the Jewish revolt against Rome. In 68 A.D. the Tenth Roman Legion reduced the buildings to ruins and ended the history of the Qumran community.

The library of this community must have been quite extensive. One of the caves near the community settlement, Cave IV, contained over four hundred manuscripts in Hebrew, Aramaic and Greek. Unfortunately, only a small part of the total library has come down to us. Among eleven caves discovered so far, the first and last have yielded complete scrolls. The rest have produced only fragments—some torn deliberately in antiquity (perhaps by Roman soldiers), many of them encrusted with dirt and bat dung, eaten by animals and insects, and worn by two thousand years of exposure to the elements. Nevertheless, these few scrolls and many fragments have proved to be one of the most important legacies left in Palestine by the people of the

Book. William F. Albright has characterized it as "the greatest manuscript discovery of modern times."

About one-quarter of the discovered manuscripts are biblical. Every book of the Hebrew Old Testament except Esther is represented. Of the deutero-canonical books (those found in the Greek Old Testament, but not in the Hebrew—accepted by Catholics, but not by Protestants and Jews), some of the original Hebrew of Sirach and the Aramaic of Tobias have appeared. Consequently, it is in the field of the Old Testament that Qumran has made some of its greatest contributions.

Before the discovery of the scrolls, the oldest Hebrew manuscripts of the biblical books dated to the ninth century A.D. Now we have, in whole or in part, over a hundred manuscripts going back to the period before Christ. In fact, some of the manuscripts are older than the Essene community itself, for instance, fragments of Exodus and Samuel from Cave IV, dating to the third century B.C. These must have been heirlooms brought by the founding fathers. One of the fragments of Daniel dates from about a half-century after Daniel was written (probably about 165 B.C.).

We must not jump to the conclusion that because the Qumran texts are older than previously known Hebrew manuscripts, they are necessarily better. At the beginning of the second century A.D. there was an organized attempt in Judaism to compare biblical manuscripts, to choose those that were carefully done, and to reject those that were textually poor. The standard Hebrew text of the Middle Ages is the heir to this tradition of critical scholarship. With the Qumran scrolls we find ourselves back in the precritical period with all sorts of textual traditions, some good and some bad. For example, two scrolls of Isaiah were found in Cave I. One was virtually identical with the Hebrew text we know; the other had a myriad of slightly variant spellings (only a few of which actually gave significantly different readings).

It should be added that even when different readings

of the biblical texts are supplied by the Qumran manuscripts, many of them are not new to scholars. After all, the medieval Hebrew texts have not been the scholars' only keys to the Old Testament. The old translations of the Bible into other languages play an important role. For instance, the Greek translation (Septuagint) of the Old Testament, made centuries before Christ, often departs considerably from the Hebrew text familiar to us. Until the Qumran discoveries, scholars could only make educated guesses at what sort of Hebrew underlay this Greek translation. Some even thought that the Septuagint was simply a free translation. The discoveries at Qumran gave us, for the first time, fragments of the variant Hebrew texts that underlay the Septuagint, and we have found that in books like Samuel and Jeremiah the Greek translators were being faithful, but to a Hebrew text quite different from the one we knew.

The scrolls, then, provide a good deal of interesting material for study of the Old Testament text as well as for auxiliary sciences such as Hebrew paleography and Hebrew pronunciation. But perhaps we should close this section by adding an assurance for the timorous of heart. The variant readings of which we speak largely concern phraseology. There will be no substantial changes in the Old Testament narrative. In fact, many modern critical translations, like our Catholic Confraternity translation of the Old Testament, have already incorporated the best of the variant readings, which were known from the Septuagint and from the Vulgate, St. Jerome's Latin translation of the Bible.

THE QUMRAN SCROLLS have also made a great contribution to the history of Judaism between the two Testaments. The Old Testament literature breaks off in the second century B.C.; the rabbinical literature has its roots in the second century A.D. Very little of the in-between literature had survived until the discovery of the scrolls. Now, from the nonbiblical Qumran writings, we can get a more realistic picture of the political, intellectual and theological ferment of the period from 150 B.C. to 70 A.D. It is true that we have to look at this period through the eyes of the Essenes. If we had some of the Pharisee and Sadducee literature, we might get a more rounded idea of the times. Yet the scrolls have given us many insights that otherwise might have been lost.

We find, for instance, that a dualistic theology had come into Judaism. Perhaps, ultimately, it came from Persia. In Essene thought, the world was divided in two, under the leadership of the angel of light and the angel of darkness. Both angels were created by God, but they were allowed to struggle in this world on an equal footing until God's final intervention on the side of light. Yearning for God's intervention produced a strong eschatological and apocalyptic interest at Qumran, and messianic expectations were especially vivid. The Essenes expected the coming of a prophet (perhaps the prophet who would be like Moses—cf. Deut. 18:15) and of *two* messiahs, one a priest, the other a king of the house of David.

The moral theology of Qumran was highly developed. There was some sharing of personal goods; amassing of wealth was severely condemned. There seems to have been an ideal of celibacy; most of the community rules concern only men. Josephus, however, speaks of celibate and marrying Essenes, and female skeletons have been found in some of the smaller Qumran cemeteries. It is not entirely clear whether the married Essenes represented a different congregation alongside the celibate community or a later relaxation in community discipline to foster membership. Fraternal charity was another virtue greatly emphasized at Qumran, far more, indeed, than in the Old Testament.

The discipline of the Qumran community is most striking as one reads the scrolls. Newcomers were submitted to a year of postulancy and a year (or two) of novitiate before being admitted to the community. In these years of trial the candidates were carefully scrutinized on behavior and adherence to the strict Qumran interpretation of the Law. Only gradually were they allowed to participate in the ritual cleansings of the sect, and only full-fledged members were admitted to the sacred community meal of bread and wine. Nothing so close to a monastic community has hitherto been found in pre-Christian Judaism.

THE INCREASED knowledge of Judaism gained from the scrolls has also given us a better insight into the background of Christianity. For instance, John the Baptist now seems less lonely in his wilderness ministry. Living and working in the same area as the Qumran Essenes, he quite conceivably had some contact with them. Like them, he insisted on repentance and a cleansing baptism; like them, he regarded his mission as preparing the way for the imminent coming of the Lord (both John and Qumran use the "voice in the desert" passage of Isaiah 40:3 to characterize their missions); like them, he tried to form a penitent nucleus in Israel. Obviously, his ministry was of wider appeal and was less legalistic, but John still stands closer to the Qumran Essenes than to any other Jewish background.

Our knowledge of Qumran thought also throws light on many of our Lord's sayings. After all, Jesus had to speak to all the Jews of his time and in their own terms, and some of the Essene doctrines may have been fairly widely known. Josephus tells us that there were Essenes scattered in various towns. Consider particularly the way Jesus speaks in St. John's Gospel. Sections like John 3:19-21 and 8:12, which emphasize the division between light and darkness, are quite intelligible against the framework of Qumran dualism. The presence of abstract dualistic terminology had led critics to deny authentic Palestinian origin to St. John's Gospel. Now, in some ways, it turns out to be the most Palestinian of all Gospels.

The scrolls have also shed light on the Epistles. A section like II Cor. 6:14-15, for example, sounds almost as if it had been copied out of a Qumran document.

The whole argument of the Epistle to the Hebrews seems clearer, thanks to Qumran. The author is probably writing to a group of converted Jewish priests who are beginning to fall away from Christianity. Yet, in the arguments he uses with them, he does not seem to have the priesthood of the Jerusalem Temple in mind, but the priestly practice of a more ancient period. Now the hypothesis has been advanced that the epistle was written to convert Essene priests. Then the comparisons drawn from the older priesthood would make more sense, for the Essenes regarded the current Jerusalem priests as usurpers. Also, the argument of the epistle that Christ is superior to every angel would be clearer against a Qumran background, in which all the sons of light were subject to a supreme angel. Finally, this hypothesis would explain the great emphasis in the epistle that Christ is a priest, even though not a Levitical priest. To satisfy Qumran expectations, Jesus would have to be the Davidic messiah and the priestly messiah all in one. We might remember the early Christian designation of Jesus Christ as prophet, priest and king. In Him, then, were to be found the three messianic expectations of Qumran.

The Qumran finds may help, also, to clarify some details of early Christian practice and organization. The ideal of a community of goods was shared by both Essenes and Christians (Acts 2:44-45). One of the Qumran officials bore a title (*mebaqqer* or *paqid*) which is the exact Hebrew equivalent of the Greek word for bishop (*episkopos*). Like the Christian *episkopos* (Acts 20:28), this Qumran official was regarded as a shepherd. His functions of teaching and guiding the small Qumran groups resembled closely the functions of the early bishops (Titus 1:7-9). Also, the large assemblies of the Qumran community were called the "sessions of *the Many*," a term very close to that used of the Christian community sessions in the Acts of the Apostles (6:2; 15:30). The Qumran practice of a daily sacred meal in which the priest blesses bread and wine has interesting parallels in the daily "breaking of the bread" in the Christian communities (Acts 2:46), although, of course, there is nothing at Qumran to suggest the body and blood of the Lord.

A final verdict as to the full significance of the Qumran discoveries is still to be achieved: the fruit of further patient investigation. Enough, however, has already been acquired to dissipate the flock of fantastic speculations occasioned by the first discoveries. Far from upsetting the credibility of the Christian narrative, the scrolls, in many cases, have offered valuable interpretations of Old and New Testament documents. We may look forward with much interest to the eventual completion of this fruitful enterprise.

New Images of Man

W. Norris Clarke

"NEW IMAGES OF MAN" is the theme of an unusually interesting and thought-provoking exhibition of contemporary sculpture and painting on view until Nov. 29 at the Museum of Modern Art in New York City. The 23 American and European artists here represented by some hundred of their works are not content with the absorption in pure abstract form which up to now has dominated about 80 per cent of contemporary exhibited art. They all feel in some way the urgent call to body forth in canvas and paint, in wood, metal and stone, their inner vision of what modern man is really like, of what has happened or is in the process of happening to him, under the veneer of his well-tailored, superficially healthy surface.

Though these artists are for the most part quite independent of each other and belong to widely different schools, there is enough unity of idea, imagery and feeling among them to make their silent judgment on our age unusually significant. The artist has always held up the mirror to his own time for those who have the eyes to see and the courage to want to see. His own eyes may not always be perfectly clear, and he may not see all that is there. But it is dangerous for the rest of us not to stop periodically in the headlong rush of our daily living and take a long attentive look at the products of his vision. This is what I would like to do in what follows, speaking not as an art critic, but as a Christian philosopher, focusing on what these men are trying to express rather than on the artistic adequacy of how they express it.

I am aware that to many the attempt to disengage any definite meaning or idea-content from a piece of modern art will seem a hopelessly arbitrary and subjective enterprise. This may well be true for most of the dominant abstract art. But one of the significant things about the present exhibition is that it marks a notable departure from pure abstractionism. The latter movement had worked out its own inner genius to such a pure degree that it was bound sooner or later, like all particular movements in art, to call forth its own counterpoint. Though the present group could not express themselves in the way they do unless they had learned the lessons of abstractionism, still I think it is undeniable that most of them are seriously concerned with giving expression to a vision or image of man that is in some recognizable sense a criticism of life. I am supported in this by the commentary of the director, the excerpts from the artists' own writings on the aims of their work, and the perceptive preface of Paul Tillich, distinguished Protestant theologian and critic of culture —all contained in the excellent book of text and photographs which serves as catalog for the exhibit. The comments of four sculptors—Leonard Baskin, Reg Butler, Balcomb Green and Theodore Roszak—on the opposite page illustrate what these artists are trying to do.

A CREATURE OF TENSIONS

What sort of images do we see in this hall of mirrors for contemporary man, so skillfully mounted for our benefit by Peter Selz, curator of painting and sculpture exhibitions at the Museum of Modern Art? The most powerful over-all impression that struck me was that nowhere here could one find man at peace, man in harmony with himself, with his fellows or with the world

79

that is supposed to be at least temporarily his home. Almost every artist lays him bare, in his own arresting way, as a creature of taut, often agonizing tension, confused as to who and what he is, painfully lonely and isolated from his brothers, the depersonalized victim of his own triumphant technology or of dark primitive forces unleashed from his own subhuman depths.

It is as though these painters and sculptors felt it was no longer safe to allow themselves or us to focus on the external beauty and grace of the human form, as earlier artists had done, mercifully concealing the machinery and obscure forces at work within. This would only flatter us, dull our sense of danger and lull us into a fatal slumber, while the poisonous gases of self-complacency and illusion continued to do their deadly work. We must rather be taken brusquely by the shoulder and vigorously shaken up, made to see what is happening to us within. Hence in most of these works all protecting surfaces are stripped away. Man is laid bare to the engine room, so to speak, with all his inner wheels, wires and power lines exposed. Even where an outer surface is apparently kept, the taut lines of bone, nerve and muscle pierce so transparently through that the veil of skin and flesh serves rather to reveal than to conceal what is underneath. This insistent urge to analysis in depth, leaving no recess of privacy unprobed, no refuge of illusion unexposed, reflects unmistakably those two typical techniques of our day for exploring the hidden dimensions of man: the X-ray camera and psychoanalysis—or rather, to be really up to date and actually more accurate— "existential" psychoanalysis. It would be hard to portray more aptly the pervasive spirit of inner tension and anxious self-analysis so characteristic of modern man in an urban technological society.

Let us now explore in detail the characteristic feelings about himself and his world that emerge from these new images of man. What are the sources of his anxiety and tension? He appears, first of all, as the depersonalized victim of his triumphant but soulless technology. Who can forget the pitiful dejected shell of the once noble hero Jason, created by the British sculptor Paolozzi out of old bits of discarded metal, wheels and tubes, pierced by jagged holes revealing the emptiness within? Or the heavy, earth-bound Icarus in the same style, stretching out helplessly his stunted (or fractured?) wings? With a little meditation one begins to share the artist's vision of man as fallen from his human dignity and his once heroic grandeur to the level of a run-down robot, a discarded reject of his own technology.

A similar theme recurs in the works of the French sculptor César, especially in his arresting figure of a female torso roughly welded of old bits of metal, whose iron blisters and amputated legs and arms, ending in jagged splinters of corroded steel, speak of human precariousness and the decay of the flesh more eloquently than any materials formed by nature herself could do. It breaks out again in a violent cry in the tremendously dynamic "Iron Throat" of the American Roszak, a great eroded skull in steel plate and tubing, its mouth open in an enormous shout—an image, one cannot help but think, of the power of communications: all voice, but dead within.

THE LONELY MAN

A second common theme is that of lonely man, isolated both from his fellows and the world of nature with which he was once in harmony. This is marvelously evoked by the "thin people" of the Swiss Giacometti, undoubtedly one of the most gifted sculptors of the group. His figures in bronze rise tall and indescribably thin, sometimes only a few inches in width for a height of six feet or more, each one infinitely withdrawn into itself and remote from every other, even when they form a group on a public square. As one critic has said, his figures may stand or walk, but they never rest; they are too tensely introspective and self-enclosed for repose. Aside from "ideology," however, for sheer artistic magic with the utmost economy of materials, his evocation of space and its mystery in the soaring figure of a "Man Pointing" has few equals that I can recall.

Something of the same impression of loneliness and isolation in the midst of the world is reflected also in the stiffly awkward couple in two of Diebenkorn's can-

81

vases. They seem to be strangely lost and out of place in the coolly indifferent and impersonal space of the room that is presumably their home. Francis Bacon's painting of the huddled figure of a man in the center of a landscape, completely alien to the nature around him and absorbed in himself, is another striking example. This lack of a sense of organic harmony and at-home-ness with his environment is one of the most frequently recurring themes of the exhibition. The same inner loneliness breaks out again in quite a different way, it seems to me, in the hectic gaiety of De Kooning's pathetically erotic women, with their desperately extrovert eyes and fixed toothy smiles. The restless, agitated flow of lines, masses and colors in these paintings achieves an admirable interpenetration of idea with artistic form.

Germaine Richier's sculptures make us pass from the isolation of man from his fellows and the world around him to his own uncertainty about himself within. Her strange, half-human figures, hovering ambiguously between man and animal or plant, like "The Grasshopper" and "Don Quixote of the Forest," seem to suggest that modern man is no longer so sure whether he is really human or only a mask for something more primitive.

A milder form of the same inner uncertainty and loss of self-identity is reflected in the blurred or faceless figures characteristic of several of the painters. Their image of man suggests someone who has lost all awareness of who and what he is, as he moves through a mysteriously opaque and alien space. Such are Bacon's two "Studies for a Portrait of Van Gogh" and Nathan Oliveira's "Man Walking" and "Standing Man with Stick," although the latter warns us: "My concern for the figure is primarily a formal one, growing out of the problems of painting itself. The implications are unconscious."

The last major theme of the exhibition, if one can call it a separate theme, I can best describe as a kind of pent-up outburst or explosion of indignation at all man's inhumanity to himself and his fellow man. The unbearable anguish of man's self-imprisonment and self-mutilation bursts out at us with brutal explicitness from

Bacon's painting of the "Man in a Blue Box," in which a powerful figure is shown rising from his chair, his whole face exploding in a vast cry of agony, his pose suggesting with sinister ambiguity that he is both judge and accused at once. Compassionate indignation flames out at man's regression to the beast in Rico Lebrun's paintings, "Buchenwald Pit" and "Study for Dachau Chamber." In a similar but more muted and resigned vein are Leon Golub's paintings of "Damaged Man," a pathetic yet heroic figure systematically dislocated in every member, and his more normal and at the same time more deeply moving "Orestes" with its strangely haunting face, the essence of bewildered pain beyond words borne with Stoic resignation and the determination to endure despite it all.

An affirmation of human dignity, however, shines through Lebrun's work in the strange light of glory that seems to radiate from his tangled heaps of limbs, some of them marking as on a clock of human history that

the time is well past high noon (the fancy struck me that it was also the time of the Passion of the Son of Man). As he himself tells us, "I wanted to express the belief that the human image, even when disfigured by the executioner, is grand in meaning. No brutality will ever cancel that meaning. Painting may increase it by changing what is disfigured into what is transfigured." Golub, too, with his "Colossal Head," points to man's heroic will to endure despite his tragic vulnerability and the enveloping threat of mutilation or annihilation. "The ambiguities of these huge forms," he writes, "indicate the stress of their vulnerability versus their capacities for endurance."

Let us call a halt reluctantly to our very incomplete inventory and reflect a moment on what we have seen. (I have had to omit many minor themes and outstanding individual works, often of stirring power and beauty, such as Roszak's "Skylark," a poem in shining steel of a figure alive with motion, half-bird, half-human, pivoting on tiptoe in the act of taking flight, with a burst of stars for a face; or the seated thinker in squared stone of the Austrian Fritz Wotruba. [*See our cover.*—Ed.])

What is the significance of it all, if any? The image of man which emerges here is certainly not one which would be greeted with enthusiasm at a Rotary Club convention. Despite the compassion behind it and the occasional flashes of hope, grandeur and beauty which break through the storm clouds, it is on the whole a somber and threatening picture: an image of inner tension, anxiety, uncertainty, depersonalization, vulnerability to powerful subhuman forces seething just beneath the thin veneer of civilization—an image, in a word, of the extreme precariousness of man's human dignity and the dangerous erosion it is undergoing on all sides.

THE ARTIST CONFUSES US

The spontaneous reaction of the non-professional, "man-in-the-street" observer tends at first to be a sense of disorientation and bafflement at being suddenly confronted with this strange and disconcerting picture of the man he thought he knew so well. This can quickly turn into a feeling of resentment—even indignation—as though he were being presented with a deliberately distorted and unflattering portrait of himself. "What have you done to our humanity?" one feels like complaining to the artist. "Why have you so mercilessly laid bare the dark corners of our soul, why have you so dislocated and dissected our natural beauty in order to reconstruct the parts in such artificial and disturbing patterns?"

But the artist can with justice throw the challenge right back at his accuser: "My images reflect what I see. What I have been watching is yourself. The real question is: What have *you* done to our common humanity, to yourself, to your brothers and to me? You accuse me of taking man and nature apart like a machine and refabricating them in unnatural new forms.

But isn't that exactly what you have done to the world around you with your technology, sometimes creating new beauty but often, with careless ruthlessness, leaving only ugliness and distortion behind as the price of utility or comfort? And isn't the taking apart and artificial reconstruction of man exactly what 'social engineering' has already done in Russia and China and threatens to do in any technological civilization?

"I am not trying to paint you as you look, or would like to look, from the outside, but as you really are, or fear that you are, within. I am not judging you, but only trying with compassion and sympathy to give expression to your own secret, inarticulate image of yourself. Do not people always experience a moment of resistance and resentment when brought face to face with their hidden selves for the first time? Yet is this not the only path to self-renewal? I am doing nothing else than what you pay your psychiatrists for doing. You should thank me, rather than blame me, for having revealed to you the hidden face of your own soul."

And I think the artist would be right. Admittedly he does not see all that is to be seen among us. Perhaps his hope and faith are not as strong and obvious as his spirit of protest. This may be his defect. But what he does see and image forth for us is certainly there. It would be wise for us not to turn away our eyes because the face in the artist's mirror is not the beautiful and inspiring one we had hoped to see, but one furrowed with agony and fear.

The people who had to live through Buchenwald and Dachau have good reason to want to bury their trauma in oblivion. But the facts behind these names, and their all too numerous counterparts in other sections of the world, remain an ineradicable scar on the face of 20th-century man which neither the historian nor the artist has the right to forget. As Tillich puts it in his perceptive preface, these makers of art

> are fighting desperately over the image of man, and by producing shock and fascination in the observer, they communicate their own concern for threatened and struggling humanity. . . . If they depict the human face, they show that it is not

simply given to us, but that its human form itself is a matter of continuous struggle.

Who of us would be willing to deny that our century will stand out in history as the one in which the ultimate precariousness and vulnerability of man were unveiled for the first time? Perhaps our artists have read the record better than we like to think.

PESSIMISM DESPITE SUCCESS

A problem remains in my own mind, however, that has long puzzled me and that the present exhibition has only brought to a head more acutely. I would like to propose it to the readers of AMERICA for their reflection and for what further light they can throw upon it. Why is it that, despite our constantly increasing prosperity, comfort and technological mastery over nature, the majority of our outstanding contemporary artists, both literary and otherwise, remain quite unimpressed by this aspect of our culture and the widespread optimism it generates, and persist in presenting a predominantly somber image of the decline and dehumanization of man? Are they merely expressing their own private resentment at their loss of prestige compared to the businessman and the scientist; at their isolation within the highly organized, team-work society around them; at their inability to feel at home and come to terms with the machine age; or possibly at their alienation from the ancient religious roots of art?

It is true that the artist, who must necessarily remain a lonely individualist in the disappearing tradition of personalized craftsmanship, and who for long has had to live as a kind of vagrant on the margins of society, has felt the pressure of the technological age more keenly than most people. He has had to face the difficulty not only of finding a place for himself within it that will not suffocate his art, but also of keeping himself from being so spiritually, imaginatively and emotionally overwhelmed by its complexity that his art itself cannot get it into focus with adequate unity, totality and depth. Could it be, then, that the artist's darkly colored portrait of lost and anxious contemporary man is really only a disguised projection of his own ambiguous and frustrated predicament?

On the other hand, his very distance from the world which nourishes him gives him a better vantage point, like that of the philosopher and the saint, for discerning what is truly significant within it. Is it he, then, who has seen more clearly than the rest of us? I do not know the answer. But while we wait for more light on the matter, it would be wise for us to give our artists the benefit of the doubt and meditate humbly on how we might provide them with the model for the happier and more harmonious image of man that we hope to find in the art exhibits of the next generation.

The Recognition of Shock

Harold C. Gardiner

THE VIRGINIA KIRKUS SERVICE is an organization that reviews books far in advance of publication as a means of alerting booksellers. Its biweekly bulletin is, as a rule, admirably outspoken in calling spades exactly spades when it is necessary to warn booksellers that such and such a book ought to be "handled with care." It is one of my chores to read the bulletin, and I have been much impressed of late by the number of books that have called forth cautions like these:

> It is, in fact, the author's concentration on bedroom/bathroom obscenities that the publishers commend as a "selling point" (*Drum and Bugle,* by Terence Fugate, Simon & Schuster).

> The lives of Amber Falls' citizens are exposed and the picture presented is one of unbridled sexuality, desperate squalor, intellectual and moral whoredom, despair nursed in alcoholism (*The Mill,* by Bradley Robinson, Random House).

> The antagonist is the occupation army . . . where boozing, wenching and brawling are the antidote

for army apathy and boredom (*A Fever for Living,* by R. A. Roripaugh, Morrow).

There she seduces him, disgusts and degrades him and finally completes his destruction. . . . It is a study in dissolution, in the fascination of the flesh which almost approaches an exaltation of evil (*The Gouffé Case,* by Joachim Maass, Harper).

The situation bursts as stickily as a lanced boil and one sniffs sensationalism at the core (*Lord Love a Duck,* by Al Hine, Atheneum).

More than the weather is steamy—there's lots of sex, cheap rather than seductive (*The Long Goodnight,* by Carl D. Burton, Morrow).

Strictures like these are common—increasingly common, I believe—in the bulletin, and though this source is hardly infallible in its estimates, its judgments do serve to spotlight a trend that is well summarized in a review of Grace Metalious' (of *Peyton Place* "fame") *The Tight White Collar* in the London *Times Literary Supplement* (January 6):

> Vulgarity informs almost every page [of this book], an account of the sex lives of the inhabitants of a small New England town hung on a plot about a witch hunt involving a teacher in the local school. The defilement of the most intimate of human experiences by language worthy of an advertisement for underwear; the prepackaging of sex in clichés so that an anxiety-ridden orthodoxy may masquerade as the ultimate in frankness and daring—these things are not uncommon in many contemporary novels, but *The Tight White Collar* is the epitome of the genre.

If these damning criticisms are accurate straws in the wind (and dozens of similar examples could be adduced), a question inescapably rises: what has happened to our contemporary recognition of shock? To pose the question in this term is to engage in an obvious play on the title of a famous book edited by Edmund Wilson in 1943, *The Shock of Recognition.* That book consisted of a series of studies by famous American writers of other equally famous literary compatriots— Melville assayed Hawthorne, for example, and Twain judged Cooper. The theme of the book was indicated in the words of Melville: "For genius, all over the world, stands hand in hand, and one shock of recognition runs the whole circle round." This tingling sense of awareness of kinship still runs like an electric current among the truly eminent of our creative writers; we find Mauriac writing probingly about Graham Greene, or, to take a very recent instance, Phyllis McGinley saluting the achievements of fellow poet Daniel Berrigan, S.J. (see AM. 10/26/57, pp. 110-111).

But the point I have in mind is concerned with the problem in reverse, so to speak. Why is it that many concerned with current literature—who, if they are not geniuses, at least ought to have basic good taste—do not recognize that much of the "shock" in that literature is a cheap and tawdry and debasing fake? Perhaps it is because, as Wilson says in his introduction to *The Shock of Recognition*, "it requires gifts just as rare—since they are just the same gifts—to be fully aware of the bad as to be fully aware of the good" (Doubleday, Doran, 1943, p. xiii).

But no—that certainly cannot be the basic reason, for it does not take "rare gifts" for readers, critics and, above all, the publishers to realize that, as critic Hannah Arendt once said: "The fact that pornography is infinitely boring is unrecognized only by the truly vulgar."

If Miss Arendt's touchstone in this matter still works, I am afraid that we are forced to the conclusion that not only a wide reading public but even more portentously a large circle of critics and a segment of publishers are persistently and perhaps all unconsciously revealing their own fundamental vulgarity. In an endeavor to attract a reading public, some publishers seem to be progressively pressing into service the tactics of the advertising agencies that whoop up the motion pictures. This particular book, like the film at your neighborhood Bijou, is sold to you (see the Kirkus statement about *Drum and Bugle* above) as "daring, frank, uninhibited, revealing, shocking," when, in sober critical fact, the excess of suggestive detail adds up to nothing more than utter boredom for the reader—unless, of course, he happens to be itching with curiosity.

WHY HAS the failure to recognize the vulgar nature of this kind of shock taken on such an unhealthy emphasis in American publishing? I believe that one clue is provided by the reference above to the influence of motion-picture promotion. It is not a large leap from the film to that other spawner of moving images that engages so much of the U. S. citizen's entertainment hours —the TV screen. Everyone has heard by now that there is much concern in all sorts of circles about the amount of violence in our TV shows. If we need anything to convince us that there is indeed a surfeit of violence, perhaps the remarks in a hardheaded business paper will do the trick. The *Wall Street Journal* (October 20, 1960), reporting a boom in the sale of U. S. television programs abroad, remarked:

> Excessive violence in some U. S. video programs has already raised difficulties abroad. Australia's censorship board last year barred 42 U. S. TV films and demanded cuts in 1,595 additional films. Sweden and Japan and other countries have also voiced objections to brawlings and shootings in American programs.

In the same issue John G. McCarthy, president of the Television Program Export Association, conceded: "There's no doubt about it; we're going to have to tone down our violence if we want to get the most out of the foreign market."

This lofty commercial motive has not yet made itself so markedly felt in the field of American publishing. Though U. S. publishers obviously do hope to pile up sales abroad, the vulgarity pays off here with American readers, where the big money is. But as the U. S. public gets more and more inured to violence through the gentle ministrations of TV (and of the "mature" films), it seems some publishers and authors feel that the only way they can attract attention is to pile violence upon violence, shock upon shock.

A most convincing case can be argued, I believe, that television violence is not merely coloring the moral attitudes of millions of younger and older viewers, but is subtly shaping the thinking of other media of entertainment as well, and not least the publishers of fiction. Alfred A. Knopf, president of the publishing firm that bears his name, has gone so far as to state recently that it's out-of-date to complain about sex in fiction today, for the simple reason that it has been replaced by sadism. Is it a mere coincidence that the replacement in fiction has followed the course of violent incidents on the television screen? Mr. Knopf may have indulged in a bit of oversimplification, but his point underlines the one I am making: we have become so used to violence in one form or another that it no longer shocks, or will shock only if dealt out in consistently larger and stronger doses.

IF THIS TYPE of shock is less and less recognized for the vulgarity it is, there is another type of shock that is—what shall I call it?—proper and human. This is the kind of impact a book makes, not through the external and sensational images of sex and sadism, but through its quality to shake the reader into a realization of what an awful, (in the proper sense of the word) creation of God any human soul is. Jean Mouroux put this succinctly and superbly in his *The Meaning of Man* (Sheed & Ward, 1948, p. 268): "On the day when, by some flash of intellectual enlightenment, or some effort at spiritual progress, we come to realize what we really are, we are seized with a kind of shiver."

And, strange to say (or apparently so), this shiver is precisely what true literature can and does provide. Arnold Bennett pointed this out many years ago in his *Literary Taste:* "The pleasure derived from a classic is never a violent pleasure—it is subtle, it will wax in intensity; it does not at all knock you down; it steals over you." In the same context Mr. Bennett observes, apropos my remarks about the vulgarity of the other kind of shock: "The artistic pleasures of an uncultivated mind are generally violent."

There is a ton more of legitimate shock—legitimate in this sense of a realization of the mystery of man—in the pages of Alan Paton's *Cry, the Beloved Country* than there is in a dozen books like James Jones' *From Here to Eternity.* Paton's great book had this quality while it was being read; the quality is there when one relives it in memory (it "waxes in intensity") and detects in one's inner self the silent canker of a pride that *could* burst out into open racial injustice. There is more genuine shock in the recently published (and churlishly received in some Catholic circles) *No Little Thing,* by Elizabeth Ann Cooper, than there is in a carload of *Peyton Places,* as one sees dramatically the Hopkinsesque "cliffs of fall" that can yawn before a priest—the fall all the more abysmal as the heights were lofty.

But susceptibility to *this* kind of shiver or shock is deadened by too frequent reception of the vulgar shock. The "daring" of a James Jones or a Grace Metalious creates such a din that the mental ear is deafened to the vibrations of the shiver set up in the soul by a Paton or a Greene, who make us face the realization that within oneself couch those awful capacities of glory and degradation—and *that* is the mystery of man.

I am inclined to think that one essential way of keeping alert to the shiver and insensitive to the shock is to preserve in one's self a deep sense of the reality of sin. I also am inclined to suspect that one basic reason why some Catholics think that the themes of a Greene or a Cooper ought not be treated rests exactly on a lack of such realization.

We would be smug indeed if we labored under the impression that we are assured of possessing this sense of sin by the mere fact that we *are* Catholics. We live in a world and are surrounded by an atmosphere in which a sense of sin has been lost. Henri Rondet, S.J., puts this well in his rich little book, *The Theology of Sin* (Fides, 1961, p. 84):

> In Kantian philosophy and its derivatives, God unfortunately remains exterior to moral obligation. Since then, morality has been secularized. In the best situations they only talk about duty, truth, justice, brotherhood and human solidarity. Moral fault is kept, responsibility remains, but whom is man responsible to? Sin as such implies a personal attitude to a personal God. In order to recover the consciousness of sin, we have also to recover the consciousness of God, His grandeur, His rights, the demands of His love.

93

In our efforts to deepen this realization of God and thereby to make more interior our "vivid consciousness of the world's sin by knowing ourselves to be sinners or capable of sin" (this is Fr. Rondet's phrase, which he applies even—perhaps I should say especially—to the saints), the shiver set pulsing by good and great books can be an immense help. But this implies that we are open to, sensitive to the shock of recognition, that we can appreciate the fact that a Greene, a Mauriac, a Cooper are not talking about "such things" for the sake of the vulgar shock that appeals to uncultivated minds, but rather for the sake of making us realize interiorly that not only genius, but our common fallen human nature stands, in Melville's phrase, "hand in hand all over the world." And if one still questions why Catholic authors have so often provided the shiver by concentrating on the theme of adultery, Fr. Rondet's words offer an explanation:

> In the Old Testament sin is disobedience, ingratitude, adultery. This last analogy, though difficult to manage, is the best. A modern man understands what an offense against love, against friendship, is. There are relations between God and man which are comparable to those between a man and his wife, between friends. Sin is an offense against a father, a friend, a spouse (p. 84).

As one grows in such a realization of the inner, personal relationship between God and one's true self, which knows the world's sin by knowing its own (either actual or potential), one grows progressively deafer to the specious attractiveness of the shock that has no basis in a sense of sin, but can do no more than flaunt violations of the conventions. Then one becomes attuned to the recognition, in the work of genuinely creative writers, of the true shock of "what we really are."

Novelists who send this shiver coursing through our fibers always make a demand on our humility and reward it. We must go to them with no thoughts that we are "not as the rest of men"; we come away from them saying to our inner selves: "Lord, be merciful to me a sinner." And if all this brings us to conclude that a spirit of prayer has something to do with really creative reading, that, at least, is a salutary shock of recognition.

Portrait of Hemingway

Charles A. Brady

"By my troth, I care not; a man can die but once; we owe God a death . . . and let it go which way it will, he that dies this year is quit for the next."—Henry IV (Ernest Hemingway's favorite quotation)

IT SEEMS a rule of human nature that one writer— and not necessarily his generation's greatest— should seize upon the imagination of his time. It was so with R. L. Stevenson yesterday, and with Byron before that. It has been the case with Ernest Hemingway in our day. In the end he, who had in life been so obsessed by death, did not have the luck to die their kind of hero death after which he had quested all his days. He died, instead, the less seemly sort of death that overtook those two other European romanticists whom he also so strangely resembled, Pushkin and Lermontov.

Because Hemingway was a man greatly loved, on the whole the newspapers have shown themselves strangely forbearing toward the not so very short and not so very happy life of a writer who, ideally speaking, should have kept his last appointment with death in Uganda or Kenya, not in Ketchum. Death came for him, not on some hot Spanish afternoon, but in a cold American dawn. His art had been dying slowly over the six or ·seven years before. While no other traveler between life and death has any right to speculate on a fellow pilgrim's passing, one cannot help but wonder if flagging creative vigor did not play its part in this final sinister installment of the Hemingway legend.

That same self-fostered legend, which occupied so disproportionate a share of our attention in Hemingway's regard, was, when all is said and done, not really so very important. Much of it was downright silly: the boxing poses for the rotogravure sections; the slangy baby talk—the "Bwana," "Papa," "Maître" business; the Byronic remarks, like the one to Fitzgerald, that his idea of heaven was a bull ring in which he owned two *barrera* seats, with a trout stream outside that no one else was allowed to fish. This was the side Lillian Ross made such savage fun of in the celebrated *New Yorker* profile. It was the side that took over far too many of Hemingway's own pages. Wolcott Gibbs used to parody it magnificently, until the inadvertent self-parody of *Across the River and Into the Trees* rendered all other lampoons forever anticlimactic —at least so far as Hemingway's art was concerned.

In everyday life the self-protective buffoonery went on right up to the time of and after 1954's near-fatal airplane crash in Africa, with the highly publicized diet of bananas and gin which Ogden Nash affectionately mocked in his calypso:

I land in the jungle by the teeth of my skeen.
Big gorilla walk up to me and talk very mean.
He put up his mits and I sock him in the cheen.
Then gorilla and me, we begin the beguine
With a bunch of bananas and a bottle of jeen.

Yet underneath all the embarrassed and embarrassing exhibitionism, beneath all the popping of corks and cocking of rifles, there was a good side to the shield. Hemingway *did* go to war, after all. Until the bitter end, Hemingway *did* escape from death like some latter-day Herakles. The admittedly too-often-

photographed hair on his chest was not ersatz at all.
The old Mohican, who, in his art, was as fastidiously
paleface as Henry James himself, was as tenderhearted
underneath as he was tough-minded on the surface.
It is always dangerous to underrate Hemingway's intelligence. Though, like Eliot in his poetry, Hemingway
deliberately filtered out all evidence of overt cerebra-
tion, he gave the game away when, in *Green Hills of
Africa*, he quoted an Australian hunter saying impa-
tiently to him: "No. Go on. Do not try to be stupid."
Reversing the little girl with the curl who, when she
was bad, she was horrid, it might be said of Ernest
Hemingway that, when he was good, he was incomparable.

For Ernest Hemingway stands as our greatest master
of fiction between the wars—and for a decade after.
I am not forgetting William Faulkner, either, who
objected so strongly to Hemingway's self-imposed classic limits (and rated Hemingway fifth, among his
contemporaries, after Thomas Wolfe, himself, John
Dos Passos and Erskine Caldwell) on the grounds of
Hemingway's esthetic cowardice: ". . . he stayed within
what he knew. He did it fine, but he didn't try for the
impossible." The difference between Hemingway and
Faulkner is the difference between a cat and a dog.
For the sake of objectivity, let me go on record as
stating that I am a cat man myself.

Nor am I overlooking those limitations of Hemingway which by any computation must be accounted
bad: his sentimentality and his adolescent bravura,
especially in the matter of sex—his heroines, for example, are really no more than the dream-boats of
an adolescent's fundamentalist reverie. But these strictures concern his attitudes, not his stylistics. Hemingway's one-finger prose music is writing like the painting of Matisse. He was far from being a naturalist.
His fiction reveals him to be a symbolist-moralist poet
"exhaling" allegory like cigarette smoke. Limitations
are one thing, limits another. He transcends and cun-
ningly makes the best of his limits to the point where,
next to James, one is inclined to rank him as our
strictest artist. This probably means that he is, at

bottom, one of the greater lesser writers, rather than one of the lesser greater writers. But there is real greatness here from many points of view.

What qualities make for this greatness? These: integrity; grace under pressure; a style whose purity produces the paradoxical impacts of obscure clarity and complex simplicity; compassion for the human condition; above all, utter honesty of intention and effect. "He could not be bought," said John Peale Bishop. "A writer's job," said Hemingway, "is to tell the truth." Needless to say, no one writer ever latches on to more than a portion of the truth. But Hemingway got hold of an astounding amount of it. He remains, moreover, the only major writer in our tradition who succeeds in sympathetically mediating the great Hispanic ethos. One must add these qualities also: a strong historical imagination; a keen sense of place and period; a power of precise observation; a subtly cadenced, stylized conversation that brings the movement of his stories curiously closer to ballet and archaic ritual than to drama; a romantic view of love which, for all its inadequacies, has the great merit, for our day, of being sexually normal.

Hemingway was a part-time expatriate with deep-seated indigenous roots. His art grew alongside one of the two main streams of the American literary tradition. He is not an Emersonian idealist, and he has no particular regard for this particular tributary of the American genius, which he once stigmatized as "the small, dried and excellent wisdom of Unitarians." No, he visits the dark side of the moon along with Twain and Crane and Poe, with Sherwood Anderson and two greater writers whom, however, he did not much admire, Hawthorne and Melville. Malcolm Cowley was right when he said of the stories: "Here are nightmares at noonday . . . a haunted and nocturnal writer."

BUT IF THE THEMES are dark, the much-praised much-maligned style is bright with the brightness of fine crystal. Once Hemingway had moved away from the first primitivist excesses of the echolalia Gertrude Stein taught her young Paris disciples, his effect justified Ford Madox Ford's famous encomium that his words like "pebbles fetched fresh from a brook . . . live and shine, each in its place." All this clean, mosaic splendor is at the service of what Hemingway himself called "the good and the bad, the remorse and sorrow, the people and the places and how the weather was," which is just about as good a definition of essential fiction as anyone has ever hit off.

Now stylistics as obtrusively tesselated as these are likely to fit the short story better than the novel. For all the poignancy of *Farewell to Arms* and the sheer brilliance of that glittering, comic *Waste Land, The Sun Also Rises,* it seems to me that the best of Hemingway's roughly sixty stories will last longer. These best must include: "Fifty Grand," "A Clean, Well-Lighted Place," "The Killers," "The Gambler, the Nun, and the Radio," "The Undefeated," "The Short Happy Life of Francis Macomber."

Hemingway has been accurately termed one of the obsessed artists. But it should never be forgotten that he was as much obsessed by God as by death, and that he was a critic rather than an apostle of nihilism, a moralist who did not approve of the rich bums and rootless expatriates who infest certain of his novels and with whom his great friend, Scott Fitzgerald, to some degree at least, was in collusion.

It has become a critical truism to describe Hemingway's muse as Catholic and Faulkner's as Protestant. (The tangled question of Hemingway's exact religious status, after his early conversion to Catholicism, might be resolved by saying that he stayed *croyant* even while his marital coil prevented his being *pratiquant.*) For once, a truism turns out to be true. Hemingway's habit of imagery is ritually Catholic. So is his sense of limits. So is his admiration for the Spanish thing, even for the Jesuit *mystique,* as we note in a reference to "the same town where Loyola got his wound that made him think." Look, moreover, at the sympathy with which nuns and priests are drawn in his fiction. If his controversial defense of the Spanish Republic comes in question here, it might be noted that this is precisely Belloc's stance where the men of 1789 are concerned.

Look also at Hemingway's fundamental admirations: courage, first and foremost, and fidelity; after these, love, generosity, chivalry, humility, a sense of the ridiculous. One remembers his credo after the announce-

ment of his Nobel award: "I do not know what Man (with a capital M) means. . . . I do know what man (with a small m) means and I hope I have learned something about men (small m) and something about women and something about animals."

Hemingway felt about animals much as the cave artists of the Dordogne felt, or a Siberian shaman. He loved them, hunted them, knew how to make them the vehicles of his deepest spiritual intuitions. Consider this, the most famous of his symbols, from the beginning of *The Snows of Kilimanjaro:*

> Kilimanjaro is a snow-covered mountain, 19,710 feet high, and it is said to be the highest mountain in Africa. Its western summit is called by the Masai "Ngaje Ngai," the House of God. Close to the western summit there is the dried and frozen carcass of a leopard. No one has explained what the leopard was seeking at that altitude.

No one has to explain. It is God, of course. As Brett said to Jake Barnes in *The Sun Also Rises:* "It's sort of what we have instead of God." And Jake replied:

"Some people have God. Quite a lot." In his own articulately inarticulate way, Hemingway's writing, too, had quite a lot of God about it.

A less famous but equally good self-symbol in animal terms occurs in the little-known "Fable of the Good Lion." The old man in *The Old Man and the Sea* used to dream of the lions on the beaches. This Good Lion, on the other hand, while he was in Africa, used to dream of the griffons, the winged lions, of St. Mark's. Only, when he got back to civilization, he went into Harry's Bar and asked for a Hindu-trader sandwich and a dry Martini made with Gordon's gin, an action which shocked all the habitués—for habitués, read Hemingway's bourgeois readers—most deliciously.

Like the Snow Leopard, the Good Lion tells us a good deal about the heraldic side of Hemingway, and about something else as well. If journalistic legends are so often false, it is partly because they mythologize too little, not too much. The Hemingway legend is already dead. The Hemingway myth is just beginning. On this higher mythic plane, he is many things: what Wyndham Lewis once called him, a "dumb ox," though "dumb" is surely the wrong adjective here; a sentient woodchuck capable of describing what he tastes and feels; even better, a medieval armorer-chronicler; a gladiator; a helmed legionary; Huckleberry Finn grown up and gone kudu-hunting; a sea-god even, a Triton or a Neptune. Above all he is a saga-hero defending a cave-mouth; a Viking gnawing his oar and well able to tell us the flavor of the bitten wood. Now that Ernest Hemingway has been released from the oar he manned in life, we can begin to view his saga-art in perspective and to discover its ultimate secret, which is precisely this: it was saga-art essentially, and that saga-art was better than any other body of writing in this grim century of ours; it was able to isolate the saga-essence, which remains one of the permanently recurring essences of man's tragic experience.

Death of a Dog
Joel Wells

ITEM: *A dog was struck and killed by a car at 9:30 last night on the highway north of town, an unidentified boy reported to the police this morning.*

The Old Dog and the Cars

after Ernest Hemingway

HE WAS AN OLD DOG and he had not chased a car for four years. But he would chase a car tonight and the boy would see him and know that he was not afraid.

He had not been afraid when he was a pup. There were not enough cars on the highway then, or big enough, or fast enough. He took them as they came, every afternoon with the sun in their eyes so they could not see him, which tripled the risk. Or at night, when there was no moon, he had run with risk so big he could feel it crowding against his flank.

There was a Sunday in his third year when he ran four Buicks within ten minutes, making three of them

103

cross the centerline and putting the last one in the ditch. The boy had given him a tire for that.

But the Ford got him the very next day, driven by a woman. She hooked to the right, catching him on the rusty bumper so that he had lost his footing and fallen before the feet of the boy. And the fear had moved up from the tip of his tail and strangled his courage as he lay there.

He did not run again and the boy got another pup. For four years he sat in the afternoon dust under the porch and listened to the applause of brakes on the highway. Tonight they would squeal for him.

Put Out More Dogs

after Evelyn Waugh

"**M**YLES! I DO WISH you could remember to steer to the right."

Lady Distraught dug bone-white fingers even deeper into the leopard-skin upholstery of the Rolls and glared at her husband.

"Damned nuisance!" His Lordship muttered, veering the wheel sharply to avoid a large moving-van already half off the road and heading for the safety of a concrete embankment.

Trust the Americans to muck up a perfectly simple thing like motoring. During the war, he remembered, some Lend-Lease-happy idiot had come before Parliament with a proposal to make England "keep to the right to make the Yanks feel at home." The fool had gone on to read a forty-minute tract about "not having to change the roads, you know, not even the center-stripes."

Well, he'd dodged more than his share of Americans at home, let them dodge him over here. And, having retrieved the hand with which he had been adjusting his tails to a better lie (they had a terrible propensity to bunch against leopard-skin), he steadied the Rolls for a fresh charge.

104

A promising petrol tanker was just settling into his sights when Lady Distraught gave one of her well-known screams and something struck the car's right fender with a thud.

"Clement Attlee!" swore His Lordship, sure that he had run up against one of those rural American types whose clothing is covered over with copper rivets and buttons, "that's bound to have marred the finish."

The Matter of a Hound

after Graham Greene

THE OVERHEAD FAN moved no faster than the minute hand of his watch. A fly clung to the underside of the blade with contemptuous ease. Through the open door the heat bulged in at him like a penny balloon filled with water. At any moment it would burst and inundate him in his own sweat. Under the porch lizards stirred and rubbed against each other with the sound of dry sticks.

Inspector Hoad opened the bottom drawer of his desk and took out a bottle. Frowning, he held it before his mouth but did not drink.

"Drunk again!" buzzed the fly, "drunk before eight in the morning."

He opened his lips and let the gin slide over his tongue and down the desolate road of his throat. Very quickly, Hoad hoped, it would reach his soul.

The balloon burst. Sweat oozed over his lashes into his eyes and the station-yard began to shimmer through the doorway like the high-noon prospect of an English country pond. Something white was moving across its surface. Slender and white: a swan? The swan stopped in the doorway surrounded by a salty nimbus.

"You killed him," the boy said. "You were drunk and you ran over him but you didn't have to stop because you are the police."

Hoad blotted his eyes on the encrusted khaki sleeve of authority. So it had come to that at last: one night you finally succeed in capturing that elusive nymph, oblivion, and in the morning a boy in a white shirt comes and stands in your door to tell you that you've killed someone.

"You have the only car," the boy said, passing sentence.

Reverently, Hoad recapped the gin and placed it back in the drawer. The moment was too sacred to spoil with sordid curiosity. This boy, Manuel, would tell it well in his careful mission-school English: "He looked at me but did not ask whom it was that he had killed."

To Hoad's lips the barrel of the pistol felt even cooler than had the bottle.

The Weakling and the Dog

after François Mauriac

IT WAS during the summer of his fourteenth year that the dog had been destroyed. His parents, in the manner of those who have only sons who are frail, ran continually to bearded physicians who assured them that all that was required was a month under the sun at X, on the coast. Heeding this divine word as faithfully as had the Israelites in going out of Egypt, they fled the city, driving their firstborn and animals before them, until they came to a halt beside the sea.

But the waters had neither divided nor performed to prescription and the boy had grown thinner while a certain cough, muted in the din of Paris, became more audible in nights as silent as a tabernacle. They baked and watered him by day and rubbed and wrapped him at night until his body became as pliable and porous as a pudding. The days ached on interminably to the drone of his mother's reading and the snuffling sounds which rose from behind his father's newspaper.

The dog had been his great consolation. On the beach it leaped and circled his chair, attacking the waves impetuously with a bravado both comic and inspiring. At night, after the droning and the snuffling in the adjoining room had merged themselves in a rich counterpoint of snoring, the dog would leap over the sill of the boy's room and settle himself in the blankets over his feet.

But doctors, as everyone knows, are able to see far more than ordinary men, and it was not long before one of them discovered dog hairs among the boy's bed-clothes.

"And you ask why he coughs and grows thinner," pronounced Zeus from the hallway outside his room.

"But he is so attached to the beast that he will never consent to being separated," his mother responded in the voice of one who knows that what they are saying is of no consequence.

"Here is what you must do. . . ." The doctor outlined his murderous plot without the least diminution of his superbly professional voice.

In the midst of a death struggle with communism, the United States carries on within itself a debate about our national purpose. In large part this debate is prompted by the moral ambiguity of our foreign policy and by worry over the image of America abroad. But even smaller nations in the West seem to float in a vacuum of national purpose. Why? Is it because the West has lost contact with the tradition of reason, the public philosophy that must underlie any fruitful dialogue on ends and means? Perhaps we cannot define our national goals until we know the real meaning of morality. Perhaps we cannot project an image that wins acceptance until we know more about ourselves. And one asks himself: can the new Peace Corps contribute modestly to America's excruciating search for self-definition in this troubled and exasperated world of ours?

East and West

Morality
and Foreign Policy

John Courtney Murray, S.J.

URING THE DECADE of the Tentative 'Fifties the course of events has thrust a number of basic issues into the forum of public argument. One of them goes under the rubric, "morality and public policy." Chiefly in question is foreign policy.

My introduction to the state of the problem took place at the outset of the decade in a conversation with a distinguished journalist who is now dead. In public affairs he was immensely knowledgeable; he was also greatly puzzled over the new issue that was being raised. His first question revealed the source of his puzzlement. What, he asked, has the Sermon on the Mount got to do with foreign policy? I was not a little taken aback by this statement of the issue. What, I asked, makes you think that morality is identical with the Sermon on the Mount? Innocently and earnestly he replied: "Isn't it?" And that in effect was the end of the conversation. We floundered a while in the shallows and miseries of mutual misunderstanding, and then changed the subject to the tactics of the war going on in Korea.

THE OLD MORALITY PASSES

I have only a fragmentary acquaintance with the growing body of literature on morality and foreign policy; the subject is outside my field. But listening, as it were, on the edges of the public argument, I have come to the conclusion that my journalist friend properly introduced me to the fundamental problem. It does not lie in the concept of policy, or even in the concretenesses of actual policies, though these matters are complex enough. It lies in the concept of morality itself. Rarely does the argument get to concrete issues of policy. And even when it does, the talk quickly turns back to the root of confusion—the question: what is morality?

The reasons for this fact lie in the history of moral theory in America. But that story is long, not to be told here (I don't think it ever has been fully told). An important event, of relatively recent occurrence, has been the recognition of the shortcomings and falsities of an older American morality that dominated the 19th century and still held sway into the 20th.

Its style was voluntarist; it sought the constitution of the moral order in the will of God. The good is good because God commands it; the evil is evil because God forbids it. The notion that certain acts are intrinsically evil or good, and therefore forbidden or commanded by God, was rejected. Rejected too was the older intellectualist tradition of ethics and its equation of morality with right reason. Reason is the dupe of interest and passion. And how is one to know, or dare to say, whose reason is right? In the search for moral principles and solutions reason can have no place.

In its sources the older morality was scriptural in a fundamentalist sense. In order to find the will of God for man it went directly to the Bible. There alone the divine precepts and prohibitions are stated. They are stated in so many words, and the words are to be taken at their immediate face value without further exegetical

110

ado. When, for instance, the Gospel tells the Christian not to resist evil but to turn the other cheek, the precept is clear and absolute. The true Christian abdicates the use of force even in the face of injury.

In its mood the old morality was subjectivist. Technically it would be called a "morality of intention." It set primary and controlling value on a sincerity of interior motive; what matters is not what you do but why you do it. And it was strong on the point that an act is moral only when its motive is altruistic—concretely, when the motive is love. If any element of self-interest creeps in, the act is corrupt and sinful.

Finally, in its whole spirit the old morality was individualistic. Not only did it reject the idea of a moral authority external to the individual conscience. It also set its single focus on the individual existence and on the moral problems that arise in interpersonal relationships. As for society, it believed in a direct transference of personal values into social life; in principle it would tolerate nothing less than Christian perfection as a social standard. Its highest assertion was there would be no moral problems in society, if only all men loved the neighbor.

Within the last generation this older morality has come under severe criticism, in itself and its later historical alliance with certain trends in secular liberal thought. The attack has centered on its simplism. The discovery was made that this morality of facile absolutes was ill-suited to cope with the growing complexity of an industrial society, domestically and in its foreign relations.

It did not go beyond the false notion that society is simply the sum of the individuals living in it, and that public morality is no more than the sum of private moralities. It did not understand the special moral problems raised by the institutionalization of human action. It did not grasp the nature of politics, the due autonomy of the political, the limiting factors of political action, or the standing of success as a political value. It had no sense of the differential character of morality and legality, no theory of jurisprudence, no idea of the distinction between private sin and public crime (witness the laws

111

that it has left on the statute books—notably the Connecticut birth control statute).

In consequence of all these shortcomings, the older morality possessed no resources for discriminating moral judgment. It tended to thrust its simple yeas and nays upon political, social and economic reality without any careful prior analysis of the realities in question. It disregarded the duly autonomous character of their lines and life. It distorted the meaning of Plato's famous dictum and understood it to say that society is the individual (not "man") writ large. In a word, what the older morality failed to understand was the nature of man himself.

ETHIC FOR A NEW WORLD

The critique of the older American morality seems to have been not only just but also successful. The older morality, though still around, is no longer dominant. This is good. It is perhaps particularly good that the older morality is still around. Doubtless it is useless against the demons that inhabit the organized structures of society and exert their sway over history from these seats of institutionalized power. On the other hand, it had a certain virtue of exorcism against the demons that dwell in the life of the individual. And it is always good that at least some demons are cast out from among us even though their departure still leaves us in combat with the "rulers of this world of darkness," whose dominion will endure until the Day of the Lord.

The avowed purpose of the newer American morality is to reckon with the full complexity of man's nature and of human affairs. Hence against the absolutism of the old morality, in which the contingent facts got lost under insistence on the absolute precept, the new morality moves towards a situationalism, in which the absoluteness of principle tends to get lost amid the contingencies of fact. Against the abstract fundamentalist literalism of the old morality the new system is consciously pragmatist; not the wording of the precept but a calculus of the consequences of the act is the decisive moral norm. Whereas the old morality saw things as so simple that moral judgment was always easy, the new morality sees things as so complicated that moral judgment becomes practically impossible. The final category of moral judgment is not "right" or "wrong"—but "ambiguous."

Finally, against the self-righteous tendency of the old morality, the new theory teaches that to act is to sin, to accept responsibilty is to incur guilt, to live at all is to stand under the judgment of God, which is uniformly adverse, since every act of moral judgment is vitiated by some hidden fallacy, and every use of human freedom is inevitably an exercise in pride.

The current argument about morality and foreign policy goes on within the climate of moral opinion created at once by the older American morality, and by the newer morality, and by the conflict between them—a conflict which does not rule out certain similarities, notably their common rejection of the whole style and structure of natural-law morality. Three basic problems, each related to the others, furnish the focus of concern.

THREE RELATED PROBLEMS

The first is the gulf between individual and collective morality. Since the day of Roger Williams and his separation of the "garden" (the Christian community) and the "wilderness" (society or "the world"), prevalent American moral theory has never found a way to bridge the chasm between the order of private life and the order of law, public policy and institutional action, especially when the question concerns the nation-state. The private life is governed by the will of God as stated in the Scriptures; it is to bear the stamp of the Christian values canonized by the Scriptures—patience, gentleness, sacrifice, forbearance, trust, compassion, humility, forgiveness of injuries and, supremely and inclusively, love.

On the other hand, it is the plainest of historical facts that the public life of the nation-state is not governed by these values. Hardly less plain is the fact that it cannot be. What, asked my journalist friend quite sensibly, has the Sermon on the Mount got to do with foreign policy? Pacifism, for instance, may be a dictate of the individual conscience, but it cannot be a public policy. What then is the will of God for the nation-state? How and where is it to be discovered? There is no charter of political morality in the Scriptures. Must one, therefore, admit that all politics is simply *Realpolitik*—the selfish pursuit of national interest in a nicely calculated play of power to which ethical norms are irrelevant?

The other two questions are consequent. First, is it not the historical fact that the nation-state acknowledges only one imperative, the dictate of national interest? And is not the fact itself also normative? Is it not right that the nation should so act? Would it not be *wrong* for the nation to act apart from the national interest, short of it, or beyond it? But if you hold this, do you not come into open conflict with the basic tenet of both the older and the new American morality, which is that self-concern is the primal sin? that the pursuit of self-interest is the pursuit of evil? that the whole function of Christian morality is to call self-interest into question, deny it all theoretical justification, and condemn it in practice? This is the moral theory. Strictly applied, it must assert that the nation is sinful and guilty in all its actions, since they are never free of the taint of interest.

Faithfully held, this theory requires that the nation should be called upon to transcend self-interest, resist its dictates, and act beyond them in a spirit of disinterested altruism. Or, since this moral call would have the ring of nonsense in the field of politics, and since morality is not supposed to sound like nonsense, one could choose not to hold and apply the theory strictly. One could fall back on the position that self-interest is a legitimate motive for the nation, even though it is an illegitimate motive for the individual. Then the question is, in the name of what theory do you make this distinction? Is this to bridge the gulf between private and collective morality, or simply to fall headlong into it?

The final issue is perhaps the most basic. It certainly is the most inclusive, since it spans all the prominent issues of the day—armaments, the politics of the Cold

War, the economics and politics of the revolution of rising expectations. It is the issue of power. As far as the sheer fact goes, most Americans seem to have finally awakened to the central relationship between foreign policy and force. But the awakening was to a state of moral bafflement and anxiety, insofar as it took place in the climate of moral opinion described above.

In the climate of this moral opinion a cold breath of evil more than faintly emanates from the very words "power" and "force." It seems to have been part of the American dream that this nation could go through history with clean hands by the simple Kantian expedient described in Péguy's genial phrase: "Kantianism has clean hands, because it has no hands." Concretely, a nation's "hands", wherewith it shapes the stuff of

history, are its instruments of power—military, economic and diplomatic power, together with the power of sheer presence and prestige. We have never wanted to have such hands, much less to get them dirty by handling any history save our own. Our historic declaration was that power-struggles were for the "barbarous" nations of Europe, not for us. Now we have become suddenly conscious of our hands—that they are sinewy beyond comparison; that they are sunk in the affairs of the world; that they are getting dirty beyond the wrists.

At least we feel them to be dirty, and the feeling is one of guilt. The United States today is an imperialism, like it or not. And we like it so little that we are even unwilling to admit the fact. The cause of our anxiety is not that there has been little in our past political experience to teach us wisely to wield the instrument of empire, which is power. It is rather that there is nothing in current American moral theories to teach us the moral quality of power itself. The prevalent teaching is simply that power is evil. The teaching, in fact, is that the evil in human nature is precisely a will to power. The will is activated as the hand closes on the thing; at that moment innocence is lost, never to be regained. To be human is bad enough; but to be powerful is to be corrupt, with a corruption that increases with each increment of power.

In what moral terms, therefore, is America to justify itself in its possession of power? And in what terms is America to justify itself to the world for its uses of its power? Can these hands be cleansed? Or must the scriptural phrase be inverted to read: Let him who is unjustified become still more unjust? The national straits are even more narrowed when·one considers that the teaching says one further devastating thing. It says that to refuse to use power is to be "irresponsible," and therefore to be more guilty yet.

ONE THING COMES CLEAR

These seem to be the basic issues involved in the current controversy about morality and foreign policy. I have found myself in a fog as I have listened intermittently, while cynics dispute with moralists, and political realists dispute with ethical idealists, and fundamentalists dispute with "ambiguists" (I apologize for the barbarism, but I must have a descriptive term for this school of thought, whose favorite word is "ambiguity").

Only one thing is clear. The real issue does not concern the moral quality of this or that element of American foreign policy. The real issue concerns the nature of morality itself, the determinants of moral action

116

(whether individual or collective), the structure of the
moral act, and the general style of moral argument.
One cannot argue moral issues until they are stated;
but what are the terms of statement of the moral issues
involved in foreign policy? One cannot come to practical solutions until one has first formulated the relevant
principles and also analyzed the factual situation in
which the principled solution is to be practiced; but
by what methods do you arrive at your principles and
establish their relevance, and what is your analysis of
the factual situation? As these issues are touched, or
as they are avoided, the whole argument flies off in
all directions.

The proper bafflers are the ambiguists. Their flashes
of insight are frequent enough; but in the end the fog
closes down. They are great ones for the facts, against
the fundamentalists, and great ones for "conscience,"
against the cynics. They insist on the values of pragmatism against the absolutists; but they resent the suggestion that they push pragmatism to the point of a relativism of moral values. My main difficulty, however, is
that I never know what, in their argument, is fact and
what is moral category (surely there is a difference),
or where the process of history ends and the moral
order begins (surely there must be such a point).

When they undertake to describe the historical-
political situation for which policy is to be framed, one
has the same feeling that comes on seeing a play by
Sartre. No human characters are on the stage, only
Sartre's philosophical categories. So, in the ambiguist
descriptions, the factual situation always appears as a
"predicament," full of "ironies," sown with "dilemmas,"
to be stated only in "paradox," and to be dealt with
only at one's "hazard," because in the situation "creative
and destructive possibilities" are inextricably mixed,
and therefore policy and action of whatever kind can
only be "morally ambiguous."

MORALITY BASED ON A PARTI PRIS

But this is to filter the facts through categories. So
far as one can see by an independent look "out there,"

the dilemmas and ironies and paradoxes are, like the beauty of the beloved, in the eye of the ambiguist beholder. They represent a doctrinaire construction of the facts in terms of an antecedent moral theory. And every set of facts is constructed in such a way as to make the moral verdict "ambiguous" a foregone conclusion.

The ambiguist rightly puts emphasis on the complexity of the situations with which foreign policy has to deal; no one could exaggerate the complexity hidden under the phrase, "the Cold War." But does the fact of complexity justify the vocabulary of description or the monotonous moral verdict? It is as if a surgeon in the midst of a gastroenterostomy were to say that the highly complex situation in front of him is so full of paradox ("The patient is at once receiving blood and

losing it"), and irony ("Half a stomach will be better than a whole one") and dilemmas ("Not too much, nor too little, anesthesia") that all surgical solutions are necessarily ambiguous. Complicated situations, surgical or moral, are merely complicated. It is for the statesman, as for the surgeon, to master the complications and minister as best he can to the health of the body, politic or physical. The work may be done deftly

118

or clumsily, intelligently or stupidly, with variant degrees of success or failure; but why call it in either case "ambiguous"? The philosophers of moral ambiguity will, of course, say that the ambiguity, properly speaking, is not in the political situation but in political man, who carries into politics the paradox, irony and ambiguous amalgam of virtue and corruption that reside in his own nature (or in the human "self," as the ambiguists prefer to say, since they have a peculiar meaning all their own for the word "nature"). There you have it.

In point of sheer method there is no reason why the ambiguist should not make use of a conceptual scheme to guide his analysis of political fact, and to furnish the terms for his statement of moral issues, and to determine the style of his argument in favor of his solutions. Every moralist does this. Every moralist has his concept of the moral order. All practical moral inquiry has theoretical presuppositions. Each moral theory has its own categories of statement and its own style of argument. And in the end every structure of moral doctrine and decision rests on a concept of the nature of man.

NO PROGRESS HERE

To this concept of man's nature the critical argument comes back. The ambiguist indicts the fundamentalist and the secular liberal for their one-dimensional views of man. But he does not recognize that the same indictment recoils on his own head. He easily disposes of all the utopianisms, both "hard" and "soft," that result from the one-dimensional fundamentalist and secular liberal views. He then spins an enormously complex analysis of the "real" nature of man in personal and political life. And at the end of it (this is the real paradox) he has again compressed the moral life of man into one dimension. Inescapably, beyond all help of divine grace—and even further beyond all help from human reason and freedom—the life of man, personally and politically, is lived in the single moral dimension of ambiguity. He who relishes irony should relish this— that the whole complicated argument against simplistic

119

theories should result in the creation of a theory that is itself simplistic; that the smashing attack on the bright and brittle illusion of utopianism should win its victory under the banner of an opposite illusion that is marshy and murky but no less an illusion.

I have outlined the argument about morality and foreign policy in a way to suggest that it is an intramural argument within the Protestant community. So it is.

The question I have asked myself is, whether and on what terms it might be possible to me to enter the argument, as a representative of the tradition of reason in moral affairs. (It is also called the tradition of natural law; but the term "natural law" is so widely misunderstood today that it gives rise not only to semantic difficulties but also to emotional antagonism.) It might be worth while to essay an answer to this question in a later issue of AMERICA.

REMARKS ON AN ARGUMENT: PART II

RECENTLY, within a segment of the American people, an argument has been set afoot concerning morality and foreign policy. The parties to it have been Protestants; hence they have defined the issues in terms of Protestant moral theory. Protestant moral theory, however, is in a sort of fluid transitional state, considerably in conflict with itself; hence the problem of defining moral issues in public policy has been further complicated by a more fundamental argument about the norms and methods of moral judgment itself.

Against an older fundamentalism that found in simple scriptural maxims the solutions to even the most complicated social and political problems—with disastrous results to the structure of the problems—a newer school has arisen. I have called it "ambiguism," because of its fondness for the word "ambiguous" as a term of moral judgment. It emphasizes the complexity of moral issues as they arise in society and state; it asserts that in all the concrete situations with which public policy has to deal the elements of right and wrong are so inextricably mixed as to be indistinguishable; hence it claims that every public action or policy is "ambiguous," that is, both right and wrong. Nonetheless, the ambiguists, like the simplists, protest their opposition to the political realists or cynics, to whom all public issues are simply issues of power in which moral judgments have no place at all.

PSEUDO-PROBLEMS

My own terms of moral definition, argument and judgment are, of course, those of the tradition of reason in moral affairs—the ancient tradition that has been sustained and developed in the Catholic Church. Consequently, listening to the public argument on morality and foreign policy, I have found it difficult to discover just what all the shooting is about. Three major issues have come to the fore. The trouble is that all three seem to me factitious. From where I sit, so to speak, in the moral universe, they are all pseudo-problems. Were I to enter the argument, this is the first point I should have to make.

The Protestant moralist is disturbed by the gulf between the morality of individual and collective man. He is forever trying somehow to close the gap. Forever he fails, not only in doing this but even in seeing how it could possibly be done. Thus he is driven back upon the simplist category of "ambiguity." Or he sadly admits an unresolvable dichotomy between moral man and immoral society.

I am obliged to say that the whole practical problem is falsely conceived in consequence of a defective theory. No such pseudo-problem arises within the tradition of reason—or, if you will, in the ethic of natural law. Society and the state are understood to be natural institutions with their own relatively autonomous ends or purposes, which are predesigned in broad outline in the social and political nature of man, as understood in its concrete completeness through reflection and historical experience. These purposes are public, not

121

private. They are therefore strictly limited. They do not transcend the temporal and terrestrial order, within which the political and social life of man is confined; and even within this order they are not coextensive with the ends of the human person as such. The obligatory public purposes of society and the state impose on these institutions a special set of obligations which, again by nature, are not coextensive with the wider and higher range of obligations that rest upon the human person (not to speak of the Christian). In a word, the imperatives of political and social morality derive from the inherent order of political and social reality itself, as the architectonic moral reason conceives this necessary order in the light of the fivefold structure of obligatory political ends—justice, freedom, security, the general welfare, and civil unity or peace (so the Preamble to the American Constitution states these ends).

It follows, then, that the morality proper to the life and action of society and the state is not univocally the morality of personal life, or even of familial life. Therefore the effort to bring the organized action of politics and the practical art of statecraft directly under the control of the Christian values that govern personal and familial life is inherently fallacious. It makes wreckage not only of public policy but also of morality itself.

Again, the Protestant moralist is deeply troubled by the fact that nations and states have the incorrigible habit of acting in their own self-interest, and thus violating the fundamental canon of morality which sees in self-concern the basic sin. Here again is a pseudo-problem. I am, of course, much troubled by the question of the national interest, but chiefly lest it be falsely identified in the concrete, thus giving rise to politically stupid policies. But since I do not subscribe to a Kantian "morality of intention," I am not at all troubled by the centrality of self-interest as the motive of national action. From the point of view of political morality, as determined by the purposes inherent in the state, this motive is both legitimate and necessary.

There is, however, one reservation. I do not want self-interest interpreted in the sense of the classic theory of *raison d'état*, which was linked to the modern concept of the absolute sovereignty of the nation-state. This latter concept imparted to the notion of national self-interest an absoluteness that was always as illegitimate as it is presently outworn. The tradition of reason requires, with particular stringency today, that national interest, remaining always valid and omnipresent as a *motive*, be given only a relative and proximate status as an *end* of national action. Political action stands always under the imperative to realize, at least in some minimal human measure, the fivefold structure of obligatory political ends. Political action by the nation-state projected in the form of foreign policy today stands with historical clarity (as it always stood with theoretical clarity in the tradition of reason) under the imperative to realize this structure of political ends in the international community, within the limits—narrow but real—of the possible. Today, in fact as in theory, the national interest must be related to this international realization, which stands higher and more ultimate in political value than itself.

No false theoretical dichotomy may be thrust in here. The national interest, rightly understood, is successfully achieved only at the interior, as it were, of the growing international order to which the pursuit of national interest can and must contribute. There is, of course, the practical problem of defining the concrete policies that will be successful at once in the national interest and in the higher interest of international order. The casuistry is endlessly difficult. In any case, one ought to spare oneself unnecessary theoretical agonies, whose roots are often in sentimentalism; as, for instance, the effort to justify foreign aid in terms of pure disinterested Christian charity. To erect some sort of inevitable opposition between the pursuit of national interest and the true imperatives of political morality is to create a pseudo-problem.

POWER AND POLITICS

The third source of Protestant moral anxiety is the problem of power. The practical problem, as put to policy, is enormously complicated in the nuclear age, in the midst of a profound historical crisis of civilization,

123

and over against an ideology of force that is also a spreading political imperialism. This, however, is surely no reason for distorting the problem by thrusting into it a set of theoretically false dilemmas—by saying, for instance, that to use power is prideful and therefore bad, and not to use it is irresponsible and therefore worse. The tradition of reason declines all such reckless simplism. It rejects the cynical dictum of Lenin that "the state is a club." On the other hand, it does not attempt to fashion the state in the image of an Eastern-seaboard "liberal" who at once abhors power and adores it (since by him, emergent from the matrix of American Protestant culture, power is unconsciously regarded as satanic). The traditional ethic starts with the assumption that, as there is no law without force to vindicate it, so there is no politics without power to promote it. All politics is power politics—up to a point.

The point is set by multiple criteria. To be drastically brief, the essential criterion is the distinction between force and violence. Force is the measure of power necessary and sufficient to uphold the valid purposes both of law and of politics. What exceeds this measure is violence, which destroys the order both of law and of politics. The distinction is teleological, in the customary style of the tradition of reason. As an instrument, force is morally neutral in itself. The standard of its use is aptitude or ineptitude for the achievement of the obligatory public purposes. Here again the casuistry is endlessly difficult, especially when the moralist's refusal to sanction too much force clashes with the soldier's

classic reluctance to use too little force. In any case, the theory is clear enough. The same criterion which governs the state in its use of coercive law for the public purposes also governs the state in its use of force, again for the public purposes. The function of law, said the Jurist, is to be useful to the community; this too is the function of force.

The community, as the Jurist knew, is neither a choir of angels nor a pack of wolves. It is simply the human community which, in proportion as it is civilized, strives to maintain itself in some small margin of safe distance from the chaos of barbarism. For this effort the only resources directly available to the community are those which first rescued it from barbarism, namely, the resources of reason, made operative chiefly through the processes of reasonable law, prudent public policies, and a discriminatingly apt use of force.

(Note here that Christianity profoundly altered the structure of politics by introducing the revolutionary idea of the two communities, two orders of law and two authorities; but it did not change the nature of politics, law and government, which still remain rational processes; to the quality of these processes Christian faith and grace contribute only indirectly, by their inner effect upon man himself, which is in part the correction and clarification of the processes of reason.)

The necessary defense against barbarism is, therefore, an apparatus of state that embodies both reason and force in a measure that is at least decently conformable with what man has learned, by rational reflection and historical experience, to be necessary and useful to sustain his striving towards the life of civility. The historical success of the civilized community in this continuing effort of the forces of reason to hold at bay the counterforces of barbarism is no more than marginal. The traditional ethic, which asserts the doctrine of the rule of reason in public affairs, does not expect that man's historical success in installing reason in its rightful rule will be much more than marginal. But the margin makes the difference.

All this is the sort of thing that the theorist of natural law would have first to say were he to enter on the ground floor, so to speak, of the controversy about morality and public policy. He could not possibly argue concrete problems of policy in the moral terms of the ambiguist. Insofar as these terms are intelligible to him at all, they seem to him questionable in themselves and creative of pseudo-problems in the field of policy. In turn, the Protestant moralist, whatever his school, cannot possibly argue questions of policy in the moral terms of the tradition of reason. The tradition is alien to him at every point—in its intellectualism, its theological emphasis, its insistence on the analogical character of the structures of life (personal, familial, political, social), its assignment of primacy to the objective end of the act over the subjective intention of the agent, and its casuistical niceties. At best, the whole theory is unintelligible; at worst it is an idolatry of reason and an evacuation of the Gospel.

It has also become customary to point out that, whatever the merits of the tradition, it is dead, in the sense of Nietzsche's dictum, "God is dead." So I was told recently. It happened that I wrote a little piece on the traditional moral doctrine on the limitations of warfare, as fashioned by the tradition of reason. A friendly critic, Prof. Julian Hartt of the Yale Divinity School, had this to say: "Father Murray has not, I believe, clearly enough come to terms with the question behind every serious consideration of limited war as a moral option, i.e., where are the ethical principles to fix the appropriate limits? *Where,* not *what*: can we make out the lineaments of the community which is the living repository (as it were) of the ethical principles relevant and efficacious to the moral determinations of the limits of warfare?" This is a fair question.

After a look around the national lot, Professor Hartt comes to the conclusion that the American community does not qualify; it is not the living repository of what the tradition of reason has said on warfare. I am compelled regretfully to agree that he is right. Such is the fact. I would further say that the American community, especially in its "clerks," who are the custodians of the public philosophy, is not the repository of the tradition of reason on any moral issue you would like to name. This ancient tradition—like the Eternal Reason of God, to which it makes its initial and final appeal—

is dead. (It lives, if you will, within the Catholic community; but this community fails to bring it into vital relation with the problems of foreign policy; there seems, in fact, to be some reason for saying that the Catholic community is not much interested in foreign affairs, beyond its contribution in sustaining the domestic mood of anti-communism.)

But if it be the fact that the tradition of natural law, once vigorous in America, is now dead, a serious question arises. What then is the moral doctrine on which America bases its national action, especially its foreign policy?

"DON'T SHOOT FIRST"

One could put the question in the first instance to the Government. It is clear that the Department of Defense and its allied agencies find sufficient moral warrant for their policies in their loyalty to the good old Western-story maxim: "Don't shoot first." With the moral issue thus summarily disposed of, they set policy under the primatial control of that powerful dyarchy, technology and the budget, which conspire to accumulate weapons that, from the moral point of view, are unshootable, no matter who shoots first. Those who are disquieted by this situation—which is not ambiguous but simply wrong—are invited to find comfort in the emanations of crypto-pacifism from the White House, which seems to hold that we shall never shoot at all. The moral argument for this unambiguous position, whose simplism rivals that of the ambiguists, is never made clear. The inquiry into the moral bases of policy would likely produce other weird and wonderful answers, if elsewhere pursued—within the Department of State, for instance, with regard to disarmament, foreign aid and diplomatic démarches among the uncommitted or emergent nations.

In any case, the question is perhaps more appropriately put to the American community at large. The theory of American government seems to be that public policies borrow, as it were, their morality from the conscience of the people. Right policies, as well as due

127

powers, derive from the consent of the governed. There-fore, on what structured concept of the moral order does the American people undertake to fulfill its traditional public moral right and duty, which is to judge, direct, correct, and then consent to, the courses of foreign policy?

There is a sentimental subjectivist scriptural fundamentalism. But this theory by definition has nothing to say about foreign policy; it is at best a theory of interpersonal relationships and therefore irrelevant to international relations, which are not interpersonal. There is also moral ambiguism. But this, in the final analysis, is not properly a moral theory. It is perhaps a technique of historical analysis, highly doctrinaire in style; but it is not an ethical philosophy. It is an interesting paradoxical structure of rhetorical categories; but it is not a normative doctrine that could base discriminating moral judgments. All norms vanish amid the multiplying paradoxes; and all discrimination is swallowed up in the cavernous interior of the constantly recurrent verdict: "This action is morally ambiguous."

The school of ambiguist thought has done some useful negative service by its corrosive critique of older types of moral simplism and political utopianism. But it has no positive constructive power to fashion purposeful public policies in an age of crisis. It can throw rocks after the event, but it can lay no cornerstones. It points out all the moral hazards, and takes none. The self-contradiction inherent in sin is indeed a massive fact of the human condition; but not for this reason, or any other, does ambiguity become a virtue in moral judgment. Ambiguism can judge no policies save those that history has already judged. It can direct no policies because it can specify no ends toward which policy should be directed. And it can correct no policies since all policies deserve by definition the same qualification, "ambiguous," and what use is it to correct one ambiguous course by substituting another? We can discard ambiguism as the moral premise of public policy.

WHAT IS LEFT?

What is there left? There is, of course, the pseudo-morality of secular liberalism, especially of the academic variety. Its basic premise is a curious version of the Socratic paradox, that knowledge is virtue; it asserts that, if only we really could get to understand everybody, our foreign policy would inevitably be good. The trouble is that the past failures of the political intelligence of secular liberalism, and its demonstrated capacities for misunderstanding, have already pretty much discredited it.

Finally, there is the ubiquitous pragmatist, whose concern is only with what will work. But he too wins no confidence, since most of us have already learned from the pragmatist source of truth, which is history, that whatever is not true will fail to work. We want to know the political truth that will base workable policies.

It would seem, therefore, that the moral footing has been eroded from beneath the political principle of consent, which has now come to designate nothing more than the technique of majority opinion as the guide of public action—a technique as apt to produce fatuity in policy and tyranny of rule as to produce wisdom and justice. It was not always so. In the constitutional theory of the West the principle of consent found its moral basis in the belief, which was presumed sufficiently to be the fact, that the people are the living repository of a moral tradition, possessed at least as a heritage of wisdom, that enables them to know what is reasonable in the action of the state—its laws, its public policies, its uses of force. The people consent because it is reasonable to consent to what, with some evidence, appears as reasonable. Today no such moral tradition lives among the American people—certainly not, as Professor Hartt suggests, the tradition of reason, which is known as the ethic of natural law. Those who seek the ironies of history should find one here, in the fact that the ethic which launched Western constitutionalism and endured long enough as a popular heritage to give essential form to the American system of government has now ceased to sustain the structure and direct the action of this constitutional commonwealth.

The situation is not such as to gladden the heart. But

at least one knows the right question in the present
matter. It is not how foreign policy is to be guided by
the norms of morality. It is, rather, what is the morality
by whose norms foreign policy is to be guided?

◆

Cassation for Good Friday

Space drains from the air between the woman and son;
They seem closer together; dark overleaps the land
As hope and memory pour into the shape of time
And time, at its full, at crux, crashes.
Silent, the woman of fifty and her sudden son
Keep their eyes empty of refusal, continue to stand,
Continue to watch the high man leap to his iron limit
In shortening thrusts.
                             Feathered by the lash
Like a love-wild bird he leaps for breath.
At the end of the dance he invents death.

At the end of the love-dance, the used-up man
Bows mystery-high in the quiet, exploded air.
The dancer hangs. The lover goes. The sky bends
Up, back and away from woman and boy, who stare
As if at each other; home, she goes to the window;
A lion lounges down the street.
The boy is waiting for friends.
Every unsuitable thing takes place.
The day never ends.   MARIE PONSOT

How We

Look to Others

George H. Dunne and C. J. McNaspy

I

THE AVERAGE AMERICAN is puzzled. He sees himself as reasonably decent, good-natured, peace-loving, generous. He has no colonies and no designs upon alien territories. He sees Fidel Castro as an unbalanced revolutionist who has betrayed the democratic hopes of many of his own followers and is busily mortgaging his revolution to the Communist powers. Why, then, should there be so much anti-Americanism about? And why, on the other hand, is Fidel Castro so popular a hero that almost every Latin American government is fearful of angering its own people should it anger him?

It would help the average American to understand if he had the gift to see both himself and Castro as others do. Two quotations from the same issue of *Time* (8/8/60) may give him just the perspective he needs. One concerns "Che" Guevara, the *éminence grise* of Castro's movement. A friend who knew him in 1953, when he held a minor job in Guatemala's agrarian program, says:

> Once we came across a group of undernourished, belly-bloated kids. We were in United Fruit land. Che went into one of his rages. He cursed everybody from God to North American "exploiters" and wound up with a frightening asthmatic attack that lasted two hours.

The other concerns an American bean grower who is fighting efforts to provide a minimum of schooling for the children of migrant farm workers who, full-fledged field hands at nine years of age, are doomed by illiteracy to dreary servitude for life. Says the bean grower: "When a migrant goes to school beyond the seventh grade, you've ruined a good bean picker."

Examine these two images side by side! The dedicated revolutionary who admits to being a Marxist, and the typical representative of American democratic society—Guevara cursing himself into asthmatic rage at the sight of underprivileged children, the American coldly assaying the value of the stunted human as bean picker.

It may be objected that the American is not typical. The objection does not stand up. He is typical of Americans who make their money growing beans or tomatoes or lettuce or fruit. The proof lies in the fact that the farm lobby bitterly opposed this same effort, as it and the ranchers' associations generally have coldly and successfully resisted for years every effort to improve the wretched lot of the migrant field hand. He is also typical of other average Americans who are unmoved by, and indifferent to, the miserable exploitation of the migrant worker.

There is more than a sermon in this confrontation of images. There is a key to an understanding of a large part of contemporary history.

There are perhaps no people more generous with their wealth than Americans, but the image which impinges itself upon world consciousness is not one of American largesse. It is rather the image of the calculating and unfeeling bean grower, and in sharp contrast to it the dramatic image of the deeply feeling, raging Guevara.

It is not the exceptional Tom Dooley, but the hard-nosed bean grower who appears to be the normal product and expression of the social and economic system which has reached its highest development in America. This, as much as anything else, is what "colonialism" and "imperialism" mean to downtrodden people everywhere. This is why there is so much anti-Americanism about.

In a recent article in the *Commonweal* (12/30/60), Fr. Albert Nevins, M.M., quotes Karl Marx:

It is quite easy for Mister Hegel in his well-heated office in Berlin to resolve the problem of man's captivity solely in speculative fashion and to explain all human contradictions on the plan of a dialectic philosophy. But while there are women and children working 18 hours a day in the mines of Manchester I cannot feel myself free.

Here is the idealism which explains the tremendous appeal Marx has had and—despite the demonstrable errors of doctrine, despite the blood which has flowed, the tyrannies which have been spawned—continues to have for millions of men everywhere.

It is basically the same idealism, the same concern, which triggered Guevara's rage. It is the same idealism, the inability himself to feel free so long as the peasant masses were poor, hungry and illiterate, that sparked Fidel Castro. Guevara sacrificed a medical career, Castro a legal career, to set the peasants free. (Recently in Rome a Benedictine priest, who only a few years ago was in British government service in Malaya, where his job was to interrogate Communist prisoners, told me what a sobering experience it was to discover how many of them were professional men who, driven by the same idealism, had sacrificed their own careers to advance the revolution. I had the feeling the experience had something to do with his decision to sacrifice his own career in government and to join the Benedictines.)

What matters is not whether Castro is destroying more than he is building, or whether the sum total of human suffering in Cuba has been halved or doubled, or whether he falls from power tomorrow. What matters is that three years ago he was a man with a handful of poorly armed followers holed up in the mountain fastnesses. Today he is master of Cuba and still a hero to the mobs which wildly cheer his interminable harangues. What matters is to understand that his strength lies in the kind of image he projects to those mobs. It matters to understand what that image is: the image of an angry man bent upon destroying a social structure which says that the poor must be kept poor and ignorant so that we will have good bean pickers.

Take the case of China. The Republican thesis which would explain Communist success in terms of an alleged American betrayal engineered by a Democratic Administration is sheer nonsense. My right to this opinion rests upon the fact that 28 years ago I was urging—there are documents to prove it—the necessity of what today would be called a "crash program" to develop an imaginative Chinese leadership to destroy the frightfully disordered social structure which produced luxury for a few and illiteracy, beggary, misery or poverty for the teeming multitudes. If this were not done, and soon, the Communists, so I argued, would inevitably come to power.

At that time the Communists were an isolated body, holed up and surrounded in the mountainous regions of south-central China. But they were the only ones who projected a sharp image of protest against an intolerable social structure. Crushed by taxes, bled white by rapacious war lords, gouged by landlords, exploited in factory and field, millions of Chinese lived in a gray misery.

One day I climbed a path to a Buddhist temple outside the city of Suchow. For half a mile I walked a solid gantlet of beggars squatting elbow to elbow, empty rice bowls held out in the age-old suppliant gesture. In Shanghai 12-year-old children worked 16 hours a day in foreign and Chinese-owned cotton mills and slept at night on their hard workbenches. What voices were raised in angry protest? Who cursed themselves into fits of asthmatic rage? Who swore not to rest until these slaves had been set free?

Missionaries, with compassion and often heroically, went about their work of corporal and spiritual mercy, but they accepted the misery as part of the structure of

Chinese life. Foreigners went to and from their factories, in and out of their offices, back and forth from their clubs, without a second thought about the misery. The great "New Life" reform movement of Chiang Kai-shek consisted largely in propaganda campaigns against the indubitably noisome habit of ubiquitous spitting.

The only ones who denied the inevitability of misery, damned the system which perpetuated it, and promised an end to it were the Communists. They were a symbol of protest and of hope. That was the source of their strength. That is why, abandoned as early as 1927 by Stalin, who neither then nor later was interested in the Chinese revolution, without outside support, surrounded by Nationalist forces, they could not be destroyed. That is why 28 years ago their ultimate triumph was inevitable—unless someone else could project an even more powerful image of protest and of hope. But no one else did, and in 18 short years, with an assist from Japanese aggression, which only advanced the date of their ultimate triumph, and in the last stages with considerable help from American military equipment delivered to them by defecting Nationalist troops or sold to them by corrupt Kuomintang politicians, they were masters of China.

This is not to say that the Communists were swept into power on a great tide of popular support. This

 would be a gross distortion of the facts. No doubt the larger mass of the people were uncommitted, but their very passivity was an asset to the Communists. Outside of its own bureaucracy, the Kuomintang had no real roots in the people. What did it offer but a perpetuation of the old ways, which is to say the old misery?

This is not a tract in support of communism. I am well aware of the mess of pottage the Chinese have bought. Many today in the communes eat their bread, if they have any, in sorrow. Probably many of the young idealists, the students of fifteen years ago, who rallied in

overwhelming numbers to the Communist cause now wonder what happened to their great dream. It would be wishful thinking, however, to believe that the dream has died for all. The hundreds of thousands of young people who turn out for parades and demonstrations do not have the look of sullen prisoners of the regime. For them the Communists still manage to project the image of promise. They are still carried along by the specter of an evil world to be destroyed and a better world to be built. It is quite likely that they still sing with gusto the song of Communist Youth:

> We are the builders.
> We build the future.
> The future world lies in our hands.

There was a time when America projected an image that enflamed imaginations, set feet jiggling and hearts dancing. That image does not come through today. What comes through is the hard-faced image of the bean growers, the hate-contorted faces of New Orleans housewives screaming epithets into the ears of a bewildered Negro child.

Far more important than guns for Laotians is the projection of a new image of America. President Kennedy in his Inaugural Address projected such an image, and it is significant and symptomatic that it stirred excitement around the globe. But it will take more than words, or even the deeds, of one man, even a President. What is needed is the image of an angry people, people who react, not blasphemously like Guevara, but like Christ in the temple, against a social structure which creates, perpetuates, or tolerates human misery anywhere.

Where will this fury arise? Is our society capable of producing this kind of rage, this kind of central concern for human beings, for other human beings? Is any society whose central concern is money and profit? These are questions which I should not presume to answer, but, on the record, I reserve the right to a certain amount of pessimism.

There have been profound changes in our social structure during the past thirty years. There will be profound changes in the next fifty years. Amidst those changes we had better find a way to do away with the bean-grower mentality and become a people who, like Karl Marx, can never feel free so long as Guatemala children or West Virginia children or Indian children have bloated bellies, or Mexican-American children slave in the fields, or Negro children are treated like medieval lepers. If we don't find a way, there will be other Castros and more surprises, and our grandchildren may learn that Khrushchev was right about them.   GEORGE H. DUNNE

II

THERE IS a Canadian proverb that says: When the wild ducks fly over, the tame ducks flutter their wings furiously—and stay on the ground. A distinguished missionary quoted this to me the other day during a discussion on the forthcoming Peace Corps. Like almost everyone else I had interviewed, he thought the Corps a splendid idea. But he wondered, and many others wonder, how many of the thousands of enthusiasts will only flutter their wings when the time comes to take off.

A random, unscientific sampling of student opinion at a recent NSA (National Student Association) meeting showed remarkable unanimity about the Peace Corps. All those questioned showed real zest for the idea. But when confronted with the question: "Are *you* going to go?" 94 per cent cleared throats and answered: "Well, no, but . . . ." Doubtless the sampling was inadequate, and possibly not even representative, but it did give one pause.

Another set of attitudes came to light during interviews with United Nations personnel (AM. 4/8, pp. 49-50). I was fortunately able to meet a number of officials from the "developing countries" (the word "underdeveloped" is no longer used), but received only standard, innocuous replies. It seemed more useful, accordingly, to talk to younger, unofficial members of these nations, with the promise that they and their nationalities would be kept anonymous. They seemed to speak freely, though how freely is hard to judge.

Rather like the NSA members, all of them assured me that they were favorably impressed by the Peace Corps ideal. They had lived in America long enough to realize personally that not all Americans are callous, dollar-mad imperialists. The real problem would lie back at home. There, innate skepticism would be hard to put down. In stately phraseology one young African warmly insisted that "America must somehow eradicate apprehensions felt by my people."

These "apprehensions" are obviously centered in the lingering dread of imperialism. Having just won political autonomy, many nations are intensely anxious to avoid anything that "savors of even the mildest form of colonialism," as one man put it. "After all," he said smiling, "we cannot forget that Americans are traditional allies of the British, Belgians and French." If the Corps is not skillfully conducted, said another, international relations will "jump from the water into the fire, as our people say."

The problems of adaptation will be even more mammoth than Americans anticipate, a young Asian assured me. "My own family back at home is rather well-to-do. An American could quite easily live with us, I believe. But it wouldn't serve the purpose of the Corps." The phrase "cultural shock" is being used to describe the probable reaction of most young Americans coming from a push-button world into many of the countries envisioned by the Corps. Even the "toughening-up" program to be given in Puerto Rico left some of my UN friends unimpressed. "They must have as stiff a training as the astronauts are getting," was the way one of them put it.

Even the very idea of the Peace Corps provokes misgivings and some skepticism. "Our countries need to be strongly reassured that the Corps is not to be given in place of money—the money we already receive or that we hope to receive," I was told more than once. For the problem in most countries, as the "locals" see it, is largely financial.

It was interesting to interview some Americans working in the UN. After our editorial "No Children's Crusade" appeared (4/8, pp. 49-50), a UN technician wrote in approval: "You are expressing an opinion strongly held 'in the house' here." It would indeed, I believe, be very wasteful if the Corps did not make extensive use of the findings of a number of the experienced UN agencies.

In this connection, several of these American experts were of the opinion that it would be more helpful in the long run to train *local* people than simply other Americans for short-term service. In any case, whatever is done, they felt, we must always avoid replacing the "locals." These developing countries vary widely: in some areas, experts are needed at high levels; but in most, the need is for intermediate workers, neither at the top nor the bottom. Above all, such practical skills need to be taught as how to sharpen tools, how to repair simple machinery, how to drain a swamp, when to plant, how to rotate crops, how to fertilize, how to improve the technique of thatching huts, how to install and use sanitary facilities, and how to maintain them. This unpretentious type of work, the UN people said, is not likely to cause unemployment nor to be interpreted as masked "neo-colonialism."

It was reassuring to learn that the Peace Corps has now released a guide insisting that the Corps is to "help, not replace," private and religious agencies already working in the field, and learn from their experience. The appointment, too, of the respected missiologist, Fr. John J. Considine, M.M., was a heartening step. I was able to meet with Fr. Joseph B. Gremillion, director of Socio-Economic Development, and F. Robert Melina, coordinator for Peace Corps, both of Catholic Relief Services—NCWC, and found that their suggestions largely corroborated the opinions of my contacts at the UN.

The executive secretary of the Mission Secretariat, Fr. Frederick A. McGuire, C.M., kindly allowed me to use answers to an important questionnaire. Inquiries had been sent out to CRS workers, to missioners and to local educators and religious leaders in the developing countries. Their answers should be of real service to Corps leaders.

THE GENERAL REACTION to the Peace Corps from this source was again favorable, but again there was the strong proviso that "the project must be properly conducted." Thomas Cardinal Tien, S.V.D., writing from Taiwan, stated: "May I suggest that this seems necessary: to train not too many young people, but very thoroughly, especially in the ethical respect." In India, Valerian Cardinal Gracias contacted heads of four important Catholic universities. Their replies echoed those of Cardinal Tien, with strong emphasis on the "need for suitable personnel, careful selection, and a cautious and slow beginning." One educator urged "something akin to military discipline, under the control of a mature officer who by previous experience knows the character and temperament of the people."

Community development is seen as a pressing need, and "social workers, medical, agricultural and veterinary graduates would be extremely useful," a missionary comments. "Let not MEDICO's experience be lost. Consult them. Use the experience of the International Voluntary Service. Discipline must be very, very rigid."

A rather blunt caution comes from another missionary: "Men with only State Department background should never be leaders, even if they have been overseas. Their attitude centers around housing, PX, motor pools and the like." More explicitly, for some areas "a person with a country background is an absolute necessity. A city-raised boy cannot adjust enough to a country living 100 years in the past."

Corpsmen must beware of chauvinistic attitudes and "must be screened for tact, understanding and psychological stability," says a native priest. "Material aid, however eagerly accepted, is suspect as to its motives." Moreover, in Americans especially there is often, he notes, an "almost inevitable sense of superiority which is resented."

A missionary in South America urges the Corps to be wary of close association with local governments. Otherwise, "the personnel will too soon become disillusioned by corruption, graft, feet-dragging, inefficiency and general lack of co-operation." Moreover, as another puts it, "the government here would take all the credit. Some 99 per cent of the people don't know anything of the millions already given by the United States."

"The hardest job," in the mind of another priest, "is to find out what the people in the host countries want. Some Corpsmen should go with the attitude and announced purpose of learning. It would be a good idea," he adds, "to invite people from other countries to participate in the Corps, and probably some young local people would like to be invited to take part."

An acute problem in the minds of several observers is how to reach the needy. "The big, already-rich land owner doesn't care about the poor. All projects should be directed at personal contact with the masses." He goes on to specify: "What I feel is needed are not the high-priced technicians living in expensive houses in the cities, but dedicated, skilled technicians living among the poor as friends and companions."

"Teaching is our prime need," notes one CRS worker. "But teaching what? Skilled trade teachers are most needed. Not highly technical teachers, but down-to-earth practical teachers." A missionary notes: "The country I am in has enough doctors and lawyers. What it needs is a skilled middle class of workers." With some acerbity he adds that "so far the United States has helped the rich man become richer. This is a point that the Communists use to the hilt."

Another missioner concentrates on the personal traits of a prospective Corpsman. "He must be sincere, sym-

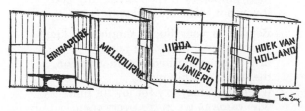

pathetic, hardworking, indifferent as to where he sleeps or what he eats, not emotional, not easily discouraged,

141

determined, rugged, the out-of-door type, self-reliant and even-tempered." This sounds very much like the qualities needed in a missioner.

After long interviews with persons differing in religion, race and background, and after reading reports from many lands, I am impressed by their overarching theme: while the ideal of a Peace Corps is unexceptionable, the hazards and problems are colossal. Knowledgeable missionaries, who have personally made the huge sacrifice of adaptation and who work with the highest motives of dedication, know this better than anyone else. Better than anyone else they know that there will be no glamour involved, that this will not be a "lark," that an inept Peace Corps would be not only futile but worse than none at all.   C. J. McNaspy

♦

Dies Sanctificatus Illuxit Nobis

This poor god shepherds once stumbled
off their hills finding and kings
groped their starred night to worship
comes to sit at the gates of the world
and judge it. And there is not one,
not one man will dare to watch
his whipped face then saying my god
it was not I that scarred your face.

Rejoice in your flocks shepherds, kings
your palaces: today the hands
of love have soothed him in a crèche,
the unimaginable word,
where men may cry for joy for him
who came to weep for every man.

JOHN KNOEPFLE

Red Bear and
Pink Toenails

George H. Dunne

PINK TOENAILS, the painted kind, have never found favor with me. They seemed a move in the wrong direction, taking us back toward a "bells on my fingers and rings on my toes" kind of primitive society. Neither have I been known as a champion of Madison Avenue. I have looked with jaundiced eye upon the world of advertising which exalts the dubious merits of products most of us could well do without and persuades people that they need a variety of things they really don't need at all. And although by nature not disposed to look kindly upon any sort of censorship, the blatant vulgarity of cheap cheesecake cover illustrations exploited by so large a part of the magazine trade has seemed to me from every point of view indefensible.

Perspective is an interesting phenomenon. There is a church in Rome, St. Ignatius, where you must stand on one spot, marked by a small mosaic, in order to see in proper perspective the rather remarkable fresco which covers the ceiling of the nave. Move even a few feet away to right or to left from that point of view and the perspective changes. Things look radically different.

When I passed through the Brandenburg Gate from West Berlin into East Berlin things immediately looked radically different. And the farther I moved into East Berlin the more different they looked. Like everyone else, I had read about the striking contrast between West and East Berlin. I had not doubted what I had read, but I had not really grasped the profound nature of that contrast. There is no use trying to describe this passage from light to shadow, from life to death. Only those who have actually themselves crossed the river Styx can understand what the passage is like. There is no reason to suppose that I could make it more understandable to others than others were able to make it to me. To grasp it, one must oneself experience the almost shattering impact of crossing from West to East Berlin.

West Berlin is a city bursting with life. East Berlin is a city which looks upon you out of the lustreless eyes of death. It is not simply a question of West Berlin's streets crowded with autos, sidewalks brimming wth people, shops crowded with customers, stores bursting with food, with furs, with clothes, with appliances. Nor is the contrast merely between the stark ruins which are still the predominant landscape of East Berlin and West Berlin's new and modern buildings, which have sprung up at an extraordinary rate where at war's end there was only rubble. For even when, as along their Stalinallee, the Communists have made a determined and deliberate attempt to match West Berlin's building display, the effort has come to naught. The buildings are there— apartments above, showcase state-owned stores below. They are massive and grandiose, each one exactly like its neighbor, all done in a heavy, stolid Soviet style, and all totally incapable of distilling a breath of life into the broad street upon which they stand. Stalinallee remains as dead as Alexanderplatz or Wilhelmstrasse, where Hitler's death bunker is now just a grass-covered mound, or Unter den Linden, where the once-famed Adlon Hotel is a rubbish heap probably inhabited by rats.

When I came back through the Brandenburg Gate into West Berlin, leaving behind the hostile faces, the uncertain faces, the fearful faces, the pathetic faces, it

144

was like a resurrection. Never had I seen anything as lovely as the painted pink toenails. Eagerly I drank in the colorful advertisements on every side: *Konditorei Mohring, Zum guten Bier, Wunderbarer Jacobs Kaffee, Imperial Weinbrand, St. John Rum!* Even the magazine kiosks with their too generous display of flesh and sex (though not as generous as those in Rome) were a welcome sight. Like the pink toenails and the flaunted ads they cried out: FREEDOM! FREEDOM! FREEDOM! and their cry seemed to make a song which echoed through the streets and, like sunlight, touched the buildings into life, dispelled darkness from alleyways, made eyes sparkle and laughter ring and hearts rejoice.

It came to me that this is what makes the contrast so shocking. It came to me that this is what our world is all about. This is what the world-wide struggle is all about. It is about freedom, and therefore about life; for the two are one. Girls free to paint their toenails every color or no color. Men free to flaunt their commercial boasts. Other men free to be foolish or not to be fooled. Free to believe or to disbelieve. Men free to follow Christ or to serve the devil.

I saw a sign ride by on the side of a beautiful double-decker bus. "*Mach mal Pause, Trink Coca-Cola, das erfrischt richtig,*" it commanded. A moment later another sign rode by with just the contrary advice: "*Dir zum Lohne trink Darbohne!*" Again it came to me that this is what it is all about. I don't have to drink Coca-Cola or Darbohne. I can take them or leave them, drink either or both or neither. This is what makes our kind of world a world worth living in. This is what makes West Berlin a city that teems with life and East Berlin a city of death.

It came to me that this is not only our kind of a world. It is God's kind of a world. God's kind of a world is a world in which men are free to make their choices, to choose heaven or to choose hell. People who say: "How could God permit sin?" don't know what they are saying. A world in which men could not sin is a world in which men are not free. If you want a microscopic view of that kind of world visit East Berlin. The tragedy of communism is that it thinks man can be brought

145

to salvation by being deprived of his freedom to choose.

I saw no pink toenails in East Berlin. Apart from dreary party propaganda slogans hanging on red banners across dismal factory fronts, I saw no advertising. I saw no vulgar magazine displays. The party has done away with the occasions of sin, the temptations to be frivolous or foolish or wicked. God's world, on the contrary, is filled with temptations and occasions of sin. That is the way He made it from the beginning when He began by planting the tree of forbidden fruit in the Garden of Eden: something to be remembered, not only by Marxists, but also by others who easily succumb to the temptation to appeal to the Inquisition or the Holy Office or the police power of the state to save us from the occasions of sin.

It is not just a matter of painted toenails. These may one day come to East Berlin. They may even be there now, unnoticed by me. But painted toenails will look entirely different in East Berlin than they do in West Berlin. No more than the imposing Georgian architecture of Stalinallee will they be able to lift the pall of death that hangs over this city from which people were fleeing at the rate of 300 a day when I was there, a figure greatly swollen since. Because, no more than the buildings on Stalinallee will they symbolize freedom. Like those buildings, they will be only another manifestation of death. For when painted toenails appear it will not be because girls have *chosen* to paint their toenails pink or green or some other outlandish color, but because a Communist bureaucrat or a faceless Communist bureau has decided that painted toenails serve the ends of the regime and are therefore to be cultivated. Freedom—and therefore life—spring not from decisions imposed, but from choices made.

I spent the evening of my return from East Berlin leaning out my open window, like a true Roman, gazing upon the human pageant in the street below: the boys and girls holding hands, the crowded sidewalk cafés, the throngs pouring out of theatres, even the silly people craning necks like Hollywood simpletons to catch a glimpse of some featherbrained movie star (it was International Film Festival Week in Berlin), the neon signs, the noisy cars, the murmur of human voices, the ready laughter, the beautiful sight of faces that were neither sullen, nor hateful, nor filled with fear.

I realized exactly what Khrushchev means when he says that West Berlin is a chicken bone stuck in his throat. How can he permit this confrontation in a Communist heartland between this small island in which men are free to make their choices and which throbs with the pulse of life worth living, and the city where death walks through the streets because freedom has died?

Fresh Look at Cuba and Castro

John Meslay

E VEN WHEN ONE IS allowed to travel freely to Havana today, it is not easy to form a judgment on the impact of communism on the life of the Cuban citizens, or to say which phase (according to Leninist theory) the Cuban revolution has presently reached.

My first impression of the scene about me, as I drove in from the airport, was its strong similarity to what I had experienced in the streets of Warsaw in Poland, several years ago: heroic-sized pictures of national or Communist heroes like Yuri Gagarin, who visited Cuba on July 26 for the anniversary of the Castro revolution; posters and slogans inviting the passer-by to work, to study and to take up his rifle.

In Havana, militiamen and militiawomen with rifles mount guard at nearly every corner, in all the public buildings and even in some private shops. The mere sight of them is enough to create a climate of fear and to give the ordinary citizen a vague feeling of guilt. Everyone is afraid of being denounced to the Committees of Defense of the Revolution (there is one on every block), or to the famous G2 (the secret police).

In such a political climate, one can hardly expect to find objectivity of judgment. Apart from some in diplomatic circles who are not directly involved, the people you meet in Havana today are either strongly favorable or strongly opposed to the regime. The dominant feelings are nervousness, anxiety and fear. The foreigner, walking through the streets of the city, cannot but notice the lack of joy and the tension plain on so many faces. The shortage of food, without yet reaching any dangerous proportions, obliges the women to queue for many hours to get meat, fruit or fish. (According to what high officials of the new regime told me, the shortage of food is not due to less production, but to the necessity of exporting agricultural produce in order to buy machines, spare parts—and weapons.) Besides, even though a great majority of Cubans were rather lax Catholics, they now resent the constant interference of the government in religious affairs. Probably Castro has not measured the strength of the religious resistance to his plans.

My visit coincided with the Literacy Congress (*Congreso de Alfabetización*) in Havana. Schoolteachers came from all over the island to comment on the literacy campaign which has been under way since April and is slated to last until January. All the boys and girls between 12 and 18 years of age who volunteered for the campaign have been sent to the villages in the country and in the mountains to help the teachers in a full-scale adult education program.

The very ambitious objective of doing away with illiteracy in Cuba by next January will most probably not be reached, but the campaign is already yielding very important psychological and ideological results. For one thing, public opinion has been mobilized on one common aim, the achievement of which is objectively good and can hardly be criticized. Moreover, the textbooks used, as well as the indoctrination campaign which goes along with the educational effort, provide the regime with a powerful instrument for the propagation of its socialist ideology.

The paradigms chosen to teach reading and writing all convey some "socialist" meaning. For instance, the very first lesson on the vowels is used for a pungent commentary on the "imperialist" character of the Organization of American States (the abbreviation in Spanish is O.E.A.).

Finally, the young people taking part in the campaign get special treatment. They appear on TV programs and on the whole seem to enjoy their work. These young persons get a sense of *participation* in the life of the country. For the first time, they enter into *communication* with the people of the remotest districts. These teen-agers appear proud of the Cuban revolution and this experience will certainly have a profound and lasting influence on their lives. On the other hand, the sending of boys and girls together without guidance for a long period of time to faraway districts is highly objectionable from the point of view of morality, and I heard many complaints about this.

There are 285 collective farms already in operation in Cuba. These are called *granjas del pueblo* and belong to the strict type of Russian *sovchoz*. I visited one of them just outside of Havana, where the administration is building nice, clean little houses for the villagers.

I also visited a co-operative plant for sugar-cane growers. Here the government is capitalizing on the agricultural progress which had been made in the preceding years by the now-expropriated big landowners. Of course, it is not quite certain that collective farming will yield better crops.

The government has not divided the land into small plots or parcels, nor handed them over to the individual peasants. Instead, it has made these people believe they are now the collective owners of the farm and has thus transformed them into employees of the government. They receive fixed wages just as they did under the preceding system. But propaganda indoctrination, which goes along with various substantial advantages, gives them, too, a higher sense of "participation." I use that word again deliberately. This class of very poor land laborers who now have their own houses will probably become the defenders of the regime. Even if the urban workers fall away from the revolution and lose their enthusiasm, these very poor peasants—together with the very many militiamen and militiawomen to whom a fixed salary and a job have been given, all over the country —will, in many cases, though by no means all, prove the best allies of the regime in case of crisis.

The weakness of Fidel Castro is that he probably no longer controls his own revolution. It is true that his figure or image is still largely popular (perhaps more so in other Latin American countries than in Cuba, where the anti-religious measures have made him lose ground). The propaganda machine leads everyone to believe that the Castro revolution has set a pattern for all future upheavals in South America. But the striking resemblance between the Cuban situation and what we have seen in other countries on the eve of a Communist takeover leaves little doubt that the uniqueness of "Fidelismo," if it was unique, has been lost, and that the point of no return on the road to "classical" communism has been reached.

Around Fidel Castro, men schooled in Soviet political doctrine are slowly preparing conditions for the final takeover, and I do not see how the trend can be reversed. It is even probable that the coming-to-power of the Communist party, as such, will coincide with some bettering of the Cuban economic situation. Cuba will receive foreign aid because, for propaganda purposes, the USSR cannot let Cuba starve.

Therefore, the chief lesson to be drawn at this moment does not concern Cuba, but the other Latin American countries. Just as the Marshall Plan in 1947 came too late to save Eastern Europe but did save Western Europe, so the Alliance for Progress program drawn up at Punta del Este last August will not pull Cuba back into the Western fold (at least not directly and immediately), but it may help some of the other Latin American nations to resist the terrible stress and temptation they are now facing.

In these nations, many influential young men are ready to overthrow their own government and to imitate Castro's shortcut to economic independence. If they are to overcome this temptation, which is so strong in the young generation of economists and left-oriented intellectuals, more will be needed than very generous and large-scale help from the United States. Required also is an effective and self-sacrificing resolution, on the part of the economic ruling class in Latin America, to take the lead in a redistribution of the national income. Unfortunately, however, this ruling class seems to think mainly in terms of military defense against the Communist threat they see embodied in the Cuban regime.

It is not difficult to find out where the secret of Fidel Castro's success lies. It is to be located in the two ideas which I have already cited: "participation" and "communication." For the first time, millions of Cubans are invited to "participate" personally in the building of their own country. And for the first time there is an effort to make all men equal and to create "communication" between the different elements of Cuban society. Lenin, with his usual shrewdness, assigned this task of making "one people" out of many elements—in sociological jargon, the function of participation and communication —to the nationalist period, which, according to his theory of the evolution of societies, was a necessary preliminary step to the Communist takeover, particularly in underdeveloped, that is to say, pre-industrial and colonial countries.

Of course, there is a great deal of bluff and lying in this pretension of building up the unity of a nation. Many people in Cuba today are neither "participating" nor "communicating." On the contrary, they live in fear of their neighbor and take no part in any public activities. The policy of the new regime has alienated a large part of the population, but these are precisely those who, in the Communist terminology, are called "the enemies of the people." In Cuban Communist slang they are termed *gusanos* (worms), that is, those who formerly lived on the work of others and who felt safe and secure through money, status and property.

As background to the panic which today seizes so many Cubans and prompts them to seek refuge abroad, a sense of insecurity is very strong. The regime will take steps to increase this insecurity for a while among this class of people who previously had status. It is an essential part of the strategy of the Communist system. In a regime like that of Castro, it is part of the usual build-up to destroy the framework of security that obtained in the old society and to build a new framework where only those who adhere to the regime feel safe and comfortable. For the others (the *parasitos*, the "worms"), life must be made impossible. They have to convert or go away.

A FTER THE ABORTIVE April invasion, religious were jailed and were there subjected to threats, but they were finally released. They were not expelled; they themselves asked to leave the country. This is what the regime wanted: to get rid of them without risking bloodshed or martyrdom.

It was only in mid-September that the government finally organized a large-scale deportation of priests. This operation included an element that is really something new in international law: the deportation of Cuban citizens without a trial, and their forcible exile to another country prior to any agreement with that country.

What set off Castro's conflict with the Church? Well, what started the present acute conflict with the United States? Is Castro's feud with the U.S.A. fully explained, as certain propagandists in Latin America and in Europe would have us believe, by long-standing U.S. abuses? The thing to remember is that when the Communists step in, *any* conflict is bound to become acute. Much the same can be said of Castro's contest with the Church. It would have been the same no matter what the Church might have done, and even if her members had always been inspired by the purest Christian spirit. A totalitarian system like communism refuses to accept any sharing of the loyalty of human beings.

Therefore, it is painfully true that the lack of justice and of love of the poor on the part of some Cuban Catholics has provided a stepping stone for Castro and for the Communist revolution. But our duty is not to judge the past. The Catholics in Cuba are now passing through the fire of persecution. If the fire consumes, it also purifies. Many formerly lax Catholics are returning to the fervent practice of their faith. Many deeds of generosity and even of heroism are performed every day. It is too early to report them because such a report would

153

endanger young men and women who have chosen to be faithful to God. The outside world cannot do much to help them except by prayer and brotherly comprehension for the difficulties and trials of this new "Church of Silence."

The most immediate duty incumbent on us is to change the social and economic conditions in those countries where communism has not yet seized political power. We must strive to do this so radically and thoroughly that the Communist system appears in a clear light for what it really is—a quick, very primitive, half-barbarian solution to social evils through brutal despotism and destructive totalitarianism: a purge which cures the illness by killing the patient.

However, let us be perfectly clear about one thing: if no other solution is proposed now—and rapidly—then the temptation to try communism will grow stronger in all the underdeveloped countries. That is exactly what makes Khrushchev so confident about the peaceful triumph of communism all over the world.

My own conclusion is that probably the situation created by the Castro regime and the Cuban power-grab by the Communist party is something that is here to stay, at least for a long time. But the drama of Castro and Cuba may yet help us to draw some badly needed lessons for other countries and other continents.

The Past Is Prologue

Leo A. Foley

F REEDOM and tyranny are common words today. They form the slogans of new nations arising out of violence, while the most tyrannical enemy of the United States calls itself a champion of freedom. In the midst of this confusion and danger, we Americans might well ask ourselves not only what we mean by freedom and tyranny, but also what our entire national purpose is. Actually, we have a model of our national purpose. We have the outstanding definition of our own aims and purposes in the American Revolution.

The American Revolution was many things. In England, it was an extension of parliamentary politics, marked by the opposition between the Tories and the Whigs. Thus it was that many such British leaders in America as Lord Howe, Lord Cornwallis and Thomas Gage were often in sympathy with the Americans simply because they themselves were influenced by Whigs in politics. In America, the revolution was not only a blow for independence. It was also a civil war, an Indian war, a vast unifying process, but always a movement proceeding from the people. In the eyes of the world, it was a startling social phenomenon soon to become the model for every progressive movement. It excited the admiration of political figures on a world-wide scale. *Democracy in America,* by Alexis de Tocqueville, is but one example.

Let there be no doubt about it, the American Revolution was one of the greatest political events since men began to congregate into tribes, cities, states and nations. Its success was, among many things, the reason for William Lyons Mackenzie's inchoate revolution in Canada, the reason why Englishmen and members of the British Commonwealth are today among the world's freest people. During these several years when we are commemorating the hundredth anniversary of the American Civil War, we might well remember that Southerners and Southern leaders would be written off as traitors and insurgents were it not for the Declaration of Independence. Yet, whereas most Americans know the Civil War almost by heart, there are many phases of the Revolution that are almost completely unknown and which by their very nature demand examination and re-examination by every American, rather, by every human being.

First, the American Revolution did not start with the killing of innocent people. It started with a document containing a theory, a statement of rights in the *Declaration of Rights and Grievances* drawn up in Philadelphia in 1765. Furthermore, the rights expressed were not those of a few. They were the rights of man given a startlingly new expression, namely, that man's rights are an extension of God's authority simply because God is God, man is man, and freedom is the manner of man's existence. The philosopher will recognize therein the position of St. Thomas, St. Robert Bellarmine, Francis Suarez and John Locke.

Nevertheless, this had been developed in a typically American manner, that of the New England town meeting. This institution was typically Protestant insofar as it was an extension of the Congregationalist Sunday meeting. The Congregationalists were Calvinists who maintained a priesthood of the people, making the layman equal in God's sight to a bishop. In civil affairs, they looked to the citizen as equal to anyone and just as capable of a correct civil decision as an appointed governor.

We Catholics should be aware of this Protestant contribution to American democracy. We have a tendency to envision St. Thomas Aquinas as the author of the Declaration of Independence. We overlook the fact that although he and Bellarmine and Suarez admired the Roman republic, they still lived in a monarchical society and wrote against the evils of tyranny by bespeaking the rights of man with a view to a benign monarchy.

The town meeting, still a potent force in many New England areas, was the model for the Continental Congress. It was also the model for co-operation of States under the Articles of Confederation and the interco-operation of citizens and States in the House of Representatives and the Senate under our present Constitution.

Europeans looked with open astonishment to see if these upstart colonists could make this sort of thing stick in a world given over to monarchy and colonialism. We Americans should be as proud as we can be of the fact that our predecessors did establish the American Republic, establish it on a sound metaphysics of man, and created a model of government that has been the ideal of every struggling state that aspired to freedom.

Second, the Revolution, as a war, was won by the people. From this emerged a new social entity, a new perfect society, a new state. Although James Otis and Samuel Adams may have breathed fire, it was the American volunteer, the minuteman, the militia man who had to learn tactics under fire, who kept that fire burning with his sacrifices. Although John Adams proposed the plan for independence and Thomas Jefferson wrote it up in the spring and summer of 1776, it was the slum dweller in New York, the waterfront scrounger in Charleston, the Philadelphia and Boston workman who fought for it. Although James Madison was to be the constitutional lawyer in the Constitutional Convention, he wrote with a background of a new nation that had begun from the ground up, from the people themselves.

Even the carrying on of the war was by amateurs. The only professional soldiers in the American army were Horatio Gates and Charles Lee. Gates was no military genius, even though later research seems to show that Gates rather than Benedict Arnold was the victor at Saratoga. Charles Lee was better at spreading dissension than at organizing troops and winning battles. Washington's outstanding generals and officers, Nathaniel Greene, Anthony Wayne, Henry Knox, Daniel Morgan, Henry Lee, Francis Marion, Israel Putnam, and the rest, were ironmasters, farmers, booksellers and amateur gentlemen—almost anything but soldiers. Yet, they rose to the challenge and defeated the finest professional army in the world by unorthodox strategy and improvised tactics. Even Washington himself had been only a colonel in the colonial forces when John Adams proposed him for command of the Continental armies.

These amateurs did a far better job than, with few exceptions, the professionals were to do during the Civil War. Military historians marvel at the audacity of a few poorly trained militiamen standing up against a professional group on Lexington Common. Yet, this was the pattern of the whole: the people against tyranny, the citizen who will not fight until he has to, and who then improvises brilliantly in order to win promptly.

There were all kinds of legalistic snarls at the beginning of hostilities. New Yorkers were fighting citizens of New Hampshire over possession of the Hampshire Grants, now the State of Vermont. When Henry Knox was dispatched by Washington to bring the cannon captured at Ticonderoga to Dorchester Heights in order to drive the British from Boston, Knox had all he could do to get any drivers to cross from one State to another. In fact, he had to depend upon local committees in township after township to supply him with relays of drivers. Yet, at the end of the war there was common unity under a Virginia Commander-in-Chief when a Rhode Islander, Nathaniel Greene, led a mixed army, and with the help of Southern officers enticed Cornwallis out of South Carolina, sniped at him, did battle with him, and nudged him into Yorktown where Greene promptly bottled him up to await the French fleet.

One cannot overestimate the unifying force of George Washington during and after the Revolution. Although he had the reputation of being a rather cold type from Virginia, he soon won the love of his men to such an extent that when Congress was unable to hold the army together, Washington fulfilled that task. When he had finally starved Sir Henry Clinton out of New York two years after Yorktown, tears flowed freely at Washington's farewell dinner for his generals.

This was the end of a process that had begun when all of the colonies leaped to the relief of beleaguered Boston. We might well remember in these days of remnants of acrimony between Northern and Southern States that Boston was indebted not only to the fiery eloquence of a Virginian, Patrick Henry, but also to two generous donations of hard, cold cash from South Carolina. In a period of 24 years, from the *Declaration of Rights and Grievances* to the adoption of the United

States Constitution in 1789, again largely because of the prestige of George Washington, a nation was born, a nation that has become the champion of the rights of man.

Today leadership of the free world is thrust upon the United States. Great nations and small nations look to the United States not only for its military strength but also because of its prestige in defense of the rights of man. We Americans can scarcely begin to appreciate the flame of hope that was lit by the American Revolution. Despite our complaints about legislation, taxes and big government, we do not know what it means not to be free.

The establishment of the United States as one nation was a process that involved the co-operation of many individuals of different opinions and viewpoints. Our forebears realized that there are limits to individual freedom. They also learned that co-operation for a common cause does not require coercion of opinion into a single, tolerated conviction. Monolithic political tenets are the property of tyrannies.

Members of the Continental Congress, colony by colony, individual by individual, disagreed, wrangled and argued before they achieved the unity of the closing of the Declaration of Independence: "And for support of this declaration, we mutually pledge to each other our lives, our fortunes, and our sacred honour."

The realization of that pledge is found in the opening words of the Constitution: "We, the people of the United States. . . ." This new unity was also preceded by discussion, argumentation and—always—serious purpose. If Herodotus is right, and if what is past is prologue, there are several truths that cannot be avoided. We cannot allow one opinion to oppress free discussion. Nor can we allow discussion to dissipate all conviction. Above all, no one can afford to identify the "American way of life" with The Great, Big, Good Time. Our existence and our future are a challenging responsibility based on a sound appreciation of our principles, our heroes and their sacrifices in their achievements.

Ireland: Vacuum of National Purpose

Gary MacEoin

THE VILLAGE in the West of Ireland seemed fuller of young people in August, 1959 than I had seen it in half a century. They were gay, well-spoken, well-dressed and free-spending. They took advantage of once-in-a-lifetime summer weather to throng to football games on Sundays, picnic at the beaches, explore the countryside in rented automobiles and see Grade-B westerns in the village movie house.

The people are living better than ever before. The selection of material comforts is sometimes odd. An electric washing machine confers prestige rather than comfort when one must carry the water half a mile in a bucket from the river. Forty years ago the village road was worn into three deep ruts, a center one by the plodding horses and donkeys, two side ones by the narrow iron rims of the cart wheels. They were so deep a cyclist could not move from one to another without dismounting. This year I drove at eighty miles an hour (Ireland has no speed limits) over its excellent macadam surface.

But the surface does not, unfortunately, tell the entire story. The prosperous young people having a gay time are home on vacation from England and the United

States. There were never so few young workers on the farms. No sooner do they finish school than they emigrate, attracted by high wages and steady work. The money they spend on vacation and the substantial remittances they send their families account in large part for the improved living conditions of those who stay.

Ireland's economy is overwhelmingly agricultural, with over 40 per cent of the population engaged in farming. The typical farm is undercapitalized and not very efficiently operated. But the currents which have affected international trade in this century have not bypassed Ireland. Agriculture, providing the bulk of its exports, must try to maintain competitive production costs, and accordingly machinery and agricultural chemicals have cut the labor input. For every four men employed on the land 20 years ago, there is room today for only three. And the trend is not yet exhausted. Output per farm worker, though up nearly 65 per cent in 20 years, is still far below that of competitors. The alternative to dwindling returns is more farm efficiency, which usually means less labor per unit of production.

Fewer farm workers, each producing substantially more, give a total agricultural output some 25 per cent higher in volume than 20 years ago. The increase is praiseworthy, though far behind the typical Western European achievement of the same period. Indeed, it is doubtful if it even pays for the higher living standards today enjoyed by the smaller farm population, for these improvements represent in substantial part the remitances from overseas of a new generation of emigrants, perhaps also some depletion of capital resources.

WHY DO THEY EMIGRATE?

What to do, however, with the people displaced from the land? This problem has been vexing the country for a century. Emigration seems the only answer.

Now the average Irish boy or girl has no wish to emigrate. The persistence of the little villages and towns, where the owners of the old-time general purpose stores and saloons lean against their doorways and wait patiently for the approach of the rare customer, shows the strength of the ties. So does the energy with which those who must go accumulate their earnings in London and New York, both to subsidize those who stay and to take frequent trips back, trips that today often pledge the future under pay-later plans.

Those who stay agree with those who go in deploring emigration. At one time, there was a lot of talk of settling more people on the land. More recently, they talk about expanding industry, and there has in fact been a small but steady annual increase in industrial employment. Yet the paradox remains. The greater the industrialization, the lower the total employment.

Today, Ireland exports automobiles to the United States and television receivers to Britain. This year its first oil refinery with an annual capacity of 2 million tons has gone on stream. A West German firm completed construction of a plant at Killarney to manufacture cranes for export. Another West German firm is building a cheese factory. Two American companies announced formation of Irish companies to produce cosmetics and chewing gum.

All of this is impressive, but it does not begin to change the basic situation. No technique has been devised to overcome the economy's persistent tendency to increase imports at a more rapid rate than exports, which is another way of saying that the Irish insist on consuming more than they produce. They settle their account by exporting their children.

I'd like to think a little about the attitude of trying to get something for nothing. Reasons exist, both technical and historical, for low labor productivity. Likewise, exposure to and facility of movement to England and the United States foster a desire for living standards which the economy does not justify. Yet, I think that the real problem is more basic. As I see it, the Irish lack a philosophy of work or, at least, a Christian philosophy of work. There is no pride of achievement, no concept of vocation, no aspiration after perfection.

This is, and is deliberately intended to be, a judgment on the moral condition of the country, for such an attitude runs counter to the whole Catholic concept of life. The existence of the attitude is, nevertheless, unquestionable. Indeed, I found no disposition to deny it. On

163

the contrary, for many this general mediocrity is a virtue, and for others a fact of life.

"Slipshod performance is universal," a priest commented, "and still you can't say that it is inherent in our character. The Irish make excellent workers abroad. My own belief is that the climate is mainly to blame."

"On the contrary," interjected a journalist friend, "this is the way God meant people to live. We in this country have found a human tempo of living. We're doing fine."

This argument—or sophistry—is frequent. It shows up, for example, in the tourist publicity, which presents the happy-go-lucky attitudes of the native as one of the country's picturesque features.

Be that as it may, optimum production is impossible with such an attitude. There is no use, for example, having modern telephone equipment, if a switchboard operator tells you there is no reply rather than make the effort to complete a call, or if the long-distance service in a major city takes ten minutes to look up a number in the London directory. I had both these experiences, and all to whom I recounted them agreed they were normal.

Similarly, there is no hope for the new industries, no matter how excellent their equipment, if the moral atmosphere considers it neither shameful nor sinful to deliver such inferior work as it is possible to get away with. And unfortunately, the monopolistic system devised—unavoidably, often—to encourage nascent industry makes it possible to get away for a time with very inferior work and haphazard schedules. However, even the worm turns. Such practices have contributed to the high mortality among new industries.

Thinking people are unhappy and frustrated with the situation. They blame the leaders for failing to devise ways of rewarding performance. Now, as I have suggested, I don't think this criticism goes deep enough, for the problem is more moral than technical. Nevertheless, it has validity. Society functions best when individual advantage and moral obligation run together. In Ireland, performance is seldom adequately rewarded.

The property structure, not only on the land, but also in commerce and industry, erects a barrier between employer and employe. The best guarantee of advancement is membership in the boss's family. The likelihood of raising oneself by honest merit out of the rank and file is so remote that the intelligent employe—and intelligence is plentiful—knows that only a fool would try.

"Longevity," as a cynic remarked, "is the country's greatest curse." Nobody is willing to relinquish authority while a breath remains in his body. Even the son slated to succeed to his father's enterprise is an old man before he can make an independent decision. By then, any enthusiasm or ideas he might once have possessed are atrophied. His concern is to enjoy his tardy power, not to risk it.

Politics suffers a like fate. Here, the men who fought in 1916 and in the War of Independence remain unchallenged in 1959. In the interval, they have concentrated all power in the central Government in Dublin and given to permanent civil servants a dictatorship over the lives and decisions of the people without parallel in the free world outside France. The county councils, which played so big a part in the struggle for independence, demonstrating at that time the possibilities for self-government existing in the country, have been made rubber stamps. The police, education, the law courts, public assistance, hospitals, transport, all are under central control. The citizen is a child who could not be depended upon to make any decisions in matters affecting the community in which he lives.

It would be strange if these trends and attitudes were not also reflected in the Church, the transcendental importance of which in Irish life is universally recognized. And when I say the Church, I am not thinking exclusively of the institutional framework of Catholicism, though obviously this is the major influence, especially in the Republic, where Catholics are 93 or more per cent of the total population. The statement may shock some, yet I am convinced that the moral values and social attitudes of the Irish Catholic and the Irish Protestant are all but indistinguishable. If only they could both see this plain fact, they would be two-thirds way toward ending Partition, and a long way, too, toward mending the rift in Christ's seamless garment.

The Church, taken in this broad sense as a social institution, offers, in the opinion of many, the most acute expression of the tendency to treat the general public as too immature to make its own decisions.

TOO MUCH BANNING

Censorship provides a good example. Here Church and State work closely together in censoring both books and films. The wording of the book-censorship legislation ("in general tendency indecent") parallels and was no doubt inspired by the Canon Law (*"ex professo obscena tractantes"*), but the application runs wildly beyond the rules laid down by the most rigid canonist. Such absurdities have resulted as the banning of a book carrying the Nihil Obstat of a Catholic archbishop and the condemnation of books as being "in general tendency indecent" because of a single objectionable incident, perhaps only a single phrase.

Censorship is also performed administratively by the Customs outside the framework of the Censorship of Publications Act. Such discretionary authority might be justified against pornography, but it is in fact used to seize shipments from reputable publishers to their dealers, and they have in practice no redress. A recent victim was Joyce Cary's novel *The Captive and the Free.* (Another Cary novel, *The Horse's Mouth,* is banned.)

Unofficial but real censorship exists at a third level, that of the public library. Twenty years ago, when I lived in Ireland, a librarian friend (now with God) used to select for me books she considered good and important. These were books withdrawn from circulation on the basis of instructions with which she disagreed but had to follow. Yet she was an exemplary and indeed militant Catholic, thoroughly qualified to make her own decisions. And the situation she endured has not improved.

The stress on book censorship would be less extraordinary if there was any substantial book-reading public. A recent survey of London publishers, the main suppliers of the Irish market, found them in agreement that 750 copies of a new hard-cover novel was the absolute top "best-seller" limit. In nonfiction, only the life of a saint or other Church personality, or a book of recent Irish political history, might exceed this figure, going as high as 4,000. The exception for recent Irish political history is not surprising, for politics remains an obsession, while social and economic thought is neglected. The other exception is curious. It suggests that the clergy attend to their spiritual reading, and no other.

Film censorship produces equal absurdities. For example, you can't have a bum tell a priest on the screen to go to hell. The phrase is cut out, even though it means that now the priest, without any provocation, knocks the bum into the gutter with a left to the jaw. One is left under the impression that it's quite proper for a priest to toss bums in the gutter, where presumably they should have been in the first instance. And, of course, you can't show a priest dressed in lay attire watching the floor show at a night club. So this scene is cut, turning what was a priest's pastoral concern for a parishioner's slightly errant daughter into a middle-aged cleric's infatuation for a starlet.

I apologize for devoting so much space to censorship, because I don't think it really makes too much difference to the average Irishman if the censors ban Kate O'Brien or cut a slice of tripe out of *Say One For Me*. But I do think that the attitude behind the censorship is important. It is the work of people who equate sanctity with the absence of sin, of a society which lacks any great purpose. It bespeaks a defensive mood, an anxiety to hold on and to conserve, not to push forward and to create. Here we are face to face with people who regard themselves as guardians of a completed society, one which has achieved its purposes and has no further great business to transact.

PATERNALISM

The civil servant, as I said, thinks of himself as guardian of a public incapable of pursuing its own welfare. But he in turn has his problems. On the plane from New York to Shannon, I spent half the night chatting with an Irish pastor returning from his first trip to the United States. Incidentally, his observations on his visit had confirmed his previous belief that the Catholic

167

Church in the United States has nothing to give the Catholic Church in Ireland but money. I mention him not for this, however, but for his exemplification of a trait frequently encountered in the clergy, a naive belief in his own omnisapience and in his absolute right and duty to override all contrary opinions. He had built several churches, he told me, without ever employing an architect (something to remember, when one is looking at Irish churches), and he always rejected Education Department school-building plans and substituted his own. The civil servants, knowing what's good for them, go along. After all, it's only public money.

The Irish, however, are not fools. They have traditionally a tremendous respect and affection for their priests, who were their only leaders and protectors during centuries of defeat. And from that historic experience they have learned great patience. But their patience is being exhausted. I was horrified by the bitter criticism of the clergy I heard everywhere I went. This is something quite different from old-time European anticlericalism, which throve mainly among nominal Catholics. Here the feeling is strongest among devout people, who remain intensely faithful to practice. And their criticism is less of the action than of the inaction of their spiritual leaders.

Ironically enough, one of the things that seems to have served to focus the dissatisfaction is a social action movement conceived and promoted by a remarkable Irish priest, the late Canon Hayes. *Muintir na Tire* (People of the Land) was designed to rejuvenate the dying countryside. The Canon effected a social and economic transformation in his own parish through the application of principles of extension education, self-help and cooperation. Others followed his urging and example, and the movement grew to national dimensions and achieved similar results wherever priests with his zeal and intelligence were willing to encourage without dominating.

And there's the rub. One wanted nothing to do with this new-fangled nonsense. Another made all the decisions. Yet another sought to channel the enthusiasm into an additional fund-raising organization for his own parish purposes. It is a pattern frequent in Ireland, where the tendency is always strong to subject the social and civic to the ecclesiastical—in effect a denial that the State no less than the Church is a perfect society seeking its own ends by its own independent means.

I should like to end on a note of optimism, but the truth is that I feel far from optimistic. The mood of the country is static and protective when the challenge calls for change and action. The danger, as always in such circumstances, is that youthful enthusiasm—lacking constructive leadership—will follow false prophets. The persistent attraction of the banned Irish Republican Army, pursuing an outmoded policy of force in keeping with the over-all arrested emotional development of the country, demonstrates alike the altruism of youth and the absence of legitimate objectives. It should serve as a warning to those who think of the present vacuum of national purpose as normality.

That American Way

Walter J. Ong

WITH THE CRISES on the international front to-day and the obvious fact that we Americans are not the most beloved people in the world, the importance of our attitude toward other nations should be brought home to us more forcefully than ever before. Is the lack of total success of our foreign policy connected with some deep-set attitude in the American psyche? What is wrong with the "American attitude" toward those of other lands?

One possible answer is, nothing at all. Other countries are inhabited by goldbrickers, who are very happy to take our money and, for compensation, to dislike us as having more than they. We are the most generous people in the world, doling out cash, credit and supplies to every other country out of the sheer goodness of our hearts without hope of any profit or reward. And what reward do we get? None. Absolutely none. It is not just. Who ever heard of anyone's being so good and unselfish as we are, looking for no reward, and then, to cap everything, getting no reward after all, not even that of being the most highly esteemed of all people on earth—a reward which would be little enough, to tell the truth, since we so richly deserve it? No wonder our blood curdles with indignation.

We have all encountered this type of reaction in others, and probably also in our own hearts. Its presence among us more than hints that there may be something wrong with our attitude toward peoples of other countries. And what is wrong with our attitude seems to be in great part a consequence of our country's history. Our own great American achievement has somehow become a positive psychological handicap. The United States has been a vast and successfully working machine for converting into ourselves persons from every nation of the world. We have met the entire human race (provided they came to the United States) and have found ourselves able to deal with them successfully. We can make anyone over into ourselves. Unfortunately, the conversion process has so far proved a one-way operation. We cannot make ourselves over, even imaginatively, into other people. As a nation and, for the most part, as individuals we have never been trained to deal with persons of other nations and cultures on their own terms.

The situation would not be so bad if we had assimilated only one or two nationalities. But our assimilation of everybody gives us illusions of grandeur. Knowing that we and all the rest of the human race can be "adjusted" to one another, we forget that the adjustment is successful, as far as we are concerned, only if all the other persons are subject to adjustment. We assume that we ourselves are born automatically "adjusted" to everybody everywhere.

THEY TALK OUR LANGUAGE

I recall a conversation a few years ago with an American businessman who stated with glowing self-satisfaction: "Whenever I do business with foreigners, I insist that they write me in English. It makes for better understanding all the way around." "Half-way around," I wanted to emend. I thought of German businessmen I had heard of who would hire Swahili-speaking clerks just to write the Bantus in their own language. But for

171

my American friend, understanding was a one-way street; if the other fellow understood *your* language, that meant that *you* were adjusted to *him* and understood him thoroughly. There is a dangerous and insulting innocence in this state of mind. Curiously enough —or disastrously enough—my American businessman friend was a genuinely devoted Catholic layman, particularly interested in international relations within the Church.

To dislike foreigners as foreigners is of course a common human failing. All nations suffer from it. Other English-speaking peoples do not particularly like foreigners. The British do not. Their insular aversion to those not like themselves, perhaps matched or surpassed only by that of the Irish, was celebrated a few years ago with infectious humor by George Mikes, who, in the role of a displaced person in Great Britain, produced the well-known book *How to Be an Alien.* Nevertheless, though foreigners feel foreign in Great Britain as elsewhere, it is no great surprise to an Englishman that a foreigner does not speak English. To an American in the United States it is a little more of a shock.

Our attitude toward the language is quite different from that of the English. The Englishman feels that the language is his and is inclined to want to keep it for himself. The American feels that it is his but should not be kept to himself. On the contrary, it ought to be enforced on everybody whether everybody likes it or not. Typically, he does not even state this feeling of his. He takes it for granted. Imposed for generations on masses of non-English-speaking immigrants, American English is a reassuring badge of conformity which —unless he watches himself—the American simply assumes the rest of the world should be made to wear.

The conviction underlying this linguistic attitude is simple: it is that everyone else in the world wants to be an American. I have some European friends who can wax amusedly eloquent at this bumptious assumption, and others who are fiercely indignant when it obtrudes on them, as it all too frequently does, from

American tourists and visitors. Again our own history has confused us. In our personal lives, or in our schoolbooks and other reading, we are familiar with non-Americans of all sorts who have come to America and want to be Americans. Our schoolbooks and folklore are full of the America-refuge-of-all-the-downtrodden-and-woebegone theme. We have inscribed this theme on the Statue of Liberty—"Give me your tired, your poor . . . wretched refuse . . ." Some present American descendants of the same "refuse" are now agitating to have the inscription changed, but most of us still relish its thought. Intoxicated with a heady appreciation of our own charity, we forget that the non-Americans—refuse or not—who did not come to America, very often did not come because, all things considered, they did not think America was worth it. When you are visiting a foreign country, you may as well face the fact that you are visiting the descendants of those who did not want to be Americans.

LOYAL NON-AMERICANS

This is truly a shattering thought for unthinking Americans. It is very difficult for an American to believe that anyone could not particularly care to be an American, and all but impossible for him to believe that anyone could coldbloodedly want not to be one. The inscription on the Statue of Liberty, and its many equivalents in all that we learn, consciously and subconsciously, at home and in school, keep our collective ego bright and shining by sheltering us from the thought of such repulsive possibilities. They condition us to the persuasion—which is something less than self-evident—that all men regard the United States as the most desirable country on earth. We are conditioned to this with such thoroughness and profundity that I am sure that by this point many readers are bristling with indignation and that some have already broken off diplomatic relations with me and are now writing to the editor.

Still, some roots stand more or less clearly exposed where even commonplace analysis can reach them. This

173

conviction that all the world wants to be American is obviously a strong defense reaction to the guilt feelings of our ancestors, which our culture subtly, but inevitably, conveys to us. We are all descended from ancestors who, if they were not violently torn from their homes as slaves or as prisoners, political or otherwise, came here, most of them if not all, by leaving behind persons to whom they were bound by affection and obligations of all sorts. Our highly commercialized folklore likes to dwell on more consoling things, making heroes of those who left but not of those who stayed. But the facts of our history which lie behind us tell a more complex story. Tearing away from family and other ties may well have been the most sensible move, but it is not accomplished without psychological hardship. The emigrant is haunted by the fact that he has abandoned others to their fate. For he has, after all, run away from home.

In such an event, to save face it was imperative that America eventually be thought of as the land of opportunity, whether it was originally so thought of or not. For the negative element in turning away from home had to be countered by building up all possible positive reasons for the abandonment. What was behind one had to be forgotten. It occasioned too much pain. One had to talk of opportunity in a new future. Much has been written about the consequent tendency of Americans to build their families around the children, who look into the future and do not know the past, even to the extent that after one generation they no longer speak the language of their fathers. Second-generation Americans become ashamed of the foreign ways of their parents—which is to say that, just as these parents felt guilty about leaving their own parents, so their own children feel guilty about them, in a different but related way. Now that the third generation is learning from sociological studies that their own parents, born in this country, felt ashamed of their parents for not having been born in this country, the third generation becomes ashamed of its parents for having felt ashamed of theirs. The American family itself, oriented through

174

the children into the future, thus becomes an instrument for smothering the past in vague and confused,
but terribly real, feelings of shame. Henry James' novels
show that such feelings are not restricted to the descendants of recent immigrant groups but go back to
the colonial beginnings of America.

Just how real this situation is was borne in on me a
few months ago when a recent immigrant to the United
States from Cen-
tral Europe, who
had just become an
American citizen,
wrote an article for
a national maga-
zine explaining his
reasons for coming
to the United
States and his reac-
tions to his new
home. In the article
he mentioned that
people in the
United States like
to think of our
country as the land
of opportunity, but

that, try as he might, he could find no such thinking
in his own decision to come here. His impulse was not
at all framed in the idea of seeking "new opportunity."
It was far more elemental. He simply wanted desperately to "get away."

Shortly after he had published this article, having the
good fortune to meet him, I mentioned my fascination
with this particular point in his account of himself.
"Yes," he said pensively, "and I had much more in the
article on the same point, but the editors suppressed
it." Did they suppress it because they did not want to
believe it? or because they did not think their readers
wanted to hear it? or perhaps for both reasons and
others, too? The Statue of Liberty approach is much
more comforting—until you get to thinking about it

too much. Strangely, then it makes you, of all things, ashamed of your ancestors again.

Are we so ashamed of immigrant ancestors, however remote, that we are repelled at the idea of further immigration into the United States? Since the recent Hungarian revolt against communism, the United States has had the lowest record of all major Western powers in the receiving of refugees. Both in terms of our total national income and in terms of our total population, we have taken in a far smaller proportion of these refugees than the war-torn countries of Germany, France and England.

Connected with our difficulties in adjusting to other persons is our extreme sensitivity to criticism of the United States by others. Again, such sensitivity to criticism of one's homeland is natural to all men. Citizens of other countries are hurt and angry when their country is the butt of criticism. But they are ordinarily not greatly surprised. The American is surprised, and his hurt and anger the greater for this. He cannot believe that there are people who would not like the United States. Otherwise, how could it be that everyone in the world wants to be an American?

AS OTHERS SEE US

I recall the Dutch priest who was a very good friend of mine and was visiting in this country. He had shared with an American priest who had lived in England the fun of *How to Be an Alien*. The American priest wanted to read anything else which George Mikes had written. With some misgivings, his Dutch friend was persuaded to lend him another book by the same man, *How to Scrape Skies: The United States Explored, Rediscovered and Explained*. The humor of this second book was lost in startled and hurt resentment on the part of the American. He could not understand how an author who poked fun at England so cleverly could so crudely and stupidly misrepresent American ways.

The American's understanding of other countries is complicated by the position of America in the technological civilization which is rapidly becoming the civilization of the entire human race. As man everywhere moves toward greater and greater technological developments, the United States is, for certain historical reasons, in the vanguard. (This judgment is made from the over-all viewpoint; in some details, of course, the United States lags behind.) Seeing others moving toward the same goal as himself, but somewhat behind him, the American is likely to believe that they have their eyes on him and his achievements rather than on the goal. This is a human misinterpretation, comforting to the ego, but hardly one which endears the American to a European, who is, also quite humanly, annoyed by the fact that the American has, by quitting home, in some matters gotten ahead of the civilization which bore him and without which he would be an impossibility.

A related complication is the superior wealth of the United States, not all of which is the product of America herself, much coming from other countries in which Americans have played the complicated role of both benefactors and exploiters—and, being human, chiefly the latter. Americans console themselves commonly here by representing other peoples as venal, panting after American wealth. They will do anything for American money, the tourist and the taxpayer thinking of foreign loans both tell themselves. This argument is a two-edged sword. Turned over, it says that the only way Americans have of getting along with other persons is money

The truth is often that Americans do attract the venal elements in other civilizations because we are venal ourselves, making up for our linguistic and other deficiencies by throwing our economic weight around. It is certain at least that we have become symbols of venality all over the face of the earth. Here are some excerpts from a recent language examination for advanced elementary education in the schools of the Netherlands: "Mr. Coppergold, a rich American, had visited several countries in western Europe. . . . And everywhere they spoke English to him, especially when he threw his dollars around." Dutch elementary school children can be expected to find translation of this passage in a foreign language not too hard. They know the elements which make up the rich American.

Monetary considerations are often operative in what has been styled the American's European neurosis. I recall the incident when a group of Americans on a tour in Italy were treated to a spectacle of a compatriot of theirs, a middle-aged woman, who without warning dramatically rose from her seat in the tourist bus and, for the benefit of all the passengers, asked the guide in a loud voice why he did not point out what was the truth, that all the rebuilding they saw in the war-torn countryside was due entirely to the kindness of Americans and to the money which "we" were sending, without any recompense, to Europe. Unfortunately, all the bus passengers were not Americans. There were a great many Italians—perhaps some whose sons' bodies were still lying under the rubble to which the American tourist so gallantly waved.

The problems of Americans in adjusting, or even in wanting to adjust, to ways of life other than their own not merely unnerve tourists but also affect our foreign policy in a myriad of ways. There are the American technical teams in foreign countries who, observers report, are, not all of them, but most of them, definitely less loved by the people around them and less in contact with these people than the corresponding Soviet teams. There is our utterly miserable record in providing diplomatic staffs able to handle the language of the country to which they are sent.

When this serious state of linguistic affairs was reported recently to President Eisenhower, his answer was that money should be raised to train suitable men. The answer certainly caused many a smile in foreign lands. There are your Americans again: more money will cure all human ills. Of course, what is needed in human relations is not just money but a certain quality of understanding and love. Money will not create in Americans the desire to learn foreign languages, which is what we so desperately need. Generally speaking, the American abroad who wants to learn the language of the people among whom he lives is the American capable of understanding and adjustment. The attitude toward the language is likely to be the attitude toward the civilization. Perhaps for this reason, in my own experience, it is Americans from academic circles, teachers and students, who do best in adjusting themselves to foreign living.

The American's difficulty in accommodating himself to others is becoming a great matter for concern in the Church as American Catholics are called on more and more to bear their share of the burdens in manning the Church's supranational offices and to share her supranational concerns. The American clergy and religious suffer of course from the same psychological disabilities as other Americans, and the results are there for everyone to see. An old-time missionary complains that newly arrived Americans take for granted that the local people will do everything the way it is done in America —or at least they *must* want to do it that way. A religious superior in Europe writes to an American religious superior with obvious concern over the fact that in the international house of which he has charge, clergy from all over the world live happily under the local regime—except the Americans, who need all sorts of preferential treatment. What can he do? He begs that some adaptable Americans be sent to the house.

It is notorious that in the Church's curia at Rome, Americans are woefully underrepresented. It is hard to get them to Rome or, once there, to keep them. We shall send money instead, and pray to be excused in our persons. The reasons for begging to be excused are by no means all invalid. Even when they wish desperately to serve, Americans are often incapacitated to do so by the fact that they are Americans. Their psychological difficulties are real and give rise even to physical disabilities.

In the present crisis of adjusting ourselves to other men, what we need to do, first of all, is to think. For it is the unthinking "innocence" of Americans vis-à-vis the rest of the world, so profoundly caught in the novels of Henry James, which dogs us, to the detriment of ourselves and others, still. Our thoughtlessness is caught in phrases such as "un-American activity" (imagine "un-British activity," "un-Irish activity," "un-Venezuelan activity," and how campaigns dealing in such slogans would impress us). It is caught in our assumptions that what we do is never chauvinistic or nationalistic, though what others do may well be. Thus, for British missionaries to teach cricket or Canadian missionaries to teach lacrosse would be chauvinistic, but for American missionaries to teach baseball is not spreading American culture but merely enabling the benighted natives to be human beings.

Our thoughtlessness is perhaps most profound in our attitude toward citizenship. For a citizen of any other country to become a citizen of the United States is admirable and a compliment to the country of his origin, which should be proud to have given birth to a potential American. For an American citizen to become a citizen of another country is quite a different thing. It is treasonable, an affront to the United States, and shows a dastardly streak which will bring no good to the adopted land. I have talked to persons who were led to suppose all through their schooling that T. S. Eliot was a treacherous and suspect character simply for having taken up residence and citizenship in England.

Such unthinking attitudes sound fantastic when they are actually formulated, but I think most of us Americans will recognize them as real, as being rooted in our hearts and not easy to eradicate, even when they have been brought to the light by explicit formulation and statement. All the picture is not black, of course. There are the bright spots of which we do not ordinarily need to be reminded and which I have consequently not shown here. There are hundreds of Americans in foreign parts who adjust thoroughly and gracefully and with love to those around them. These are the cultural ambassadors of whom we can well be proud, and the response to their behavior on the part of those of other countries is gratifying beyond expectation. For most persons do wish to like Americans. We are, after all, not more repulsive than other people in the world, provided we try not to be.

But we have to try, not only in ways we like, but in ways others like. We must work hard for acceptance by others if we want to be loved. We cannot assume that we are accepted, and scream when we are not, as the immature do. We must understand that while our psychological h e r i - tage breeds in us an overpowering desire for acceptance, to se- cure acceptance in foreign lands we need more than wishful w a r m - heartedness and candy bars for the children. We need imagination and hu- man sympathy f o r those unlike us. And a b o v e all—a hard thing for us Ameri- cans—we have great need to understand ourselves.

Back in 1948 President Truman felt that it was high time for the churches to man the parapets of civilization against the menace of godlessness. But he failed in his efforts to have the Christian forces in the West make common cause for the Kingdom of Christ against the Kingdom of Satan. Such is the scandal of Christian disunity in the face of the monolithic solidarity of communism. Ecumenism is still in its unstable infancy and has a rugged road ahead. Catholics like to lay the blame for division upon our separated brethren. Too often, however, the fault lies in our own stars. Too many Catholics still lurk in the ghetto, fearing to be "engaged," not daring to become "involved." God grant that the Second Vatican Council may show us how to open the path to Christian unity before it is too late.

Christian Unity

In a Catholic Ghetto?

Robert I. Gannon

"GHETTO" has been a fighting word for centuries—like "Inquisition." It refers to the formerly restricted parts of certain European cities where, by law, all Jews were compelled to live. Some insisted that as an institution it was well meant, but, no matter what could be said of it as assuring the safety of the Jews and promoting their solidarity, it was recognized as a symbol of Gentile aversion and fear.

In the last few years, the word has become almost as popular in some Catholic circles as "dialogue" and "pluralistic." Every time a good long convention is held where lots and lots of papers are read, someone is sure to condemn "the ghetto mentality," which is said to have resulted among us from the fact that we were so long an ostracized minority. They point out that in Colonial days we were like the *gens lucifuga* of Penal England—the "people who shunned the light"—and although there was a breather after the American Revolution, the Nativist movement, following the heavy immigration of underprivileged Catholics, kept us on the defense for generations, and forced us to live in a more or less restricted circle. Now we are taken to task for continuing a situation which we did not create—a situation which has become a symbol of Protestant fear to some and of Catholic solidarity to others. "Ghetto," therefore, is today not only a fighting word among us, but a very confusing one as well.

There is, first, a false sense in which the word is used by apologetic Catholics. These are the brethren, with us almost from Apostolic times, who shrink from identification with a Church that prides itself on being the Church of the poor and the humble. They would like to remove their identification tags, the last differences that set them apart from the non-Catholic world around them. To these people the parish is a social ghetto. They resent getting permission for anything from a pastor, pride themselves on the fact that they have never met any of the priests, and, of course, would not be caught dead at a meeting of the Bona Mors. The reading of Catholic books and periodicals creates, they say, a mental ghetto. But it takes the Catholic schools to combine the social ghetto with the mental. When children are confined in that sort of atmosphere too long, they will believe for the rest of their lives that there is only one true faith and that the Pope is the Vicar of Christ. They may even want to go on to a Catholic college, and eventually to marry their own.

This sort of thinking is encouraged by a wave of exaggerated self-criticism that is spreading from our uninhibited juniors to the ranks of our neglected intellectuals. It is considered the healthy thing to belittle accomplishment, to ignore the handicaps which have been so nobly overcome and to shout our defects from a sound truck in Times Square. The whole world must hear that there is not one single St. John Chrysostom to be heard in St. Patrick's Cathedral, not a single Oppenheimer among Notre Dame's alumni, not even a Rodgers or a Hammerstein, and the reason is that American Catholics are living in a ghetto.

It may be embarrassing to be caught agreeing with this type of coreligionist, but wrongheaded as they are, there is something in what they say. We are in a sort of ghetto, but not the sort that they like to talk about. Our parishes are not ghettos. They are the extension of our homes, and being identified with them is like fostering the family spirit. Our Catholic press is not a mental ghetto designed for isolation. It is a constant reminder that ours is an eternal faith that is eternally adapting itself to a changing world, and this makes us all the more eager to study contemporary society. But no one with a day's work to do can keep up with the modern problems of the Church unless he learns what the best-informed Catholics are thinking about them. As for our schools, far from being ghettos, they are lighthouses in these tempestuous times, and the Protestants and the Jews who attend them appreciate what they mean for the future of America. Like the parishes and the press, they are symbols, not of fear, but of Catholic solidarity.

WHAT OUR GHETTO IS

What, then, is the ghetto that too many of us *do* cultivate? It is the ghetto of social aloofness. Too many of us are living in the past, nursing slights of another generation, aloof from our fellow Americans, where no aloofness is called for.

It is true that not too many years ago Catholics were excluded from ordinary neighborhood activities and snubbed by an intolerant majority. Unless these new arrivals would give up the Mass, they were not welcome on any committees that worked for the community as a whole. Today, however, in the urbanized United States this attitude is exceptional. A new generation of Protestants has matured, and the aspect of their maturity that interests us is the fact that they no longer believe what their fathers believed about the Roman Catholic Church.

A professor of the University of Virginia once told, at the end of a faculty forum, how his father had warned him as late as 1910 that Catholics were dangerous only when they started to be friendly, because every Catholic was pledged to make friends with one Protestant and kill him. Tales like that have retreated to the hills, and ordinary Protestants can joke about them now. They still find many of our practices and assumptions irritating, and they are not completely reassured about the future; but it is worth noting that the uneasiness of some admirable ministers of the gospel in the present campaign reflects their reading not of *The Awful Disclosures of Maria Monk* but of commentaries on what Leo XIII wrote in his *Immortale Dei* (1885) about the Christian constitution of states. That indicates considerable progress. The professors of philosophy and religion at Colgate University last spring were troubled about this same encyclical, the only one they were studying in their classes, but they were eager to get to the bottom of the misunderstandings.

The deep South looks different from up here in New York and some of the blasts that are coming from the Texas Baptists sound like the roaring 1840's, but even in Dallas itself a Jesuit was asked only last year to open the drive for the Community Chest. He spoke at a traditional luncheon that had formerly been restricted to Protestant ministers and their congregations, and afterwards he was congratulated in the good old Protestant way. The line formed to the right, and one after another each said to him: "Thank you, Reverend, for your message!" Later he spoke to 14 old-fashioned preachers who sat in the front row of a series of meetings in Tuscaloosa, Alabama, and had the leader, a venerable Methodist, say at a little reception afterwards: "Reverend, I had no idea that a Roman Catholic priest could preach such a good Methodist sermon." Needless to say, it was not a Methodist sermon. It was a Christian sermon on the matter found in the first two pages of the catechism, and it worked. It stressed the fact that there was much common ground to be shared by all of us, and its selection was in line with the conciliation that has characterized the reign of "Good Pope John."

This conciliation involves no compromise in any essential, still less any risk of indifferentism. The dictionary says that conciliation is "the gaining of good will by pleasing acts." Madison Avenue might call it public relations, but the Pope would call it in the present circumstances spreading the charity of Christ by imitating the amiability of Christ. See how the Holy Father has gone out of his way to change the wording in several passages of the liturgy, not because they were untrue, but because they were unnecessarily harsh and offensive to people who would like to be friendly.

Our minimum program, then, in removing the bitterness that wells to the surface of our national life when the spirit of politics gets out of hand, is to make sure that intelligent Americans of good will, Protestant, Jewish and whatever, know us for what we are. If we are real Catholics we are worth knowing; and our character answers more objections than any sermon would. And yet, out of habit we keep on being clannish; we cling to our ghetto. We still hesitate to join non-Catholics in social, charitable and recreational movements that are city-wide and even national. What leadership do we give the Red Cross, for example, or the Boy Scouts, or the public library, the art galleries, symphony orchestras and movements for better parks? And yet, what is more conducive to seeing God in my neighbor and having my neighbor see God in me, than to work with my neighbor when we are both at our best working for others?

What public-spirited people do for the community is good, but the fact that they do it together is even better. By working shoulder to shoulder for the public weal, men of every religion and no religion can build up what we need more than ever in the gathering chaos—American unity.

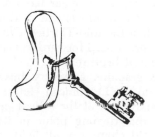

Perspectives for the Council

Yves M.-J. Congar, O.P.

P ERE CONGAR, distinguished European theologian, is
a consultor to the Theological Commission of the
coming Ecumenical Council. Among his many
significant books, *Lay People in the Church* is perhaps
best known to the English-speaking world as an outstanding contribution to the theology of the laity. For a
quarter of a century Fr. Congar has been in the fore-
front of the ecumenical movement in European circles.
His pioneer work of the mid-Thirties, *Chrétiens désunis*,
is still cited as a great leap forward in Catholic thinking
about the separated brethren.

Recently this gracious and very busy man invited
the interviewer to the Dominican House in Strasbourg,
France, to discuss some of the salient problems facing
the Church at this turning point in its history. The
astoundingly rapid change in all fields today compels

us to reconsider the role of "the little Church in a vast world." The quest for the most effective ways of witnessing to the truth that is in us forms the supreme challenge for all Catholics who are living in this new age a-borning.

* *

Q. *You have mentioned in your writings that the coming Ecumenical Council should rethink the role of bishops in the Church. Would you like to develop this idea?*

A. I think many bishops desire that the Roman Curia be more international, with better representation in Rome for the various national churches. One could imagine a number of ways of achieving this, for example, enlargement of the College of Cardinals, or perhaps the setting up of a new permanent commission.

Even more important is the question of the theology of the episcopal college. You are aware that, from a theological point of view, the bishops are successors of the apostles, not in the sense that each bishop succeeds a determined apostle, but in the sense that the college of bishops as such succeeds the college of apostles as such. This is important, because the apostles had a universal jurisdiction; thus, the college of bishops has in itself a universal jurisdiction in the Church.

Each individual bishop is limited *de facto* in his authority to a determined territory, but as a member of the episcopal college he possesses a certain power and obligation in regard to the whole Church. This power and duty to the whole Church is exercised mainly in an Ecumenical Council. The Council is the perfect realization of the episcopal college. When the bishops are dispersed in their own dioceses, each bishop individually is not infallible. But when the bishops universally preach a doctrine, it becomes a matter of faith, a matter of infallible teaching. This was the chief argument of Pius XII for declaring the dogma of the Assumption.

Furthermore, each bishop in his individual diocese should show an interest in the universal Church. He shows this interest, first, in the administration of his own diocese.

A diocese is only part of the whole Church, but it carries within itself the nature of the whole Church. Thus, for a diocese to be truly catholic, its bishop must govern it not as an independent unit, but rather as a portion of the universal Church. This implies that the bishop actualize in his diocese all the great causes of the universal Church.

Moreover, a bishop should interest himself in questions that are distant from his proper domain. Pius XII strongly underlined this in his encyclical *Fidei Donum*, in 1957, when he said that bishops are missionaries for the whole world. You have a striking example of that in Louvain, where you live. Cardinal Van Roey and the Belgian bishops constructed there a seminary for Latin America. The Belgian bishops felt that they had a certain charge of South America, where, as you know, the pastoral needs are very great. Of course, the Belgian bishops will not intervene in the diocesan affairs of Brazil or Chile, but they are pastorally interested in these dioceses, as the creation of this seminary signifies.

Q. *What attitudes do you think the Council will take regarding the mission countries?*

A. I feel that the Church of the 20th century has already begun to have a completely new outlook on the missions. This is to be understood in two ways.

First, the Church today understands better that it is missionary by nature, missionary in all the aspects of its life. There is no role or position in the Catholic Church that is not missionary. This should be strongly marked in the Council. And when I say that the Church is missionary in all its members, I am thinking especially of the laity.

Recent pontifical documents and the recent conventions of laymen in Rome have all expressed this thought.

Second, the Church has had to revise its attitude toward the missions because of the gradual collapse of colonialism. The 19th century was the most glorious missionary epoch of history, but it was also a century of colonialism. It was the century of the geographical discovery of the world, of Africa and Oceania in particular. Missionaries were sent on the heels of the explorers, and under the protection of Western governments. This protection seemed necessary at the time. But it linked the missions to the West, to colonial interests and to a certain form of imperialism. Today this is absolutely impossible. Wherever the missions are tied to Western powers, they are meeting serious obstacles. The missions remain alive and vigorous only in places where for at least a decade they have separated themselves from too much dependence on Western powers.

Today there are native hierarchies and native clergies. Of course, these indigenous churches have not been very active in theology, but I think they will bring to the Council their authentic problems. Instead of being represented at the Council by Western powers, as was the case in 1869, the mission countries will be represented by men who have firsthand knowledge and intense interest in the evangelization of their own lands. This should bring to the Council an extraordinary broadening of perspective in matters of canon law, Catholic Action, liturgy, and maybe even in the formulation of certain doctrines.

Q. *What position do you think the Council can take on the thorny question of non-Catholic groups evangelizing in previously Catholic areas?*

A. I do not believe that the Church can directly declare the right of non-Catholic Christians to evangelize in already Catholic areas. After all, the

Church must be honest with its own conscience; it must be loyal to its most profound beliefs. What it can do, and should do, I feel, is to make a firm declaration of tolerance, of respect for the religious liberties of other consciences.

In place of bitter rivalry, a disagreeable and very negative thing, the Church should seek an amicable understanding with the separated Christians, especially in missionary areas. Such understanding, I am happy to say, is coming more and more into evidence.

This question of tolerance is very important. The World Council of Churches in Geneva expressly requests that we take up the subject. In May we brought together 12 Catholic and 12 non-Catholic theologians (appointed by the WCC). This semi-official meeting was to discuss religious tolerance and freedom of conscience.

Today this problem of tolerance must be seen as forming an indivisible unit in the world. I mean that it is impossible to demand tolerance in one country and not practice it in another. The question cannot be divided like that. A few years ago the Catholic Church in a certain country desired that Protestant citizens support the Catholic stand for religious liberty in Hungary. The Protestants did not want to co-operate; they said that they suffered from injustices in Catholic countries— Spain and areas of Latin America. This was not overly generous on their part; it would have been better had they not raised that matter. But one can certainly see their point of view. The example shows that this question of religious liberty must be viewed on an undivided, world-wide basis.

Q. *You have said that a whole new chapter on the laity could be written today, that the Church must go beyond the canonical notion of the laity and see the place of laymen in the sacral order of the Church. Would you like to comment on this?*

A. Yes. When the word "layman" is used, it makes

some theologians think of a canonical distinction which says that a layman is neither a cleric nor a monk. Actually there is much more to it than that.

The laity would be very disappointed if the Council just says that laymen are neither clerics nor monks. They already know that. They want to know just what they are, and this in a positive way, not only as regards their rights, but also concerning their duties. The laymen are anxious to take on the responsibilities of the Church, but they ask to know just how intimately they belong to the Church and form the Church.

On these points we have reached a sort of consensus since 1950, thanks to two world congresses of the lay apostolate, to the discourses of the Pope, and to a number of books. I think it is quite possible today to determine the ecclesiological role of lay people in a positive and constructive way.

Of course, there are some delicate points, such as the obligations of married people and the role of spouses in the Mystical Body (which is not made up of individuals but of families). Marriage must be seen as a Christian state in and of the Church. Pius XII, you know, speaks of such a state in his encyclical *Mystici Corporis.*

The role of lay people in the liturgy constitutes a considerable problem. First, there is the question of a more active lay participation in the liturgy as it is today. Much good work has been done. We have just recently celebrated the Holy Week and paschal ceremonies. I found them very impressive when the lay people had been properly prepared. But, in my opinion, there is a much more fundamental problem facing our liturgy. I doubt whether the Council will be able to take it up, since it will demand much time, work and gradual experimentation. I refer to the problem of a less clerical or monastic liturgy than our present one.

Let me give an example. Take the "Exsultet" preface of the paschal vigil. As a cleric who knows

Latin, who has an ecclesiastical, patristic and monastic background, I find myself relatively at home in this prayer. I recognize a quantity of poetical and biblical and traditional allusions in the song. It says something to me; it profoundly touches my religious spirit. But I imagine that for laymen, who have none of this Latin, traditional, clerical formation, it must have much less meaning. Many of the images say nothing at all to them; their formation has in many cases been just the opposite of a Greco-Latin traditional schooling.

I really think that this is *the* liturgical problem for some time to come. The difficulties will first present themselves in the far-off mission lands, like India and Africa, which totally lack the Western tradition. It is evident that in France our whole culture has a Latin foundation; the problem is, therefore, somewhat less urgent here than on the missions. But this liturgical "gap" will become more and more pronounced even in Western Europe. It will be necessary to envisage forms of expression that are more accessible to the laity, if we do not want to see the liturgy restricted to a specially trained elite. Such a restricted liturgy would have no contact with the masses.

I am very much impressed, having followed closely the French liturgical movement, by the fact that the liturgy did not really become popular until the singing of psalms in the vernacular was introduced, thanks to the translation of the Bible of Jerusalem and to the melodies of Père Gelineau. The movement is encouraging, but it is not an ideal solution. It is rather a paraliturgical, or peripheral, solution. I regret that in our student Masses we do not sing the Mass itself but, instead, various psalms distributed throughout.

Q. *In your major work on the laity you explain how lay people participate in the priestly, prophetic and royal roles of the Church. Which of these functions do you think is the most important for the layman?*

A. I would hesitate to say that one of these roles is more important than another. I think that the priestly function of the layman is the most comprehensive, that is, if one understands it, as I did, in conformity with the text of the Epistle to the Romans: ". . . offer up your bodies as a living sacrifice, consecrated to God and worthy of his acceptance; this is the worship due from you as rational creatures" (12:1). All the life of a layman is included in this complete offering to God.

The prophetic role is also very important. This is what the apostle Peter underlines in his first letter: ". . . that you may declare the virtues of him who has called you out of darkness into his wonderful light" (2:9). It is the prophetic people of God, the Church, like Israel of old, giving witness before the world of the existence of God, of His greatness and the need for serving Him.

And concerning the interior life, the royal role is very important. It is the domination of one's self, the conquest of one's liberty. I think that we could formulate all of Christian morality in terms of the conquest of liberty. It would then appear as a baptismal and paschal morality, that is, the escaping from the slavery to sin to live in the spiritual liberty of the sons of God.

Q. *Would you like to say something about the relation between the priest's and the layman's roles in the Church's mission of evangelization?*

A. If we understand "evangelization" in the sense of preaching the gospel, it is clear that this role is exercised mainly by priests. Under certain circumstances the priesthood could perhaps suffice for this mission. One would have to find out, though, whether the priests could establish contact with all the parts of a given population, and whether these clerics could express themselves in a language understandable to certain groups. Here I would insist on the need for lay people to evangelize among the workers and especially in the proletarian milieu.

But there is a special mission of the Church which the priest simply cannot fulfill adequately, and which becomes the special domain of the layman. This task is the influencing of the temporal directly.

By the nature of his calling, the priest cannot place himself completely in the political and professional order. This area is one of competition and of rivalry; the priest must remain the man of God, the man of charity. The priest in the Latin Church has no family; usually he doesn't have a secular occupation, except in an accidental way as teacher or researcher, etc. Thus, it is the layman who actually carries the cause of the Church, the cause of the "consecration of the world" of which Pius XII spoke.

This consecration does not mean a ceremonial consecration of the world to the Sacred Heart, which can be a very external affair. This consecration consists in an orientation of the world toward God. It cannot be achieved unless the world is reformed according to God, unless we reform the sordid structures of money, of sex, of egoism toward structures of service, fraternity, justice, truth and love. That is what it means to remake the world according to God. It is the laymen's role to carry forward this task of the Church. The priest must second their efforts by forming them spiritually, by advising (and being advised by) them.

Q. *I noticed in one of your articles that you hoped the Council would refrain, in the interests of Church unity, from making any Marian declarations. Why do you feel this way?*

A. I think that if the Council made some of the Marian declarations that are talked about in various theological publications and congresses, it would constitute an almost definitive obstacle to the unity of Christians.

196

It is a fact that the definition of the Assumption has created a new obstacle that is very difficult to overcome not only for Protestants—that's too clear —but even for the Orthodox, who nevertheless admit the Assumption. From the very day that it was dogmatized unilaterally by Rome, the Orthodox said: Ah no, we can't go along with that; it's not the same thing.

If we were to make dogmas of subjects not yet mature in the conscience of the Church, such as the co-redemption or the universal mediation of Mary, I think it would create an almost insurmountable obstacle to unity. Fr. R. Leiber, as you know, was a close adviser of Pius XII. In his memoirs of Pius XII, he stated that the Pope considered these questions not yet mature in the conscience of the Church, not sufficiently clear to be the subjects of a dogmatic declaration. I would ask those who want to make a dogma out of the co-redemption of Mary to give me first a good theology of the redemption. I'm sure they would have to admit that it is not an easy thing to do.

Q. *There seems to be a desire among many Christians to see a greater simplicity and modernity in the Church's modes of expression and in the ways of her ecclesiastics. What is your thought on this?*

A. I think the Council must express itself in language that will be understood by the men of today, that is, in clear, pastoral, and nonacademic words which have a truly evangelical and religious tone. As far as I can see from my personal contacts, this is the formal desire of the bishops. They absolutely do not want to propose to the world a theological dissertation. They want to address a pastoral message. The Church must become more and more aware that it is speaking not only to a believing but also to an unbelieving world.

On the question of external forms of clerical life in the Church, yes, it would be well if we simplified our ways. But what is of much more importance, and I say this from personal experience, is that men be able to express themselves freely. Men are not happy unless they can express themselves. In fact they do express themselves. But where? In their family lives, their diversions, their work—not in church.

Now it seems to me that because of the pomp that surrounds and imprisons them, bishops and Cardinals are practically never in occasions and places where men freely express themselves. They are with men, yes, but in formal ceremonies where men do not express themselves, where they enter into a ready-made rite that is perhaps too solemn. These high dignitaries have a certain contact with men, but this contact is fenced around with protocol and the marks of respect. In these circumstances men are careful not to say too much.

Thus, members of the upper hierarchy generally encounter artificial rather than real situations. One would wish that they had contact with men in those domains in which the latter express themselves freely. How to achieve such contact practically, you ask. One suggestion would be that members of the upper hierarchy four times a year take a worker's train for half an hour during the rush period. Of course, they would have to present themselves in a way that would not obstruct the liberty of expression of the men around them.

Q. *You have been engaged in the ecumenical movement for many years, Father. What do you think are some of the best means for furthering Christian unity?*

A. One must first understand that the ecumenical movement is a process of long duration. Success cannot be achieved by a resourceful five-year plan, as in the economic order.

On both sides we need a profound change of attitude, a revision of many ideas—not only prejudices, but also certain formulations of doctrine that could be improved. This will take a lot of time.

This gradual movement toward unity is advanced by numerous means. The first step is to see that there is correct and adequate information about one another in our schools and press. Second, we must all convert our hearts and abandon egoistic ideas of self-righteousness, of possessiveness, of a sort of religious imperialism. In short, it means becoming truly Christian. One of the principal means for this is prayer.

As you know, prayer does not always obtain what we pray for, but it always makes our heart more true, more fraternal, more humble. A prayer in common with our separated brethren is particularly significant. The directive of 1949, *Ecclesia Catholica*, approves the reciting together of the Our Father or any other prayer approved by the Church. Thus, for example, Compline (Evening Prayer) could be recited together. What is forbidden, of course, is liturgical worship together: Mass, Holy Communion, etc. Also, Catholics cannot ordinarily participate in non-Catholic services. This should not be looked upon as a spirit of opposition, but rather as a fundamental honesty and respect for one's own conscience.

Common prayer within the limits that I have mentioned is a very important and profound means of furthering unity. If something has already been accomplished in the Western countries, it is in large measure due to prayer, especially during the unity octave in January. I have had the privilege of preaching this octave for the last 25 years, and I must say that it is during this time that I find the audiences in the most attentive mood.

There are many other ways of preparing the ground for unity: historical studies, informal as well as theological discussions, and, yes, even the practical suggestion of M. Oscar Cullman about mutual collections, which are a good means of disinterestedness. All this has a role to play; nothing must be neglected. It is a world-wide process that is gradually leading us to something that only God foreknows. What this will be, we must not try to learn too early.

What we must try to develop more fully as an essential condition for dialogue is confidence in our separated brothers. This implies that we treat "the others" as persons, as subjects and not as objects—objects of study, objects of sanctions, objects of repulsion, objects even of interest and of solicitude. This demands more than an exterior attitude, a certain form of politeness; it must come from within, from a conversion of heart. Yes, there is an intimate kinship between the dialogue and a confidence in others that treats them as persons.

Q. *Has the study of the ecumenical movement made much headway in European schools?*

A. It is a fact that ecumenism is becoming a subject of study in universities, seminaries and other institutes. I have drawn up a list of 17 universities which have one or more chairs dedicated to ecumenical problems. Last year I gave a course on the subject at the University of Strasbourg. As you know, the Holy See asks that a certain period of time be devoted to Orthodox theology in our seminaries; the same should certainly be done for the various forms of Protestantism. The important thing in this is that a few key ideas be developed well; it is not a question of covering much ground, but of delivering what is authentic.

In other words, one learns more in one hour of authentic conversation with a separated Christian than in years of artificial controversy. What counts is not to have at hand many facts and figures, but to have grasped the genius of the Orient and the inspiration of the Reformation. For this we must seek the authentic, whether that means inviting a minister into our seminaries from time to time or some other means. Surely we must cast about for new avenues of rapport if we are to be faithful to an inspiration for reunion that comes from the Holy Spirit.

Pope John XXIII:
Teacher

Philip S. Land, S.J.

A RECENT issue of AMERICA (9/30, p. 820) found it
necessary to return once more to the defense of
Pope John's encyclical on "Christianity and Social
Progress." The most recent occasion for this defense had
arisen in the form of a letter to AMERICA from one of
National Review's editors, Frank S. Meyer (AM. 9/30, p.
813).

Much more moderate in tone than *National Review*'s
now notorious editorial on the encyclical, Mr. Meyer's
letter no less firmly rejects the authority of the Pope as a
teacher on social, economic or political matters. Commenting on the Meyer letter, AMERICA deplores the fact
that, whereas we ought to be getting on with the work
of interpreting and applying *Mater et Magistra*, it is dis-
couragingly necessary to continue to defend the very
right of the Pope to be heard on questions which involve
both the temporal and the spiritual orders.

Why does AMERICA continue to stress this right of the
Pope to act as teacher? Are its editors excessively preoccupied with this problem?

It assuredly does not seem so to me. I say this after
three months of lecturing on and discussing the new

encyclical in U.S. cities from coast to coast. Repeatedly I met the challenge: "What right has the Pope to talk about social questions?" and "Why doesn't the Pope just stick to preaching what's in the Gospels?"

Such questions, needless to say, come from people who are disgruntled because the Pope has said things in *Mater et Magistra* that they don't like. Such questioners were—it will be no surprise to learn—all members of the so-called conservative Right. Many of them declared themselves to be avid readers of *National Review*, from which they said they draw their personal social creed.

This challenge to papal authority in social matters was not exactly new to me. I had already been familiar with it in Europe while engaged for seven years in teaching and writing on Catholic social thought, principally in Rome. Moreover, I have had frequent contact with social developments elsewhere on the Continent. Thus, I have come to know Europe's strong element of *laïcisme*, which repudiates papal authority in all social matters. Says the laicist: Let the Church confine her exhortations to the Gospels and to the law of love. These are things of the spirit, and it is only things of the spirit that are or should be properly the concern of the Church. The task of civilization is purely temporal, wholly secular, and the Church has nothing at all to say about it.

In Europe, this laicist mentality finds its principal expression on the left wing of society. For there it is the left-wing Catholic who wants freedom—in his case, freedom to turn to what he believes to be the only vital source of social thought today, that is, to Hegel—and to Hegel as expounded by the Marxists.

During a long summer of travel in the United States this year, I continually met up with a strange variant of this European mentality. The only difference is that here in the United States I everywhere encountered it among extreme right-wing Catholics, notably those identifying themselves with the *National Review*. Aiding and abetting this mentality, if only unconsciously, were a small number of priests who—themselves also followers of *National Review*—are totally out of sympathy with Pope John's new encyclical, and in consequence are inclined to give it the silent treatment.

While these priests do not *ex professo* reject the Pope's right to teach in these fields, their silent treatment amounts in practice to just that; and it is so interpreted by some members of the laity who, in their encounters with these priests, find encouragement for their own rejection of papal authority in social questions.

AMERICA's editorial of September 30 devoted itself to the task of bringing about a better understanding of papal authority. The editorial is a clear statement, backed up by the weight of a century of the Church's own express teaching. It should be quite enough to convince loyal Catholics of what their attitude ought to be.

However, there is another approach that could be taken to bring us to the same conclusion—an approach that I made frequent use of this past summer. This is to point out a simple fact always missed by the dissenters: *there is in fact an already existing body of Catholic social thought*. The implication of this idea will escape no well-instructed Catholic. What the Church *in fact* has done and does today with regard to matters of faith and morals—namely, teach—she *rightly* (*de jure*) may do. And this is obviously no less true in matters of social morality than in those of personal morality.

I have made use of this approach because the current debate often goes on as though Pope John were the first Pope to try his hand at writing a social encyclical. The fact is that to an already existing and considerable body of Catholic social thought, one Pope after another throughout the past century has continually been adding more and more material. Most notable, of course, are the writings of Pope John's predecessors, Leo XIII, Pius XI and Pius XII. But others, too, made contributions of considerable importance.

Furthermore, all of these Popes recognized that they were contributing to an already existing body of Catholic doctrine, which, if it needed organization, growth and adaptation, nevertheless constituted a true *corpus doctrinae ecclesiae*. Each Pope in turn relies upon his predecessors; directs attention to what they have already written; insists that it is truly and properly the Church that is teaching; asserts that he himself is merely applying the existing body of Church doctrine in a new situation and covering a field hitherto left untouched (*Mater et Magistra's* extended treatment of agriculture, for example), or that he is resolving a disputed question.

There is no more striking evidence of the truth of what I have just been saying than the encyclical *Mater et Magistra* itself. There Pope John devotes practically one-fourth of a lengthy document to the point of recalling the teachings of his predecessors and to reaffirming their authority.

In brief, centuries of papal practice refute the supposition of those who dissent from *Mater et Magistra,* claiming that the Popes do not have a right to speak to the Christian conscience in questions of social policy. If one cannot understand the limpid reasoning with which AMERICA explained to Mr. Meyer the theological basis of the Church's right—and duty—so to speak, surely this formidable fact of the Church presuming over centuries to do precisely what Mr. Meyer rejects, that is, to inform the Christian conscience in social as well as personal morality, should make the Catholic's dissent seem awesomely brash.

I repeat: *in social as well as personal matters.* What a singular prejudice it is which reduces morality to not cheating, not telling lies, not fornicating, not eating meat on Fridays, while dismissing from the realm of morality all questions that touch justice in economic life. And just such a rejection of Catholic tradition is involved in denying to the Pope an authoritative voice in social policy. Evidently, it cannot be repeated too often that wherever moral values are at stake, there the Church rightfully speaks in the name of Him who entrusted to her the task of guiding men to their eternal destiny.

In asserting thus vigorously the Catholic's obligation to heed any message propounded in a social encyclical, I do so in full awareness of the fact that it is not the competence of any individual, but that of the Church, to lay down norms for the proper fulfillment of that obedience. The Church, it should be remembered, is a loving mother who will not bruise the bent reed. Much less will she impose anything that is against reason.

It would take many more paragraphs than I have space for in this brief article to set forth the Church's norms for the reading and interpreting and applying of an encyclical. But those acquainted with them will recognize how eminently satisfactory they are.

It has been reported that Pope John has manifested a desire that Catholics engage in serious but loyal discussion of what he has written. He wants commentaries to be written and questions to be proposed. There is nothing at all surprising in this, because each of the social encyclicals has inevitably been followed by such discussion. The Pope recognizes that one may legitimately question the prudence of his decision not to enter into a detailed condemnation of the economic systems of socialism and communism. If this question is raised, the Pope would like the opportunity to answer it. He recognizes, too, that in some countries his use of the word "socialization" has given trouble. He welcomes clarifying comment on the topic.

The Holy Father realizes that the half-dozen general lines of "socialization" that particularly interest him will need both comment and careful application. He is keenly aware that the application of these policies calls for great discrimination, in view of the differing situations in which many nations find themselves—some well advanced along lines of sociative activity, others scarcely embarked upon any sort of organized group enterprises for the common good.

The Vicar of Christ acknowledges that "socialization" can and does have negative aspects. He will not consider it a disloyalty if some, commenting on "socialization" in local situations which they know well, wish to insist upon these aspects more than he did.

However, in the preparation of all such commentary and in any application that may be made to his principles, the Pope hopes that we will attend most carefully to his nuances of language, to his emphases, to what he states as certain and to what he sets down as reputable and authoritative opinion. Finally, he would certainly want a manifest distinction made between statements of what is truth, generally and everywhere applicable, and what is rather exhortation to a line of action which, according to circumstances, is mandatory or advisable or less advisable in one part of the world or another.

Among the dissenters I met this summer some few justified their slowness in accepting the new encyclical on the ground of a certain familiarity with the normal procedures by which encyclicals are prepared. They begin with the fact that a Pope must rely upon collaborators who have the technical competence he may lack in certain purely economic aspects of some questions, questions on whose social and moral implications he believes it timely to give guidance. These dissenters also recognize—and argue from the fact—that the Pope freely seeks help in setting forth accurately and fully those elements of Catholic social thought which he intends to bring to bear upon the questions at issue in his encyclical.

If Pope John had relied on more conservative collaborators—so these hesitant ones argue—a different (and to them more acceptable?) encyclical might have emerged. If, therefore, such a different teaching might have existed, how, they argue, can they be forced to accept the presently oriented document?

It is not altogether easy to answer this question, put in this way. And certainly, something must be conceded to the questioner of good will. It is perfectly true that the Pope's collaborators can be more liberal or more conservative in orientation. Indeed, the same could be said of Popes themselves. Hence, certain inevitable emphases will be reflected in what they write or in the manner in which they express themselves.

At this point, however, ends the comfort which the so-called "Catholic Right" can draw from the above consideration. For the collaborators of a Pope, whether liberal or conservative, would certainly know that a corpus of Catholic social doctrine exists. And this body of doctrine necessarily channels and confines whatever line their individual conservative or liberal tendencies might seek to take.

Moreover, as one whose life work has been the assimilation and the teaching of Catholic social doctrine, I can say with absolute certainty that *no* collaborators Pope John might have turned to—European or American or other—could have or would have prepared an encyclical that would be acceptable to the editors of the *National Review*.

An Appeal for
Christian Unity

Robert Pell

T HAT distinguished lawyer and businessman, Myron C. Taylor, who dedicated his latter years to
diplomacy in the highest sense of the term, has
quietly "set sail," as he was wont to say, "for that other
shore." He has left behind with those who knew him
and worked with him memories of unflagging loyalty
to principles and men, of tireless devotion to what he
conceived to be his duty and his unquenchable Christian faith. I had the privilege of working with Ambassador Taylor from 1938, when I was assigned by
President Roosevelt as his Assistant at the Evian Conference on Political Refugees, until he left public life
in 1950 upon resigning as the Personal Representative
of the President of the United States to His Holiness the
Pope. I can think of many episodes which would illustrate his high qualities, but none more poignant than
his conversations in 1948 with leading churchmen to
promote unity—among Christians and others who believe in God—in the face of the mounting threat of international atheism and materialism, and in the interest
of peace.

Early in 1948 President Harry S. Truman decided

that the time had come to appeal for the cooperation of the religious forces of the world to resist communism and turn the influence of all those who believed in God toward the positive preservation of peace. The President had held informal preparatory exchanges to this end with Pope Pius XII through the mediation of Mr. Taylor. He had discussed this "thought," as it was regularly described, with Protestant leaders, notably Bishop G. Bromley Oxnam and Dr. Samuel McCrea Cavert. He had consulted outstanding Jewish personalities and was in communication with Muslim leaders.

By March of 1948, the President decided that the time had come to translate his "thought" into action. Accordingly, he called Ambassador Taylor to the White House in the week of March 20 for talks about his plan. This review of the possibilities led to the President's decision to send Mr. Taylor on a "voyage of exploration," in which the Ambassador would discuss the President's "thought" with the outstanding churchmen of the Christian world. Then, if agreement could be reached among Christians on a simple, positive formula,

others who believed in God would be brought into the conversations. Accordingly, on March 23, President Truman gave Mr. Taylor a commission appointing him as his Personal Representative, with the rank of Ambassador, for these exploratory conversations. At Mr. Taylor's request, the President designated me to accompany Mr. Taylor, with the rank of Minister.

At once, Mr. Taylor set off for his post at Rome, as his point of departure, by way of Lisbon and Madrid.

In Lisbon, Mr. Taylor conferred at length with the Catholic Patriarch of Lisbon, Cardinal Emanuel Gonçalves Cerejeira, to whom were confided the revelations of the three children who had the vision at Fatima. On the basis of these revelations, the Cardinal was confident that the tide of godless communism would recede when faith was restored in the Christian world. Mr. Taylor, fortified by the Cardinal's confidence, then flew to Madrid where, at the suggestion of highly situated persons in the Vatican, he had an appointment to meet Cardinal Pedro Segura y Saenz, Archbishop of Seville, who was reported to be unbending with regard to any contact between the Catholic Church and the Protestants.

Ambassador Taylor met Cardinal Segura in the Episcopal Palace in Madrid, with Cardinal Pla y Deniel, Archbishop of Toledo and Primate of Spain, acting as host. The ensuing conversations between Ambassador Taylor and the Cardinal were on a high plane of objectivity, although it was clear that Cardinal Segura had serious doubts about the feasibility of joint action by Catholics and Protestants. Nevertheless, he seemed to be won to sympathize with Mr. Taylor's moving message in behalf of President Truman. Those of us who were present concluded that, at the very least, the Cardinal would not raise his voice in opposition to the President's appeal to the conscience of the world.

Thereupon, Mr. Taylor continued his flight to Rome, which he reached on April 4. The next day he visited the Vatican for a protracted review of the President's "thought" with Pope Pius. Mr. Taylor carried a brief, courteous letter from the President, but the main body of the message was oral, developed at some length by Mr. Taylor in a series of meetings with His Holiness, which took place on the balcony outside Pope Pius' private apartments.

Pope Pius XII was not only favorable to President Truman's "thought," but was prepared to support it in a letter to the President. This letter His Holiness handed to Mr. Taylor some days later, and it was considered by experts with a broad knowledge of Vatican practice and history as unusually liberal. Moreover, His Holiness gave Mr. Taylor permission to show a copy of the letter to other Christian leaders in the course of his exploratory tour.

President Truman thereupon directed the Commander-in-Chief in Germany to place his "Potsdam plane" at the disposal of Mr. Taylor. The Ambassador took off from Rome in the second week in April—I was with him and so was Miss Bushwaller, his confidential secretary—with Geneva as the first stop.

PROTESTANT REACTIONS

The object of Ambassador Taylor's visit to Geneva was to confer with the Provisional Executive Committee of the World Council of [Protestant] Churches, notably with Mr. Visser 't Hooft, its secretary. Immediately upon our arrival in Geneva Mr. Taylor sent me to notify Mr. 't Hooft that he wished to confer with the members of the executive committee on a matter of grave importance affecting the Christian world, and an appointment was arranged for the same afternoon. The American Consul General had advised the committee previously that Mr. Taylor would come to Geneva on a special mission, so that the Ambassador's arrival was not altogether unexpected. At all events, Mr. 't Hooft stressed to me that the committee was committed irrevocably to the principle of the separation of Church and State. For that reason it could not receive Mr. Taylor as the representative of a temporal sovereign. The members of the committee, as a courtesy, would talk with him, however, as a prominent Episcopalian layman who had important things to say.

At the appointed hour, Mr. Taylor and I presented ourselves at the headquarters of the executive committee. At the gate we were met by a spokesman for that austere body who repeated once more that the committee would not receive Mr. Taylor as the representative of a temporal sovereign. However, the members would be pleased to talk with Mr. Taylor in his capacity as a prominent Anglican, or Episcopal, layman. Mr. Taylor agreed to these terms and was ushered into the Council Chamber, where he was received coldly but with dignity and seated in a stiff-backed chair— I beside him.

Ambassador Taylor then outlined his mission, stressing the vital importance of union among Christians in the cause of world peace, threatened as they were by international atheism and materialism, and urging agreement on a positive statement of their intention of standing together under the banner of Christ. The members of the committee who were present heard Mr. Taylor through politely but without comment. However, they declined to read his Commission from the President or the copy of the letter which he carried from Pope Pius. Indeed, Secretary 't Hooft suggested that a wiser approach would be for the Protestant Churches to come together in the first instance with the President's "thought" in mind. Then, when they had agreed on a formula, approaches might be made to the Orthodox and Roman Catholic Churches. The secretary of the executive committee observed, moreover, that the Protestant Churches were planning to meet in the near future in Ecumenical Conference at Amsterdam. This would be the appropriate place and time to give further consideration to the possibilities of cooperation of the religious and moral forces of the world. He stressed, however, that this meeting would be held completely free from all governmental pressures. There would be no place at Amsterdam for Ambassadors or Personal Representatives of Heads of State. Mr. 't Hooft added that the World Council embraced over 150 Protestant communions in five continents. The Council, and as a consequence its proposed reunion at Amsterdam, was as inclusive as it had been possible to make it by several years of constant endeavor. There did not seem to be any possibility of widening its membership and scope at that late hour, despite Mr. Taylor's appeal to make the meeting inclusive of all Christianity.

I should like to take this opportunity to underscore that Mr. Taylor did not propose on this occasion, as some publications suggested later, that he should go to Amsterdam. Nor did he try to complicate "with the apparatus of diplomacy" what had already been accomplished. He confined his presentation to a solemn and very fervent appeal for contact between *all* Christian leaders, and those who believed in God, in the interest of peace and in defense against the inroads of international godlessness. He urged that this action be taken before it was too late.

From Geneva Mr. Taylor flew to London for a conversation with the Archbishop of Canterbury, Dr. Geoffrey Fisher, in his capacity as head, under the sovereign, of the English State Church. The Archbishop, accompanied by a Coadjutor and an Under Secretary of State at the Foreign Office, joined Mr. Taylor at luncheon in a private dining room at Claridge's Hotel. Dr. Fisher was most cordial but at the same time most cautious. He read President Truman's Commission without comment and expressed pleasure with Pope Pius' letter, which he read over twice. He indicated that the President's initiative was worthy of careful consideration, but indicated that there might be obstacles. However, they could be overcome, and he gave Mr. Taylor his full blessing and expressed the hope that his mission would lead to a concrete result. He would give much earnest consideration and prayer to President Truman's "thought" and would remain in contact with Mr. Taylor through diplomatic channels.

Paris was Mr. Taylor's next destination, where the ground had been prepared for conversations with Pastor Boegner, speaking for the French Protestant, or Huguenot, Church. Mr. Taylor sent me to sound out the Pastor preparatory to a meeting, and I found him far from conciliatory. Indeed, at first he was not sure that it would be possible for him to meet Mr. Taylor at all. Finally, with marked reluctance, he agreed to lunch with Mr. Taylor in a private dining room at the Ritz Hotel. However, he said over and over again that he could not see his way to coming to agreement with the Vatican and that the only form of cooperation which was conceivable, or desirable, was that based on a reaffirmation of their faith by the Protestant Churches, with which the Orthodox Churches might be associated, including the Patriarch of Moscow.

In any event, Pastor Boegner kept his luncheon date with Mr. Taylor, but merely reiterated his previous position. He was uncompromising with regard to cooperation with the Catholic Church, and those who were present could not help recalling the stand taken some weeks previously in Madrid by the Archbishop of Seville. They hoped, moreover, that the Pastor would unbend as far as the Archbishop had.

Finally, Mr. Taylor asked the Pastor point blank if he was to regard the statement of his views as closing the door to any form of further conversation with regard to President Truman's "thought." The Pastor replied that Mr. Taylor might consider his statement in that light. Mr. Taylor, a final time, asked Pastor Boegner to hold the door open at least a cranny while he continued further explorations, but the Pastor replied that he could not in all conscience do this. At least he would interpose no obstacle during the early stages of the talks and would be interested to hear how they progressed. The following day Mr. Taylor wrote Pastor Boegner, renewing his appeal. The Pastor replied with a letter firmly closing the door to further exchanges.

Thereafter, Mr. Taylor met in Paris with representatives of the German and Scandinavian Lutheran Churches. They were less obdurate than Pastor Boegner, but cannot be said to have been encouraging. The Ambassador also talked with the new Patriarch of Constantinople, Athenagoras I, an American citizen who had been obliged to assume Turkish nationality in order to take possession of his see. He was on his way to Constantinople and was most gracious in his meeting with Mr. Taylor. He expressed profound interest in the conversations initiated by President Truman, al-

214

though he stressed most honestly that he foresaw that many obstacles would have to be overcome. For instance, he must not break irrevocably with the Patriarch of Moscow.

REPORTING TO WASHINGTON

It was now June, and Mr. Taylor believed that he should make a full report to President Truman before entering upon a further phase of the conversations. Above all, Mr. Taylor wished to avoid anything which might appear to be a break. Accordingly, he suggested to the President that it might be wise to have a breathing spell, while emotions cooled. The groundwork had been laid. The Amsterdam Conference of the Protestant Church would be opening shortly. It might be well to wait and see what transpired there. Later, when the atmosphere was more propitious and there was a greater possibility of reaching a concrete accord, the conversations might be resumed and further considerations given to President Truman's "thought."

Mr. Taylor thereupon prepared a lengthy report for the President, to which he joined the memoranda of all his conversations and such documents as might be useful as background. Mr. Taylor instructed me to fly to Washington with this report and place it directly in the hands of the President. He would return to Rome, the seat of his Embassy, from where he could piece together the various skeins and threads of his preliminary talks. Moreover, Mr. Taylor directed me to go to Washington via London, where I was to send word to the Archbishop of Canterbury through the Foreign Office of what had transpired.

I reached Washington, after completing my mission in London, on June 13. On the 15th I went through what was then called "the State Department passage" to the White House and placed Mr. Taylor's report in President Truman's hands. Three days of deliberation followed, and on June 18 the President handed me a letter which he instructed me to deliver personally to Mr. Taylor in Rome. In his letter, Mr. Truman, after congratulating Mr. Taylor on the manner in which he had conducted his delicate mission, agreed that the conversations should be suspended for a time. He said it was deplorable that the Amsterdam Conference would embrace only a section of Christendom and reaffirmed that both State and Church had a solemn obligation to man the parapets of Christian civilization.

In the meantime, plans for the Conference of the Protestant Churches at Amsterdam had been completed. It had been announced, moreover, that invitations had been extended to leaders of the Orthodox Churches. Indeed, it has been especially stressed that these invitations had included one to the Patriarch of Moscow, who had accepted. Finally, the Congregation of the Holy Office on June 5 had warned Catholics that they would not be permitted to attend the meeting. Every day the prospects of agreement among the leaders of Christendom seemed to be more remote. Nevertheless, in the thinking of Mr. Taylor, with which I understand President Truman agreed, the conversations were merely suspended. At some time in the future they would be carried forward from the point where Mr. Taylor, to his profound disappointment and undying regret, left off.

In conclusion, it is my duty to stress that Ambassador Myron C. Taylor's negotiation in the spring of 1948, seeking unity among Christians in the interest of the preservation of peace, was conceived in the highest idealism and carried out step by step with profound sincerity. Mr. Taylor spoke as one modest, faithful Christian to other Christians, in moving, simple language straight from his heart. He tried earnestly to lift the conversations from the slough of politics. He offered modestly to serve as a simple workman in building even one bridge to cross the ravines which separated the churches. He strove mightily to help them find together a few positive words to express their unity in the interest of peace and in their rejection of international godlessness. He may have seemed to falter at one split second in the eternal course of history. But surely Mr. Truman's message, and Mr. Taylor's, still and small though it may seem amid the thunder of jets and missiles, lives on. It is not yet too late to heed it. But it may be, very soon.

A Catholic-Protestant Conversation

Thomas F. Stransky, C.S.P.,

and Claud D. Nelson

STRANSKY: Dr. Nelson, would you please tell me something about the response of non-Catholic Christians to the fact that the Catholic Church is going to hold an ecumenical council?

NELSON: Well, Father, there was the inevitable first reaction of surprise, a second reaction of gratification, and perhaps, in the first few weeks, a certain amount of over-optimism, particularly since the question of unity and even of union was on the lips of so many who spoke about the council. My own reactions were—"well, it's about time!" And then: "What will come of it?" In a word, profound encouragement and limited expectations. These reactions have not changed substantially. The fact of the council—the fact that Pope John feels the need of a council and sees the possibility of something constructive from its work—remains an event of fundamental importance.

STRANSKY: What do you mean by "It's about time"?

NELSON: The discouragement of non-Catholics by what was done in the unfinished council of nearly 100 years ago has been profound and, of course, prolonged. My impression of the First Vatican Council is one of fellow-Christians outside being completely ignored. If anything can be done to lighten that darkness, it is important that it be done, particularly since we are in a period when Christian unity, not only in a spiritual and emotional sense, but in a practical and co-operative sense, is certainly more important than it has ever been.

STRANSKY: Pope John hopes to lighten that darkness you speak about. So often he stresses the necessity of bringing the Church up-to-date—of an *aggiornamento*. If the council can clarify some of the doctrines misunderstood by non-Catholics, if it can inspire a more intensified Christian life among Catholics, and modernize many of the Church's external disciplines, I am sure the Catholic Church's image will become clearer to our separated brethren.

That, after all, is the best task the Church can perform at this stage of the ecumenical movement. Neither the Protestants, the Orthodox—nor, I'm afraid, Roman Catholics—are prepared for a "reunion" council, with the hope of reaching some general agreement.

Our council is a domestic affair. Yet, in its preparations, the Secretariat for Promoting Christian Unity is trying to help guide the council's theological and pastoral decisions in such a way as to aid our separated brethren to see the Church as the Church believes herself to be.

Furthermore, the very fact that there will be a council shows that the Church is conscious of its many imperfections on the human side and is, therefore, conscious of the need for much self-purification.

NELSON: I would like to underline one of your comments and to express deep gratification over another. Even those Protestants (and I believe also Eastern Orthodox) who are engaged actively in ecumenical or unity movements, apart from Rome, are certainly not ready to discuss union with Rome. They are not yet aware of what they must do to unify themseves, Protestants with Protestants, Protestants with Orthodox.

But it seems to me that the Roman Catholic confession of humility in the face of our human imperfections and of the situation of our divided churches is one of the surest and most necessary signs of an ecumenical spirit. Its very expression, both by the Pope's calling of the council, as you said, and by statements of other persons in positions of responsibility, draws us together in a way that an older attitude—one that sometimes has seemed to be self-satisfaction, almost arrogance—assuredly did not do.

 It seems to me that the calling of the council is not only a factor in the development toward Christian unity, but it also gives evidence that the time is ripe for some actual advance in that direction.

I think of all the signs within the Roman Church: the fact that two former Popes have either thought of calling a council or have been urged to do so; I think of the creation of the World Council of Churches and similar councils of smaller scope over the last fifty years; then there is the new surge toward unity within Orthodoxy. All this together indicates that the Holy Spirit is working among many, many Christians.

I wonder to what extent you would connect this movement of the Holy Spirit and the widespread movement toward Christian unity with renewed interest in the Bible?

STRANSKY: It is always difficult, often presumptuous, to trace God's handwriting in specific historical events. Yet, as I read the Holy Office's Letter on the Ecumenical Movement (1949), I don't think it presumptuous of the Catholic Church to see the workings of the Holy Spirit both in the mood of non-Catholic Christians and in Catholic desires and efforts to promote Christian unity.

Christ works through the Holy Spirit, and one can hear Christ's voice and feel the breath of His Spirit in the growing and ever firmer conviction that a divided Christianity is against Christ's will and a scandal, and that His prayer "that all may be one" is also a command for our own prayers and actions.

NELSON: Consider the biblical revival. The Bible, after all, is the Book of all us Christians, and by searching into God's mind and will as revealed in the Bible, Catholics better understand not only themselves but their separated brethren.

Catholic and non-Catholic biblical scholars are following more closely and more objectively each other's exegetical work (which, in large part, is itself becoming more objective). For example, recent studies on the Epistle to the Romans prove to be a fruitful ground for discussions with Protestants, for around that epistle, as we know, centered much of the Reformation theology of Luther, Calvin and Melancthon. And a developing biblical theology—expressing doctrine in biblical categories—finds closer rapprochement with Eastern theology.

STRANSKY: Interestingly enough, Cardinal Bea, president of our Secretariat, had his first serious theological discussions with non-Catholics in the 1920's at congresses of biblical studies, meetings at which scholars of various confessions discussed their common Book on objective, scientific grounds.

NELSON: Indeed, we do well to bear in mind, with reference to Christ's prayer for unity, that our unity is not simply important in itself, as evidence of our faith—for our own reassurance, guidance and salvation. It is more than that. Jesus seems to have held it as indispensable evidence to the non-believer. Only if we are united will the world believe that He was sent by the Father.

STRANSKY: The division of Christianity certainly makes it very difficult for the rest of the world to believe. It is a tragedy that we have had to pass through this bitter experience of history in order to prove Christ's words.

NELSON: The experience of Protestants bears you out. It is the missionaries who have been most eloquent and most impressive in saying that a divided Christianity cannot bear effective witness among those who have never heard of Christ. In fact, most Protestants date the concrete beginnings of the modern ecumenical movement from the International Missionary Congress in 1910 at Edinburgh. At first there was concern only for correcting the scandal caused by diverse mission methods, but soon Protestant missionaries had to face honestly the blunt judgment of Bishop Brent of Edinburgh: the division of the churches is itself the fundamental cause of the scandal.

STRANSKY: From another angle, I would like to paraphrase a remark of Bishop Newbigin, General Secretary of the International Missionary Council. It is not surprising to the non-Christian, he said, that Christians are concerned with promoting Christian unity; what is surprising is that they are *content* to be divided.

NELSON: One of the things I am sure many Protestants would hope will be encouraged by the council is more sustained and intelligent activity on the part of laymen. My own experience has been largely through the Y.M.C.A. Surely, the work of laymen in the Y.M.C.A., which is not necessarily nor exclusively Protestant, indicates how laymen may be effective in the Christian cause, without interfering in any way with ecclesiastical prerogatives and responsibilities.

 STRANSKY: When Pope Pius XII said that "the Church *is* the laity," I suspected many Catholics, including priests, were shocked that a Pope could sound so "Protestant." The statement implies that the very authority Christ gives to Pope,

bishop and priest is given that they may act under the same title the Pope has assumed for centuries—"servant of the servants of God." And the laymen—the servants of God—have a specific role to play in the Church's mission to the world—something far more than supplying for the lack of priests.

Just what the layman's role is, and its consequences in modern pastoral practices—well, the council will handle that, through the Theological Commission, the Commission for the Lay Apostolate and our own Secretariat.

But no matter what the council decides, I hope that non-Catholics, as well as Catholics, will be patient. It's hard enough to teach a small child to walk without stumbling, and it is much more difficult to teach that full-grown giant of a child—the laity—to take his first mature steps. Furthermore, many of the laity's authorized teachers—bishops and priests—are not yet mature enough in this field to guide them. We are still in the era of *talking about* the laity, surely not yet in the era of the laity. And it will take a long time, I fear, before the whole Church will work out, in practice, the layman's apostolate.

The council would issue only guiding principles on the layman's role. It will not decree their minute applications, for that depends on the varied circumstances of nation, diocese, parish, etc. Even then, conciliar decrees require a long time before their implications filter into the conscious spiritual life and activity of the Church's members. (Look how long it took for the Council of Trent's decrees on seminary training to shape a somewhat uniform practice throughout the world!)

Yet, this "filtering process" can be hurried along if every Catholic—not just the Pope and the bishops who will gather in Rome—considers the calling of the council as a summons to the whole Church to commence a vast spiritual retreat. As we prepare for it, our prayers should be for the enlightened guidance of those who will formulate and approve the council's conclusions, and for all the laity, priests, religious and bishops, that they may accept these conclusions in faith and love, and live them and practice them, in that same faith and love.

NELSON: Is it foreseeable that after the council more Catholic laymen will feel competent and confident enough to engage in dialogue about religious matters with their Protestant neighbors who are willing and interested?

STRANSKY: The Council will try to make the layman more conscious of his duty and privilege of witnessing the Catholic faith by word and action. Of course, no matter what his competency, the U.S. layman must challenge that prevalent and enigmatic commandment of the American way of life, namely, that religion is not a topic for conversation. Besides, a layman is often afraid to talk about his faith because he doesn't trust his own information. Many non-Catholics have been referred to me by Catholic laymen; the issues raised in those conversations have been so simple that I failed to see why the average Catholic could not have answered them on his own, in his own language, and in the circumstances which provoked the questions originally.

As for serious dialogue, we have some very competent laymen. Often the layman can do much better than a priest in discussing Christian issues in the educational, political, social and economic fields. There he earns his daily bread and butter, and, if observant, he is quite conscious of their religious implications.

NELSON: My reference to "dialogue" probably put too much emphasis on its formal aspects. Of course, I think that's where we are at the present moment. "Dialogue," however, is also the proper word for that natural spontaneous conversation which ought to take place whenever Christian brethren—laymen or priests—have occasion to discuss common problems, or simply to meet and talk with each other. At present, however, in many, many cases, the occasion has to be created and even made formal before real conversation—any real dialogue, any truly attentive listening and honest speaking —can take place between Catholics and Protestants, at least in this country.

There must be a lot of casual, spontaneous conversation going on which is not recorded or formalized. I'm sure there is. However, I'm also sure that there are a good many Protestants who feel it really isn't worth while to talk with a Catholic layman, or perhaps even with a local priest, because they don't feel he is free. We feel that he is somehow restricted.

Sometimes that impression is due to what we read in the press, or in an encyclical meant for a certain specific situation, but we don't know which situation, and, therefore, read it as a general thing. I think of the common understanding of prohibitions for membership in the Y.M.C.A. or the Rotary Club, and also of the fact that there are areas where Catholics participate in the National Conference of Christians and Jews, and other areas where they do not.

 STRANSKY: Divine Truth is binding, and every Catholic, be he pope or layman, believes that Christ reveals His Word through His Church. In this sense, we are all "restricted" in what we believe. A Protestant who will talk to a Catholic layman or the local priest only if they are allowed to speak apart from the authority of the Church—well, he doesn't want to talk to a Catholic, but to a fellow-Protestant.

The authority of the Church is understandably a crucial, divisive issue between Roman Catholics and others. (Here I may mention that much of this problem of "Catholic authority," at least in America, is not theological but philosophical—what is authority and its relation to human liberty?)

Your mention of Catholic universal prohibitions is ticklish. Christian organizations, such as the Y.M.C.A. and the World Student Christian Federation are beginning to catch an ecumenical spirit to which Catholics could subscribe. I say "beginning to catch." How widespread this spirit is, how reflected it is in policy and action, requires much more honest and common study by Catholic and non-Catholic authorities. The question for us will be whether or not universal prohibitions or approbations for Catholic memberships in such organizations are the most prudent and practical, or whether these decisions should be left up to local hierarchies.

Even a prohibition does not exclude all cooperation or dialogue. Last May at Louvain, I took part in a most fruitful discussion between thirty leaders of Pax Romana and the World Student Christian Federation on the provocative theme: "Technology and Christian Faith."

NELSON: You refer to certain practical fields—fields in which practical understanding and cooperation are necessary. Certainly we are agreed that conversation in such fields is necessary. Would you mention two or three of those fields and what specifically one ought to hope for from this coming council or within a short range of time?

STRANSKY: There are abundant opportunities for fruitful discussion and action, and they are already within reach, if we but open our eyes and catch an authentic ecumenical spirit. I think, first of all, of our mission lands. Faced with the increasing influence of communism and materialism, with the new problems of the scientific, economic and national revolutions, Christians as such in many African and Asian countries should realize their obligation to defend and promote their common deposit of truth and goods in the Christian patrimony. With all the shared social and political problems of a Christian minority, Christians can meet to discuss advisable ways and means of defending the fundamental principles of the natural law and the Christian religion and of re-establishing a sound social order.

What providential occasions for common discussion and much concrete cooperation are the relations we have as Christians with national governments and international organizations (e.g., Unesco, FAO); questions of mission schools and the religious orientation of public schools; questions of racism, just wages, the dignity of marriage and womanhood; problems of over-population, urbanization, immigration, alcoholism, etc.

Then there are the more properly theological issues that must be discussed together: Church-State relations, religious tolerance, false proselytism, catechetics and liturgical worship, mixed marriages, the question of a common Bible translation, problems confronting the personal and social life of Christian ministers and priests, etc. Of course, you can see that most of these problems are not confined to the mission lands.

What can the council do? The 1949 Holy Office Letter on Ecumenism recommended the lines of action I've already mentioned. Such cooperation is implied, too, in Pope John's latest encyclical, *Mater et Magistra*. I doubt whether the council should do anything more than reinforce and clarify these general principles. Then it will be up to the local churches or international organizations to work out the details.

 NELSON: We've been speaking of dialogue, and certainly one of the things we both hope will be encouraged by the council is the mutual respect, confidence, trust and tolerance between Catholics and non-Catholics, which is in one sense a condition of Christian conversation.

It seems to me that a very important foundation in the field of tolerance was laid by Pope Pius XII when —without recognizing the religious pluralism which exists, not only within countries but internationally (I mean what we call Christian pluralism)—without recognizing it ecclesiastically—the Pope indicated to Catholic statesmen, in Catholic countries, that they must respect it as a political fact.

That statement seems to me to have represented, at least in the attitude toward Protestants, a change in policy, not in doctrine, which may be very significant. I have been encouraged to hear Cardinal Bea say, and others not usually thought of as being quite so ecumenical-minded as the Cardinal, that the council will not go back on that change of direction, will not reverse it.

Would you have further comments in this whole field of religious liberty, particularly as to how the activities of the Secretariat for Promoting Christian Unity are affecting and may affect mutual respect and tolerance and, in a civic sense, the protection of religious liberty?

STRANSKY: Yes. The Secretariat itself, as well as its members throughout the world, have been receiving so many similar questions, that I am convinced that Catholic relations with other Christian communities cannot be securely founded until a Catholic doctrine of tolerance is fully developed.

A growing school of Catholic theologians is pleading for more than a "policy change," that is, for a deepening and development of our doctrine. Such a doctrine of tolerance cannot be based on mere social or political expedience, or on the intellectual abstraction that "error has no right," but rather on the nature of human liberty itself and of divine faith. Of course, we cannot foresee how much time the council will spend on the question, or what it will decide.

NELSON: It seems to me that we have to realize—and we are beginning to realize quite generally—that tolerance, religious liberty, is something not conceded, either by the person who has power, or the person who has truth, to one who, in his opinion, does not have truth and who obviously does not have power. There is a mutuality about Christian tolerance which is essential to it.

In the various fields of which we have been speaking it seems to me that the creation of the Secretariat for Promoting Christian Unity and what it has done so far are of tremendous significance. For one thing, it has made possible two-way conversation. Indeed, without too much hesitation as to who speaks first, such conversations are going on between groups assembled jointly. For example, these conversations are being held between the World Council of Churches and the Secretariat. The mutuality we've been speaking of certainly exists in these encounters. This is very desirable.

I can think of nothing more significant likely to come out of the Council than the encouragement and development of the Secretariat and the process which it typifies. It seems to me to represent a willingness to talk on the part of Roman Catholics, which heretofore has not been evident, or so officially evident, one might say. This is so necessary if we are ever to arrive at a practical Christian unity.

STRANSKY: All I can say to that is that the Council itself will determine if and how the Secretariat should continue as a permanent official organ in the Church. But even as it stands now, the Secretariat gives witness that the Church is taking an official position on the dialogue. We treat our separated brethren more as *brethren* than as "separated"—not as adversaries to be conquered, but as brothers who ask questions that deserve an answer.

The short experience of the Secretariat has already shown how the Holy Spirit speaks to us Catholics through our separated brethren in many of their criticisms of the Church of 1961. As a Catholic, I believe that the Roman Catholic Church is the Church of Christ. Her structure cannot change. But, as Pope John has said, we should try to eliminate as much as possible the human obstacles that obstruct the work of showing the Church to be what she is. Many non-Catholic Christian criticisms, positive and negative, that have been received by the Secretariat, have given us a clearer insight into what these obstacles are.

Pope John has asked us Catholics not to rest until we have overcome old habits of thought, our old prejudices, the use of expressions that are anything but courteous. Thus, we are to try to create a climate favorable to the reconciliation to which we look forward. Thus, too, we must strive to cooperate with the work of grace.

The union of all Christians in the Church of Christ

will be a work of God's grace. If and when He will work His wonder, I do not know. But this I do know: we must begin now, in a very humble and patient way, to prepare for His gift. I hope you and all the rest of our separated brethren may look on our council as just a step in this preparation. So pray for us. I know you will.

NELSON: As you must for us. *Fiat Voluntas Dei!*

In the view of Teilhard de Chardin, evolution became self-conscious in man and the vital process ultimately converges toward a sort of divine togetherness which he called the Omega Point. We will let Mother Church judge the validity of this challenging synthesis of the phenomenon of man. But there is no doubt about our human heritage under God's providence and revelation. He is our Father. We are His children, born spiritually from the loins of Abraham. We must love not only Him, but all those who bear His image. One can wonder at the quality of that neighborly love, when he weighs our attitudes toward the poor, or even when he thinks about our treatment of criminals. Yes, man has come a long way from his origins, but he is still a far distance from any conceivable Omega Point.

All God's Children

Ethics at the
Shelter Doorway

L. C. McHugh
S.J.

THE AMERICAN people are burrowing underground in a grassroots movement for survival; the shelter business is booming. Civil defense officials have already noted that many citizens are very furtive about building a modest haven in the cellar or yard. The more secret the nuclear hideaway, the less likely they are to be troubled by panicky neighbors at the shelter door when the bombs start falling.

Some rugged householders are not banking on mere secrecy to insure their families a fair chance of survival. *Time*, on August 18, cited a Chicago suburbanite who intended to mount a machine gun at his shelter in order to keep unwelcome strangers out, and it also quoted a Texas businessman who was ready to evict unbidden guests with tear gas if any such occupied his shelter before his family did. Inevitably, *Time* raised the question: what do the guardians of the Christian ethic have to say about the pros and cons of gunning one's neighbor at the shelter door? *Time* got some strange answers

232

in its brief clerical poll, and one was rather remarkable:

> If someone wanted to use the shelter, then you
> yourself should get out and let him use it. That's
> not what would happen, but that's the strict Chris-
> tian application.

I cannot accept that statement as it stands. It argues
that we must love our neighbor, not as ourselves, but
more than ourselves. It implies that the Christian law
runs counter to the instinct of self-preservation that is
written in the human frame. If I am right, then the
American people need more than blueprints for shelter
construction. They also need a little instruction in the
grim guidelines of essential morality at the shelter
hatchway. Are there any moral constants that apply
when unprepared or merely luckless neighbors and
strangers start milling around the sanctuary where you
and your family have built a refuge against atomic fire,
blast and fallout?

This aspect of nuclear warfare has received no attention, but its relevant principles were the common
property of Catholic moralists long before Hiroshima.
They are generally treated under the discussion of what
kind of activity is permissible when one's life is under
attack. These principles are worth reviewing, if only to
show that sound Christian morality does not solve
moral problems by appeal to sentiment, but by the cold
light of reason.

Each of us has a natural right to life and the essential goods, such as liberty and food, without which life
is brutish—or impossible.

The right to life and its equivalent goods is a curtain
of inviolability drawn around the human personality.
But if that curtain is torn aside by unreasonable interference with one's freedom, nature still provides a
second line of defense against injustice. This is the right
to use violence as a last resort or emergency measure
for securing the just needs of the human person. This
right to employ violence, which the moralists call "coactivity," is a *limited* grant of power, just like the rights
for whose protection it is given. Nobody enjoys unlimited rights, simply because no human being can have

unlimited needs.

Working from this basis, Catholic moralists teach that the use of violence to defend life and its equivalent goods is justifiable, when certain conditions are met, even if the violent defense entails the death of the aggressor. It will be very enlightening to reflect on what these conditions are.

1. The situation is such that violence is the last available recourse of the aggrieved party. Either you take desperate action now or, in your best judgment, you are going to be done in.

2. The violence used is employed at the time of assault. It is not vengeance for a deed already done, neither is it a preventive against a merely projected assault. The violence is leveled against an attack which, in the prudent estimation of the victim, has been actually initiated. How is one to determine that an attack upon one's life has truly begun? Sometimes the intent to work deadly harm is obvious, as when a known killer runs at me with a drawn gun or an unsheathed knife. But this is not always the case. In the actual circumstances of life, a man under assault has no time for academic niceties. He is not a logical machine but an excited and harried individual. He cannot be blamed if his use of violent defense is based on nothing more than a quick and honest judgment that he will suffer most grievous harm unless he resists promptly.

3. The third condition is that the violence is employed against an attack that is unjust. In other words, the violence is used to ward off an unwarranted invasion of one's undoubted rights. In the technical vocabulary of moralists, the assailant is called an unjust aggressor, but the term refers to an objective situation, not to a state of soul. The delirious madman who thrusts at me with a rapier may be incapable of moral guilt at the moment, but his invasion of my basic immunities is as objectively unjust as if he were a paid emissary of Murder, Inc.

4. Finally, when one uses violence to defend his essential rights, he may employ no more violence than is needed to protect himself. Coactivity is thus a marginal grant, strictly tailored to the end it serves. Moral-

ists have argued for centuries whether this grant ever allows one to intend the death of his adversary. We do not need to resolve this scholastic dispute here. What moralists agree on is that a man under grave attack may take those emergency measures which will effectively terminate the assault, even if they include the death of the assailant. Moreover, common sense tells us that men under attack seldom have a nice discrimination of weapons to employ in self-defense. They must use the means at hand, rough and ready as they are. Unless they use them in the surest way, they are likely to come out on the worse end of an unequal contest.

So much for the general conditions that cover the use of violence in defense of life. Two more observations are very much in order before the picture is complete.

To say that one has the right to employ violence in defense of life is not to say that one has the duty to do so. Indeed, in the Christian view, there is great merit in turning the other cheek and bearing evils patiently out of the love of God.

But it should be noted that people who consistently manifest this exalted brand of supernatural motivation are deservedly called heroic Christians. Their conduct reveals a dedication to the full Christian ethic that is far above what God requires under pain of eternal loss in the way of the Ten Commandments. Nowhere in traditional Catholic morality does one read that Christ, in counseling nonresistance to evil, rescinded the right of self-defense which is granted by nature and recognized in the legal system of all nations.

Again, we must observe that because of special responsibilities the individual bears to other members of society, circumstances may easily arise in which it is positively immoral to turn the other cheek: one may have a positive duty to employ violence in his own behalf and/or for the sake of others. Secret Service agents are bound in justice not to bear ills patiently when the President is set upon by assassins. So, too, a well-armed hunter who surprises three hoodlums attacking a lonely woman in the forest cannot absolve his grave and obvious duty in charity with a shocked "tsk-tsk" and a resolution to inform the State police when he gets to the nearest telephone. . . . More relevant to our immediate interest, we ought to note that the father of a family is tied to his wife and children by bonds of both love and justice. His every normal instinct prompts him to nourish and protect his dependents. He cannot carelessly squander their essential welfare for the needy stranger and call this irresponsibility an act of charity. He may not idly stand by while his brood is robbed of what is necessary for life and then explain that his cowardice is actually a wholehearted obedience to the Biblical injunction to overcome evil by good.

I think that this review of some constants in the general morality of human survival has an obvious relevance to the questions that are raised in the mind of the cautious householder when he thinks about building a family shelter, and wonders how he can insure its availability, in the moment of greatest need, for those in whose behalf it was intended. But I would prefer that every man apply the principles to his own set of circumstances, although I am not averse to setting out a few practical norms that I think would be broadly acceptable to Catholic teachers of morality.

What is your family shelter? It is more than a piece of property that should be secure against trespass. It is a property of a most vital kind. When the bombs start falling, it is likely to be the one material good in your family's environment which is equivalent to life itself. The shelter is your ultimate line of defense against fire, blast, radiation and residual fallout. Moreover, because of its strictly limited resources (space, food, medical supplies, etc.), its use must be carefully regulated if it is to guarantee even marginal opportunity for survival over a protracted period. If you go underground with just one occupant above the maximum number for which the shelter was designed, the survival value of the shelter diminishes for all that take refuge in it.

If a man builds a shelter for his family, then it is the family that has the first right to use it. The right becomes empty if a misguided charity prompts a pitying householder to crowd his haven to the hatch in the hour of peril; for this conduct makes sure that no one will survive. And I consider it the height of nonsense to say that the Christian ethic demands or even permits a man to thrust his family into the rain of fallout when unsheltered neighbors plead for entrance. On the other hand, I doubt that any Catholic moralist would condemn the man who used available violence to repel panicky aggressors plying crowbars at the shelter door, or who took strong measures to evict trespassers who locked themselves in the family shelter before his own family had a chance to find sanctuary therein.

I shall even go so far as to offer a partial code of essential shelter morality. This will offend those who dread to think that the points could conceivably have serious bearing on human survival within the next few months. I am more interested, however, in finding what response such a code might have among readers and "guardians of the Christian ethic."

1. If you are an unattached individual and wish to yield your shelter space to others, God bless you. You can show no greater love for your neighbors.

2. Think twice before you rashly give your family shelter space to friends and neighbors or to the passing stranger. Do your dependents go along with this heroic self-sacrifice? If they do, and you have not yet built a shelter, don't bother to do so. Go next door and build one for your neighbor. In an emergency, he can take refuge there more quickly if it is on his own property instead of yours.

3. When you have sheltered your family, you may make a prudent judgment as to whether you may admit any others to your sanctuary without undue risk to the essential welfare of those who are most closely bound to you in justice and charity. It would be hard to

prove that you have any grave obligation to do so.

4. If you are already secured in your shelter and others try to break in, they may be treated as unjust aggressors and repelled with whatever means will effectively deter their assault. If others steal your family shelter space before you get there, you may also use whatever means will recover your sanctuary intact.

5. The careful husbandman who has no heroic aspirations will take precautions *now* so that his shelter will be available for those for whose safety it was built. If it is marginally equipped, it would be a normal exercise of prudence to conceal the entrance, if feasible, or make it inaccessible except to the members of the family. Does prudence also dictate that you have some "protective devices" in your survival kit, e.g. a revolver for breaking up traffic jams at your shelter door? That's for you to decide, in the light of your personal circumstances. But as Civil Defense Coordinator Keith Dwyer said in the *Time* story: "There's nothing in the Christian ethic which denies one's right to protect oneself and one's family."

—More on the Shelter Question—

MY ARTICLE on "Ethics at the Shelter Doorway" (9/30, p. 824) produced quite a bit of critical comment. This week's State of the Question (p. 288) reflects only a minute sampling of the varied reactions which greeted the article at home and abroad.

I grant that the article took a somewhat technical approach to a crisis of conscience that could arise for thousands of householders under thermonuclear attack, if our families had to rely, for survival, on the minimal protection offered by a do-it-yourself fallout shelter program.

Viewed in this light, my treatment of the principles of self-defense generated much emotional revulsion, especially among rabbis, Protestant clergymen and the gentle souls who make too simple appeals to the "unequivocal ethic" of the Sermon on the Mount.

Theologian Reinhold Niebuhr, for instance, said that I had given a "new and horrendous" twist to the shelter debate by "justifying the murder [*sic*] of anyone who dared to usurp" one's family refuge. Few indeed conceded, with Dr. Paul Ramsey of Princeton University, that my code of emergency morality might be the one to which we are impelled, "not by the cold light of reason, but by the warm light of Christian charity."

I do not apologize for the article, even though it posed some very unpleasant choices for a national community in panic and threatened with that shocking relaxation of normal social bonds which must be anticipated in any realistic forecast of thermonuclear war. Indeed, I am happy that my controversial discussion, by evoking the unwelcome thought that some of us may be driven to liquidating our neighbor even before Mr. Khrushchev's bombs can incinerate him, helped to highlight the essentially moral aspects of the Great Shelter Debate that is now engaging our attention.

For as *Look* will say in its December 5 issue (p. 21), "A moral debate with few parallels in American history is sweeping across the land. . . . Should every American prepare to claw his own hole in the ground in a lone effort to escape the lethal radiation of a possible nuclear war?"

The chief elements of this debate are well worth noting, and after following its development for more than a month, I believe they can be expressed in a series of rather excruciating questions:

■ Are all forms of fallout shelter a cruel governmental hoax foisted upon our people? Are they a fraudulent type of insurance against what must be an unbounded catastrophe? Does reliance on shelters encourage acceptance of nuclear war, even though such a war involves the death of civil society and the survival of no more than a barbarous tribe of mutants for whom life is short, brutish and raw? It is amazing how many people, paralyzed by despair and the fear that any nuclear war must end in the total ruin of all that they value, greet every proposed shelter program with the cry: "Noah's Ark? Include me out!"

■ Are all shelter precautions "immoral" because they either siphon off the energies that should be devoted wholeheartedly to the search for peace, or because they upset some delicate "balance of terror" and thereby bring on the inevitable holocaust that must terminate the open-ended arms race? Many are expressing the perilous view that a vote for shelters is a vote for nuclear war.

■ Is the family shelter in particular an immoral device, because it puts survival on a competitive basis that favors the affluent and leaves the poor with nothing more than the hope of seeing God sooner? Again and again, in the shelter debate, the note is sounded that private indulgence of the instinct of self-preservation is not just un-Christian but subhuman, and that the people who build private shel-

240

ters are an inferior breed who cannot possess the pioneer virtues that would be needed on the radioactive frontiers of a post-Armageddon Era.

■ If shelters do give some assurance that we can maximize human survival in a moderate nuclear assault, or at least insure the continuance of a germinal community after a saturation attack, then where does the primary duty of building shelters lie? Is a vast investment in community shelters, planned and controlled by government, the only program that can be reconciled with the demands of distributive justice? Many are talking as though shelter construction is an unjust and discriminative prescription for survival, unless it shows a common effort to meet the common defense—something that sounds suspiciously like saying that until Uncle Sam controls the weather, one cannot buy an umbrella without doing an injury to his neighbors.

THESE ISSUES are of pressing importance. Their universality far transcends the narrow problem to which my recent article was addressed, and the confusion that exists about them calls loudly for informed discussion, especially if, in the uncertain years ahead, shelters, like indoor plumbing, must become part of our way of life.

So far, we have had little more than emotional reactions to the rationale of what may come to be called the Subterranean Society. For that matter, the Administration has not yet developed a meaningful shelter policy, although it is four months since the President took note of the urgency of the present crisis.   L. C. McHUGH

God Bless America!

George H. Dunne

ROME, MAY 23——This is one of those days which make one proud to be an American. "The Stars and Stripes Forever. . . DA DA-DA-DA DA-DA-DA." "God bless America, land that I love." "Columbia, the gem of the ocean." Land of the free and home of the brave! You can see the Statue of Liberty standing there welcoming to her shore—urging them to come—the poor, the outcast, of every color, land and creed. The melting pot of the world! Irish and English, French and Germans, Dutch and Indonesians, Algerians, Liberians, Laotians, Cambodians, Burmese and Vietnamese. "Send me your poor from every land!" What a proud boast. You can hear the bands playing, the crowds cheering; see the flags flying, the orators orating. "God bless America, my home sweet home!" This is a day to make an American hold his head high and step out lively, conscious of the eyes of the world upon him.

The eyes of the world are upon him. You know it from the moment you step out the front door here in Rome. There is the story in every paper; the great black headlines screaming from every *chiosco*; the photographs on every front page, arresting the eye, making the heart beat faster, thrusting the proud American chin up higher. Oh, it's a great day to be an American! Not just in *L'Unità*, the Communist paper; but in *Il Messaggero,* the middle-of-the-road paper, and in *Tempo,* the right-of-center paper, and in *Il Popolo,* the Christian Democratic paper, and in *Il Quotidiano,* the Catholic Action paper, and in *Osservatore Romano,* the Vatican City paper. Not just in the papers in Rome and in Italy, but in the papers in France and in Germany and in Spain and in Portugal, the papers all over Europe, the papers all over the world—rejoicing our enemies, dismaying our friends, alienating the undecided.

Let the Russians boast of their sputniks and their shots around the moon and their astronauts circling the globe. Cheap publicity stunts! Propaganda flares! When it comes to capturing headlines, Americans can give them hearts and spades and beat them every time. Look at those headlines: RACE RIOTS IN ALABAMA! . . . WHITE MOB ASSAULTS BUSLOAD OF NEGRO AND WHITE YOUTHS. . . . . TWO GIRLS AMONG INJURED TAKEN TO HOSPITAL. Thank God, that vaunted Southern chivalry lives on, come hell, high water or the day of doom!

Look at those great photographs! A young white man, blood streaming down his face, dyeing his white shirt red. Red, white and blue! Oh, it's a grand thing to be an American! He'd been riding around the country in the Freedom Bus with this crowd of young colored people, being a friend. Well, that'll show him. That'll show the world how red-, white- and blue-blooded Americans handle that kind of thing. Look at this other picture! A young Negro man knocked to the sidewalk and three big, tough Alabama cops closing in to grab him and give him the works. That should teach those Baluba tribesmen down in the Congo a thing or two. They think they're primitives. It'll take them a few hundred years to catch up with us. They don't even know what being primitive means. Look at this picture here. Brave, frightened Negroes watching the long wakeful night hours through, besieged in a church by a howling mob of whites outside. A young woman, face contorted by sobs, broken up, broken to pieces by the pounding waves of hatred beating in through the windows, through the doors, through the walls.

I know what she feels, the tearing sickness of feeling yourself hated for nothing that you have done or could undo, but only for what you are. I stood on a debating platform once in Boston and felt waves of hatred coming up at me out of the audience. They came from only part of the audience, a small part, from two or three hundred people who hated me because of what I was—a Catholic and a priest defending the Catholicism they hated. Nothing I could say or do could reach them or touch their hearts or stop those waves of hatred which were tangible, which I could feel and almost taste. When it was over, I went back to New York and into a hotel and was sick. I was sick for three days. So I know what makes this girl sob her heart out. Mine, however, was an isolated, single experience. She has had to live with hatred all her life, surrounded by it, engulfed by it, and tonight it howls about this church where she is at bay.

Yes, sir, those Communists think they are real good haters. "Imperialists," they call us. "Capitalists." Sticks and stones. We can give anybody lessons in hate. Remember those magnificent pictures a few months back that made the newspapers all over the world? Those New Orleans housewives, good solid American housewives, their faces grotesquely contorted with hate, screaming epithets at a bewildered little Negro child and her quietly courageous father? There was an orgasm of hate that even Hitler might have envied.

We have erected and maintained for a hundred years a segregated social system based upon hatred; hatred and pride, a (literally) God-damned pride in the whiteness of our own skin. A kind of skinolatry. That's what it means to be an American. Always out there in front. Firstest with the bestest. Maybe not in sputniks, shots to the moons, men in orbit, but first in the things that count, the things that *really* make the world sit up and take notice.

"There's a great day a-comin'." Yes sir, today is one of those great days, a come-and-get-it day. A day when

244

it feels great to be an American.

Feels great to be a Catholic, too. Associated with those Catholic racists in New Orleans—frequenters of the sacraments, the papers said—who keep the fires of hatred burning bright. "Keep the home fires burning." Great to be a priest, along with those priests who encouraged their racist friends to defy their Archbishop and who quietly sabotaged his efforts to destroy the white-skinned calf in his own diocese.

A day indeed for a Catholic to be proud. A day for all Christians to be proud. For Alabama, like Mississippi, like Georgia, like all of the Deep South, is God-fearing, go-to-meeting country. You may be sure most of that howling mob besieging the church and trying to set it on fire to make a holocaust to the Lord of those who huddled within will be in their own Baptist and Methodist and Evangelical churches next Sunday, piously and proudly shouting their Christian hymns.

Some years ago a Jesuit missionary, after twenty-five years in India, said to me: "Stay in America and fight racial segregation. You can do more for the missions and for the cause of freedom that way than by returning to China, because the example of racial segregation in America is the biggest obstacle we have to overcome in trying to sell people on the merits of both Christianity and democracy."

We were looking out the window upon a Marian procession of young college men and women, of surpliced priests, and a resplendently robed bishop. "As long as they can gather a few hundred well-scrubbed boys and pretty girls to walk in a procession carrying banners on a sunny day, they think God's in His heaven and all's right with the world," he said. "And all the time the earth is crumbling beneath their feet."

Since that day China has gone, Eastern Europe has gone, Southeast Asia is tottering, Africa is threatening, Latin America is in doubt. And we still talk about "prudence" and "gradualism" and reduce to silence the voice that is raised in warning or in protest and say we must tread quietly here and walk softly there.

But this is such a complicated question! As complicated as charity. No doubt Christianity has become quite complicated. But the essence of the Christian way of life remains as simple as Christ said it was: love of God and love of neighbor, summing up all of the law and all of the prophets. And unless we have this, we do nothing more than tinkle the brass and sound the cymbals. Love is a hard saying and, because in two thousand years we have not learned to practice it, the thunder gathers on the left while we sing our pious hymns and walk in pleasant processions.

But enough of these melancholy thoughts. This is a day to exult in. Unfurl the flags! Strike up the bands! Thrust our your chests! Lift up your chins! "I pledge allegiance to the flag and to the Republic for which it stands, one nation, indivisible, with (come on now, shout it out) liberty and justice FOR ALL!" That's it! Now all together: "God bless America, land that I love . . . From the mountains, to the ocean. . . ." Hold it! Hold it! Wait a minute! How did that nigger back there get in? Ushers! grab that jigaboo and throw him out on his ear. . . . That's the way to do it. Wait'll *those* pictures hit the papers. That'll make them sit up and take notice. Now—let's hear *EVERYBODY* join in. "Onward, Christian soldiers, DA, DA, DA, DA, DA. . . ."

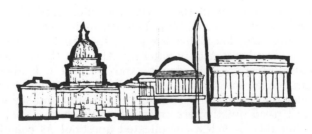

Freedom Now

I ASKED a young Freedom Rider in New Orleans last week: "How long have you felt the way you do now?" She smiled politely at the simplicity of my question. "Why, ever since I realized I was a Negro and was treated as something different."

Some thirty faculty members—priest, religious and lay—in New Orleans's two Catholic universities gave a variety of answers to my questions. Almost unanimously, they felt that the Freedom Rides were a good and probably a necessary thing. "Thank God," said one, "these youngsters aren't throwing bombs!" There was some division on the matter of timing: was this the best step here and now, or would it set interracial progress back? However, such division as existed followed lines of age, not race.

When I spoke to Negro university students, no difference of opinion could be found. It was always: "How much longer do we have to wait to be treated as human beings?" I asked about gradualism. "Our grandparents could wait a hundred years. We can't. We *have* to act." What had brought the change about? "It's simple. We read the same books as other Catholic Americans —St. Thomas, the Constitution, Lincoln. We want the same things. No more, no less." Had the explosion in Africa affected them? "Of course. At last we could feel that we didn't have to have white blood to want freedom and even to die for it."

With no pose or trace of braggadocio, one of them said quietly: "I don't want to get killed, but I will go on protesting even if it does mean I get killed." He had spent time in jail for taking part in a sit-in and had been beaten by an Alabama mob. His buddy nodded and explained: "We don't want freedom in 1963. We want it in 1961."

C. J. McNaspy

The Old Neighborhood

Katharine Byrne

E RIC LARRABEE has observed that one of the unanswerable arguments in the American conversational repertory is "for the children's sake." These are the words which may yet kill the City. This is the inscription on the banner which leads the "leavers," though there is, among them, occasional evidence of a faint regret.

Whenever I see my friend Marcia, she asks: "How are things in the old neighborhood?" In a way I think she misses it despite the blatant charm of life in far-out Homogeny. Somewhere in our conversation she is likely to ask: "Does Miss Claussen still unpack her trunkful of dolls each Christmas and ask the children in to drink chocolate out of those cups her mother brought from Sweden?" Or: "Does Mrs. Calvin still follow the garbage truck down the alley with a broom and a dustpan?" Or: "Does old Mr. Stein still wear his Prince Albert to temple on Saturday morning?" Life on a square block where there is no child over ten and no parent over forty, has made my friend lonely for some of our aging eccentrics.

I assure her that things are pretty much as they were

when she lived here, except that Mrs. Calvin is now too old to tidy up the alley, and must content herself with hollering out the window at the garbage men.

"I really should bring the children back sometime," she often says. Perhaps she has read the recommendation of the social scientist who notes that many children are growing up in great metropolitan concentrations without ever seeing the City. He recommends that thoughtful parents plan an occasional excursion back into town, presumably so that the children of this generation may enjoy the exotic sights and smells, the varied colors and customs of those who live here. (I have also heard, but do not believe, the story about the earnest PTA member who petitioned her Human Relations Chairman to find a way to import a busload of "less fortunate" children from the City to share the toys and treasures of the local kindergarteners. "How can our children learn to share," she wondered, "when all of them have the same things?")

We live in Marcia's old neighborhood, but it is really not old, even as areas are dated and outdated in a profligate City which uses up and abandons in such a hurry. It's only middle-aged, and bears up fairly well, especially in summer, which covers soot-colored stucco with bright Boston ivy, and makes green archways of the elms and maples. The streets are quiet. The alleys are where the noise is. There are collapsible pools in some of the yards, but out in the alley it's more fun to jump off an old kitchen chair into a garbage can of icy water, or play a peer-directed brand of baseball which can accommodate any child between the ages of six and twelve. Forty-year-old lilac and mock orange bushes push through the fences and over the gateways. The peonies and roses were planted by Nordic Protestant burghers who moved in when Taft was President.

When summer is gone, the sagging fences show their years and weather cracks the seams of stucco slipcovers. Only the handful of really handsome houses can stand the stark glare of winter. The Old Settlers will not be seen again until spring, unless one of them dies. Otherwise they are locked in until an April primary brings them out to vote the straight Republican ticket.

All but Mr. Ramsey, who hasn't voted for anyone since Norman Thomas stopped running. To their homes come the grocery boys, balancing on their handlebars the orders called in by the arthritic and weatherbound. The butcher's wife listens patiently. ("I'm sorry about your back, Mrs. Wesley. Will there be anything else today? Yes, I'm sure George has a nice knucklebone for Queenie. I'll send Jerry right over with it.")

Very early yesterday morning, I saw two of my neighbors leaving home. I am *not* peeking at people from behind drawn drapes at six o'clock. It's this early-rising baby beagle that has me prowling the streets and alleys at dawn. With a houseful of loving friends at all other times, at this hour he is mine alone. First we met Dr. Newland, whose wife was about to drive him to the airport. He was off to address a world congress of neuro-ophthalmologists meeting in Barcelona to discuss the control of glaucoma. A few houses later it was Mrs. Lipman who greeted us. She was off to the early shift at the girdle factory. A Polish DP, she has a five-figure tattoo on her left forearm, and other concentration camp mementos. Once in a while my eight-year-old travels around the block with hers, offering girdle "seconds" at half-price to any lady who will answer her back doorbell. And he wants to know why *his* mother doesn't produce anything a boy could sell from door to door.

This is a neighborhood of contrast and diversity. On a warm day with the windows open the incense which is wafted heavenward during the high Mass may mingle with the insistent smells of the bagel-and-onionroll bakery nearby. We have the lady who bakes cookies and loves children, and we have Mrs. Marshall. She's the one who turns the hose on the man-next-door if he dares to place a ladder a few inches over her property line when he wants to clean the leaves out of his garage gutter. We have the little girl whose clothes come from Best's and the children whose mother has discovered the joy of browsing in the Thrift Shop's perpetual rummage sale. Bobby Richardson's grandmother still walks him to school and calls for him, but there is another six-year-old who is on his own all day and may appear at your kitchen door with a breakfast quarter which he wants to trade for some orange juice and a bowl of cereal. If you are looking for a Finnish bath, a German-language movie, or a slice of almond-filled, honey-dripping Greek *baklava,* all are close by and available.

In American urban society clever and ambitious elements of the population are often able to move from

a poor ghetto to a plush ghetto in three generations. Here we seem to have bogged-down permanently in heterogeneous Phase Two. The voting roll reads: Bantsolas, Batka, Becker, Binkowski, Blumfield, Bongiovanni, Borge, Brooke, Brye, Byrne. At the other end of the line are the Yoshimuras, who always win the garden contest. There are other evidences that we are obsolete. We have a paperboy whose father is dead and whose mother *needs* the money he earns. Our backyards have fences between them, and neighbors who have known one another for years call each other by their last names. Dogs wear leashes. The other day I saw a Davy Crockett sweatshirt.

Harry Golden, in one of his syndicated slices of nostalgia, speaks in admiring reminiscence of the lower-East-Side mother of other years who used to keep the delicatessen open 16 hours a day and literally put her boy through law school with the contents of her pickle barrel. She isn't dead. Now she lives a few blocks from us. At a busy intersection, early and late, she's selling newspapers. Depending on the season, she wears a flowered housedress, a patrol boy's slicker or army boots and a combat jacket. Her boy will enter medical school this fall.

Sociologists will tell you that a good community spirit implies standards to which each member of a group feels forced to conform. Here some of us score low. While the general level for the care and feeding of front lawns is fairly high, no one gets excited because Mr. Richter, who teaches violin lessons in his basement studio, attacks his grass with a sickle once or twice a summer, and no oftener.

The evening paper features a daily column on the women's page called "Split-Level Living" or something. It's a serial chronicle about life in a certain bucolia on the edge of our metropolitan area. The author recounted an interesting episode recently. It seems that on a warm spring afternoon all the well-slacked and loose-shirt-tailed girls on her block were surprised to find one of their number out of uniform. Unaccountably, she appeared on her driveway dressed in the manner of a young woman about to be taken out to dinner. They asked her for an explanation. She told them that it was her birthday, that her husband had apparently forgotten the fact, and that she hoped that this subtle hint, when she picked him up at the station, would inspire him, et cetera, et cetera.

Now her motive and method are routine and incontestable. The fascinating point, to an outsider, is that all her neighbors knew what she was wearing, that they cared, and that they demanded and received an explanation for her wearing it. Fierce old Mrs. McCorkle, in a wide-skirted flannel nightgown and with an iron gray braid hanging to her waist, has been seen weeding her moss roses in the early hours. On our block the item is not considered newsworthy. And if it were, I wouldn't want to be the one to ask why she doesn't wear a more suitable gardening garment.

Not far away lie the newly-hardened arteries, the expressways built to move the hot-and-cursing or the cold-and-cursing multitudes who daily fight their way to the City's edge and far beyond. What has kept us from joining them? Is it social inertia, an unwillingness or an inability even to approach the starting line for the great push which engages the best energies of so many able people? Or is it because the woman of this house cannot face the prospect of cleaning out an attic crowded with tropical fish apparatus, the operating room of an abandoned doll "hosbidel," all those little pens from the time the guinea pigs had guinea pigs, fifty pounds of moldering raw data from one doctoral dissertation, who-knows-how-many Halloween costumes, and the boots which keep accumulating because no one ever seems to be the right size at the right time of the year?

THE TRUTH IS, of course, we like it here. And this in spite of the fact the City never had a good press. As far back as the days of Dick Whittington, it has been beaten down in print as a source and a magnet for the forces of evil. In our day, advertising, the great status-conferral instrument of our culture, has done nothing to improve the image. Who ever saw a boy in a Crest commercial come bounding up the steps of the ancestral two-flat to tell his mother he had no cavities? Who ever saw a beautiful young wife toss the laundry in with the sponsor's product and walk carefree from the basement of a 35-year-old octagon-front bungalow?

The City, "like a patient etherized upon a table," lies naked to the probings of the attending social theorists. What is the prognosis for one of its well-preserved but aging parts? The mature community, as they call it, cannot long prosper on the limited commitment of the elderly, watching life with waning interest from the sidelines of their screened front porches. It needs bright and scrappy young parents, interested, involved. People who will fight with the alderman or petition the pastor, if necessary, and whose move in this direction is not an interim expedient, but a positive choice made —and why not?—"for the children's sake."

If they come back to the Old Neighborhood (and I do not see them in great numbers), it will be because they find something of value here in spite of embarrassing deficiencies. (Eighth grade graduation parties are backyard barbecues with little brothers and sisters hanging around; 16-year-olds walk to school or take a bus; there is no local of the John Birch Society.) They will come, if they do, as independents bucking a dominant mobility pattern. Not as self-consciously different, perhaps, as the society column's "young marrieds," buying up the central city's "divine old houses with the magnificent leaded panes in the bay windows," but different, nevertheless. Candidates for the last place in the list of those whom the predictors suspect may ultimately inherit the City: "the very rich, the very poor, and the slightly odd."

On Being Slightly Odd

ANYONE who has lived through a generation which gave rise to successive epidemics of Davy Crockett hats, Jack Paar viewing and collegiate conservatism must be aware of the deep-rooted American compulsion to conform at all costs. Then, too, what contemporary parent hasn't wilted before the lament: "Aw, gee, all the kids are doing it"? From the cradle on, the pressure bears down on one—conform or (the horror of it!) be considered "slightly odd."

Now conformism in the matter of coonskin caps presents no serious problem for most people. The same goes for many of the other manias that capture the people's fancy and defy common sense. But for American Catholics the urge to be part of the faceless throng can be a real threat at times. Take, for instance, some prevailing family-life patterns in the typical American suburb.

The rub here, as Fr. John L. Thomas, S.J., leading Catholic student of family-life sociology, put it in an address at the recent 23rd annual convention of the American Catholic Sociological Society, is that "the Catholic family is clearly defined from the point of view of norms—its philosophy and theology."

History shows only too clearly that these norms have faced challenges from the general culture in the past. But for American Catholics, particularly for young parents, the difficulty lies in the fact that the move to suburbia puts them for the first time in a position of finding their "way of life in sharp conflict with what is considered normal by other Americans." It's a question that deserves close attention.

In this issue of AMERICA (p. 41), Katharine Byrne points the finger, in slightly less direct fashion, at the same sore spot. Speaking of self-exiled former residents of a big city's "Old Neighborhood," she suggests that a visit back there will help them to "find something of value" in the community life they abandoned when they fled over the expressway to cook-out land.

What could be more quaint, indeed, from a split-level viewpoint, than celebrating eighth-grade graduations by "backyard barbecues with little brothers and sisters hanging around"? Or more primitive, in the eyes of some Crestview Manor, with its fierce parking problem at the local high school, than the sight of 16-year-olds who "walk to school or take the bus"?

Neither Mrs. Byrne nor Father Thomas, we feel sure, is plugging a back-to-the-farm or even a back-to-the-metropolis movement for their fellow Catholics. What both hint at, however, is that there may be some virtue, not to mention necessity, about being "slightly odd" in a pluralistic society such as ours. Their observations make it clear, too, that the time has come to pose a few questions about the impact of so powerful a social force as advertising, "the great status-conferral instrument of our culture."

Whether it be a question of eating fish on Friday when everyone else is biting into prime beef, or the more serious issues of early dating and other matters subject to family decision, the Catholic has to recognize that living up to his faith inevitably demands on occasion that he be *different*. What makes the demand more painful for many of today's Catholics is the fact that they are more than ever cast as members of a minority "trying to preserve its values when the rest of society is not."

Angular Catholicism for the sake of angularity has nothing in particular to recommend it. But the risk of being labeled "slightly odd" by the conformist herd is a price worth paying for the sake of basic Catholic ideals.

On Loving the Poor

William A. Schumacher

S T. VINCENT DE PAUL once said: "The poor are our
most demanding masters; we must love them so
they will forgive us the bread that we give them."
We rarely think of him as a master of paradox, yet in
that epigram St. Vincent struck, three centuries ago,
a profound truth about Christian charity. It seems
strange at first glance to think of the poor as our masters; we tend, rather, to regard them as our inferiors,
the recipients of our liberality.

But St. Vincent was looking to the real fruit of Christian charity, the spiritual benefit of the giver. It is
axiomatic in theology that pure, disinterested love of
God is impossible; the virtue of charity—that is, the
love of God for His own sake—will always have overtones of our own spiritual profit from the very act of
loving God. This is also true when the same virtue of
charity is directed towards its other proper object—
namely, the children of God, who are to be loved for
His sake. When we truly love the poor, we stand to
profit thereby. The poor, then, are really our benefactors, our masters, who have the right to demand love

in God's name, and who confer on those who love them an eternal benefit.

A VISIT TO THE POOR

However, we cannot love what we do not know. True love is intensely personal; it cannot be accomplished by remote control. We cannot love the poor merely through a check or a gift; we must first come to know them. Come with me into the homes of the poor, here in the affluent United States, 1960. Whatever their race or color; whatever their religion or lack of it; whether they speak broken English, sloppy Spanish, or in the heavily accented tones of the Southern hills and fields—all the poor are very much alike. Climb the steps to the third-floor rear of a tenement, making your way past abandoned baby buggies, old tire casings and discarded clothing. Be careful to step over gaps where boards are missing from the stairs. (When there are no locks on the doors, removing a few planks from the back steps makes an effective barrier against surreptitious entry into a third-floor flat.) Don't forget to look for the poor living in so-called cottages, shacks located on the backs of lots which seem to be occupied completely by small factories and filling stations. You will find the poor there, too, down that littered path and behind that broken fence—living where you would never expect to find anyone at all.

Wherever they live, you will find the poor enveloped in an all-embracing odor, with mingled scents of unwashed bodies, moldy clothes, strong cooking smells, escaping gas and, in winter, the piercing reek of fuel oil from their stoves. The buildings in which the poor live were erected some fifty to eighty years ago; time and neglect have taken their toll in broken door jambs, sagging floors, peeling walls, leaky pipes and balky plumbing. Be sure to bring a flashlight if you make this trip at night; it will spare you many a bad fall and embarrassing situation.

Beyond learning where the poor live, we must meet them face to face. From a natural viewpoint they are not very attractive people. Defeated by life, old before their time, they seem suspicious at first acquaintance.

When you come to know them better, you will learn that this is a deep-seated shrewdness born of years of experience in detecting phonies and frauds. Surrounded by crudity, living in the midst of every sort of degradation, accustomed to disease and corruption, grown used to all varieties of sin from childhood, these people cannot be fooled or tricked. All the effects of original sin are well known to the poor; the weaknesses and shortcomings of human nature are drilled into them every day of their lives. Hammer blows of human cruelty and selfishness have fallen on their hearts, uncushioned by the conventions that protect the rest of us. What wonder, then, that something twisted and misshapen has occasionally resulted.

These, then, are the poor who are our most demanding masters; these are the people whom we must love if we are to love God. But what is lovable about them? Only the image of God in their souls is an adequate object of true charity. The whole problem lies in seeing this image in such a place. Philosophy tells us that God is Goodness, Truth and Beauty. His image is hard to find in those whose lives are choked with evil, lies and ugliness. Only the virtue of faith, the acceptance of God's word that these are, indeed, His beloved children and the brothers of His Son—only such faith makes this kind of love possible. In the world's eyes they are worthless, yet the price which God paid for each one of them was the death of His own Son on the cross. Only the crucifix is the correct price tag for each of the souls of the poor.

But once we have come to know the poor, then we must learn to love them. More than this, we must learn how to love them in such a way that the expression of our love will not corrupt its spiritual fruits and make a hollow mockery of the love itself.

We cannot truly love the poor with "charity" in the do-gooder, Lady Bountiful sense, because this sort of charity is often only a disguised selfishness. Those "charitable Catholics" whose acts of love consist of graciously bestowing a gift on some waif with a photographer in attendance remind me always of Christ's story of the Pharisee, who gave his alms to the accompaniment of blaring trumpets. In His words, "they have had their reward." If they have had their reward in public praise, whether by a fanfare of trumpets in the streets of Jerusalem or by a picture on the society page, there is no spiritual reward to be gained. What could have been profitable for eternal life has been wasted in attempting to buy the fickle praise of men.

This love of God in His own poor will never be carried out in practice merely by writing a check, even for those whom we have come to know. What young lady would take a lover seriously who made his proposal by mail, money order enclosed? If we claim to love the poor, it must be with the qualities of a lover—a burning personal interest in their happiness both here and hereafter, a concern for their welfare in every sense, a compassion for their weaknesses, an understanding of their all too human frailties.

It would be easy to love the poor if they would respond in kind; it is very difficult to offer love to those who take it for granted and spit in your eye. Yet only this kind of love is worthy to be called by the noble

name of Christian charity, since Christ our Lord loved us in this way. As St. Paul reminds us: "It is hard enough to find anyone who will die on behalf of a just man, although perhaps there may be those who will face death for one so deserving. But here, as if God meant to prove how well He loves us, it was while we were still sinners that Christ, in His own appointed time, died for us" (Romans 5: 8-9).

The poor are shrewd, and all our protestations of

interest and concern, all our well-meaning compassion and understanding will be rejected as a sham and a fraud unless it is backed up by self-sacrifice. This is the ultimate test of love, the willingness to put oneself out— in time, in convenience, in cold cash until it hurts. The poor are realists; they know from bitter experience that pious words do not put food in children's mouths, nor do beautiful sentiments put shoes on their feet.

But if merely writing a check is not enough to love the poor, how then are we to do it? Did you ever think of learning about a poor family and then visiting them, literally feeding and clothing them? An experience of this sort can teach us more about poverty and its effects than all the sociology texts ever written. It can also be quite humbling and thought-provoking.

WHEN LOVE IS PRACTICAL

Could you pick up a poor expectant mother and drive her to the prenatal clinic, saving her a long walk in the cold? Could you visit some of the most pathetic poor people of all, the old ones whose lives are leaking away in a dark, forgotten room, surrounded by the pitiful relics of a lifetime of suffering? Could you stomach a trip to a county hospital, to an old people's home, to a Catholic hospital for unmarried mothers—not just a visit, a tour of inspection, but hours of heart-rending suffering perhaps, just sitting there and listening to the sick poor pour out the tragedy of their lives? Would you have the love and patience necessary to teach the catechism to a child not quite mentally deficient enough to be placed in an institution? No one who has never tried these works of mercy can know what bittersweet experiences they can be, how our very understandable natural loathing can be turned into a spiritual joy by the alchemy of God's grace.

Loving the poor is not something optional for a Christian; it is of the very essence of his faith, and it makes sense only in the light of that faith. It is easy to love lepers in Asia, when someone else changes their bandages; it is easy to love orphans in our own city, when devoted Sisters care for them day and night. But it is repulsive, it is awkward, it is sometimes em-

260

barrassing, it can make us uncomfortable to think about really and truly loving the poor in person.

When the followers of John came to Christ to ask if He were the Messiah, our Lord made the preaching of the Gospel to the poor the sure sign of His office as Saviour. Love of the poor has always remained the certain mark of His true followers, of His real lovers. As His beloved Apostle, St. John, remarks: "Yes, we must love God; He gave us His love first. If a man boasts of loving God, while he hates his own brother, he is a liar. He has seen his brother, and has no love for him; what love can he have for the God he has never seen? No, this is the divine command that has been given us; the man who loves God must be the one who loves his brother as well" (I John 4: 20-21).

Christ Himself has given us a very practical reason for this personal love of the poor. On the most momentous day of our lives, He will say to the elect: "I was hungry, and you gave me food, thirsty, and you gave me drink; I was a stranger, and you brought me home, naked, and you clothed me, sick, and you cared for me, a prisoner, and you came to see me. . . . Believe me, when you did it to one of the least of my brethren, you did it to Me." And others will hear, to their eternal sorrow, that the contrary of this is also true: ". . . when you refused it to one of the least of my brethren, you refused it to me" (Matthew 25: 35-46).

The Phenomenon
of Man

Francois Russo

S. J.

CONSIDERABLE ATTENTION has been given recently
to a book translated from French into English
and entitled *The Phenomenon of Man* (Harper).
It is the most important of the works that came from
the pen of a man who was both a member of a religious order and a well-known scientist. Born in France
in 1881, in the Province of Auvergne, Pierre Teilhard
de Chardin entered the Society of Jesus in 1899 after
completing his secondary education at the Jesuit College of Mongré, near Lyons. He was ordained to the
priesthood in 1911. From that time on he dedicated
himself to the so-called earth sciences, and above all
to paleontology. His scientific work made it necessary
for him to spend long years in China, and to travel
widely. He returned to Paris in 1946, and then in 1951
went to New York, where he died suddenly on Easter
Sunday, April 10, 1955.

Fr. Teilhard was a member of the French Academy
of Sciences. During his latter years in the United States,
he was attached to the Wenner-Gren Foundation for
Anthropological Research.

The author of *The Phenomenon of Man* was a scientist of international repute, to whom are owed several

extremely important contributions in the field of human paleontology. Moreover, Fr. Teilhard was a man haunted by the apostolic needs of his time. All through his life he struggled to understand the human and religious meaning that lies behind the advance of the sciences and particularly of biology. Over and above his directly scientific works, he is to be credited with numerous writings of a general character that treat large philosophical and religious problems, especially in so far as they touch upon the sciences. The *Phenomenon of Man* is but the first volume of these works; four other volumes have already been published in France (Le Seuil, Paris). These writings of Fr. Teilhard often seem rather to resemble sketches, or the presentation of a piece of research as it might be proposed to intimate friends and not to the larger public, than the expression of fully elaborated thought. But *The Phenomenon of Man* is one work that Fr. Teilhard most explicitly wanted to see published. In its pages he brought together the essential and guiding ideas of a world-view which he kept progressively sharpening and reformulating all during his life, and whose basic insights date from 1916. This book, which was finished in 1947, was published only in 1955, a few months after his death.

One can distinguish in *The Phenomenon of Man* two levels of its author's considerations: on one hand, a *description* of the evolution of the world, where certain elements of hypothesis are already present; on the other, an attempt at an *explanation* of this evolutionary process.

TWO ASPECTS

For Fr. Teilhard the evolution of the world is viewed in four stages: pre-life (*la prévie*), life, thought, and hyper-life or a stage beyond life (*la survie*).

During the period of pre-life, matter passes from an undifferentiated state to that of organized forms which represent, first of all, the elements, and then bodies composed of a more and more complex structure. All this while, the universe in its totality is setting in motion a process of expansion which it will continue to pursue all through its history.

The appearance of life represents a passage through some sort of threshold. From this point on there begins a continuing development of living forms that mount up into more and more organized species—a process that ends in animal forms very close to that of man. Life has certain well-determined "manners," and its qualitative progression is clearly measured by means of what Fr. Teilhard calls a "parameter of cephalization." This means that the farther we advance in evolution, the more the complexity of the brain increases, at least for certain phyla. Among these phyla, there is one of "pure and direct cerebralization," which leads to man.

We find here, traced by the hand of a master, an admirable tableau of the story of life. These precise and stirring pages constitute, beyond all doubt, that part of Père Teilhard's work which is the most scientific in nature and also the least debatable.

After the appearance of life, the emergence of man constitutes a new and critical transformation. It is "the momentary pause of reflection." It is an extremely important moment in the history of life, and we shall perhaps never succeed in elucidating the phenomenology of its process with total clarity. The author writes: "At that point earth is reborn, or better, it finds its soul." This part of the book contains a masterly treatment of the paleontological problem of the origins of man—an exposé that requires only a few retouchings here and there in order to be brought up-to-date with discoveries which have taken place since it was written. These stirring analyses truly make us understand that "man is the ascending arrow of the great biological synthesis."

In a final stage, we confront the prolongation of evolution in the development of humanity. This development is presented essentially as a phenomenon of *convergence*. By this is meant a "coming together" of humanity—a process which in no way tends to form an impersonal all, but rather to effect a union which allows for differentiation and distinction within an ever more narrow solidarity and through a ceaseless interpenetration—the end-product of consciousness being constantly enlarged and augmented.

AN ULTIMATE CONVERGENCE

Thus, we see that the synthesis that Fr. Teilhard presents to us has evolution as its basic principle of explanation. Recognized at first in the realm of living forms, evolution is extended to the whole ensemble of the universe. Thus, there was already an evolutionary process at work before the appearance of life, and then, after the emergence of man, evolution goes on following its course from the very start in man's biological form. Finally, evolution is more and more evident in the course of the progress of humanity.

Each of these great stages of evolution, of course, has its own proper character, but the same fundamental dynamism runs through and sustains them all. Thus, what is to be has already been "fore-announced" by that which is and by that which has been. The evolution of matter somehow prefigures life. In fact, this is precisely why Fr. Teilhard calls it pre-life. To speak in general, "nothing could one day emerge as final on any of the divers thresholds—no matter how crucial these thresholds are in the successive leaps of evolution —which was not first of all obscurely primordial." Hence, we are in the presence here of a "cosmogenesis" whose "tendrils insist on reaching up and being prolonged in us. . . . Evolution is on its way into the realm of the psyche. . . . Thus, thought has part with evolution." Making his own the formula used by Julian Huxley, Fr. Teilhard dares to declare that "man discovers he is nothing other than evolution become self-conscious."

With respect to the convergence of humanity, this phenomenon appears as "a subtle shifting of the onward flow of evolution." Humanity, in its first movements of expansion over the earth, tended to branch off; now, however, it converges toward a point that constitutes the "superior pole" of all evolution—what Fr. Teilhard proposes that we call *Omega*. This final stage of evolution is of an "ultrapersonal" nature; at this point persons come together in unity and self-realization. (We have here a conception that is completely opposed to the totalitarianism of the Marxist when he discusses the "end" of humanity, a destiny wherein we discover in only a very feeble degree anything like care for the individual human person and his dignity.)

In order to clarify and give an account of this "upflow" of life to achievement in *Omega,* Fr. Teilhard appeals to two pairs of notions: that of "complexity-consciousness" and that of tangential and radial energies.

Growth in the complexity of beings is accompanied—explicitly, at least, in the realm of animal life—by progress in psychic development. This fact leads the author to lay down as a general principle the notion that consciousness and complexity are but the "two faces"—one internal, the other external—of a single phenomenon that keeps showing itself all through the course of evolution. Hence, all things in a certain sense have a kind of "inwardness"; even inert matter has this character. The degree of complexity-consciousness constitutes the fundamental parameter which allows us to measure progress in evolution and the dignity of the beings that evolve. Thus, man, although quite lost in the immensity of space and time, appears as the summit of the cosmos, as its final achievement and its résumé, because he is the most conscious and the most organized of beings.

From the pair of notions that we have named complexity-consciousness, we pass on to the second pair, that of tangential energy and radial energy. Thus, the dynamism which, in certain areas of the universe, tends to go along certain privileged lines of development, ever at work to assure the growing complexity of being, cannot be understood except as springing from an energy of arrangement, called radial energy. We say "radial" energy but it is a dynamism which assures a progression of development. This energy comes to be associated with energy in its classical sense, the energy that Fr. Teilhard calls tangential. He calls it tangential because he conceives it to be only an element of the conservation of beings, and to be unable to assure the forward march of evolutionary progress.

NEED FOR SUCH A VIEW

It would be impossible to overemphasize how opportune is the effort at synthesis that is outlined in *The Phenomenon of Man*. New scientific data, especially in the realm of biology, have up till now remained quite unrelated. This makes it all the more necessary to bring these data together into a coherent and total picture. Such precisely is the task which so preoccupied Fr. Teilhard, and it is to his great credit that he was able to put himself in a frame of reference which—transcending the narrow scope of cosmologies that were completely exterior to man, and with which we have been too easily content up until now—is ample enough to permit the integration of man himself into cosmology. Such an enterprise is by no means a mere intellectual game or some sort of fancy speculation. It answers a pressing need of our era. Today, looking beyond the partial and disjointed world-views which have up till now been offered us, we seek a global insight. We want such a view to be founded on general principles, of course, but it must also take into account all the riches of scientific findings.

Fr. Teilhard was blessed with exceptional gifts as he set out to accomplish this purpose. He had great powers of analysis, wide culture, a forceful imagination, an aptitude for synthesis and, above all, that largeness of mind which scorns the shabby and the second-rate and puts no limit to its ambition to understand.

True, certain of Fr. Teilhard's descriptions and formulas have a certain poetic attraction about them, but it must be kept in mind that for him poetry constituted a genuine method of the phenomenological study of reality. I repeat that despite its appearances of poetic form the thought of Fr. Teilhard should not be ranged alongside the thought of authors of those numerous cosmogonies that are more or less literary or philosophical in nature, but which are also so careless in taking account of details of reality. Teilhard's formulations are at times daring, but the views he expresses in *The Phenomenon of Man* have their origin in the realm of fact. These views are deeply rooted in reality. His master ideas were born and matured in the womb of a real world with which the author's long career in research made him intimately familiar. Moreover, although these ideas may evoke criticism, they cannot be lightly pushed aside; for they embody a very remarkable attempt to answer the fundamental cosmological problems which preoccupy those gifted enough to possess at once a feeling for research and a desire to master difficult problems independently of immediate results.

Moreover, despite his highly personal style, we ought not be afraid to compare the work of Fr. Teilhard with those of a Descartes or a Leibniz. Like these men, Fr. Teilhard was haunted by the desire to grasp reality in a truly profound way, and to account for and unify that multiplicity of facts that has been brought into evidence by science. In his case as in theirs we find profundity wedded to true mastery of thought. He knows what he wants to say; he is not satisfied with words nor content with pseudo-explanations. What he affirms in his book is not the result of some totally conceptual deduction, but is intimately bound to an "experience" in reference to which he understands how to give an account  of the real and intimate structure of things.

We are led quite naturally to ask whether views that are so ample and so rich really belong in any properly exclusive sense to the field of science. As he expresses them, are they not already filled with philosophy, and even with ideological considerations?

It is not easy to answer this question. Like all great creative minds, Teilhard was not slowed down by methodological discussions of any kind. Such exercises in the minutiae of scholarship seem never to have greatly interested him. Nevertheless, he was anxious to emphasize that his research was based exclusively on real phenomena, and that it was not dealing with either the principles of metaphysics or the data of revelation. Metaphysics and revelation certainly had a part to play in his thinking, but the views that we find expressed in *The Phenomenon of Man* are developed above all on the basis of the reality of phenomena that we can observe and understand with the help of the scientific method. In all truth, Fr. Teilhard could say that "in so far as man can distinguish in himself different levels of knowledge, it is not the believer, but rather the naturalist, who speaks in this book and asks to be understood."

Be that as it may, the views found in *The Phenomenon of Man* do not seem to be of the same type that one is accustomed to encounter in properly scientific treatises. Moreover, even if Fr. Teilhard should seem to us to have turned aside from science, this is not because he poses daring hypotheses for our consideration—every great scholar and savant does the same— but rather because his hypotheses introduce notions and considerations which lack that mark of objectivity and precision to which science is commonly accustomed. The "inwardness" of things, the radial energy of arrangement—concepts which for Teilhard are fundamental themes—are not truly scientific notions at all.

Must we therefore brush them aside as though they had no value? Not at all. Bear in mind that in the realm of physics the distinction between science and philosophy can be made without too much difficulty, but that it is not so easy to make the same distinction in the field of biology, in spite of the progress that has been made in methodology. Particular difficulties arise in the study of evolution, the field to which the aforesaid concepts have reference.

It is by no means proven that one can neatly distinguish the "scientific" and the "philosophical" in a study that sets out to understand life itself. If we attempt to imprison ourselves in too rigid or too scholarly a separation of these two orders of knowledge, we run the risk of impoverishing and stunting our investigation. Numerous treatises written in this narrow and rigid spirit give proof that such impoverishment is not only possible but real.

Some will object that Fr. Teilhard, by transcending positive science, is really engaged in philosophy. If so, then he should have concerned himself to a greater extent with principles and notions that have been tested. There can be no quarrel with this. But it must also be remembered that precisely here, in the field of cosmology, our classical views and concepts, valid as they may be fundamentally, today demand reappraisal. A work like that of Fr. Teilhard can be of great help in this regard. At first blush, of course, notions such as the "inwardness" of things and radial energy appear to be somewhat baffling, but he who reflects even a bit seriously on the world as made known to us by contemporary science will find them illuminating.

The author of the work in question was a philosopher in action rather than a professional philosopher. There can be no disputing the fact that Fr. Teilhard lacks the technical sureness of touch of a trained metaphysician. Yet his cosmological analyses are not for that reason of diminished importance. We may dare to criticize them or to revamp them, but if we are really anxious to enlarge the scope of cosmology in order to integrate into it all recent scientific knowledge, we must take his views into consideration. It is impossible to imagine how anyone could continue to teach cosmology without making generous allowance for the views of the author of *The Phenomenon of Man.*

CRITICISM OF TEILHARD

Thus, we see that the perspective or frame of reference in which the work of Fr. Teilhard is situated can be completely justified. Moreover, we may safely concede the riches and the profundity of the synthesis he proposes. Nevertheless, his thought calls for criticism on several points.

On the plane of positive cosmology, and quite independently of any philosophical or religious views to which it might seem to lead us, Fr. Teilhard's synthesis seems at fault in giving too great a place to evolution, and in presenting views as totally valid which in fact arc still under discussion. Let us consider this point with particular respect to the realm of the physical world. Up to the present, while the thesis of the general evolution of the universe has by no means been set aside, it is equally true that it has not met unanimous acceptance in competent scientific circles. This general theory has as its most precise formulation the theory of the expanding universe, but today there are some who would oppose to the theory of the expanding universe a quite different theory—a theory of a pulsating universe, or a theory of irregular evolution; these, like the theory to which they are opposed, also rest on plausible foundations. Nevertheless, so far as the formation of the elements is concerned, the idea of the evolution of matter upward from an undifferentiated state toward more and more highly structured stages, which are represented by chemical elements of increasing weights, seems probable enough.

Moreover, the tight link which is asserted by Fr. Teilhard to exist between the world of pre-life and that of life in the general movement of evolution is not so solidly established as he affirms it to be. The "juncture" between the physical world and the world of life remains a problem which Fr. Teilhard has clarified less satisfactorily than he seems to think.

When he deals with questions related to the development of civilization, Teilhard appears to rely a little too exclusively on evolution. Evolution may be quite an essential aspect of human history, but it is difficult to go along with the seeming evaluation of Fr. Teilhard that evolution is the *only* principle by which to explain the dynamism of civilization. To insist too much on this one principle is to seem to fail to recognize realms like the history of art or the history of philosophy, which are likewise very important. Evolution, with its principle of linear growth, is not able fully to account for them. Moreover, in these fields we come upon patterns of evolution of a quite different type—cycles, regressions, etc.

When we regard it from the point of view of philosophy and religion, *The Phenomenon of Man* is again found to be other than completely satisfying. Let us be very clear on this point. For the reasons outlined above, we ought not see in *The Phenomenon of Man* a metaphysical or theological work, and hence should not attempt to judge it as such. As is obvious, Fr. Teilhard has propounded certain views of man and his destiny which have a bearing on philosophical and religious problems. In fact, in a final chapter of his book, he has a treatise on the "Christian phenomenon." But here again he is dealing with considerations of *fact* and not with properly *doctrinal* affirmations.

In a word, Fr. Teilhard has in no way attempted in *The Phenomenon of Man* to give us the *total truth* regarding man. Such truth can come only from metaphysics and, especially, from revelation. Teilhard said this explicitly in his preface: "Let no one seek here a final explanation of things, a metaphysics." And again: "The book that I have written should not be read as one might read a work of metaphysics, and much less as one might peruse some sort of theological essay."

However, despite the care that the author took to write only within the framework of the phenomenal order, and to restrict what he has to say to that order, Fr. Teilhard—precisely because of his remarkable ability at synthesis—has presented us with a set of views on man and the world which to some little degree gives the impression of being *self-sufficient*. This, of course, is a problem that he himself recognized, for he said: "It is impossible to attempt a general interpretation of the universe without seeming to explain it from top to bottom."

The reader must not fail to note this problem about the work in question. One might wish that Teilhard's synthesis were *more open, more unfinished*. If it had been, the book would then be more useful in the work of building up a complete vision of man and the universe. This vision would have been one which, since it summoned up other sources than mere scientific experience, would have found a place for the very mystery of our freedom—for that mystery of evil which is not simply the mystery of the evil that we undergo (a problem which cosmology can discuss)—but also of that evil which is the product of our wills. This sort of evil is not very directly treated in *The Phenomenon of Man,* even though Fr. Teilhard was at pains to note that the fortunate evolution of the world that he envisaged depended on the free decision of man.

One should not try to find things in *The Phenomenon of Man* that are contrary to the faith. Even where Fr. Teilhard's views leap beyond the realm of strict positive evidence, they always allow complete scope for the teaching of the Church. It is impossible for us, here within the limits of this article, to examine this point in detail. Let us look at several essential observations.

The immanence and the spontaneity of evolution by no means rule out the transcendence of God. After all, evolution is only the phenomenal aspect of divine creation. Moreover, God's transcendence is demanded by the very exigencies of human action which, with its eternal striving for immortality, must constantly refer itself to a God who transcends the world. We have here one of the fundamental points of the thinking of Fr. Teilhard, and one that in other of his writings he calls the dialectical meaning of his thought.

Still other critics have felt obliged to note that in *The Phenomenon of Man* an excessive primacy is given to continuity in the development of beings. But this is to fail to recognize in Fr. Teilhard the notion of transformation, where continuity and discontinuity are indissolubly complementary to one another. Such critics also fail to recognize that the particular intervention of God, who creates the human soul, cannot be fully expressed on the phenomenal level, where we can find nothing more than a reflection of the mystery of creation.

Furthermore, so far as the mystery of original sin is concerned, is there any need to add that we would be guilty of a grave error of methodology if we reproached Fr. Teilhard for not having given this mystery an organic place in the cosmological synthesis he was making? In fact, it cannot be recognized in experience, because original sin is something which, in the last analysis, we know through revelation.

When examined attentively, the synthesis of Fr. Teilhard is seen to be less complete and less satisfactory than it might appear at first glance. Moreover, it is proper to read a book like this with a mind forewarned in regard to certain points of interpretation. But be all this as it may, Fr. Teilhard's synthesis is nonetheless a major contribution to our thinking about the world. Moreover, we must remember that for a world fascinated by the progress of science, a synthesis like that of Fr. Teilhard opens great windows to faith, and this in an age when such approaches are more than ever indispensable. One other notable achievement of this synthesis is that today, when the Marxists are attempting to monopolize the field of science, Fr. Teilhard demonstrates that science cannot find its fulfillment except in a spiritual dimension.

Let us therefore take up this book with the same simplicity of heart which caused its author to write: "I might well have deceived myself on a number of points. Let others, then, try to do better. All that I wanted to accomplish is to have made people feel the difficulty and the urgency of the problem, to comprehend its immense magnitude, and to perceive the form in which a solution must inevitably be found."

Should Men Hang?

Donald R. Campion

O N OCTOBER 21, 1959, Caryl Chessman won his seventh legal stay in an eleven-year battle to avoid California's gas chamber. The convict-author of the autobiographical best-seller *Cell 2245, Death Row* first heard the death sentence pronounced over his head in the Superior Court of Los Angeles on June 25, 1948. Now, two days before the U. S. Supreme Court intervened, Gov. Edmund G. Brown (despite his personal opposition to capital punishment) had turned down a plea for clemency submitted by Chessman's attorney. The execution had been scheduled for October 23.

Chessman's success in staying the law's avenging hand by complex legal maneuvers had long ago aroused international interest in his case. His book—a tract against the death penalty—won recruits to the abolition movement in California and around the globe. More than one prominent voice in other lands was raised on his behalf. *Osservatore Romano,* the Vatican City daily, recently labeled Chessman's "eleven years of awaiting the gas chamber" as an example of "immovable justice which has become supreme injustice."

Over the centuries the law has created a series of devices designed to protect the rights of any condemned person. The extreme uses to which Chessman and others have been allowed to push them serve to dramatize society's inner hesitancy about taking a life in payment for any crime. In a sense, then, the fate of the highly articulate prisoner in San Quentin has simply brought to the surface many already existing doubts about the wisdom of the death penalty. Should men continue to be executed?

Similar misgivings arose during the recent Truscott trial in Ontario. In the Canadian town of Goderich, on September 30, 1959, a jury found Steven Murray Truscott guilty of murdering a 12-year-old girl. Under Ontario law the death sentence is mandatory in such a case. What drew widespread attention to the trial was the age of the defendant—a boy of 14. In all likelihood, public opinion will never permit Truscott to be hanged. Indeed, many are now beginning to wonder whether a law which allows for the possibility of such a sentence is any longer socially desirable.

Dr. Benjamin Rush, pioneer psychiatrist, patriot and leading medical man of his day, fired the opening shot in the American campaign against capital punishment. Before a meeting of his fellow Philadelphians gathered in Ben Franklin's home in 1787, Rush read a paper entitled "An Enquiry Into the Effects of Public Punishment Upon Criminals and Upon Society." The substance, if not the words, of his case against hanging echoed as recently as a year ago in legislative discussions on the death penalty in California, Connecticut, Illinois, Indiana, Missouri, New Jersey, New York, Ohio and Tennessee.

A TREND TOWARD ABOLITION

Today, in the United States, 41 States and the District of Columbia still retain the gallows or its equivalent. When Gov. J. Caleb Boggs, on April 2, 1958, put his signature to a bill abolishing the death penalty in Delaware, he added that State to Michigan, Rhode Island, Wisconsin, Minnesota, North Dakota and Maine in the abolitionist ranks. Since that date, the admission of Alaska and Hawaii to statehood raised their number to nine.

Most of the States, then, carry the death penalty on their books. But actual executions have become rarer each year. In 1958 the U. S. total was 48, the lowest on record since a national count was begun in 1930. Even more indicative of the trend are the annual averages for each of the last three decades. Executions averaged 167 a year from 1930 to 1939. During the years 1940-1949, the yearly average was 128. The figure was down to 74 for the nine-year period of 1950-1958. Thus, though the opponents of capital punishment have made little headway in fighting for its legal abolition, the death penalty seems destined to disappear at least by a process of increasing disuse.

Outside the United States, the picture varies. Four nations that reintroduced the death penalty for collaborators at the end of World War II quickly reverted to its abolition by law (Denmark, Norway, the Netherlands) or by custom (Belgium). Currently, most of the non-Communist countries of Europe, with the notable exceptions of England, France and Spain, do not execute criminal offenders. The death penalty has been dropped also in a majority of the Latin American republics. Closer to home, Mexico does not execute. But Canada, true to what promises to become peculiarly an Anglo-Saxon folkway, still clings to the noose.

Though England and Canada both retain the death penalty, it is not for want of abolitionist sentiment in these countries. Parliament in London has rung repeatedly with heated debates on abolition. Both in the 'thirties and again in 1950, Royal Commissions conducted extensive hearings on the subject. More recently, on this side of the Atlantic, the Canadian Parliament set up a Joint Committee on Capital and Corporal Punishment and Lotteries. The minutes of the committee's hearings constitute one of the best sources for up-to-date evidence and arguments for and against retaining the death sentence in a criminal code.

The story behind much of the drive to do away with capital punishment is one of gradual change in public sentiment throughout the Western world. In time this change made itself felt in our criminal law. Some 350 capital crimes existed in English law, for example, as late as 1780. These narrowed to about 220 by 1825. Today, murder remains the one crime for which an execution, in practice, will take place. Moreover, in addition to these modifications in the law, prosecutors insist that juries often mitigate the law's severity on their own in trials involving capital offenses. This accounts, it is said, for the difficulty encountered in getting a verdict of guilty where the death sentence would be mandatory.

Possibly the greatest change resulting from the shift in public opinion, however, has been in the way executions are handled. Today the authorities take great care to keep publicity to a minimum. In the not too distant past, on the contrary, the law appointed that a hanging or beheading be carried out in the most public manner possible. This often meant before a crowd of both sexes, young and old, gathered in a holiday mood and setting.

What accounts for the change in popular feeling and, in turn, in the law? To a great extent this reaction stemmed from a tide of humanitarian sentiment which swept Europe and the Americas in the 19th century. Church groups commonly proved to be the earliest and strongest supporters of the movement. Thus, many Quakers opposed capital punishment with the same intensity with which they struggled to free the slaves. Much of the present strength of the abolition movement, it is interesting to note, comes from the same sources. The following sample of resolutions taken by U. S. church bodies in the past year tells this story.

On October 17, 1958, the 59th triennial convention of the Protestant Episcopal Church went on record as strongly opposing capital punishment. Then, on January 2, 1959, the president and general secretary of the Massachusetts Council of Churches recommended that capital punishment be abolished immediately in that State.

Similar action was called for in Pennsylvania and New Jersey by the Philadelphia Yearly Meeting of Friends (representing some one hundred Monthly Meetings in the two States) in its session of April 1,

1959. Later that same month, on April 14, the 72nd annual assembly of the Texas Convention of Christian Churches (Disciples) denounced capital punishment as "cruel, inhuman and unjust." The assembly requested local churches to urge the Texas Legislature to abolish the electric chair.

Back once more in Pennsylvania, the Protestant Episcopal Diocese of Pennsylvania in its 175th annual convention (May 7) condemned capital punishment and asked that it be outlawed in the Commonwealth. Four days later, at Oak Lawn, Ill., the 116th annual meeting of the Congregational Christian Conference voted support of a proposal pending in the Illinois Legislature to suspend the death penalty for a six-year period by way of experiment. Outright abolition, however, was urged by the Colorado Congregational Conference on May 25—shortly after a similar move made by the Colorado Episcopal Diocese. Again in Illinois, at the end of June, the Methodist Rock River Conference summoned all Christians to take "deliberate and appropriate" action to end the death penalty. And in the East, the New York Yearly Meeting of the Religious Society of Friends, with members in New York, New Jersey, Connecticut and Vermont, on August 12 appealed to the Governors of those States to do away with capital sentences.

It will be noted that all of the above actions arose in Protestant or Quaker groups. Catholics and Jews in the United States seem to have had little to say on the subject in the past year. However, Rabbi Maurice N. Eisendrath, president of the Union of American Hebrew Congregations (an association representing one million Reform Jews in 565 synagogues), announced that the union would consider a resolution on capital punishment in its biennial assembly, at Miami Beach, Fla. in late November of 1959.

Despite the widespread support of abolition by church groups and other organizations, defenders of the death penalty still claim a solid measure of public support for it. Though philosophers, theologians, social scientists, the police and legal authorities have all had their say on the subject, the debate still goes on. Most current discussions, however, center around two basic issues. One is whether executions are necessary to satisfy justice; the other, whether such a penalty is necessary or desirable as a deterrent from serious crimes.

REASONS FAVORING CAPITAL PUNISHMENT

"A life for a life"—this remains the simplest version of the argument that justice demands death in some cases. Over the centuries, however, this rule of thumb has all but vanished from sight. It lies buried beneath a forest of distinctions on the morality of exacting such a price even for murder.

It may be of interest to note one corollary of the proposition that justice demands the death penalty. The claim is made that death should be meted out, at least for unusually heinous crimes, to satisfy in a legal manner the public's sense of outrage and its desire for vengeance. One wonders, however, whether the revolt against public executions in the old style does not point to a hopeful lessening of the cry for revenge.

Paradoxically, defenders of the death penalty sometimes also insist that an execution, in addition to being an economy measure for the State, is an act of kindness to the prisoner. The basis for this argument is the judgment that life imprisonment is a crueler fate than swift death on the gallows. Whether or not it costs less to hang a man by the neck than to keep him in prison for life would seem to depend on the price you attach to a great many intangibles. And as for weighing the relative humaneness of any mode of execution now in use, again your judgment will be influenced by deep-rooted factors in your own personality. In any event, the task of devising an objective method of making such estimates will not be easy.

Take, by way of example, the history of a recent experiment in Maryland. A few years ago that State substituted the gas chamber for the gallows. The change was widely hailed as a move to eliminate pain or torture from an execution. Yet, the late Fr. Joseph J. Ayd, S.J. (veteran sociologist at Baltimore's Loyola College and companion, over several decades, to all Maryland condemned men who walked the last mile to

280

the gallows), could write soon after in the Baltimore *Sun*:

Had Arthur Koestler witnessed the recent gruesome gassing of a mentally underprivileged human (I.Q. 62) at the Maryland Penitentiary, he would have had added fuel for his remarkable book, *Reflections on Hanging,* and, maybe, some unkind words about Maryland's new gas chamber would have been brilliantly expressed.

IS IT A DETERRENT?

In the final analysis, however, the case for capital punishment today is based mainly on the claim that the retention of the death penalty affords society an effective protection by deterring potential evildoers. Advocates of capital punishment go on to insist that a comparison of crime rates in countries having and those not having the death penalty will support this claim. For example, this was the point of testimony by Walter H. Mulligan, president of the Chief Constables Association of Canada, before the Canadian joint committee on April 17, 1954. Speaking of policemen killed in the line of duty in those parts of the world where capital punishment has been abolished, he stated:

I submit that it will be found that the number is much higher than in those countries where the death penalty is still in effect, and this point is the main one in our submission that our Government should retain capital punishment as a form of security.

He then told the committee that his extensive contacts in police circles led him to believe that his view reflected the "general opinion among police officers on the North American continent."

In view of the forthright nature of Mr. Mulligan's statement, and because my own research on capital punishment convinced me that it was typical of the primary argument advanced everywhere in favor of retaining capital punishment, I attempted a minor test of its validity. The records on which the investigation was based came from the State police chiefs in 24

States (including the six in which the death penalty was abolished). They reported on all State police killed in these States during the years 1905 to 1954. From my analysis of these figures I concluded, in a report submitted to the Canadian joint committee in 1955, that the available data did not square with Constable Mulligan's claim.

In referring to this study—restricted as it is in scope—I do not pretend that it says the last word on the subject in itself. Its significance derives, rather, from the fact that it agrees with a series of other investigations of the anti-abolition claim. As Prof. Karl F. Schuessler of Indiana University wrote earlier in the November, 1952 issue of the *Annals* of the American Academy of Political and Social Science:

> Statistical findings and case studies converge to disprove the claim that the death penalty has any special deterrent value. The belief in the death penalty as a deterrent is repudiated by statistical studies, since they consistently demonstrate that differences in homicide rates are in no way correlated with differences in the use of the death penalty.

As this article is written, the Chessman and Truscott cases still hang fire. Bizarre elements in both these dramas will keep alive the discussion of questions raised in these pages. Moreover, on October 26, representatives of Sweden, Austria, Ceylon, Uruguay, Ecuador and Venezuela called on the United Nations to underwrite a study of capital punishment with a view to its universal abolition. This proposal, it is interesting to note, calls specifically for a study of the effect of abolition on the rate of criminality. Finally, it is most likely that in the coming year several State Legislatures will once again take up proposals to abandon the death penalty. What will be the outcome of these moves?

Certainly, the over-all trend is toward the abolition of capital punishment. Many factors lend support to this movement. Of fundamental importance is the influence of a more Christian or, at least, more humanitarian attitude toward the criminal. For a time this old-yet-new spirit seemed doomed to disappear in the bloodthirsty era of the 16th and 17th centuries. Today, to be sure, we may not be ready as yet to resume the medieval practice of sending a murderer on pilgrimage to Jerusalem or Compostella in punishment for his crime and for his personal rehabilitation. Still, a wide sector of public opinion now inclines to the view that the execution of a prisoner is an unnecessary confession of defeat and an act of self-degradation on the part of modern society.

One final factor has, of late, won new adherents to the abolitionist cause. It is the growth of a critical attitude toward claims long made on behalf of the death penalty. Further research into the validity of these claims, particularly those concerning the alleged deterrent effect of this punishment, may well hasten the abandonment of the death penalty everywhere in civilized society. Then the gallows, the electric chair, the guillotine and the gas chamber can be relegated to our museums. There they will take their appointed places alongside the rack, the thumbscrew and other rightfully discarded instruments of earlier systems of justice.

God Knows Best

Pat Somers Cronin

EARLY TRINITY SUNDAY, some two months ahead of
schedule, I was wheeled into Mercy Hospital for
the birth of our seventh child. . . .

What is it about a Catholic hospital that makes it
seem that some peculiar protectiveness slips about you
even as you enter the doors? Perhaps any nurse anywhere would have greeted you as sympathetically, and
any orderly anywhere would have wheeled your chair
as gently. But would there have been anywhere, save
in a Catholic hospital, the statue of Christ at the Labor
Room door? And would there have been in His beloved
hands that prophetic scroll, "If you love Me, come
follow me"?

The moments preceeding birth, particularly premature birth, are emotion-packed for any woman. But
again, in a Catholic hospital, when all during labor you
have only to look at the crucifix and think over and
over through each pain, "If you love Me, come follow
Me," what is possible but complete resignation to the
divine Will?

However, life does mean hope and even a little boy
will fight for survival. "What is his name?" the doctor

284

asked, "Father is waiting to baptize him." "Eileen Marie," you said, having planned on another girl. "That won't do," the doctor smiled. Still groggy, you murmured, "Joseph Anthony," and only then could you relax. Father was waiting and little Tony joined the Church Militant, moments after his abrupt arrival into life.

Going back to your room, you questioned the attendant: "He is baptized?" "Yes, yes," you were assured, again gently, and I suppose gentleness is to be found in many, many secular hospitals too. At the nursery window you looked into the little incubator where Tony would spend only God knew how much time. But you and your weary husband were at peace; the matter was in God's hands now. What a blessing baptism is, surely a sacrament to inspire courage. Three A.M., Trinity Sunday: with all of Heaven rejoicing, little Tony had had quite a magnificent birthday!

In your room, still overcome by the actual fact of the baby's arrival, you faced up to his uncertain earthly future, discussing it in whispers with your husband. Two and one-half pounds, the doctor estimated; there had been no time for the usual accurate weight and height routine. But again, because you were a Catholic in a Catholic hospital, living at this dramatic moment under the roof with the Blessed Sacrament, you had only to turn to heaven and say to our Lord, "Dear God, it is in Your far-seeing Will; we have done our part, the rest is up to You." And because little Tony was named for your personal favorite, St. Anthony, who has aided you again and again, you turned to him, asking him to intercede for you, not for Tony's life, but for the manifestation of the divine Will.

During the long hours from three A.M., when sleep simply would not come, your mind dwelt on the phrase, "If you love Me, come follow Me." Surely, surely the baby would follow immediately; God had been most generous to give you such a direct sign.

But little Tony continued his valiant struggle, although Sister would never admit to more than a "pretty good, honey." "And," she added, "God sees so far ahead; sometimes He knows a soul may be lost unless He takes it right away." Where but at a Catholic hospital, you wondered gratefully, would the head nurse of the maternity department offer sage spiritual consolation as well as physical comfort?

The hours of Trinity Sunday dragged on, each adding to Tony's chances for survival. Over and over you said the rosary, content to rest and let God act. A few phone calls to your favorite convents and literally thousands of prayers were being said. What a wonderful thing, the Mystical Body of Christ!

Sunday evening your husband wheeled you down to see Tony, all pink and lovely, kicking gaily, even opening his tiny unseeing blue eyes. Toothpicks for arms and legs, you thought, but the little heart was beating and that little soul was ready. And here you were, in league with God, waiting, waiting for His word; how powerful God makes mothers and fathers; what a privilege to be parents!

Finally, after your nursery visit, you could succumb to that body-encompassing sleep that follows birth. The job is done; the body can rest. And when you awakened next morning, because this was a Catholic hospital, you had only to lie in bed, waiting for our Lord. This is something you will never accept as commonplace, that every single morning our Lord, preceded by a white-robed sister sounding a small tinkling bell, comes to any patient who wishes Him.

As you anticipated our Lord's arrival, you thought too of little Tony down the hall. He must have survived the night at least. Almost for him, you received our Lord, made your thanksgiving gratefully and then were amazed to hear the doctor's "Good-morning." The report was not good and something, surely another bit of divine generosity, propelled you from bed to take another peek at this littlest son. And what did this tiny bundle do? He curled his small fist and somehow managed a pint-sized wave. You smiled and tucked in your heart for all time that simple gesture.

Again in your room, you picked up the phone to prepare your husband for any news, but divine Providence and the hospital staff had raced ahead of you. Even as you held the phone in your hand, your dear strong husband walked in to give you the final news. In that short moment since your return from the nursery, little Tony, some twenty-eight hours old, had returned to our Lord.

Surely there are times of triumph for every mother and father, but what can ever equal this: to know with sublime certainty that a son is safely with our Lord, always and forever. Truly, He had spoken: "If you love Me, come follow Me."

Man is a toolmaker. Therefore his efforts to tame the refractory material world befit his rational nature: the accelerating pace of technology is a good thing, so long as man fits his mastery of things into a proper hierarchy of values. But many questions may be raised about the real value of our technological triumphs. They have not begun to solve the problem of giving all men real access to the goods necessary for human development. And there is a real danger that science may foster unhealthy pride in the heart of man and deaden in his heart the sense of sin. Does all this mean that there is some rationale behind that hostility to the works of science which sometimes exists among Catholics? Surely, in a world that belongs to God, there is some way of baptizing the science and technology that are changing its shape.

Men and Things

Christians

Confront

W. Norris Clarke
Gustave Weigel
Walter J. Ong

Technology

S.J.

I

BOTH THE NEED and the aptitude for technology are rooted deep in human nature. Man simply cannot live with security and dignity in this material environment of physically superior forces unless, by means of tools, he extends and intensifies the power of his own comparatively feeble body so that it can cope with nature. Thus we find that as far back as the presence of man is discernible in history it is accompanied by the use of tools. And the use of even the simplest tools is a rudimentary technology, differing only in degree from the technology of the great machines of today.

Nor is it merely a question of decent survival. If man is made as a living image of God, he must imitate in his own way the life of his Father in heaven. Now in relation to the material universe God is both its thinker and its maker—the supreme Artist. Hence man, His adopted son, must first strive to rethink the handiwork that his Father has thought up. (This is the fundamental justification for all pure science.) He must then strive to remake or transform creatively, by his own God-given powers, the world that his Father has made for him out of nothing and given him as his workshop. (This is the fundamental justification for all technology and art.)

The goal of all technology, then, is to make matter serve the human spirit as the most pliant instrument possible for authentic human growth. This involves both the gradual liberation of man from all inhumane, degrading, purely animal-like drudgery and subservience to matter, and also the positive transformation of matter to express man's own spiritual vision of the meaning and purpose of the universe.

Secondly, the unfolding of this basic aptitude is of its nature dynamic and progressive. Each successive achievement in technology builds upon what has gone before and opens up in turn new developments beyond it. As the range of possibilities constantly expands, the rate of development is able to advance more and more rapidly. This dynamic dialog of man with nature is inherent in the very presence of man as a rational animal in a material universe. To attempt to freeze this process, which by its very nature is developmental, at some supposed point of ideal equilibrium (determined by whom and on what evidence?) would be like trying to immobilize the growth of art or of human intelligence itself—indeed of any living thing, which dies when it can no longer put forth new fruits. The conclusion of all this is that progressive technology in some form or other is both natural and essential—hence basically good—for man if he is to fulfill his God-given destiny in the universe.

So much for the credit side of the account. On the debit side it must also be admitted that there are grave dangers inherent in the pursuit of technology, as in the use of all natural human aptitudes. These perils arise, it seems to me, from two main sources: first, lack of subordination to the higher spiritual good of man; second, lack of the proper rational control of the rate and timing of technological development.

The danger of the first is that what should be a mere

means, at the service of the spiritual growth of mankind, may become so all-absorbing that it will upset the proper hierarchy of values and reduce the spiritual intelligence of man to the role of a mere servant of technological progress pursued for its own sake. Technology would then become like an overdeveloped organ or the runaway growth of cancerous cells—a threat to the basic cultural and spiritual health of mankind.

The second main danger is that the tempo of technological development be allowed to follow unchecked its own inner dynamism, independently of its relation to the balanced over-all good of the people it is supposed to serve. Too rapid a tempo of change can produce an atmosphere of such constant flux and severe social dislocation that the people subjected to it will be in grave danger of becoming culturally rootless and deprived of all fixed landmarks as they are whirled hectically along by the racing current of "progress." Thus the ever increasing mobility made possible by the automobile (though not so much, strange to say, by the airplane) has so far proved to be a very mixed blessing, whose dissipating effects we have not yet learned to control.

Another form of the same danger is the so-called "enslavement of man to the machine." The harsh rhythms of the machine and its artificial environment will, it is said, dominate or destroy the healthy natural rhythms of the living body in harmony with nature as God made it. The example of the assembly line, with its impoverishment of creative ability and subjection of the workers to a monotonous, repetitive routine, is sufficient warning of where technology can lead.

All of these dangers are real and serious, in addition to the very special and obvious perils connected with the use of the immense power now at our disposal. I have no intention of trying to conjure them away by general optimistic affirmations of the "inevitable forward march of progress." The latter is a dangerous modern myth, a secularized distortion of Christian hope. Uncontrolled technology can certainly bring down disaster, perhaps irreparable, on our race. The only protection against it is a growth in man's spiritual and moral maturity proportionate to his growth in technical skill and power. Either we grow in both dimensions or we perish, like the overgrown monsters of our prehistoric past. But this is already a law in the development of every individual personality (for example, scientific geniuses with a child's knowledge of religion and morality). If individuals can solve it there is no reason why people generally cannot either. Actually, it seems to me that there is already a rapidly growing recognition on the part of both scientists and political leaders—who are also the ones most able to do something about it—of the urgent necessity of greater moral control over the exploitation of scientific and technological advances.

Furthermore, as technological development proceeds along its course among a people still endowed with basic biological, social and moral vitality—as I believe our people still is—certain laws of equilibrium and self-correction seem to be constantly and unobtrusively at work. Thus, losses in one area are compensated for by gains in another, or exaggeration in one direction generates its own counterreaction in the other direction. Thus the very mobility which seems, at least temporarily, to be weakening our roots in the family and the local community is at the same time strengthening our bonds with the rest of the world. The very increase in perfection of the means of communication at a distance, as in television, may eventually make it neither necessary nor desirable to move about so feverishly on a small scale as we now do. We may end up by visiting our friends and clients relaxedly on a two-way television circuit rather than by transporting ourselves physically to them along overcrowded highways or airways. Or by the mysterious providence of God it may happen that the insistent challenge of outer mobility may succeed more effectively than pulpit sermons in making us turn within and discover that it is possible to achieve a sense of permanence, self-identity and rootedness in more interior, supramaterial and universal values than we now believe capable of winning our allegiance.

What of the threatened enslavement of man to the machine? The danger is real. But I am convinced that it is limited by the very inner logic and laws of equilibrium of technology itself to certain transitory types of techniques and local or temporary abuses during periods of transition. The whole innate drive of technology is to substitute machines for man in all areas where monotonous, repetitive actions are the rule, and to leave man free for more intelligent creative or supervisory work. The supposed threat of a constantly increasing slavery to the machine, as some kind of inexorable drift inherent in the process of technology itself, seems to me to be largely a myth, without solid historical, psychological or sociological foundation. The true danger lies in the moral dispositions of those who use technology. The far greater peril is that men may become slaves to their fellow men rather than to their machines.

I believe, therefore, that if we have the courage to assume with full moral and intellectual maturity the responsibility of actively guiding and controlling the mighty power of technology that is now in our hands, far from being ruined by it, we shall be able rather to turn it into a profoundly beneficent instrument for the authentic growth of the human family. And no one has greater inner resources for rising to this challenge, nor more urgent motives for doing so, than the Christian.

W. Norris Clarke

I HAVE BEEN asked to propose a counter-thesis to the essay of Father Clarke concerning a Christian judgment on technology. Fortunately I cannot make any substantial objection to the preceding article. For this I am grateful because I never like to disagree with the brilliant Father Clarke, and when the necessity arises it causes me pain. Hence most gladly I make my own the thoughts of the first essay in this trilogy. Perhaps, however, it will be possible to add some shadings to the reflections made by the preceding author.

The Prophet Ezechiel (28:26), speaking of the house

and family of Jacob, says: "Securely it shall dwell there, build houses, plant vineyards, fear no attack." St. Thomas, writing of what man can do without the help of grace (S.T., 1-2, 109, 2), says: "In the state of corrupt nature man can indeed act to achieve some particular good by the power of his own nature, as for example, the building of houses, the planting of vineyards and things of this kind." The employment of atomic power, which is today in the power of man, seems to be a far cry from the building of houses and the planting of vineyards. Yet it is more than pure whimsy to think that St. Thomas would include such a feat in his phrase, "and things of this kind." Technological progress derives from the cumulative knowledge of the human family. It is the fruit of the continuous struggle to possess the earth and to subdue it, which was the destiny divinely imposed on man according to the words in Genesis.

The malaise of moralists in the presence of the increased technological capacity of man has always struck me as unwarranted and irrelevant. Evolution seems to be the very basic pattern of the universe. Hence it seems only logical that technology among men should evolve into ever more complicated structures of power capable of ever wider efficiency. This evolution of technology should be welcomed, not lamented. The people who cry alarm because of the dehumanizing effects supposed to be inherent in the better tools and instruments that human intelligence has devised for our aid, rarely see that the high human activities they so praise were possible in past times only because vast masses of the human family were condemned by a humanly erected social system to be mere hewers of wood and carriers of water. The degree of humanism achieved by this numberless proletariat was certainly not high. In those days, in a vast sea of drab human misery, there were indeed little green isles of comfort and luxurious leisure where the few beneficiaries of the iniquitous system could discourse eloquently and elegantly on art, history, philosophy and the patterns of nature. To stand in rightful awe before the brilliant contributions of these few men and women, while overlooking the dull and tedious existence of the countless surrounding majority of their contemporaries, seems hardly humane.

Those who nostalgically praise those days of old usually do so because they wittingly or unwittingly share Plato's conception of mankind. In this vision there are a few individuals who are really men. They are the *aristoi*, the best, who in turn breed other *aristoi*. The rest by very nature must be treated as a good husbandman treats his horses and asses. He gives them what is necessary so that they can survive and be able to use their muscles to serve the master. Their own aspirations and desires are not consulted. They are expendable. Even their procreating capacities are exploited to assure the owner a continuance of an efficient physical force at his disposal.

In this cruel and un-Christian view of mankind, our human family is deterministically divided into two unequal groups: the "haves" and the "have-nots." By some kind of weird and unverified assumption, the real men are "haves" who beget real men who will be "haves," while the "have-nots" are subhuman and can only engender more subhuman "have-nots." Technological evolution endangers the privileged position of the hereditary *aristoi* because the lesser breed finally has an opportunity to enjoy leisure. Now they too will discourse about life and eternity. But since they are the great un-

washed, it is asserted that they will vulgarize the arts and sciences. In this contemplation, the Platonic lover of aristocracy can only shudder.

For my part, I let him shudder—without any felt need to shudder with him. The Christian understanding of man, and indeed history itself, find no room for the postulate that human genius and human creativity come forth by some law of biological determinism. The genius is a glorious sport showing up in the most unexpected places. Improved technology gives the sport who happens to rise out of the masses the opportunity to develop his genius richly for the benefit of all. Improved technology gives leisure to all men. That very many will use it for creative ends is more than we can hope for. Yet it is true that some will do so and their contribution to culture and civilization will be a blessing.

Having made this basic recognition of the goodness of technology, we must now consider its ultimate value. Optimists, whether they be Christian or naturalist, usually look no further than the goods rooted in creation. They overlook, or at least do not take seriously, the dogma of original sin. They are always looking for man's better moment in the future. According to faith there will indeed be a golden eon at a point beyond history. However, before that point is reached, the ever-evolving human situation under the sickening influence of original sin will not change substantially.

Now technology is a phase of creation. But in this creation man himself will always be faced by his own sin and the consequences thereof. Selfishness, guilt, ignorance, anxiety and frustration will not be removed because of technological progress. The improvement of human know-how will not increase the amount of virtue in the world. The kingdom of God is not the fruit of the natural evolution of man. Only the catastrophic intervention of God, unmerited and undemanded by human efforts, will bring it about—after history ceases. In this evolving world, through the power of God and not through natural evolution, the Mystical Body of Christ gradually achieves its full stature. The growth of natural human potential is the setting in which the whole Christ matures. The setting does not enter causally into the process of maturation. Increased human efficacy furthers the designs of God's natural creation. However, the salvific will of God is distinct and only occasionally related to the creative drama. The heightening of man's creative power can readily also be the occasion of man's moral deterioration. It need not be so. Yet, whether it is or is not is a matter of indifference to the salvific designs of God. St. Paul says that where sin abounds, grace abounds more. It is not sin which produces grace. Grace is freely and independently given by a gracious God who uses the unfolding of nature as the framework for His saving bounty.

To put it briefly, the expansion of technological prowess, since it is a manifestation of God's creative will, is like all of God's creation—good. It should be hailed as such by all men. But it is *not salvific*. Of itself it neither hinders salvation nor does it hasten it. Salvation comes not from nature, no matter how effective it be. It comes from grace given on God's mysterious initiative. Before the kingdom of God arises, nature lies under the dominion of the Prince of this World. Salvation will not be of it. Only at the end will the order of creation and the order of salvation be identical. Although technology is a good thing, it is helpless to make man righteous. Man's hope can not be founded on increased human power, good though that be. Only faith and trust in the world to come can give substance to human hope. GUSTAVE WEIGEL

THE OBSERVATIONS on technology by Fathers Clarke and Weigel seem to me eminently sound. Not knowing how to take exception to them, nor wanting to do so, and yet being constrained by a kind but persevering (that's why he has the job) editor to produce some sort of counterstatement, I can do nothing better than use their observations as points of departure, examining the implications of those observations in a more extended field.

The problem confronting the Catholic mind today is not the problem of tolerating the technological age, of living with it. This we do well enough. The problem is rather that of participating unselfishly in it, of contributing to it. Our present difficulties, philosophical, theological and other, vis-à-vis the age we live in seem to be in great part our inability to conceive of it positively and imaginatively.

We badly need a cosmology and a Christology sufficiently developed to enable us to lay hold of the technological age, and all it implies, with inspiration and vision. This means a cosmology and a Christology which will enable us to conceive of our entire evolutionary universe with a positive intellectual humility and enthusiasm. For our problem with the technological age is only one facet of a much larger problem—our problem with an evolving cosmos. Our technological age cannot be grasped as the reality which it is outside the framework of cosmic and intellectual and social evolution in which and as a result of which the technological age has come into being and achieved its own self-consciousness.

Our vision and enthusiasm should be without illusions, of course. As Father Weigel says so well, the expansion of technological prowess is not salvific. Men are not individually closer to God because of it. Yet, as he also says, this expansion is a good. There is in many quarters a growing awareness of this fact, but the Catholic consciousness as a whole, in the United States and elsewhere, still labors under a certain disposition not to welcome technological growth or the over-all movement of cosmic evolution as a good in its view of the over-all scheme of things. Evolution is not of course presented as an evil in most Catholics' world outlook. In fact, in this country at least, for the most part, it is not presented at all. A few articles and lectures are putting in their appearance now. But corporately we still react as though evolutionary processes were not there.

The unfolding of the basic aptitude of man to transform material reality, Father Clarke has well observed, "is of its nature dynamic and progressive." And Father Weigel adds the more general theorem: "Evolution seems to be the very basic pattern of the universe." No one can today adequately understand the material world, or those reaches of the spiritual world enmeshed in the material, such as human thought, without immersing himself in evolutionary studies. Courses in the details of evolution are a standard part of any biological training, but familiarity with the developmental pattern of God's universe is necessary as well for geology and astronomy and semantics and linguistics and an increasing number of other fields. Yet, strangely enough, in many quarters it is still thought possible to develop a Catholic world view with no serious attention to evolutionary fact or, at the maximum, with a permissive, tolerant gesture toward this great "basic pattern of the universe." How much of our corporate thinking about divine revelation gives any sustained attention at all to the possible meaning of the Incarnation for the real evolving cosmos as we know it, in which technology has now come to be?

One cannot fully understand a fact which one accepts reluctantly. When Marxist thought has stolen the march on Catholic thought, it has often done so by enthusiastically making its own a fact which Catholics have not had the courage or humility to warm up to. Father Weigel expresses a sympathetic interest in the massive and miserable proletariat of Plato's republic, the "have-nots" as contrasted with the "haves" who were Plato's *aristoi*. He does this in order to explain the forces at work in our technological age. A Marxist could urge, quite legitimately, that Marx had used this type of analysis and had felt the sympathy which it entails more than a hundred years ago: the "haves" and the "have-nots" form a perfect dialectical dyad for a dialectical penetration and understanding of the historical processes. Father Weigel is right in his analysis and in insisting that the lot of the proletariat ought to be changed, but

Marx was right first. He was right before some Catholics were, not because of his basic principles nor of his over-all aims, which were wrong, but because in this he was building upon a keen and enthusiastic awareness of the fact of change which some Catholics lacked. He seized imaginatively and creatively upon the emergence of a new technological order and upon the evolutionary nature of the universe and of human society, at a time when all too many Catholics were meeting the challenge of the age by talk of being satisfied with one's "state in life" (that is, with a static view of human society).

Certainly, we must not abandon what we already know and teach. But you are sure to lose if you are so cautious about keeping what you have that you feel no real enthusiasm for what is new and unknown.

The time is past when we can afford a merely permissive reaction to evolution or to such things as this technological age to which an evolving universe has brought us. We cannot even begin to understand technology or to make it our own and Christ's own if we regard it and other massive developments in a changing universe as incidental items, inconsequential or even unintelligible phenomena to which one need pay no great attention as human life creaks along in its supposedly unchangeable ruts. The age of technology is an era in the history of the universe, coming into being at a certain time just as dinosaurs and the first mammals and man himself did. Technology is a shape which material reality takes at a certain stage in its development, a stage at which it arrives when it has ripened over a period of some five or ten billion years for man's appearance and when human culture thereupon, over a period of probably some several hundred thousand years, has developed a store of experience such as we have at our command at present. The age we live in is a shape all things have today, different from—yet comparable to—other and earlier shapes.

What is the over-all Catholic educational effort doing to interpret with positive and creative vision this shape of things? What are we doing to interpret philosophically and theologically, imaginatively and inspiringly, a universe which everyone knows is in full evolutionary career? A retouched medieval cosmology is hopeless and inevitably false. If evolution is "the very basic pattern of the universe," it is not a thing to be discussed charily, handled gingerly, touched on incidentally. Although its details are even now not fully worked out, and probably never will be, cosmic evolution has been known to thinking men as a fact for generations. As early as *The Descent of Man* (1871), the sage and cautious Darwin notes that the principle of evolution is already "admitted by the majority of rising men" among natural scientists. This was almost a hundred years ago. The hour is getting late.

As long as change in the world was thought to be a simple cycle of life and death complicated only by the vagaries of human history stemming from unpredictable human decisions, one could enjoy a world view in which change was treated simply by contrast with non-change, as becoming is contrasted with being. Once we know, however, that created material being changes over vast eons in a directional pattern, in which more advanced forms of being appear only after less advanced forms, we cannot be so offhand. The study of the supposedly cyclic generation and corruption postulated by ancient philosophies no longer suffices. Knowing a vast number of details concerning the movement of matter through time and its greater and greater complexification, if only to understand ourselves better we need to puzzle out the *sequence* of changes which represent an activity proper to material being of which earlier man had been unaware. Protein molecules far postdate a world of simple inorganic molecules and atoms. Early geological deposits yield only simple forms of life, later deposits alone yield higher forms. And these become more and more advanced as time moves on. Moreover, species vary from place to place, as well as from age to age. The flora and fauna of the Western Hemisphere are quite different from those of the Eastern. And there is a long and complex sequence in human cultures, too.

301

Such facts were of course inaccessible to earlier "wisdom lovers" or philosophers, who lived before man had been able to accumulate all the vast store of observation and reflection on which we can draw today. But now that we know these facts, we cannot afford to disregard them in constructing our world view. Any wisdom, natural or supernatural, worthy of the name now has to deal with the fact of linear development in the history of material being and of human culture, in which the present grows out of a past different from itself and opens into a future different from both present and past. This is particularly true if we are to try to interpret the meaning of the technological age. For this age is most strikingly something which has not existed before and yet something into which the preceding history of human society and of human thought has led, as well as something pregnant with a future sure to be different from both present and past.

For the technological age is not only an epoch in the totality of cosmic evolution—all ages have been that—but it is also keenly conscious of the fact that it is an epoch in a total process. To interpret it to itself we must work out a world view which is thoroughly Christian and which also operates in close, imaginative, sympathetic, creative association with the facts of cosmic development, for this development is God's work and something humbly to be revered.

WALTER J. ONG

Sin, Sickness and Psychiatry

John R. Connery
S. J.

ARE SIN and mental sickness two faces of the same
coin? If they are, psychiatry, presuming that it
can cure the illness, can make saints of us all.
Psychiatry, of course, makes no claim to produce sanctity. But there are psychiatrists who will say: "There
are no sinners, there are just sick people." This is an
extreme position, and more philosophical than psychiatric, but it illustrates the needless conflict that has in
the past divided the theologian and the moralist from
the psychiatrist.

The various aspects of this conflict are best illustrated in the different opinions advanced regarding
the relationship between sin and mental sickness. Besides the opinion that would regard sin as sickness
rather than an expression of the will of the sinner,
there is a second opinion that, surprisingly, reverses
the relationship. According to this opinion, sin, far from
being the result, is really the cause of mental illness.
A third opinion refuses to consider sin either as cause
or effect of mental illness, but tends to regard it as a
cure. Fortunately, these opinions, at least to the extent
that they are objectionable to either theologian or

psychiatrist, cannot today claim any large number of adherents. Nevertheless, it will perhaps be profitable to explore them, to understand more precisely just what is the relation between sin and mental disease.

THE PURPOSE OF CONFESSION

If sin were merely a symptom of mental disease, the psychiatric clinic could, and should, be substituted for the confessional. Sacramental absolution is not intended to be a cure for disease. The Church has a sacrament for the sick, but it is the sacrament of extreme unction, not the sacrament of penance.

It is quite true that the act of confessing one's sins, the acceptance of the priest and of his pastoral counsel can have some therapeutic value, but this is not the primary purpose for which the sacrament was instituted. The sacrament of penance is primarily a sacrament of forgiveness. Consequently, it presumes responsibility for sin and has no meaning outside of a context of responsibility. An opinion that reduces sin to sickness, then, is incompatible with the existence of a sacrament of forgiveness; it is unacceptable to those who hold the sacrament of penance to be an integral part of the Christian economy for fallen man.

There is, of course, a sense in which all of us are sick—we are all victims of original sin. But Catholic theology has never admitted that this sickness destroys human freedom. Responsibility for personal sin cannot be shifted entirely to the shoulders of Adam. Moreover, this sickness is not something that will yield to psychiatric treatment. Nothing short of a restoration of man to the state of original justice will remove it.

WEAKNESS OR MALICE?

Yet we must admit that we do not enjoy the freedom of Adam. Original sin has introduced a certain dissociation into our nature, a conflict between our instincts and appetites and right reason. We do not have the control that a perfectly integrated nature would have. Passions and emotions leading to sin arise spontaneously and often continue to make their demands beyond the intervention of the will. And when man does sin, it

304

is more often because he yields to these passions than because he deliberately arouses them. As theologians say, his sins are very frequently of weakness rather than of malice.

It is a strange paradox, too, that we do not begin to function freely until in a sense our freedom is already prejudiced. If the child could function freely from birth, he would be more his own master. But the child's early formation is directed by his parents. By the time he gets to the age where he can gradually take over personal responsibility for his conduct, he is no longer the same pliable person he was at birth. He has already been given a certain *formation*. It would indeed be incongruous if this early formation were incompatible with personal autonomy, but there can be no doubt that the child's life has already been given a distinctive moral direction which will not easily be changed. If the child has been properly formed, he will have certain moral assets to draw upon. Otherwise, he will begin his moral life with definite liabilities.

There are also other elements to be taken into consideration in estimating the extent of individual freedom. Even when a child begins to direct his own moral life, every action he performs influences his future. This is true of good actions, but it is true as well, and perhaps even more so, of sinful actions. No act can be isolated or erased from our past. It leaves its mark on our personalities and becomes a factor in our next decision. This will be particularly true if the act has become a habit. As has often been said, we *are* our past. If that past has taken a certain direction, it will become increasingly difficult to change it as time goes on. In common parlance, it will take greater will power.

This continuity with our past is admittedly a handicap where that past has been sinful, but it is also essential to moral progress. If one is to grow morally, he must be able to profit from past good conduct. Habitual good conduct in the past should endow him with a certain moral facility. Once such ease of operation is acquired in one moral area, the will can be released for conquest in other areas. This is the way moral progress is made. It is, if you will, the economy of moral development. It is not desirable, of course, to reduce moral conduct to a purely mechanical process. But it is not desirable, either, that good conduct should always be difficult. We do not necessarily measure the goodness of an act by the difficulty it involves. But a certain facility in performing moral acts is certainly consistent with liberty, especially if this facility is itself a testimony to constructive use of freedom in a person's past life.

At any rate, when the will goes into action, it can be confronted with an initial inclination toward good or evil. In fact, in the beginner the tendency toward evil may be more frequent. But whether this tendency is the result of original sin or early formation or previous habit, it will not necessarily be compelling in the sense that it will rule out human freedom. Even when it amounts to a pathological force, although there will be some impairment of will, freedom will not be ruled out completely. Sin, then, like good conduct, can be explained to some extent in terms of a person's past, but this does not prevent an actual commitment of the will in the present. To what extent there is actual commitment will depend on the circumstances not only of the person but also of the individual act.

GUILT FEELINGS

If it is an oversimplification to reduce sin to sickness, it is even more naive to consider sin a cure for mental illness. Fortunately, the reputable psychiatrist will not subscribe to any general theory of sin therapy. But it is not difficult to see how the temptation arises to resort to such therapy, particularly in the realm of sex. In his professional work the psychiatrist comes into contact with guilt feelings that not only disturb emotional balance but actually paralyze action. The curious aspect of these feelings is that they plague the innocent; the hardened sinner fails to experience them. It is easy for a psychiatrist whose theological or religious background is deficient to conclude that guilt feelings are the result of moral striving and that, consequently, the remedy consists in abandoning such striving.

The Church, of course, has always considered a

healthy sense of sin a moral asset, not a liability. The recognition of the self as a sinner is the beginning of conversion. It is quite true that this sense of sin grows as the Christian grows in sanctity. The Christian attitude, no matter how far along the road to perfection one may have traveled, always remains that of the publican rather than the pharisee, and the saints have always considered themselves the worst of sinners. One might be tempted to consider this a pious exaggeration; or worse, one might conclude from it to the pathological nature of guilt feelings. Either judgment would be a mistake.

The sense of sin is linked to the sense of God. As the creature comes into more intimate contact with infinite Sanctity and Goodness, it is quite natural that awareness of his own defects should be sharpened and deepened. Remoteness from God, on the contrary, removes the contrast necessary for the sinner to recognize his true condition. The only contrast available to him is that with other creatures, which is immeasurably less revealing. The sense of sin that the saints experience is neither insincere nor pathological. It is a genuine and healthy experience of the evil in themselves — however slight it may seem to  others—against a background of infinite Sanctity.

There is certainly a sense of sin that is not religiously profitable, but a real handicap. It makes psychological cripples of its victims. This sense of guilt has nothing to do with religious truth but, rather, resists it stubbornly. For instance, it is not affected by sacramental absolution, and it turns a closed ear to the appeal of divine mercy. This is the sense of guilt which manifests itself in the scrupulous conscience; psychiatrists classify it as a form of obsessive neurosis. While the obsessive neurosis itself is no more common among Catholics than among others, the anxiety characteristic of it seems more often related to the sphere of morality among Catholics. The scrupulous conscience is a kind of "Catholic disease."

It is their contact with this type of guilt that has stimulated the attacks of some psychiatrists on the doctrine of sin. In a desperate attempt to alleviate these feelings they may even be tempted to have recourse to sin therapy. Yet most of those who take issue with the Catholic doctrine of sin and punishment are realistic enough to recognize that wrongdoing is not the cure for neurotic guilt feelings. But they would like to see the concept of sin disappear. In their opinion the world of obedience, sin and guilt is the world of the child, the world of the primitive, which the adult must outgrow and which adult society must leave behind. The adult, they feel, should be motivated by more positive and altruistic feelings.

The Catholic theologian certainly recognizes the importance of loyalty and love in the realm of motivation and he approves wholeheartedly of a morality founded upon charity. But it is difficult to see how a morality of love is possible without a morality of sin. How deep can that love be which does not recognize betrayal or experience remorse about it? One would rather expect that the deeper the love goes, the more profound the feelings of remorse will be. Guilt feelings are, indeed, the inverse expression of love. The loss of a sense of sin, therefore, far from being an indication of psychological maturity, must be traced to a loss of love, a loss of the sense of God. As such, the loss of a sense of sin must be considered a psychological handicap, tantamount to a loss of the sense of pain in the physical order. As long as sin is an objective reality, there must be a sense of sin. The whole concept of salvation and redemption in the Christian economy presumes this reality. It assuredly has no meaning without it.

There is, unquestionably, a great difference between a healthy sense of sin and a morbid one. The healthy sense of sin is linked to genuine fault; the morbid sense is unfounded. The origin of this morbidity is not always easy to detect, but one can point to certain factors in the early training of the child that can foster a scrupulous type of conscience. Overanxious parents, for instance, can easily pass their morbid fear of sin on to their children. Just by their association with them, the children can copy the anxious attitude of the parents toward sin. Similarly, overprotective parents, who attempt to lead their children's lives instead of teaching them personal responsibility, can deprive the children of the education they need in decision making. Since life will eventually demand such decision making, they will be overwhelmed with anxiety whenever they are faced with the task. Parents who are perfectionists, who constantly nag at children and make no allowance for the failings natural to children, can generate a kind of nagging conscience in the children themselves. Finally, parents who motivate their children primarily through fear and punishment, and, when it is necessary to bolster their own authority, introduce God to the children as a kind of superpoliceman, can easily cause the anxious conscience of the scrupulous person.

FRIENDSHIP WITH CHRIST

The child's first contacts with religion and moral duty should not be dominated by fear. It is clearly unfair to threaten with divine punishment a child who is as yet incapable of formal sin. The Christian must indeed have a sense of sin, but it should not be linked primarily with a fear for self and a dread of punishment; it should have that reference to love and to Christ that is basic to it. Sin consists primarily in breaking off a relationship of friendship and love with Christ. Similarly, recourse to the sacrament of penance and genuine penitence are not aimed primarily at personal security but at restoring this all-important friendship. Fear of punishment, which is at best emergency motivation, should not dominate the Christian's moral life. If the conflict between psychiatry and religion succeeds in correcting a negative approach of this kind to the Christian moral life, it will have been worth-while.

A third source of conflict between religion and psychiatry originates in the opinion that neurosis must be traced to sin. According to this opinion, it is because the patient has repressed his conscience, rather than because he has repressed his instinctive drives, that he develops a neurosis. Although this position is held by some first-rate psychologists and psychiatrists, it is vigorously opposed by equally competent representatives of these fields. From a religious viewpoint it does not have the objectionable features of the opinions already considered, but it is not necessarily true because it is religiously more agreeable. There are, indeed, good reasons for questioning it. First of all, there are many habitual sinners who to all appearances have average mental and emotional balance. Secondly, the scrupulous conscience, which usually must be classified as neurotic to some degree, does not ordinarily have its origin in past sin. For the most part, the scrupulous individual is as remote from sin as one can get.

THE CONFESSOR'S PROVINCE

But whatever may be said for the truth of the opinion that neurosis must be traced to sin, it would be a mistake to conclude from it that mental sickness is a simple moral problem. It is just as much a mistake to attempt to substitute the confessional for the psychiatric clinic as it is to substitute the psychiatric clinic for the confessional. Whatever may be the cause of the serious neurosis, it is not the province of the confessor to deal with it. The confessor will have to deal with the neurotic and psychotic penitent, but the disturbance itself is not his concern, any more than sin is the concern of the psychiatrist. Even if one were to grant that the neurosis or psychosis in a particular case originated in some moral failure, it has gone beyond the stage where it will yield to the ordinary moral solution. I suppose the best illustration of this fact will be found in the case of the alcoholic, although it is by no means clear whether one is an alcoholic because he drinks to excess or drinks to excess because he is an alcoholic. At any rate, ordinary excessive drinking will yield to a moral solution. But the alcoholic seems to have gone beyond the point of moral return. Sacramental confession will not cure the alcoholic, any more than it will cure tuberculosis. As we have already pointed out in another connection, it is not a cure for disease. Mental disease is no more the province of the priest than physical disease.

A conflict between psychiatry and religion is just as needless as a conflict between medicine and religion. It is encouraging to see this conflict being resolved today in the recognition on both sides that each has its place and that neither can be reduced to the other. Religion will not eliminate the need for psychiatry, and psychiatry will not serve as a substitute for religion. The Christian needs the ministrations of the priest; he may also need the services of the psychiatrist. He must not expect either to do the work of the other.

311

The Vanishing Hero

Andrew M. Greeley

K ENNETH S. DAVIS ends his tragic biography of Charles A. Lindbergh with an account of how the motion picture, *The Spirit of St. Louis*, was a box-office failure because few moviegoers under forty either knew or cared about Lindbergh. In the closing words of the book, Davis describes the famous *New Yorker* cartoon in which a boy says to his father, as they walk away from a theatre: "If everyone thought what he did was so marvelous, how come he never got famous?"

Davis sees this as the unhappy ending of a great hero's romance with the American public. However, there is more involved. Even those young Americans who can grasp that flying the Atlantic alone in a 110-mile-per-hour Ryan monoplane was an amazing feat find it impossible to understand the tremendous ovation which this accomplishment earned for the Lone Eagle. That Lindbergh could have become one of the truly important men in our land, offering counsel to the powerful and the wealthy, and arguing in public with the President, is beyond the belief of the present generation. As one young man put it to me, "So he flew the Atlantic. So what?" It is not merely that flying the Atlantic has become a commonplace, but that heroism —or at least hero worship—has become a thing of the past. The Lindbergh phenomenon simply could not happen in 1959. The first man to the moon—in the unlikely supposition that he is an American—will never capture the public imagination as Lindbergh did in 1927.

One searches the national scene in vain for any trace of a hero. The attempt to build up Dr. Jonas Salk as a latter-day Edison had barely gotten off the ground when confusion over the safety of his vaccine and governmental bungling in its distribution destroyed his chances. Mickey Mantle (or, last year, Luis Aparicio) does not occupy in the imagination of the young anywhere near the place that Ruth or Gehrig or even Di Maggio did. College football stars, if we are to believe a recent article in *Harper's*, are no longer the toast of Big Ten campuses; the student body regards them as mere paid professionals. The "astronauts" are favored with complete coverage by *Life*, but few think they will beat some nameless Russian into orbit, and not many people really care. Tom Mix, Buck Jones, Gene Autry and Roy Rogers have been succeeded by the very nonheroic but indisputably "cool" Maverick Brothers. No longer do boys collect cards with pictures of home run kings or galloping halfbacks. Now, on the contrary, they save and trade pictures of ghouls, vampires, werewolves, two-headed monsters, dancing skeletons and assorted odd creatures from outer space—all furnished by the obliging manufacturers of bubble gum.

THE TIMES ARE SOPHISTICATED

Hardly anyone can remember the last time a leading character in a major novel was in any noticeable way an admirable person. The idols of campus radicals (such as they are) are no longer a Dewey or a Tugwell, but rather a Kerouac or a Corso or a Ginsberg. The typical labor leader in the public mind is not a crusading John L. Lewis or Phil Murray, but a smooth though tough negotiator like David McDonald. The efforts of several Catholic organizations to promote

313

prominent entertainers as model Catholics have ended disastrously. Even in the political world no leaders excite the admiration or the dislike caused by an F.D.R. —or even by an H.S.T. President Eisenhower was unquestionably a hero eight years ago, but it is dubious whether he remains one. A father figure or a "personality" perhaps, Dwight D. Eisenhower might well be called the last of the national heroes.

There are many reasons for the disappearance of the hero from the American environment. World War II was a disillusioning experience, and the disillusionment was compounded by the Korean debacle. We were sated with heroes during the war and found out later that many of the heroes had been created by publicity men for "morale purposes"—whatever that might mean. There was, of course, much bravery in both conflicts, but in mechanized, electronic wars bravery is often irrelevant and not infrequently useless. Most Americans, therefore, tend to feel that heroism is pointless and that a hero may be stupid or mad, or else that he is a chap who is running away from personal problems, or is incredibly naive. An alleged hero is greeted with scepticism and even distrust, but hardly ever with admiration.

Besides being disillusioned, we are also extremely sophisticated. A good number of us know enough about the techniques of publicity build-ups to be able to spot a smart public-relations man's handiwork a mile away. We are well aware that the world we live in is anything but an honest place, and we suspect that everyone who has done something outstanding or is alleged to be doing something outstanding has an "angle" somewhere that will emerge if we just wait around long enough. We know that big-time athletics is a business, not a sport—and if we happen not to know it, the antics of major-league baseball owners would certainly convince us. We know that major scientific discoveries are no longer made by isolated geniuses but rather by well-equipped teams working for big corporations or affluent foundations. We realize that the entertainment world is a land of what would charitably be called make-believe or, uncharitably, phoniness.

314

Hence, we at least pretend not to be surprised by the quiz-show scandals. (Despite the anguished moralizing of editorial writers, average citizens seemed remarkably unperturbed by the whole sorry spectacle. Most of them seemed to feel that if they were in the same circumstances as the hapless contestants, they would do the same thing themselves.)

We know that politics is a profession of eternal compromising where even the most shining knight must soon learn to "make deals" and ignore the many shady doings of his underlings. We know that "personalities" can be "merchandised" and "packaged" and that any connection between the public image of a "personality" and his real personality is entirely accidental. So we take our heroes with a grain of salt.

There seems to exist a kind of Gresham's law of publicity, according to which authentic greatness is corrupted by association with pseudo-greatness. Heroes are somehow tainted by association with "personalities." If the Lindbergh feat happened today instead of in the 1920's, he would appear on the Ed Sullivan show, have his picture on the cover of *Time,* be the subject of a serial biography in the *Saturday Evening Post,* star in a highly fictionalized movie account of his flight and trade wise cracks with Jack Paar. All of these "stunts" are harmless in themselves, but taken together they constitute the same type of treatment accorded to quiz-show winners, tennis stars, rising young "sick" comedians, the newest teen-age "singing" (the term is used loosely) sensation, beatnik poets and any other odd characters that temporarily awaken public interest in our blasé society.

THE RADIO AND TV DID IT

There may, of course, be a real distinction between celebrity and heroism, but such nuances are hard to catch in a mass-culture world. The nation is strongly tempted to lump all public figures into the same two general types of phenomena—the amusing and the diverting. If the chair next to Jack Paar is occupied one night by a medical missionary from Laos and the next night by a brash and clever starlet, there is little chance

315

of the missionary's raising public respect for the starlet, but considerable danger that the starlet will subtly taint the public image of a real hero. (Obviously, I'm talking about Dr. Thomas Dooley; he seems to realize the chance that he is taking and is willing to run the risk. Perhaps there is no greater heroism than to endanger a reputation for genuine personal heroism in the service of something more important.)

The mass media are the great equalizers. They reduce everything to a homogenized common denominator. They have room for the bizarre and the peculiar, but no room for the truly different. "This is Your Life" and "Person to Person" carry this equalizing function to its logical absurdity. The extraordinary person is made to look ordinary. The hero, who is always just a little bit larger than human, is reduced to the commonplace, the banal, the trivial. And the public yawns and looks for someone else to stimulate its dulled sensitiveness. Heroism vanishes not in disgrace but in boredom.

Yet, there are other reasons for the vanishing hero than the equalizing influence of the mass media. For there are still heroes, even if we do not see them or try to convert them into something different from what they are. A year that has seen a Thomas Dooley and a Shirley O'Neill, the girl who tried to save a man from an attacking shark, can hardly be said to be a year with-

out heroism. The human spirit is still capable of bravery against incredible odds. Man is still capable of breaking out of the ordinary run of everyday activities and doing the startling, the remarkable. There are still a few Americans who are willing and eager to dedicate their energies and even their lives to the service of their fellows. There is still a handful of young people searching for ways to make a significant contribution to human progress and happiness no matter what personal inconveniences, discomforts and even sacrifices are required. It is not heroism that has vanished but hero worship. If we Americans have allowed the mass media to taint heroism and even to destroy the notion of it in the market place of common ideas, the reason is not the absence of heroes but rather our fear of them.

SOMETHING GOOD HAS GONE

For the hero is a reproach to the rest of us. We seek eagerly for his feet of clay so as to find an excuse for our own laziness and indifference. In the late 1950's we Americans do not want heroes. Reaction to Dr. Dooley is a perfect example of our contemporary indifference to the hero. This "splendid American," in many ways so like Lindbergh, has stirred a completely different public reaction. As a young friend of mine observed: "If Dooley should die, a lot of people will weep, but there won't be very many who will want to take his place." In 1927 the public identified itself with Lindbergh. The young could see themselves doing the same thing when they got older. The old would have liked to see their children accomplish something like what the Lone Eagle had done. But precious few Americans want to go to Laos to heal the sick; in fact, practically no one would even give it a thought. Most parents would be gravely displeased if their offspring even suggested that such a vocation might be worthwhile. As a nation we admired Lindbergh and wanted to imitate him; today we are fascinated by Tom Dooley and cheerfully contribute money to his cause. And so the hero vanishes and is soon forgotten.

The vanishing of the hero is then part of the whole complex malaise which has settled upon our nation.

We want to be left alone so we can enjoy our Good Life free from social responsibilities. We want to "play it cool." Life is too complicated and involved, too confused and dangerous for us to depart from the comfortable routine we have carved out for ourselves from the mess around us. The hero and the heroine refuse to subscribe to our National Compromise with indifference. Hence, we have no room for them.

Such a state of affairs should be disturbing to a religion which has based much of its traditional pedagogy on hero worship of Christ, the Blessed Mother and the saints. In a world determined to "play it cool," one wonders how much relevance these models of the Christian life still have. The ascetical dictum *"admirandus sed non imitandus"* ("to be admired but not imitated"), used at times of certain of the more spectacular rigors of the saints, has at best only a limited area of proper application; yet American Catholics are sorely tempted to make it a guiding principle of their spiritual life, and to place it side by side with that glittering absolute of other-directed man—"Everybody is doing it."

The decline of public idealism between the age of Lindbergh and the era of Dooley should be even more troubling to Catholics. The healing of the sick in Southeast Asia is far more in line with the traditional Christian notion of heroism than the spanning of the Atlantic. There still exists in our land some depth of interest in scientific achievement, but the corporal works of mercy occasion only passing fancy. It is not a question of our trying to defend these principles of charity from direct attack. That would be relatively easy. It is rather a problem of maintaining them as vital parts of the life of our Catholic people when all the world around us acts as though heroic virtue were irrelevant.

When someone writes a pessimistic little piece like this one, it is normally demanded of him that he provide answers. A writer who attempts to describe a critical problem is apparently thought to incur some sort of obligation of coming up with a pat, easy answer. But there are no easy answers. There never have been and there never will be. If our nation is to survive we must once again have popular heroes who will inspire us to greatness through times of crises. The mass media must bear much of the blame for corrupting the image of heroism in the public eye. The media can be reformed, but only if the public wants this reform badly enough. The public will get authentic heroes again only when it feels the need for them. But there seems to be no way in which the public can be persuaded to face the reproach which heroism involves. Yet must not some way be found—and soon? For, as the atomic scientists keep warning us, it is two minutes to midnight.

Science and the
Catholic Tradition

Ernan McMullin

IT IS A STRIKING fact, and one often commented on, that
the underlying attitude of the average Catholic toward
the whole enterprise of theoretic science is usually assumed to be one of hostility. He is the inheritor of a
sad tradition of misunderstanding and misjudgment in
this matter, one which goes back more than three centuries to the period in which physical science, as we
know it today, was just beginning to take shape.

The Church's condemnation of Galileo marked—
though the Church's spokesmen could not realize it—
a moment of grave decision. Galileo had been eloquently (if not always convincingly) contending for
the freedom of the "new science" from theological
control. In his forthright *Letter to Castelli*, he argued
that

> the authority of the Sacred Scriptures has as its sole
> aim to convince men of those truths which are nec-
> essary for their salvation. . . . But that the same
> God who has endowed us with senses, reason and
> understanding should not wish us to use them and
> should desire to impart to us by another means
> knowledge which we have it in our power to ac-

quire by their use—this is a thing which I do not think I am bound to believe.

He was on the right track, as it turned out, but the theologians were not going to be easily convinced of this. Their supervisory competence in matters physical had too long gone uncontested for them to take kindly to a man who warned them bluntly that "professors of theology should not arrogate to themselves the authority to decide on controversies in professions they have neither studied nor practiced." Forgetting the caution of Augustine, who had written that "the Holy Spirit did not desire that men should learn things [from the Bible] that are useful to no one for salvation," and ignoring the lesson of Aquinas, whose successful efforts to establish the autonomy of philosophy had met with vehement opposition from the Augustinian theologians of his day, the theologians declared that the theory that the earth moved was formally heretical. This was equivalent to outlawing the upstart new science of the "mathematicians"; it was, in fact, a declaration of war.

It is easy to find excuses for the theologians. The die-hard Aristotelians whom Galileo had confounded were the ones who actually initiated the campaign to have his novel views declared heretical. As the denunciation to the Holy Office phrased it, the Galileans "were treading under foot the entire philosophy of Aristotle, which had been of such service to scholastic theology." Strong measures of repression were therefore demanded by the beleaguered philosophers, who were, as Galileo wryly put it, "unable to withstand assault on their own." In addition, the theologians themselves had been growing increasingly sensitive about questions involving the interpretation of Scripture; the Protestant challenge had made them much less receptive to developments in the concept of inspiration than their predecessors had been. The Church's reaction to the 17th-century crisis was thus very different from her reaction to the very similar crisis in the mid-13th century, when Aristotle's novel views in physics and psychology seemed to threaten the intellectual foundations of Christendom. That Galileo failed in his attempt to provide a way for the Church to assimilate the new

321

learning where Aquinas had succeeded, is not so much an indication of a difference between the two men (though Galileo with his scorching polemic and soaring vanity stirred up a fierceness of personal opposition that the serene Aquinas never had to contend with) or between their aims, as it is an indication of the failure of nerve that followed the Counter Reformation, a failure which all but paralyzed intellectual initiative in the Church for two centuries.

A SUSPICIOUSNESS TOWARD SCIENCE

Physical science had not originated in the East nor in the Arab world, even though mathematics and technology were already relatively advanced in these areas, but rather in the Christian West, whose belief in the orderliness and "creatureliness" of the universe encouraged a constructive and theoretical approach to the problems of nature. In the early growth of this science, bishops like Grosseteste, Albert of Saxony and Oresme, as well as priests like Albertus Magnus, Bradwardine and Buridan had played a decisive part. In Galileo's own day, the contributions to science of priests like Copernicus and Mersenne were known to all; indeed, Galileo's strongest support came from priest-friends—Foscarini, the Carmelite provincial; Castelli, the Benedictine professor of philosophy; Ciampoli, secretary to Urban VIII; Dini, archbishop of Fermo. But within a generation all this had changed. Reactionism set in, and the intellectual forces of the Church gradually withdrew from the fields of secular learning which they had dominated for so long, back to the seemingly secure fortifications of a tried and true traditionalism. For the next two hundred years scarcely a single theologian, philosopher or scientist of the caliber of a Cajetan or a Copernicus was to appear behind those fortifications. Their places were taken by the preacher and the casuist. During the time when the "modern" mind was being molded by intellectual giants like Newton, Leibnitz and Kant, the Church was voiceless and intellectually almost impotent; she could take no part in directing the flood of new ideas. The age of Newman and Mercier and Lemaître was still a long way off.

322

But this unhappy story reveals only one of the causes of the latent hostility between scientists and Christian theologians that sometimes, even today, breaks out into open warfare. In England, the national church at first greeted the new science rapturously, seeing in it a wonderful manner of discovering (in Robert Boyle's phrase) "the footsteps and impressions and perfections of the Creator," of scrutinizing "the vast library of creation" (John Ray). But a nagging doubt soon arose. Did not the new physics lead to a mechanistic world-view which was incompatible with belief in the Christian God? And so the Anglican apologists began to find themselves on the defensive against allegations that they were promoting atheism (Bishop Bentley's reproach against Newton); they set about writing treatises like *The Darkness of Atheism Dispelled by the Light of Nature* (Charleston) and *The Wisdom of God Manifested in the Works of Creation* (Boyle). These works tried to provide a rational basis for a belief in God which would be acceptable to scientists. They searched for "such principles as might work with considering men for the belief in a deity" (Newton in a letter to Bentley). But their natural religion, with its emphasis on physical law and design, took little account of the supernatural side of the Christian message; instead of shoring up Christianity, natural religion soon began to displace it. As a recent writer puts it, "although their absorption in natural religion and the external manifestations of divine power did not dispute or deny any specific Christian doctrine, it did more to undermine Christianity than any conclusion of natural science ever did" (Westfall).

In the century that followed, Anglican theology was buried in the deist landslide. The Watchmaker God of the Newtonian world-machine (who had to intervene in the universe to keep planetary systems stable) became less and less believable as time wore on and scientific explanation was seen to be total in its own order. We do not *need* the hypothesis of God in physics, as Laplace explained patiently to Napoleon. Neither do we need it in biology, as Darwin's century was soon to discover. The diversity of species, the marvelous in-

323

tricacy of organisms, the adaptation of living beings to their environment—all these and many other striking facts about the living world which had formerly been thought to be explicable only in terms of special divine "interventions," now appeared as natural corollaries of the all-embracing scientific concept of evolution. But the Anglican Church felt that the time had come to make a stand (American Protestant groups like the Baptists were later to come to the same decision), and so they joined battle with the "atheist" defenders of the new theory—Huxley, Haeckel, Tyndall and the rest. History repeated itself—except that the Anglicans lacked the canonical sword Galileo's opponents had been able to wield—and the results were just as disastrous as before.

And so, both ways were tried. The Catholic Church, after an auspicious start, lost confidence and treated the new science with suspicion as a potential competitor of theology. This led for a time to the eclipse of science within the Church and to a near rupture of the harmonious, but always precarious balance between faith and natural reason which had prevailed since the time of Aquinas. The Anglican churchmen welcomed the new science as a helpmate of theology; but it ultimately proved a Trojan horse to them. Here, then, is the background we must keep in mind when we talk of the "conflict" between science and Christianity. It will be seen that the tension is due to two main factors— the past efforts of theologians to regiment science and to extend their competence considerably beyond its proper limits, and the growing "Caesarism" of science, which seems to explain everything and to make supernatural modes of thought appear hopelessly old-fashioned. The "Catholic" attitude toward science has been strongly affected by these factors. The peculiar combination of bad conscience and inferiority complex they can give rise to is vividly illustrated, for example, by the statements about science and scientists that one sometimes finds in certain sections of the Catholic press.

CAN THE GAP BE BRIDGED?

There is a real problem to be faced here. Science has

unquestionably encouraged the spread of irreligion. Are we then to retreat from it, as some leading Protestant and Jewish thinkers seem to advocate? Are we to set science over against religion as an alien and hostile fact, or are we to incorporate it in our world-view as part of that total intelligibility which it is given to man alone to discover in God's universe? Does the advance of science pose a threat to the Christian's theological understanding of the universe?

To answer this last question, we must see something of what is meant by "explanation"—scientific, philosophical and theological. Each discipline proceeds in a different way to "explain" the same thing; each has, if you will, a different idea of what constitutes "explanation." For the scientist, the death of a dog will be "explained" in terms of a virus; for the philosopher, it will be "explained" in terms of matter and form; for the theologian, it will be "explained" as part of God's providence. Now the exponent of any one of these modes of explanation is quite liable to regard the other modes as being trivial or even spurious. For instance, the scientist may protest to the theologian: "Your statement that God is good is applicable to every contingency that can arise. Therefore, its truth cannot be tested and it explains nothing." Nor does it, in the scientific sense of the word, for science requires that an explanation be specific and thus, conceivably, falsifiable.

Of course, this objection will not be taken seriously by anyone who has first grasped the difference between these quite diverse ideals of "explanation" and has then satisfied himself of the legitimacy of each. Furthermore, an analysis of the methods of procedure followed in these disciplines shows that each of the orders of explanation is autonomous and—in principle, at least—*complete* in its own domain. This means, for example, that there cannot ordinarily be any question of science happening on "something it can never explain." Science is capable of explaining any repeatable physical phenomenon, according to its *own* sense of the word "explain."

These points are understood much better today than they used to be. It was once assumed that each kind of being (a stone, a gnat, a star) was fashioned separately by God; this implied that the *only* explanation of why things are as they are is a theological one. The projection of a scientific framework back into the prehistory of our universe—in the theory of biological evolution, for example—indicates that the matter is much more complicated than this. We are beginning to see a continuous line of descent from the primitive nebula almost to man himself. Science has discovered the laws by which the stars evolved; we know why the earth is the sort of planet it is; we know a little, at least, of what brought about those genetic changes which produced more and more complicated animal nervous systems until finally one may have been sufficiently developed for God to infuse into it the breath of rationality, the human soul.

It is a majestic picture that science here presents to us, one after the heart of Augustine, who saw the whole plan of God's creation contained germinally in the original desolation. God is not "intervening" at every moment, as Platonic teleology and Aristotelian physics led medieval thinkers to assume. God is *transcendent;* He is no demiurge or watchmaker. His creation and conservation are one and the same, timeless and all-wise. There are no last-moment additions to His plan: everything is allowed for in the original blueprint.

SCIENCE, TOO, IS GOD'S

Science now begins to emerge as an essential component in our understanding of God's plan. Man is still the pivotal point in the universe. The scientist realizes better than most of us that the rationality on which science depends can be found only in a single creature, and that this creature alone has the incredible power of encompassing the universe in the sweep of his mind, of becoming "potentially all things," in Aquinas's happy phrase. The whole history of the world has led up to this creature. The Aristotelian of Galileo's *Dialog* objected to the vast empty spaces between planets and stars in Copernicus's model of the universe as "vain and superfluous," because they did not serve man. Galileo's spokesman tartly replied that it is not for *us* to say what is vain. Nowadays we can answer this objection even more effectively. No longer is man's key role as the "crown of the universe" to be insured by the crude device of making his dwelling its physical center. We can understand that enormous spaces are necessary if somewhere, by the ordinary laws of physics, planetary systems are to appear. We can appreciate the fantastic time-scale that is required if on one of those planets an incredibly improbable and complex grouping of atoms is to occur, and if a new kind of matter which will have the power of adaptation and development is to appear. We can realize that a profusion of living creatures must try their luck in the struggle for existence if some day one of them is to be found a suitable abode for an immortal spirit. The advance of science does *not* therefore involve any weakening of the plausibility of theological ways of thinking, if it be properly understood. On the

hapgood.

contrary, it notably deepens our theological appreciation of the grandeur and nobility of God's plan for man and the universe.

It has often been said that Catholicism is lacking in a theology of the temporal order, that its otherworldliness and preoccupation with the absolute have made it dilatory in evolving a true humanism. That there is an element of truth in this allegation cannot be denied. Catholic thinkers, like Maritain, Thils, Dondeyne, Mouroux and Norris Clarke have been trying hard to remedy the fault. They have sought to validate proximate temporal ends for man. Among other things, they point out that advances in technology are liberating man more and more from his slavery to matter—to disease, hunger, climatic extremes, exhausting labor and so on. Man has tapped sources of untold energy. He is about to conquer the barriers of space. The range of his brain is being enormously extended by intricate electronic computers. He will soon be able to communicate almost instantaneously with any one of his fellows. His whole conception of work and of society is gradually being transformed. Three hundred years ago, science scarcely touched the daily life of man; the range of his mind and body, the energies at his disposition, even his means of communication were almost as physically limited as those of his primitive ancestors tens of thousands of years before. Today man's whole relation to the universe is changing at a pace that few have really grasped. Yet we are still only on the threshold of undreamed-of changes. Biology and psychology are at approximately the level of theoretical development that physics was in the time of Newton. What the next century will bring only God knows.

There are two points of view among Christians regarding this upheaval. One is that technology is a "bad thing," that it replaces God's image in creation by that of man, that it is a manifestation of the "original sin" of man's nature, a sign of his inordinate desire for knowledge and greed for power. This attitude (which is exemplified in the work of some few American Catholic writers, like Wilhelmsen and Carol Jackson) is partly Manichean, but is chiefly rooted in a "good-old-days" mentality which Catholicism somehow seems to foster among some of its adherents. The opposite view is that technology is man's way of obeying God's command to Adam to "dominate the earth;" it "transforms matter by imposing upon it traces, as it were, of rationality, of spirituality, even of humanity" (Thils). In remaking the universe in the image of man, science brings it nearer the image of Christ, too, thus helping it participate in its own way in the work of redemption (Dubarle, Teilhard de Chardin). It may even be (as D'Arcy, Malevez and others suggest) that in transforming the earth, science is preparing for that "glorified earth" on which the resurrection of the body is to take place and which will be entirely dominated by man.

Be this as it may, it is certain, at least, that the technological transformation of the world is in itself good. But it unquestionably poses man with the gravest challenge he has ever faced. The more man's capacities are magnified, the greater the dangers to which the weakness of his still human nature exposes him and his world. It is imperative, then, that man's spiritual growth match the increase of his physical potentialities. It is not merely that man needs new wisdom and new restraint as he gains command of nuclear and genetic forces. It it not just that the terrible new capacity he possesses for dominating his fellow man may turn his head and put an end to freedom on the earth. It is above all the fact of *knowledge,* the feeling of omnipotent reach, that can make the mind of man so swell with pride that it may set itself over against God.

In this growing crisis, the troubled scientist can find little solace in either of the great philosophical orientations which have, between them, dominated the intellectual arena for over three centuries. The tradition which stretches back from Nagel and Ayer through Mach and Hume to Locke and Hobbes, and which in its various manifestations has been labeled "empiricist," "phenomenalist" and "positivist," could never find room for the notion of a transcendent Creator. Within the limits of its own categories and starting-point, the most it could hope to achieve was either a rational affirmation of a limited God (the deism of the 18th century is an instructive example of where such an effort is bound to fail) or else an incompletely rational affirmation of a Creator. (The fideism of the later medieval nominalists or of many of the leading Protestant thinkers of today might here be quoted as examples.) On the other hand, the idealist tradition, represented in our own day by such great scientific figures as Einstein and Eddington, deifies the human mind, and therefore tends to accentuate rather than diminish the peril in which the scientist now finds himself. Existentialism may be regarded as an anguished attempt not so much to resolve as to underline this dilemma bequeathed to the modern world by Descartes. It recalls man to a sense of his own contingency, to wonder at the *fact* and not merely the modalities of his existence.

But this recall is not enough. If we refuse to move beyond the level of human frailty and fleetingness, the human condition becomes a horrid absurdity without a history or a meaning. It must be seen as demanding a total creative Cause as its existential ground. This requires acute metaphysical analysis of the kind in which Aristotle and Aquinas excelled and which contemporary thought is beginning to master again after many centuries of neglect. Here is where the realist metaphysical inheritance of the Christian can be of such immense service. Yet the very strength and complexity of this inheritance poses a great danger, too— the danger of treating the words of a master philosopher as things, of converting philosophy into history. Metaphysics more than any other part of philosophy requires an utter integrity, to which the memorizing of "theses" and the division of the history of philosophy into the "good guys" and the "bad guys," in the manner of a juvenile western, are altogether alien.

A KEY TO UNDERSTANDING NATURE

The passionate desire for understanding and comprehensiveness which distinguishes the true metaphysician can be seen in the best of contemporary Catholic thought—in the works of Maréchal, Maritain, Lonergan and de Raeymaeker. It is here (so it seems to me) that the scientist may find a way of relating man to nature and to his Creator. Man is seen as totally dependent upon God at every moment of his existence. Everything that he is or makes himself to be, the very fact that he is at all, all these find their ultimate ground in the Creator. Man accepts things as they are, tries to understand them and modify them. But the whole order of space and time, of man and materials, takes its being from One who stands alone, who conserves the world in being and does not merely modify it.

The scientific quest itself takes on its full significance only within this context of *creation*, of the universe as God's handiwork. This insight is probably the principal legacy that the Judeo-Christian tradition has bequeathed to philosophy, as Pieper and other historians of medieval philosophy have often stressed. Natural science is, then, in its own way a searching out of the intelligible imprint that the Creator has impressed upon His work. It is true that some have held that science tells us nothing of real structure; it is to be regarded, they say, simply as a convenient way of cataloguing phenomena. This was Bellarmine's contention against Galileo, and it echoed a distrust of the "mathematicians" which was common at that day among philosophers and theologians. It is the view of modern positivism, too—a view which, partly through the influence of Duhem, has tinged the whole of contemporary Catholic thinking on the nature of science. Though possessing a certain plausibility in the context of descriptive theories of motion, this view is utterly and demonstrably inadequate as an account of what science in general is doing. It makes the scientist either a collector of curiosities or a technologist.

Science, then, must be taken to disclose in some sense, however oblique, the hidden structures of the real. We can now begin to realize the true dignity of the scientist's vocation. It is he who is charged with interpreting the Book of Nature in which God reveals Himself no less surely—though much less clearly—than in the Bible. What the scientist finds is what God Himself has put there, the intelligible structures which are the proper objects of man's God-given intellect. Next to love of God and neighbor in the scale of values governing those spiritual activities which make us the "image of God" comes intellectual understanding. And foremost among the goals of that understanding is the scientific grasp of that created world to which God has fitted man's sensory powers. If we are to prepare for eternal life by the development, both natural and supernatural, of our faculties of intellect and will, it would seem that scientific understanding is among the highest of natural activities, one that every Christian should hold in the most profound esteem.

The struggle which convulses the world today can, in a certain sense, be regarded as a conflict between two competing theologies of science. One theology assigns to science a messianic role in bringing about the millennium here on earth; secularists of the West and Communists of the East agree in making scientific progress the supreme norm for man. Christianity, however, sees its millennium elsewhere, and declares that prayerful union with God is more important to human destiny than is scientific research. The scientific exploration of the universe, as our late Holy Father so often emphasized, is good, but it attains its full significance only when it reverently respects God's overlordship. It must be carried out with humility; there must be a real ascesis of knowledge, of the kind that the Incarnation not only dramatizes and symbolizes, but also makes possible for the individual Christian. The Word of God freely limited Himself by assuming our condition, and even gave Himself over to suffering and death for the love of men. The scientist who looks upon nature and sees in it the imprint of this same Word cannot fail to realize that his commitment to truth must involve an equal commitment to love, and that the discovery of truth must lead man to humility, not to pride.

How Affluent
Are We?

Benjamin L. Masse

O UR COLLEGE hosts had done their homework; they had planned exceedingly well. What more timely discussion topic could one have desired than "Our Affluent Society"? Wasn't Professor Galbraith's *The Affluent Society* on all the best-seller lists? Wasn't it stirring economists to angry debate on half the campuses in the country? Wasn't it jolting complacent businessmen and sending labor leaders into mental tailspins?

Nor was the topic merely timely. It raised a host of important questions, not only for economists and statesmen, but for theologians, philosophers, sociologists and psychiatrists as well. What was the affluent society doing to the one and only absolute commitment man has here below—the commitment to serve God and save his immortal soul? What was it doing to the character of individuals? What was its impact on family life and on the community? Freed at last from the grinding struggle to overcome scarcity, were we falling victim to a siren world of abundance? Were we, in short, going the decadent way of the ancient Romans, masters of the world but slaves to sense appetite?

What about all the economic problems our affluence was generating? Were we working too hard and too long? Were we building too many autos, refrigerators and television sets, and not enough roads and schools, parks and hospitals? Were individuals taking too much of the national product and the state too little? How could it take a bigger share without taxing us to death and coming to dominate our lives?

Surely, that was the sort of chow to feed the troops (the troops in this case being an invited group of businessmen, union leaders and public officials). Our academic hosts were right. If the social documents of the Popes are to have significance for our times, such are certainly some of the questions that must be posed and pondered.

But the play didn't run according to script. The discussion began well enough and Galbraith's blueprint for a happier future was being subjected to the sort of lively crossfire one expects whenever scholars and businessmen gather for dialog. Before the discussion reached the point, however, when it might be expected to continue under its own momentum, one of the professors gained the floor and proceeded to execute the maneuver known as the diversionary tactic. He announced that the Galbraith tome wasn't worth the paper it was expensively printed on. It was as phony as a three-dollar bill. It assumed that ours was an affluent society and then went gaily on from there. But was ours truly an affluent society? How many in the room, the speaker demanded, waving an eloquent finger in a sweeping semicircle, felt notably affluent?

That did it, of course. Our academic hosts remained perfect gentlemen. There was not the slightest sign of dismay over what was happening to the agenda they had so carefully planned. But from that point onward, the argument was not about the consequences of an affluent society; it raged over the bald fact of its existence.

Now from several points of view, this switch in the discussion might seem to have been a waste of time. Certainly, compared with other countries, American society is not merely affluent, but garishly so. What other

334

country can pay its farmers not to produce? Or permit its industry to· practice a wasteful policy of planned obsolescence? What other country can finance a World War and at the same time raise the standard of living of its people? What other country can over a ten-year period lend or give to other countries more than $60 billion?

Nor is the United States well-to-do merely in reference to other lands. By any criterion of material wealth, it is simply and absolutely rich—period. It has vast natural resources. It has the greatest industrial machine known to man. Its citizens possess in abundance those goods and services which are recognized as evidences of wealth—telephones, automobiles, indoor plumbing, electric refrigerators, radio and television sets, motor boats, schools and hospitals. They spend a fortune annually on liquor, tobacco and cosmetics. In their search for pleasure and diversion they travel to the ends of the world. Never before has there been a society in which so many workers enjoy so much of what are commonly regarded as the perquisites of wealth. How can anyone seriously quarrel with Mr. Galbraith when he writes that "as an economic and social goal, inequality has been declining in urgency"; or when he notes that, subject to a few qualifications— including the major qualification that depressions be prevented—the age-old "preoccupation with economic security is largely in the past."

It is only when one ignores the forest and looks at the trees, so to speak, that doubts about the Galbraith thesis arise. It was the sudden descent from the global and national to the individual that detoured the conferees. Only those who did feel affluent—and they appeared to be a small minority—frowned when the professor suggested that all of us might take a long look at family-income distribution and then make our own estimate of the affluence of the U.S.A. He had taken such a look, he affirmed, and so far as he was concerned, his articulate colleague from Harvard had laid an academic egg

But did Dr. Galbraith lay an egg, albeit a very lucrative one? Let us together have a look at the most recent estimates of national income distribution and make up our own minds about this business. (These estimates are the work of Mrs. Selma F. Goldsmith of the Office of Business Economics of the U. S. Department of Commerce, and appear in the April *Survey of Current Business*.)

In 1958, total pretax income of families and unattached individuals (consumer units) increased, despite the recession, by $6 billion over 1957. The total was $338 billion. However, since the number of consumer units also increased, from 53.5 to 54.3 million, average family income remained stationary at $6,220. The median income stood at $5,050, that is, half the country's consumer units had incomes above $5,050, half had incomes below that amount. The following table shows the distribution of income among consumer units:

Family Income	No. of Families (Millions)
Under $2,000	7.6
$2,000- $3,999	12.1
$4,000- $5,999	13.6
$6,000- $7,999	9.2
$8,000- $9,999	5.0
$10,000-$14,999	4.3
$15,000 and over	2.5

Over the status of the 7.6 million consumer units with an annual income of $2,000 or less, there won't be much argument. They scarcely consider themselves affluent, nor are they so regarded by popular estimate. In 1958, as in the preceding year, they represented 14 per cent of all consumer units. Since in large part these units are made up of farm families and unattached individuals, their economic status, while unenviable, is not quite so hopeless as the bald figures suggest.

Nor would families in the $2,000-$3,999 bracket look upon themselves as well-to-do, or be so described by others. If there are children in the household, these consumer units are barely able to get by. Even a single person earning $76 a week before taxes—the highest level in this bracket—is hardly in a position to indulge in riotous living. More than a fifth of all consumer units arc in this category. Combining these two lowest brackets, we find that 36 per cent of U. S. consumer units might reasonably wonder what Dr. Galbraith is talking about.

It would take a Gallup Poll to find out whether the large number of consumer units in the next bracket—a fourth of all consumer units—consider themselves affluent, at least in the restricted sense of being reasonably well off. In this bracket, where incomes range from $4,000 to $5,999, are many of the nation's unionized and better paid workers. The bracket is also heavily populated by school teachers and government workers. At the risk of exposing the Editor-in-Chief to a flood of irate letters, I would guess that at least those in the upper half of this bracket, especially the unattached individuals and small families, have some feeling of being part of an affluent society. They can't send their children to college—unless the children work their way through—but they do enjoy many of the amenities of modern life. They can contribute to the week-end jam on our highways (in a second-hand car) and (with the aid of a long-term mortgage) swell the number of Americans who own their own home.

If Professor Galbraith's thesis makes any sense at all, all the remaining consumer units, those making $6,000 and over, must be deemed in some sense affluent. In this group are 21 million consumer units, or 39 per cent of all units. These are the families, notably the 6.8 million among them with incomes over $10,000, who set the affluent tone of American society. They are the ones for whom the big slick magazines are mainly published. They are the ones on whom Madison Avenue concentrates much of its frenetic energy. They are the darlings of Fifth Avenue and Detroit, and of all the centers of fashions and big-ticket consumer durables in the land. And, significantly for the Galbraith thesis, they are a growing percentage of all our family units. In terms of 1958 dollars, only 27 per cent of U. S. families had incomes of $6,000 or more in 1947. Last year, as wo have seen, nearly two-fifths of our families had such incomes.

The late Hilaire Belloc, as I recall, used to argue that a dominant minority customarily establishes the character and sets the tone of a society. If that criterion has validity, the sizable minority with incomes of $6,000 and more—or even, perhaps, the smaller minority with incomes in the $8,000-plus class—stamps American society as signally affluent.

One reason some of Dr. Galbraith's critics are loath to accept this fact is the growing institution of the working wife. In 1940, less than 15 per cent of the wives of family heads were in the work-force. By 1958 the percentage had jumped to 28. Still more astonishing, the increase was as marked among the top two-fifths of our families—rated according to income—as it was among the other three-fifths. In fact, the closer one looks at the figures, the clearer it becomes that in many cases the dividing line between affluent and nonaffluent families is determined by the presence or absence of a working wife. In 1957, 41.2 per cent of the wives of working husbands in the top fifth of our families were in the paid labor force. In the next fifth, the percentage was 33.4. Those figures contrast sharply with the 16 per cent of working wives in the lowest fifth of our income classes, and with the 22.3 per cent in the next lowest fifth.

But, however great the sociological significance of this phenomenon, it doesn't change the economic fact of affluence—except in one respect. To the extent that many of our affluent families have sacrificed leisure for income, by sending the wife out to work, they are less affluent than the size of their billfolds suggests.

The conclusion would seem to be that Professor Galbraith was correct in positing an affluent society; and that, as a consequence, anyone who wishes to think realistically and creatively about our society must start from that fact. Among other things, the fact of affluence has enormous importance for our conduct of the Cold War. It means that if our leaders should decide that more strenuous efforts are needed to turn aside the Communist threat, they can safely demand much greater sacrifices than the American people are presently making. They can forget, in other words, the nonsense that the Kremlin, by keeping the Cold War going, is forcing us to spend ourselves into bankruptcy. That such a belief has gained some currency in the country, even in high councils, is itself, perhaps, another proof of our affluence. For too often one of the fruits of affluence is a certain softness of character which shrinks from sacrifice, and which, if indulged, spells the doom of men and nations. It was this softness which Sen. J. W. Fulbright had in mind when, rising in the Senate on July 8 and looking on the wreckage of the Foreign Relations Committee's mutual-security bill, he said bitterly that "we are determined to end up the richest, fattest, smuggest and most complacent people who ever failed to meet the test of survival." God grant the Senator is wrong.

Where Do We Go From Togetherness?

Katharine M. Byrne

THE WORDS "Magazine of Togetherness" are no longer to be found on the cover of *McCall's*. What does the withdrawal of this familiar subtitle mean? Is this the end of another era? Is each man now to go his way alone, picking out his own underwear, pursuing a destiny apart from the other members of his bowling team? What happens to the American housewife of the ad writer's dream—the girl who, as Philip Roth has it, "couldn't buy a roll of ScotTissue unless her companion-helper-husband went along to help select the color"? And our children? Are they to sit around reading books or looking into space, or to gather in haphazard little groups composed entirely of other children? Will we abandon them to their unorganized devices?

These are serious questions to which precipitate answers will not suffice. But perhaps we should back up a bit. What is "Togetherness," anyway, and what could possibly be wrong with it? At first glance the word appears to describe an admirable ideal of shared familial experiences. It also suggests any number of pleasant pictures, often in four colors and on slick paper. The pictures usually include Mother, Father, Sister and Brother. Sometimes, with a nod to the national birth-boom, Brother and Sister are joined by Baby. If it is spring, Father may be planting a young tree just beyond the last flagstone of the patio. Brother holds the shovel, Sister carries the bag of fertilizer, and Mother stands by, decorous and deferential, with a tray of frosty lemonade. This scene offers only one of a number of heartwarming variations on the theme: parents and children engaged in mutually satisfying projects. Abroad, this companionable group might be seen eating off the tailgate of the station wagon. At home or away, they usually move as a unit.

If the mother does leave the house alone, it is to attend a PTA meeting, to act as a canister-bearer for a worthy childhood disease, to solicit money for new band uniforms or to set out the casseroles for the Cub Scout dinner. Father will venture beyond the range of the television set long enough to coach young ballplayers, to participate in a symposium on "Our Turbulent Teens," or to pressure local authorities for some significant community project, perhaps a new swimming pool for the children.

This is a world of children, and, as someone might viciously suggest, of adults who have never been anything but children first and parents afterward. It is an egalitarian world of family councils in which a four-year-old's vote is as good as his father's, and of schools where everyone sits in a circle and the teacher just happens to be bigger than the other members of the group.

FAREWELL TO PRIVACY

From the earliest days of the Togetherness movement, some there were who did not like that word. Hardly was the ink dry on the first "Togetherness" issue of *McCall's*, when an articulate opposition moved into print. It was, of course, among persons who had never purchased a copy of the magazine—and were now even less likely to do so—that the wheels of high-minded protest were set in motion. Joseph Wood Krutch, for one, writing in *The American Scholar,* announced that at the first sight of the word "Togetherness" he felt a "little wave of nausea" come over him.

Thus, from the beginning, there were dissenting voices: those of persons who persisted in thinking of families as composed of individual members, each with a separate as well as a corporate destiny. These objectors saw in Togetherness a leveling process, an invitation to mediocrity, a cutting down to a least common denominator. And they set out to plead the case for apartness, separateness, or individuality; the right to spiritual and physical withdrawal, at least on a part-time basis.

Not that either condition is readily achieved in our time or in our places of residence. What are the possibilities for physical separation from those we love? Invasions of privacy are usually discussed as though they consisted only of wire-tapping, bugging, assailing of captive commuters by canned commercials, or the impudent questionings of a personnel department. But actually, parenthood, in our culture, is probably the greatest single invasion of the right to be let alone. What has the Fourth Amendment, guaranteeing the rights of the people to "be secure in their persons, houses, papers and effects," to say to the parents of a number of young children?

Physically, we are all in this together. New-house space is dearly purchased, if at all, and the old houses are running out. There is no place to hide. The French, in other times, built into their homes a *boudoir* (Fr., *bouder*: to pout), to which an adult female might retire in times of stress. Today's lady of the house will have to stand up and scream right in the middle of the living-dining-and-hobby area. Even "mommy's room" is likely to be shared by the ironing board, an electric train running around and under the bed, and enough Lincoln Logs to reconstruct Abe's birthplace.

Domestic privacy, the right to be alone now and then, used to be purchasable at low rates. Our mothers could summon to their aid the laundress, the nursemaid, the indigent and grateful relative. Even today, in other climes, the middle-class mother has the assistance of amahs, ayahs, scullions and nannies. It wasn't long ago

342

that the British family of *Mary Poppins* operated with "Mrs. Brill to cook for them, and Ellen to lay the table, and Robertson Ay to cut the lawn and clean the knives and polish the shoes. And, of course, there was usually a nanny to look after the children." In most of the households we know the mother plays all these supporting roles herself, including that of Robertson Ay.

The world of hired help and adequate living space is gone, but not without being missed. Even Mary Drone, John Keats' wretched, unheroic heroine of *The Crack in the Picture Window*, knew that the life of community Togetherness in cracker-box identicality, with endless talk of tots and toilet-training, was wrong, but she didn't know what to do about it. Several of her more literate sisters could have told her, and in no uncertain terms.

MESDAMES OF THE OPPOSITION

First among a recent group of Togetherness's detractors was Anne Lindbergh, happy to be tied by the bonds of family love and necessity, but not willing to be strangled by them. Her gentle protest, in *Gift from the Sea,* is the classic plaint of a mother giving love and time freely—but with a small reservation, asking to be removed in time and space, occasionally, from the exigencies of hems to be let down, items to be purchased, schedules to be met, and the endless tyranny of telephones. Hers is a plea for the chance to replenish the ever pouring pitcher, to preserve a small corner of life untrammeled. Moreover, she is certain that, given time out, she will return to the fray a better and more understanding parent.

A more raucous voice on a shriller note is that of Jean Kerr. In her anti-Togetherness tract, *Please Don't Eat the Daisies,* Mrs. Kerr makes the basic assumption that a parent is entitled to a considerable part of her own life, even during the years of her children's early childhood, a period of active and unavoidable Togetherness. Mrs. Lindbergh is gentle, tentative, inquiring; Mrs. Kerr's cry is loud and clear. It says, very simply, "lemme-out-of-here." This girl wants to sleep late and to write, and has been doing both with considerable success. She loves each cereal-smeared little face, but not exclusively.

About housing, Jean Kerr could tell Mary Drone a few things. "I wanted," she writes, "a house that would have four bedrooms for the boys, all of them located some distance from the living room—say in the next county."

Of the period during which she and her husband were working on the musical, *Goldilocks,* she writes: "First we would locate the children and threaten them with violence and sudden death if they came near us. Then we'd take coffee and cigarettes and hide in the den." Lines like these must give us pause. As a writer who is a Catholic, what is Mrs. Kerr doing to our Mother Machree Fantasy? (This is the illusion which haunts one segment of Catholic writing—that unless the dear fingers are toilworn on a 24-hour-a-day basis, the mother doesn't really love her children.)

Another assault, this time two-pronged, was launched by a man and his wife named Smith. Elinor Goulding Smith is the author of books with such remarkable titles as *The Complete Book of Absolutely Perfect House-keeping* and *The Complete Book of Absolutely Perfect Child Care.* (This last contains the Absolutely Perfect Chapter Title: "The Child, from One to Forty.") Mrs. Smith's answer to the "groupiness" of organized, adult-directed childhood is a prescription for developing an "ingoing" child. Her formula calls for no fuss, no muss, nor any procession of uniformed and mud-footed affiliates. "Just drop the child off at a library," she suggests, "or buy him another stamp catalog or a scrapbook for his bug collection, or a little formaldehyde, or a tub of tropical fish, or even a book, and he's safely and economically disposed of."

Now the literate females of the community may be accused of selfishness: turning their backs on their children because of the desire for a more ego-satisfying hour or two out of life. And, it must be admitted, they all are spiritual sisters of Joan, one of A. C. Spectorsky's *Exurbanites.* You remember Joan. She is the one who asks herself why, with her brains and good taste, she shouldn't start a small business, or write a book, or "put up my onion soup with sherry that everybody always

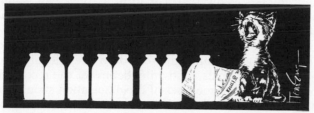

raves about, and sell it by mail?" But not Mrs. Smith's husband, Paul Robert, who, in *"Where Did You Go?" "Out." "What Did You Do?" "Nothing."* sets up all his arguments for adult apartness from the *child's* point of view.

He is the one who comes right out and says that "parents are the natural enemies of children." And before you get excited about this premise, please note that it is just Mr. Smith's way of postulating a dual family-world, as opposed to the social monism of Togetherness. The title of his book is a reference to the mutually impenetrable curtain which hangs between a child's world and that of his parents. The book itself is a plea for the part-time freedom of growing children from the encroachments of interfering adults, with a special emphasis on the right of a child to breathe in and out without direction, and make his own way among his peers.

By last spring the pendulum was arcing wide with savage swipes. There was a Mercury ad featuring three crowded and uncomfortable males packed into the rear seat of a competitor's automobile, and captioned, "Are you suffering from too much Togetherness?" There were admonishing magazine articles and hortatory headlines. "Don't Be a Pal to Your Son!" cried the *Reader's Digest.* "What child wants a 40-year-old man for a friend?" sneered Al Capp in a New York *Times* interview.

Yes, it seemed only yesterday that the last father had been wrested from his beer can and television set, and pushed over to the third-base foul-line with his boy's team at bat. But now the order was being countermanded. He was being warned to get back into his house-slippers, and to cultivate a stern look of authority; and advised coldly that a man in a chef's cap leaning over a hot barbecue pit does not present the proper "father image" to the young.

Someone even suggested an organization called "Fathers Anonymous," dedicated to the rehabilitation of reformed parents, erstwhile victims of the recent unpleasantness. If a member of the association should feel a cook-out or an all-family beach excursion coming on, he is advised to telephone another father, one who will speak firmly to him, strengthen his resolve, or even come over and hold his hand until the crisis passes.

Another straw in the wind came out of the Middle West when the leader of a Cub Scout Pack in Winnetka, a man described in newspaper accounts as a "former Cubmaster who is also an opinion research executive," announced that his youthful affiliates were being placed on inactive duty. This man, you must note, is not only a Cubmaster, but a poll-taker, sensitive to the wavering seismograph of buyer-reactions. And he has sent forth a ukase: "It's time that the kids started their own baseball games, or just went outside and kicked a can."

One can almost see the little IBM wheels turning madly in the ex-Cubmaster's eager, probing mind. (And in our world that "ex" pays highly: whose word is more impressive than that of an ex-narcotic, -Communist, or -drunkard?) This man is probably giving us the word from our alternate sponsor, with next week's prescription for family living. If the organized child is on the wane, let's start working on the Child Apart, the Pale, Skinny Kid with Glasses, reading a book, or the Lonely Scientist with his "600-power microscope with reversible plano-concave mirror with illuminator, 5-hole revolving diaphragm and zippered, simulated-leather carrying case." Or even, absolutely new and in time for the Christmas trade, a package of "non-toxic, non-allergenic and highly educational KANS FOR KIDS. Hours of Healthful, Solitary Fun. Just for Kicks."

If an epitaph to an era is now in order, let it be written by the small boy in a cartoon in this morning's paper. He is standing in his backyard with a firm grip on his own football, and he is saying to the puzzled adult who is signaling for a pass: "I'd like to have you for a pal, Dad, but you're just not my type."

The John Birch Society

Robert A. Graham

L IKE THE influenza epidemic now predicted for next spring on the basis of known cycles, a new onset of such extreme "ultra" movements seems at hand and may well be already upon us. One writer claims there are today 2,000 right-wing organizations in the country, with a membership or following of eight million. No one knows how many persons are really active in these bodies, but it is already clear enough that they are in a position to make their influence felt. Is there reason to be alarmed about this revival of extremism?

In the opinion of many observers, the recrudescence of what is a chronic malady of American democracy will soon be cast off by healthy antibodies in the nation. Others, less complacent, point to perhaps lasting ill effects on the political system of this country. While most of the right-wing organizations now proliferating claim the highest patriotic purposes, the fact is that they weaken the fight against communism by indiscriminate charges against the most respected leaders of government and public opinion. The resulting loss of public confidence in our national leadership, it is feared, will seriously compromise the United States' ability to emerge from the current world crisis with its honor and its prestige intact.

Many of the new organizations now recruiting the extremists of the country in their ranks are small and insignificant both in numbers and in the quality of their leadership. This, however, cannot be said of the John Birch Society. This is no movement led by an obvious paranoid like the self-proclaimed Nazi, George Lincoln Rockwell, or by opportunists with a police record, such as the promoters of the Minute Men in Southern California. The JBS boasts of a national council comprising leading industrialists, military men and former government officials, as well as several well-known Catholic figures, one of whom is a priest. A fairly large number of Catholics seem to have been drawn to the society. Without accepting the estimate of 50 per cent of the total membership, we can believe society president Robert Welch's Catholic secretary, who told the editor of the Rochester *Courier-Journal* that there is "a high percentage of Catholics in the John Birch organization."

Yet for this organization of presumably responsible and conscientious men, Sen. Thomas J. Dodd (D., Conn.) has had words of sharpest criticism. Alluding to the JBS-propagated charges against such respected public leaders as President Eisenhower and the late John Foster Dulles, he spoke of an "affront to both decency and intelligence." This sort of charge, he said in a speech last March 31, "brings our leaders and our institutions into disrepute, it sows division, it makes it easier for the Communists and the 'ultraliberals' to equate opposition to communism with political lunacy." Senator Dodd's vigorous, practical anticommunism needs neither apology nor proof. This will not necessarily guarantee him against attack by the Birchites as a Communist.

What is this organization that called forth such a cry of protest from so respected and forceful an anti-Communist as Senator Dodd?

John Birch, after whom the society is named, was a young Georgia-born fundamentalist Baptist missionary in China when America entered the second World War. He become a captain in the Office of Strategic Services, and appears to have been instrumental in preparing the escape route of Gen. Jimmy Doolittle, who landed in China following the successful bombing of Tokyo. Killed by Red Chinese troops ten days after V-J Day, Birch is hailed as "the first victim of the Cold War."

On December 9, 1958, Robert Welch, a retired New England candy manufacturer, founded the John Birch Society at a meeting in Indianapolis. Its purpose: to cope with "the threat of the Communist conspiracy." Unless, he said, "we reverse forces which now seem inexorable in their movement, you have only a few more years before the country in which you live will become four separate provinces in a world-wide Communist dominion ruled by police-state methods from the Kremlin."

Growth of the new society was rapid, and it is now described by observers as "the base-organization of the extreme Right." It is so much to the right politically that it treats with disdain such conservatives as Sen. Barry Goldwater, former Vice President Richard M. Nixon and the late Sen. Robert Taft. According to the Attorney General of California, its membership consists primarily of "wealthy businessmen, retired army officers and little old ladies in tennis shoes." Nevertheless, it has aroused public attention and concern. It spearheads a campaign to impeach Supreme Court Chief Justice Earl Warren. Its material was used by Gen. Edwin A. Walker in his controversial right-wing activities in the U.S. armed forces in Germany.

The JBS is perhaps most known for its founder's sweeping charges against President Eisenhower, whom Mr. Welch has described as a conscious Red agent. Relying less on mass membership than on disciplined and organized (and perhaps often camouflaged) small groups scattered around the country, the John Birch Society seems capable of making its influence felt, if only by co-ordinating the efforts of other existing extremist groups that await direction.

Included among the estimated 60,000 members of

the society are, no doubt, a large number of ordinary citizens who seek to "do something" to combat communism. They are usually surprised and puzzled that an organization led by such a God-fearing man as Robert Welch, who patriotically aims to save the country from what they are convinced is imminent danger, should be subjected to denunciations and complaint. The attacks on the JBS they regard only as confirmation of the gravity of our national peril and proof of the extent of Red influence. These persons exhibit an amazing naïveté about the real issues at stake and what they are getting into.

Typical in this respect is the view expressed in a letter sent by a pious and apostolic Catholic woman and brought to this writer's attention. She wrote: "Critics of JBS all speak vaguely of extremism, questionable methods, fascism, etc., and they all sound as if they haven't even read the Blue Book. I read the Blue Book first two years ago, thought it made good sense and was impressed with Welch's spirituality. I read it again this spring when all the controversy arose and still feel the same way. I read his monthly bulletins fairly regularly. The only specific objection I have read about is Welch's contention that 'government is the enemy of the people' or something to that effect." For the writer of that letter, as for many others, the criticisms seem trivial compared to the great task undertaken by the society in fighting communism.

It is perhaps true that much of the criticism leveled against the John Birch Society at the beginning was based on instinctive fear of any right-wing organization and not on a detailed knowledge of the creation of Mr. Welch. After all, these movements are not new. They all tend to follow a common pattern, since they recruit from much the same reservoir of human temperament and political bias. They are usually anti-Semitic and anti-Negro. Many are also anti-Catholic. All are anti-Communist, but it usually turns out that they use the Communist issue as a cloak to cover such less admirable features as racism, bigotry and intolerance, both religious and social. Even the best of them invariably attract a host of dubious personalities all too ready to ride the crest of popular hysteria. The general public cannot easily forget the Ku Klux Klan or the Christian Nationalists of Gerald L. K. Smith.

The John Birch Society seems to have avoided the cruder manifestations of the extreme rightwingism of its predecessors. Accusations leveled against it on the score of anti-Semitism seem based on suspicion rather than on the record. While it opposes desegregation, it refrains from anti-Negro bias. Nevertheless, the JBS inevitably falls heir to the heavy legacy of distrust and revulsion that sad experience has left in the public's mind. The feeling is almost unavoidable that the old forces of bigotry are latent in the society, ready to burst forth at the right moment.

Critics of the society have found more concrete grounds for their attacks in the contents of the Blue Book already referred to. This is the "bible" of the society and consists of the talks Robert Welch delivered to the charter members in Indianapolis in 1958. The Blue Book is not just a collection of speeches; it is the acknowledged chief source of orientation on the principles, attitudes and program of the John Birch Society.

Many of the individual aphorisms in the rambling Welchian prose of the Blue Book are not particularly incendiary unless given an extreme and literal meaning. Mr. Welch believes, for instance, in "less government and more individual responsibility." A much-publicized slogan runs like this: "This is a republic, not a democracy. Let's keep it that way." Other statements are found which, though disturbing by their connotations, are capable of an innocuous interpretation, or are at least not as sinister as sometimes made out. For instance: "The greatest enemy of man is, and always has been, government." Democracy is defined by Mr. Welch as "the unbridled rule of demagogic men." It is "merely a descriptive phrase, a weapon of demagoguery and a perennial fraud." Taken in their ensemble, however, these and other statements of governmental philosophy reek strongly of distrust of the democratic principles upon whose validity the American system rests.

The authoritarian structure of the society is another object of criticism. In the Blue Book the founder leaves no doubt that it is he who will decide things. The society, he said, is to be "a monolithic body." It is not to be run on representative principles or on the principle of local autonomy: "The John Birch Society will operate under completely authoritative control at all levels." He declared also: "We can allow for honest differences of opinion. . . . But whenever differences of opinion become translated into lack of loyal support, we shall have short cuts for eliminating both without going through any congress of so-called democratic processes." Needless to say, these remarks confirm the picture of the society as an antidemocratic body.

There is no doubt that Mr. Welch is not only the founder but the main force of the society. He stresses repeatedly in the Blue Book the leadership principle, that recalls the grim Nazi precedent: "The men who join the John Birch Society . . . are going to be doing so primarily because they believe in me and what I am doing and are willing to accept my leadership anyway." His justification is the following pragmatic one:

353

"We simply are not going to be able to save our country from either the immediate threat of communism or the long-range threat of socialism by organizational leadership. Our only possible chance is dynamic personal leadership."

The methods proposed in the Blue Book provide another reason for criticism, by seeming to sanction the use of any means, fair or foul, in combating communism. "The front business," said Mr. Welch, "like a lot of techniques the Communists use, can be made to cut both ways." He advocates the pillorying of suspected Communists by techniques which he admits are "mean and dirty."

Such arguments against the John Birch Society, based on its antidemocratic principles, its authoritarian structure and its avowed readiness to use "short cuts" to achieve its ends, fail to impress, much less embarrass, the average Bircher. These features, they say, are necessary to prevent the society from degenerating into an ineffective debating society or from being infiltrated by Communist disrupters. Membership is voluntary,

and those who don't like the founder's ideas are perfectly free to take their leave. The accusations, in the Birchers' eyes, are mere caviling on secondary issues at the very moment when the nation faces mortal danger from communism.

The John Birch Society is certainly no model for a movement in the American democratic tradition. From this standpoint it deserves the castigation it has already received, and one must regret that its members treat so lightly these unpatriotic aspects of a supposedly patriotic organization. But the real danger of the society lies not in its organization and methods but in its impact upon the American national life at this moment. It is proving to be an instrument of division and a threat to the national morale. For, to justify its claim that the country is infiltrated by Communists, the John Birch Society has embarked on an unprecedented and arrogant campaign against almost all our leaders, Democrats or Republicans, liberals or conservatives. A man for whom President Eisenhower himself is a "dedicated, conscious agent of the Communist conspiracy," can only be said to use the Communist issue for his own purposes and to define "communism" in a sense all his own.

This now famous statement about President Eisenhower is dismissed by Birchites as antedating the foundation of the society and therefore as not being a JBS document. The statement was not repudiated, however, by its author. It still reflects, so far as the evidence shows, the frame of mind of Mr. Welch. The citation deserves quotation at greater length:

> While I too think that Milton Eisenhower is a Communist and has been for thirty years, this opinion is based largely on the general circumstances of his conduct. But my firm belief that Dwight Eisenhower is a dedicated, conscious agent of the Communist conspiracy is based on an accumulation of evidence so extensive and so palpable that it seems to put this conviction beyond any reasonable doubt.

This frightful distortion should suffice to damn the allegedly anti-Communist movement headed by Robert Welch.

Yet the society prospers. This is an uncomfortable fact that requires explanation. One element in the mushrooming of the John Birch Society is no doubt the widespread anxiety of the general public over the world situation, marked by what appears as a succession of defeats on the international scene. There is a strain in the body politic. To many, the history of the postwar years is, to quote one Birchite, a story of "defeatism, humiliation, incompetence, surrender and treason." The only explanation of the setbacks, for many a perplexed citizen, is Communist infiltration. A new "stab-in-the-back" legend is in the process of creation.

Robert Welch was formerly active in the National Association of Manufacturers and his views represent the extreme right wing of that right-wing group. His program expresses resistance to the trends of the past thirty years toward more government control in the economic life of the nation as well as the "welfare" phases of contemporary governmental policy. This is the "communism" he is fighting, not the communism of Marx or the Leninism of Moscow. Of late years, however, a new factor has entered upon the scene in the form of an upsurge of anti-Communist activity by fundamentalist Protestants.

It was not so long ago that Catholics were regarded as the most active foes of communism. This can no longer be said today. Dr. Fred C. Schwarz' Anti-Communist Christian Crusade is of predominantly Baptist inspiration. The National Educational Program of Dr. George Stuart Benson of Searcy, Ark., is another fundamentalist operation. It is no accident that the key centers of the John Birch Society are in the fundamentalist South and Southwest, and that Welch himself stems from this background. John Birch, be it recalled, was a fundamentalist.

To these Protestants, the charge of widespread Communist infiltration seems to make sense when they look at the National Council of the Churches of Christ in the United States. For years they have regarded with distrust the efforts of their NCCC brethren in the urbanized regions of the country to adapt Christianity to the social problems of modern times. This is "modernism," or "liberalism." It is socialism or communism, and they will have none of it.

Related to this tension within the Protestant body in this country was the uproar over the Air Force training manual last year. In that text, the author (who drew his material from one of the fundamentalist anti-Communist organizations) charged that the Communists had infiltrated the Protestant clergy. The National Council reacted with vigor to these charges and succeeded in having the manual withdrawn by the Defense Department.

The charge of Red infiltration of the Protestant clergy remains in the air, however, and Mr. Welch explicitly refers to it in his talks. As he himself reports, his remarks are usually matched by the query as to whether the Reds have also infiltrated the Catholic clergy as well. In a talk at Garden City, N.Y., October 9, 1961, Welch said that about one-half of one per cent of Catholic priests are "comsymps." When challenged on this statement by the Boston *Pilot*, Welch admitted that his figure was "simply pulled out of a hat, as a complete guess, and without any substantiation even being claimed." In the same letter, however, he revealed that by "comsymp" he means not merely a priest whose sympathies may happen to be turned toward "socialism," but, actually, a Communist who became a priest. So far does Robert Welch push his theory of Red infiltration.

Reference has already been made to the participation of Catholics in the John Birch Society. This is hardly the kind of Catholic-Protestant dialogue encouraged by the ecumenical movement. But it would be a serious mistake to contend that the Communist danger does not greatly worry Catholics. One Protestant writer, Dr. David M. Baxter, whose notes on the John Birch Society greatly aided this writer in the preparation of the present report, testifies that many Catholics, like others, were stampeded into the Welch camp by an hysterical fear of a Communist takeover. Dr. Baxter's own article on the society in the Catholic monthly magazine *Extension* brought this sample of confused thinking from a reader:

If America and Holy Mother Church are to survive in the face of our Catholic Action, our Catholic President and too much of our Catholic press, we sorely need the John Birch Society to keep us from destruction. As long as good people continue to give God their first allegiance, as Robert Welch does, there is hope that our Lady's Son will be appeased and we will enjoy an era of peace.

The writer of those lines is no doubt a very religious and sincere person. It evidently does not strike her as strange that Catholic Action (which is under the bishops' supervision) and President Kennedy (a practicing Catholic) and the Catholic press (also under episcopal supervision) should be a menace to our country and our Church and that only Robert Welch stands between us and disaster.

As A. A. Berle Jr. remarked recently at a dinner honoring Christopher Emmet, the key to the survival and influence of the U.S. Communist party is its success in focusing the issue not on itself but elsewhere. An attack on communism is automatically transformed by them into an assault on precious values of our society—civil liberties, social progress, democracy, even peace.

To date no remedy has been found to cope with this tactic of dodging and camouflage. The John Birch Society has made this problem more difficult than ever before. If allowed to proceed with its work of confusion and division, it will destroy the basis of a solid effective fight against communism carried on by such organs as the FBI. As it stands today, communism, which as an enemy of freedom and order should be the common foe of both liberals and conservatives, eludes (perhaps indefinitely, so far as our action is concerned) its final accounting. Indeed, applying Welch's own principle of "inversion," a perfect case can be made out for the thesis that Welch himself is a Communist, so much has he helped the Red cause in the United States.

After this lengthy analysis of the sad role of extreme rightwingers in the fight against communism, a word about the responsibility of the liberals is in order. These are not at all without blame for the situation that drives so many anti-Communists to extreme positions. The liberals seem to have adopted the old European slogan, "No enemies to the left." Not only have they predictably defended Communists in the name of civil liberties, but they have lacked both courage and consistency in carrying out their self-proclaimed devotion to freedom. There are few outspoken anti-Communist liberals. Outside of such persons as Berle and Emmet, and such organizations as Freedom House, few have applied to the Communist totalitarian danger (internal or external) the same zeal they deployed for years against the Nazi totalitarianism and Nazi-like movements in this country. Until the liberals adopt a less ambiguous position in relation to communism, extremist groups like the John Birch Society will continue to win ready credence among average citizens for their scattergun accusations of infiltration and subversion.

On Turning Seventy
John LaFarge

F OR MOST PEOPLE the seventieth birthday comes in
silently, like Carl Sandburg's San Francisco fog.
The usual reaction to congratulations is: Why this
fuss *today*? Nothing new has happened. In fact, you
would hardly have noticed the event if you hadn't received retirement papers or read the passage in Psalm
89 (90) which defines the usual limits to human age.
That psalm, incidentally, was composed long before our
present days of regulated diet, vitamin tablets and other
aids to longevity.

This very unobtrusiveness of the 70th year, however,
inclines me to believe one should not wait until decrepi-
tude sets in before formulating some plan for life's last
lap. I don't believe that one should just trust to luck—
or the good Lord's merciful indulgence—so as to enjoy
various happy and pleasant latter-day moments. I don't
think it's enough to reread Cicero's *De Senectute* and
be persuaded that age is not such a bad thing after all.
Age, it is true, can be a time of harvest, of selection
and wisdom, with wonderful compensations all its own.
But for you, perhaps, it isn't quite so commendable.

Let it be clear to us that our Christian and Catholic faith is not content just to console us in old age, to shield us a bit from the rude blasts of the time-wind. Our faith can transform old age, just as it can transform each period of our life upon earth, and it is of supreme importance that we act in our faith's full consciousness: that we do not leave matters to mere hazard, but that we approach this period fortified with a definite purpose and program.

Old age, after all, is a natural phase of life. The burden of human weakness that we have inherited from our father Adam lies heavily upon it, both physically and morally. Yet it is endowed with a right, a nobility of its own. Since it is a humanly natural phase, one wants to make the best use of it. The latter years, in various ways, can repair the mistakes of youth or of the middle years. But who can repair old age when it has been wasted?

For this reason I propose a conscious and deliberate choice. People want something they can remember, something that will stick by them and serve as a rallying point when their minds become a little foggy. The words I propose will carry very different meanings to people in different states of life. They will mean one thing to a priest, another to a layman or woman. Yet I think they are applicable to all. Close to a dozen years of latter-day pondering have strengthened my conviction of their validity.

Old Age Is a Time of Prayer

I am not suggesting anything unusual, anything like the good old medieval idea when the conscience-stricken knight or the too-comfortably-cushioned prelate decided it was high time to retire to a monastery or hermitage and begin to "prepare for death." If, of course, you have drifted along to your 70th birthday without ever having made a spiritual retreat—or at least a really devout one—now, obviously, is the time to repair that particular lacuna. I am speaking of a deepening of value, rather than of any extension to our time schedule.

The latter years are a time when we simply allow ourselves to become more familiar with God and with His saints in heaven. We should let ourselves grow closer to that source of life, that ocean of love, toward which we are inexorably moving, just as the water-borne traveler on a great river begins to scent the first tang of the mighty sea to which the current is noise-lessly carrying him. It means talking much to God: to our Father in Heaven, to His Son, our Redeemer, to the Holy Spirit, who is our invisible and ever-working companion, and to Christ's Blessed Mother Mary.

We don't delude ourselves. Our minds may wander more readily in the later years. Troublesome memories of the past may obtrude, if we don't banish them at their first appearance. We may even forget prayers we used to know by heart.

But the main thing is that prayer become more and more a part of the texture of our lives. We dwell a little longer in meditating. We spend a little longer time in a visit to the Blessed Sacrament. We refer things to God more naturally and frequently. We do much praying for the Church, for the See of Peter, for all the body of faithful, for souls akin to us outside the visible Church, for so many great intentions. In our later years we become more conscious that we do not pray alone. The Church is praying with us and in us—the whole Mystical Body of Christ. With this, many things take on a new and fresh meaning, such as the prayers of the Mass and the Divine Office, the rosary and its mysteries. They are more specifically our prayers—especially if we have cherished since our youth some pious practice like the Angelus or Morning Offering or the *De Profundis.*

These latter years are a time for *listening,* listening for that Voice which could not make itself heard so well in the clamor of busier years. Now that voice begins to converse with us in the cool of the evening, speaking to us of what it all means and what we should really be thinking of.

We can become habitually thankful in our prayer, making the latter years a period of constant thanksgiving, in union with the divine Eucharistic thanksgiving each day in the holy Mass. We thank the Creator for these years, present and past. We think of all those who have not been privileged to enjoy them, those who died in infancy, in youth, in midstream. The *oblata* each morning are placed in our hands. Shall we not offer them consciously and joyously? Each day is a gift, ever more precious, mysteriously gathering up in itself all the value of all the days that have preceded.

Moreover, we pray silently and informally with all other *viatores*—fellow pilgrims, *sputniki*, as the Russians call them. For the curiously twisted thing about the Bolshevist jargon is its use of an honored and an-

cient term—the fellow-traveler on a sacred pilgrimage —to express the opposite of all that is holy: the Communist "fellow-traveler" here on earth, and the Soviet satellite, the "fellow-traveler" of the skies.

Let us pray for our fellow pilgrims here and in the world to come. In the latter years we live, or should live, in closer relationship with those who have already completed their term and are waiting to rejoin us in eternity.

Time, as it duly unfolds, becomes familiar with eternity. In our childhood we felt eternity's cool freshness, its bright simplicity. We return to it today, the time of the *Heimholung,* the ingathering into the Father's house. By the same token, it is the time of remaking one's life by reparation. Of that, a little later.

With every additional year the Church's traditional worship takes on new meaning. The Church is praying *in us and with us.* Our prayer becomes less purely private and is more absorbed into the great sacramental intercession of the Mystical Body of Christ.

Do I pray only for myself, or for my associates and friends, or for all mankind? It makes little difference. The Holy Spirit prays within me; it is His voice speaking through my own.

Largire lumen vespere
Quo vita nusquam decidat.
("Bestow upon us Thy light in the evening [of life], that our life may never languish."

—Hymn for None)
Is it my own personal life, or is it also the life of the whole creation that the Church prays for? It is either or both. It is the fulfillment of God's plan.

Old Age Is a Time of Charity

The second element in the program is that of charity, which is the new life into which we are reborn in baptism.

In the latter years you cannot, as St. Jerome says, practice anything like what you once could in the way of strenuous works for your neighbor. The area in which you can operate becomes gradually smaller. Your greatest hope is seeing that others carry on your works, perhaps much better than you could hope to do, for they can profit by your experience; perhaps less effectively, at least from your particular point of view. But with all this, the latter years offer countless opportunities for charity, many of which are appropriate to that very time. In those years you watch over your tongue, and see that it does not yield to the particular temptation for elders to listen to gossip and to disparage others. Old age is the time for *hidden* charity: a good word spoken here and there, a quiet service performed, visits to those in suffering, visits to others of the same age period whose predicament you can understand.

Most of us, unless we have had a particularly unhappy youth, remember how much we valued the kindness then shown us by older people, sometimes by the aged. The world of conversation with the young is a world each older person must construct for his or her particular self. There are infinite differences as to what each of us can contribute. Yet *give we can,* though apparently it may amount to no more than listening to what younger people say and providing for them the audience they naturally crave.

Even where age has taken the heaviest toll of physical strength and alertness, there is always something the older person can do to make life tolerable for those with whom he lives in daily contact. There are always ways and means to aid even those at a distance. It may be no more than a letter of congratulation or condolence or attendance at a wedding, a wake or a funeral; or greetings to newlyweds, or a word of encouragement to a struggling student, or a good word for the worker in some intellectual specialty who feels isolated and misunderstood. It may be the answering of questions out of our somewhat fading store of knowledge. Or the ordering of our affairs for our successor. Or the bit of counsel given to those who need it and whom we can really advise. Or encouragement to true creative effort—oh, this is a great *opus* for the aged! Or taking on the companionship of someone in the community or neighborhood whom one might, perhaps, instinctively avoid.

Nothing indeed can extinguish our capacity for love as long as the breath of life remains within us, and that love *is life,* its very breath and current. One of the deepest inspirations to one's life as a priest is to see those wonderful instances where a married couple have nourished a flame of true love that neither age nor custom can wither, that endures until the very end; indeed, beyond the grave itself. . . .

Love for the living and love for the dead. For our

faith teaches us that we can show our love for the faithful departed in a tangible manner. There are so many, so very many, in your later years, who are inscribed in your roll of remembrance. *Ipsis, Domine, et omnibus in Christo quiescentibus*: for those (whom I have named) and for all who rest in Christ. How much meaning those words gather!

Charity for those who are spiritually kin but separated from us by misunderstanding, by honest differences and, perhaps, by prejudice. Indeed, do we ourselves understand them as we ought? Certainly it is not mere chance that one has remained in close friendship with friends of one's youth throughout the vicissitudes of life, and that we can still forgather. Such things are God's working, the working of His Holy Spirit, in ways that we know not.

And there must be charity for the spiritually dead, for our enemies, if there are such; even for God's enemies, as long as they struggle in the uncertainties of this life. In my old age I can still, perhaps, represent for them the charity of Christ as I could not in younger and more active years. When age bears heavily upon them, they can more reasonably tolerate my own differences, more readily concur in those points of attraction or rejection wherein we coincide.

Old Age Is a Time for Courage

The two preceding items in our latter-years' program are more or less obvious. The third is less so and may seem to contradict our idea of age, belonging specially to youth. Yet it is, in a way, the most essential of all. This is the factor of courage.

Courage, of course, is what we elders demand and expect of the young. We tell them how brave we were in our own time, conveniently passing over various episodes of moral or physical cowardice. The middle-aged are to be guided by common sense, and the elders by the divine virtue of caution. Age is the time for prudence, for if's and maybe's, for looking before you step out, for seizing the railing when you step downstairs, for slipping a coat over your shoulders of a chilly evening, for going easy on cholesterol in your diet, for avoiding contentious argument; in other words, for sailing close to the wind and keeping your hand on the tiller.

Now there's plenty of good reasoning to be displayed for each and every one of these precautions. None the less, the latter phase of your life is not just a calamity to which you must somehow accommodate yourself. It is an adventure—your greatest adventure.

It is the time when you bet your final stake upon the great battle of life itself: when you place all your life, hopes and being in the hands of Him who gave His all for you; who hoped in you and for you, when you had forfeited any such grounds for His hope. And it's a precious privilege to enjoy this experience.

There are ample motives, then, for courage in our old

age: courage from a sense of destiny; from a sense, too, of the strength that can flow to others from your own example; for the multitude of mankind have no more respect for the cowardly aged than they have for the timid youth. The courage of old age is the courage of an exalted vision. Not the dreams of a visionary, but the glimpse of the reality and the foundations of things that is given by our faith, and the interpretation that our faith offers of the meaning of the universe—the sense of the truly sacred, of the genuinely and infinitely precious elements of our existence.

The fact that so little of life still remains is the added reason for being, as it were, prodigal of that little. Thus it was put by the patriarch Eleazar when, at the age of 90, he refused to listen to the kind advice of his cautious friends and went to torture and death, because, as he said in very blunt language, he'd "go to hell" before violating the holy law of God by eating the forbidden food.

The courage of old age is not an added decoration, a mantle of heroism in which we dress up and masquerade. The *material*, the occasion for courage, floats to our hand in the circumstances of old age itself.

The latter years are a time of diminution, no matter how you look upon them. They preface the total diminution which ends life itself. "Does the road wind upward all the way? Yes, to the very end" (Christina Rossetti). Our *habitual* courage consists in accepting this diminution, in living with it. The sense of achievement is rudely lessened. The circle of associates grows smaller. We find fewer capabilities for exciting adventures. We grow steadily smaller; after a while even our bodily stature may shrink. There is less we can say, less attention we can command, fewer opportunities to make our ideas and opinions known, greater perils of error and blunders, less control over our own functions, of mind or body, less ability to concentrate. We are becoming steadily—let us hope, gracefully—less.

Courage, then, consists in accepting this diminution just as it comes; not only accepting it, but welcoming it, going forth to meet it and embracing it with simple honesty, as the Christlike pastor of an old German parish in Ohio told me he and his parishioners would go forth and welcome the first Negro family to put in an appearance as residents.

It takes courage, therefore, to face, day by day, the manifold pinpricks of diminution. Indeed, one becomes accustomed to it, as a new rhythm of life—if in point of fact it *is* new, and is not a mere emphasis on diminutions that wafted their chilly breath upon you long ago in less propitious moments. It can steal over you in so many ways, as when you clear out your desk drawers or files and wonder how you happened to accumulate so much junk, and why you have such reluctance in disposing of it. You stroll in the library and realize finally (what common sense should have taught you long ago) how many areas of knowledge must remain forever unexplored, how many intellectual or imaginative byways you can hardly ever hope to investigate. You are grateful for honors or congratulations if any such have come your way; but now they point only to the past. They no longer presage any particular surprises still to come.

As I said, there is a courage of diminution, and it has its worthy place in the program of old age. It precludes certain conventional types of adventure. Yet the approach of the latter years is a vast liberation, not an imprisonment. It is the time now to speak frankly, since you no longer have anything to lose or anything earthly to gain. You may not speak at length, and your voice is croaky, but you can express the outcome of a life's experience—modestly, of course, but also confidently.

The natural difficulties of communicating with one's neighbor should not drive us in upon ourselves. On the contrary, it is the time for going forth and meeting people, for talking, a bit confusedly, perhaps, but in the depth of human warmth, with those whom you alone can reach. For the older one grows in the service of God, the more accessible one becomes for the young, the troubled, the doubting, the despairing and perplexed. Your advice, your warning, your prophecy, may be lost nearly a hundred times. But that hundredth time, like the slim notches of a Yale-lock key, may mean for another human heart the unlocking of life itself.

Old age is traditionally the time for exceptions and exemptions. Some of these, of course, must be. You need a bit more sleep, a bit more rest, less worry and responsibility. But may it also not be the time for a somewhat greater exactness, greater fidelity, in matters that do lie within its scope? In such a way, one can take a revenge, as it were, on the tyranny of time. You can make a little more room for eternity just when time has tightened its grasp.

Most of all, we remember that old age has a special work to do. With the merely natural apparatus that time offers us, we cannot turn back the hands of the clock nor erase by one iota what the Moving Finger may write. But with the help of the Eternal we *can* turn the clock back: we can rewrite our lives, and even rewrite the lives of others. Old age is pre-eminently the time of reparation. Into that great work of reparation enter all three factors that I have just mentioned: prayer, love and courage.

I have never been able to determine just where thanksgiving ceases and reparation begins. The two great actions are inextricably blended in the great prayer of the Church's sacrifice. We give thanks by offering, on our part, the sacrifice of atonement. We introduce the sacrifice of atonement by a solemn preface of thanksgiving. We give thanks at the end, after the great offering has been made, by partaking of the Victim's Body and Blood in reparation for the world's burden of sin and alienation.

This is the great *opus* of old age. This is its wonderful work, its glorious Amen; the response to the divine "Amen, I say to you. . . ." That Amen can never be in any way diminished as long as we keep it in our hearts and minds. In these years we are always busy with the great work of reparation: each day offering ourselves and offering the divine Victim—sacramentally or just spiritually—as an atonement for all the sins of the world. The facets of reparation are innumerable, yet they all blend into one great harmonious whole: the offering that is consummated with the offering of life itself. In our days of relative strength, while the light does still shine (*lucerna Dei antequam exstingueretur*), we can offer that final sacrifice, accept its manner and mode and circumstances wholly and entirely from the hands of the Creator, pray that it may be meet and just, fitting and salutary.

"The Old"

"WHEN THERE IS QUESTION of blood-relationship, three ancient traditions—the merely human, the Judaic and the Christian—agree that children are obliged to render concrete assistance to their needy parents. We must note that in one sharp controversy with his faithful enemies our Saviour explicitly and heartily condemned the failure to discharge this duty. . . . An impersonal government, paternal as it may wish to be or seem, cannot altogether assume the filial obligations that exist in the highly personal family."

Vincent P. McCorry, S.J., on the November intention of the Apostleship of Prayer.

There is a close link between classroom and commonwealth. It is in our schools that men are trained to take part in that civilized dialogue which leads to the rational consensus so vital to a healthy democracy. But what is happening in our schools today, particularly in our institutions of higher learning? The campuses are in political ferment, to be sure, but is that necessarily a healthy sign? Are our philosophers dutifully guiding youth's search for truth? And by the way, what is the responsibility of government toward all schools in our democracy? Should the state, for instance, assume the basic costs of education? Can it do so, without demanding a dangerous control of the whole process, to the ultimate harm of truth and human liberty?

School and Society

Only Higher Education, Mr. President?

Charles M. Whelan

PRESIDENT JOHN F. KENNEDY's education message to Congress contained several surprises, but none more disappointing than the statement on parochial schools:

> In accordance with the clear prohibition of the Constitution, no elementary or secondary school funds are allocated for constructing church schools or paying church school teachers' salaries; and thus non-public school children are rightfully not counted in determining the funds each state will receive for its public schools.

With one sentence Mr. Kennedy disposes of a problem which has vexed the best constitutional lawyers in the nation. Who's being dogmatic now?

In fairness to the President, we could hardly have expected an elaborate constitutional argument to the Congress. But in fairness to the people we could and did expect a silence respectful of the complexity of the problem and the traditional modes of constitutional adjudication. As an American, Mr. Kennedy is entitled both to hold and express his personal opinions. As a candidate, he was compelled to detail his position on Church and

State. But as President of the United States, speaking to the Congress of the United States, he should avoid unnecessary pronouncements on delicate constitutional issues.

Mr. Kennedy's warrant for the "clear prohibition" of the Constitution is undoubtedly the Church-State theory elaborated by the Supreme Court in the *Everson, Mc-Collum,* and *Zorach* cases. In the first of these cases, decided in 1947, the court affirmed that neither a State nor the Federal Government

> can pass laws which aid one religion, aid all religions, or prefer one religion over another. . . . No tax in any amount, large or small, can be levied to support any religious activities or institutions, whatever they may be called, or whatever form they may adopt to teach or practice religion.

One year later the court reaffirmed this position in *Mc-Collum* and held religious instruction on public school premises during public school hours unconstitutional. In 1952 the court sustained a released-time program off public school premises in the *Zorach* case, but did not explicitly temper the language of *Everson* and *McCollum.*

In view of these statements from the nation's highest tribunal, Mr. Kennedy has a prima facie case. He also enjoys the unquestioned support of distinguished constitutional lawyers like Leo Pfeffer, counsel for the American Jewish Congress and author of the best one-volume work on Church and State in America. The President can count on the active political support of such eminent and powerful organizations as the National Educational Association and the National Council of Churches. He cannot fairly be blamed for having incurred the endorsement of POAU and similar extreme groups.

What complaint, then, have the advocates of church schools, save that Mr. Kennedy has given the coup de grâce to a cause already lost? What refuge have they, short of saying the Supreme Court was wrong and thus conceding the President's proposition?

The answer is short and decisive. Mr. Kennedy's statement was erroneous, inopportune and unnecessary.

It is no disparagement to say that neither the President nor his Attorney General enjoy the reputation of great constitutional lawyers. If Mr. Kennedy were speaking simply as a private citizen, his opinion would not be recorded in the law journals of the land. The fact that he is President lends only political, not persuasive, force to his assertion.

A short survey of the professional legal journals would demonstrate to any inquirer that there is no clarity of expert legal opinion on the constitutionality of Federal aid to education through limited grants to church schools. It is the fashion of opponents of such aid to pretend that there is no question what the Supreme Court would do with such a program, and that the only people who think there is a serious argument for the constitutionality of such aid are Roman Catholics. The publications of such eminent non-Catholic legal authorities as Mark DeWolfe Howe and Wilbur Katz are simply ignored.

A GREAT amount of confusion could be saved if all parties to the controversy would agree to this statement of the question: May the Federal Government bear part of the cost of educating all of the nation's children in literacy, science and general human culture, regardless of the benefits which result from the program to religion or religious groups? Everyone is talking at present as though it were the name on the Federal check which determined the issue of constitutionality: if it is P.S. 101, it is constitutional; but if it is St. Mary's, it is a betrayal of the American way of life. I submit that it is an incontrovertible principle of American constitutional law that it is the purpose, and not the payee, that determines the check's constitutionality.

The only constitutional issue in Federal aid to education is whether the First Amendment means that Congress may not advance the level of national education by any method which even indirectly results in benefit to religious bodies. There is no decision of the Supreme Court on this precise point. All that the *McCollum* case decided was that government may not use its revenues and its coercive force to co-operate with the teaching of religion in public schools. *Everson* sustained the use of tax funds to provide transportation to church schools, and *Zorach* relied explicitly on the right of government to accommodate its programs to the religious interests of the citizens it serves.

Moreover, there are two significant facts in our constitutional history which militate for the legitimacy of Federal aid to education, regardless of the religious consequences. First, the Blaine Amendment was not adopted. In spirit and in letter, it would have forbidden every form of Federal aid to education which would result in direct or indirect benefit to church schools. This amendment repeatedly failed of enough votes in Congress to be proposed to the States for ratification. The failure of its adoption by the nation is the precise reason why it was adopted in New York and many other States.

The second fact supporting Federal use of church schools to advance general education is that the *Cochran* case of 1930 was not overruled in *Everson, McCollum* or *Zorach*. As is well known, *Cochran* sustained the constitutionality of state-supplied textbooks to church-school children. Opponents of Federal aid airily dismiss this case with the comment that the First Amendment issue was never argued. This is true only in form. The whole point and substance of the objection overruled by the Court was that the educational program could not be public because it resulted in benefit to a religious, and therefore necessarily private, group. The side effects of the program, the Court held, did not nullify its public character.

Cochran involved the due process clause of the Fourteenth Amendment; so did *Everson, McCollum* and *Zorach*. What puzzles the experts is the coexistence of these four cases with one another and with our total constitutional history. It is a brave prophet, indeed, who confidently predicts what the Supreme Court will decide.

In the clarity of his constitutional vision, however, Mr. Kennedy has succeeded in adding to the confusion. On what principle of law will he justify the enormous chasm in his program between higher and lower education? An admirable scholarship program, buttressed with supplementary grants to the institutions chosen, apparently raises no First Amendment problems; but once you descend to the high school or elementary level, there is no possibility of aid through church schools. Is it because lower education is compulsory that aid would be unconstitutional? Surely that argument cuts both ways; religious freedom becomes a joke when government itself makes the price too high. Is it because direct and unrestricted grants for construction or teachers' salaries are the only conceivable means of aid to grammar and high schools? What about scholarships, tuition payments, tax credits or deductions, with supplementary grants to the institutions of parental choice? "Education must remain a matter of state and local control," says Mr. Kennedy, "and higher education a matter of individual choice." Only higher education, Mr. President?

The Administration, however, has double reason for gratitude to its leader. Not only has Mr. Kennedy illumined the experts; he has saved his advisers the trouble of further thought. The speed with which he has done so should give them particular consolation and great hope. Less than two months after assuming his grave responsibilities, he has settled for them the position they must publicly take on an intricate constitutional problem. That is, they will take it or risk their jobs. With the necessity of uniformity in the Administration, I have no quarrel; but why should uniformity have been imposed, when it was not necessary to say anything at all?

No one would have been surprised if Mr. Kennedy had simply submitted his proposals to Congress without defining the unconstitutionality of a proposal he was not submitting. The basic models of the President's program are the Thompson and McNamara bills of last year; each passed its own House with scarcely a whisper of Church-State debate. For all practical purposes, the question of aid to education through limited co-operation with church schools was simply hushed up and ignored. Senator Morse was the one exception, and he made a speech in favor of its constitutionality.

The President has solid political reasons for refusing to champion aid through church schools. I affirm the propriety and the wisdom of his standing on those reasons. More importantly, I concede that, wholly independently of Mr. Kennedy's position, there are sound arguments against the desirability of Federal aid through church schools. The dangers of excessive Federal control, of disproportionate complexity in administration, of imbalance between the public and church-school systems, and particularly of the use of such funds for segregated schools, are not mere chimeras. These arguments I understand and respect, even though I do not think they outweigh the contrary arguments in favor of the desirability of Federal aid once such aid is given to public schools. But what I cannot understand is why a President who had such excellent political justification for the omission of private schools from his program has chosen the deeper and swampier ground of constitutional law on which to take a stand.

The reason he gave Congress, of course, was to explain why non-public school children would not be counted in determining the funds each State would receive for its public schools. An equally good and perhaps more candid reason would be that he had decided in the national interest not to help them. Or was it that Mr. Kennedy did not dare to say that? Did he recognize that the discrimination would be too manifest unless hidden in constitutional clarity?

The Philosopher's Responsibility

James Collins

W E COULD DEAL very abruptly with the theme of the Catholic philosopher's responsibility in our world by saying that he has the same responsibility as anyone else in the guild—simply to do a good job at philosophizing. Nothing more need be said after that, provided we really understand what it takes to do the job. There are some disagreements about what it does take, however. It may be useful, therefore, to spell out some of the philosopher's responsibilities.

We have, for example, a share in the responsibility for long-range preparation of college students who show an interest in becoming philosophers. It seems to me we could improve our counseling of such students when we talk to them about the study of modern languages and the choice of a major subject.

TRAINING THE PHILOSOPHER

In order to avoid wasting valuable time at the graduate level, and to insure a real use of language resources in later life, the undergraduate who plans to work in philosophy should master both French and German.

For students of the sciences, translation of new research is systematically planned; the main objective is simply to make information available. Philosophical reading, however, demands something more than the transfer of information; the reader must eventually make a close study of sources. Norman Kemp Smith has provided fine renditions of Descartes and Kant, but the student who wants to become an effective philosopher must be able to consult the original French and German—and Latin—of Descartes and Kant. He cannot be fully confident about the bases of modern thought without attending to the pattern and nuance of Descartes and Kant, which elude even a well-turned translation.

More is at stake here than background erudition. Language deficiencies often limit the course of studies and thus hamper the comparative work which Catholic thinkers should be doing in the contemporary field. Often, a student will embrace some current trend in British or American analytic philosophy simply because he lacks the linguistic skill to explore any other regions. And then he goes along with the trend of many analytic philosophers to become rapidly impoverished in thought content, because they fail to consult other ways of thinking which are not reducible to questions about English usage or formal systems. A counterpart case is furnished by the student who knows French well but is vague about German; he then spends his graduate program in assimilating only the French approach to such philosophies as phenomenology and existentialism. His understanding of the roots of these movements is likely to remain forever slightly out of focus; he cannot make the thorough evaluation of them for which we are looking.

Another domestic responsibility we have is to counsel prospective philosophers about the proper value of non-philosophical studies. I would prefer to see these students do their major undergraduate work in something other than philosophy and develop a lifelong habit of cultivating other fields of interest. There is nothing so barren as a mind which has tried to specialize too early and too exclusively in philosophy, especially in the School philosophies, which introduce one into an autonomous world of terms and syllogistic rules.

The fact is that it takes time and experience to ripen the philosophical temper of mind, and to date there is no way available for artificially stepping up the process of maturation. Hence, while the undergraduate and graduate student is learning the rudiments of philosophy, he should also be feeding his mind with large doses of literature and the sciences, history and the arts. In doing so, he will discover his own way of overcoming the isolation that initially separates the Catholic philosophical mind from contemporary modes of inquiry. Sometimes it is suggested that desperate measures must be taken to remove the isolation. Some, for example, recommend sloughing off the entire philosophical tradition in philosophy and simply combining one's Catholic faith with some currently important tendency in philosophy. I am convinced that this procedure results in more of a net loss than a gain, particularly when the individual is asked to compare the merits of the philosophy of his choice with other alternatives. What we should do is, rather, to try the indirect route of deepening our hold on both the philosophical tradition and the liberal heritage in the sciences and humanities, to the point where we begin to feel confident about bringing them to bear upon some present issues. This is a very steep demand to make of the maturing philosopher, but the road to philosophical competence is an unusually steep one in any case.

There is one definite advantage in cultivating interplay between one's philosophy and the other cultural components. It gives the Catholic philosopher sufficient familiarity with problems in other fields to enable him to speak with wisdom about them.

The complaint is often heard today from jurists, directors of foundations and other administrators that they consult the philosophers in their midst for guidance on major human issues—but do so in vain. The analytic philosophers, who are in the majority, are quite adept at describing the logic of courses of action already taken and at pointing out flaws in all plans for future action. But they regard even their work in the moral area as a meta-ethics, which usually means a theoretical analysis that does not result in any definite practical recommendation or norms of choice.

On the other hand, the Catholic philosopher is apt to be embarrassingly rich in general recommendations which he does not know how to make concretely applicable to the particular shape of the problem confronting the practical man. Here, the individual thinker has to rely upon his own fund of experience drawn from his personal studies in the relevant natural and humane sciences.

THE PHILOSOPHER'S ROLE TODAY

An example of my point is furnished by a session of this year's meeting of the Western Division of the American Philosophical Association, which was devoted to the theme of the public interest. This theme was suggested by the urgent need to set up some working standards for determining when there is a genuine conflict between the private connections of a public officer and his official responsibilities. As the assembled philosophers probed into this matter, it became increasingly clear that much more than a procedural issue was at stake. It was important to settle the rules of adjudication, but there was, at bottom, a good deal of murkiness about the meaning of the public interest and about the grounds for upholding and promoting it, even at cost of sacrificing a political administrator's private advantage. To clarify the difficulties, analyses were given from many different philosophical standpoints. The official printed report mentioned St. Thomas and Yves Simon, but it was noticeable that no Catholic thinker had been invited to give an independent development in the light of present legal and political circumstances.

The pertinent point here is that some independent rethinking of the "conflict of interests" issue is required, and it must be done with resources drawn both from political philosophy and from acquaintance with relevant materials in political history and present-day economic situations. The theme of the common good may be more productive in this inquiry than the general topic of natural law. There is no easy equivalence between the common good and the public interest, any more than there is between moral good in general and moral value. But there is a definite relevance of the doctrine about the common good to what underlies our present talk about the public interest and the obligation to uphold it. One of the Catholic philosopher's prime tasks is to search out the points of insertion and make his findings available in public conference and readable prose.

We are beginning to realize that in our pluralist world more is required than agreement on some practical courses of action. We see a need for dialog concerning widely divergent general interpretations of human existence and the roots of obligation. I think, however, that the time is overdue for Catholic philosophers to perform their critical function in respect to the theme of dialog itself. There is a danger that dialog may be praised out of all proportion, and hence a danger that we may demand more of it than we can reasonably expect it to yield.

THE NATURE OF DIALOG

Recent probes into the role of religion in American society show the need not only of carrying on a persistent dialog, but also of clarifying the nature of dialog. Suppose that we want to understand and develop our human nature to the fullest. Can we do so if we consider only the individual man in isolation or as related solely to the world of impersonal things? A negative answer is required because isolated man merely talks to himself, and man, the shaper of things, does not profit by independent criticism. For full use of his capacities and for full understanding of what they are, the individual man has to enter into the society of persons. Only then can a man open himself to another (as well as to himself) on a person-with-person basis. This interpersonal relation constitutes the situation of the dialog, which is the mainspring for humanizing our social bonds.

For over a century now, philosophers have been examining this dialogal relation among men. From the time of Feuerbach and Kierkegaard down to Buber and Marcel, a remarkably broad zone of agreement has been developing among otherwise quite divergent thinkers. They refer to man's attitude toward nonpersonal things as an I-and-it relation, and they observe that this nondialogal relation is a powerful but one-sided realization, since it leaves out the challenge of a free response from another person. The human individual is made for something more than dominion over things; he looks for a sharing in the goods of interpersonal life. This is possible only in the form of an I-and-Thou bond among persons engaged in dialog, whether the other person be a fellow man or God. Dialogal analysis is the philosophical study of the traits that unfold when a man reaches out to other persons and seeks to produce conditions of mutual understanding, critical integrity and, perhaps, love. The dialog itself is the interlacing of several mutual interpersonal efforts.

One of the lasting achievements of our century is to uncover the values of interpersonal communication, both among men and between man and God. But I would like to see the basic findings of existentialism and psychiatry about the unique reality of the I-Thou relation and dialogal intercourse kept distinct from the mass of loose variations and overextensions of these insights. At least two critical limitations should be respected. First, although a dialogal relationship establishes the conditions most favorable for studying questions about the human person and the acts of freedom, it does not automatically generate the evidence which will lead to sound conclusions about them or to firm agreement among the participants in the dialogal situation. Second, there is a wide range of significant philosophical issues, especially those concerning scientific methods and the domain of nature, which must be investigated by non-dialogal means. Philosophy itself reaches farther than analysis of the dialogal situation, some of whose implications can be followed through only with the aid of other philosophical approaches.

A CASE IN POINT

In last year's televised study of Western ideals which was sponsored by the Rockefeller Foundation, two of the participants, Arnold Toynbee and Reinhold Niebuhr, described the West's commitment to respect for personal integrity, moral courage and the constant search after justice in society. The discussion became highly charged, however, when an observation was quietly introduced by the third participant, the only professional philosopher in the group, Prof. Charles Frankel of Columbia University.

While approving of the dialogally congenial traits of the Western mind that Toynbee and Niebuhr described, Professor Frankel said that he missed any mention of another outstanding trait—the search for truth and the readiness to accept objective evidence, regardless of where it leads. Frankel then pressed home the point that one of the triumphs of the Western mind is its investigation of the structures rationally attainable in nature. He left the impression that Greek and Judeo-Christian humanisms contained some strong convictions, but that it was left for the Enlightenment and the scientific method to sift these convictions, identify the sound ones and eliminate any views conflicting with a scientifically grounded humanism.

We may readily sympathize with the judgment, expressed in Crane Brinton's recent *History of Western Morals* (Harcourt, Brace; 1959), that Frankel takes too simplistic a view of the relation between the scientific mind and rational structures, as well as between the Enlightenment and Christianity. There is much more constructural activity, and hence more need for decision and a basis of intelligibility, on the part of the scientist than Frankel admits. There is also a much broader basis of intellectual content and independent, self-critical activity on the part of the Christian mind than he allows.

The main criticism which suggests itself to me, however, tells against all the participants in the discussion: it is unfortunate that anyone should be permitted to pit the values of dialogal existence against the truths about nature as embodied in a scientific ordering. The Catholic philosopher's acceptance of God as the creative source of both personal and nonpersonal realms of being, as well as his acceptance of the value of all our sources of experience, requires him to overcome this

385

dichotomy.

The Catholic philosopher is clearly obliged to keep open the meaning of philosophy, so that it can include the method of object-analysis as well as that of dialogal analysis, the world of nature as well as the society of personal centers. He may well specialize in one of these modes of philosophizing and be thorough in working out its interpretation of human experience. But he should not succumb to the tendency to absolutize only one of these approaches and to reduce the other to a subordinate function.

The Catholic philosopher's respect for nature as well as man fits him for healing the split which has made intellectual strangers out of the existentialists and the naturalists, the phenomenologists and the analysts. His critical integrity will prevent him from following the momentary fashion of pretending that these two groups are saying the same thing in different ways. The problem of bringing unity and order into our philosophical world is much too genuine and complex to tolerate such an illusion for very long. But the Catholic philosopher can work deliberately and steadily to bring these two approaches within range for a mutual study of evidence that will benefit philosophy as a whole. This search for wholeness in the midst of a jagged world of natural events and interpersonal relations is the beacon of his professional labors.

Throughout this article I have referred to "the Catholic philosopher," without specifying any of the schools to which the individual thinker may belong. My reason for this usage is that at least a few things can be said in common about the present responsibilities of Catholic philosophers, before getting into further points of method and doctrine which separate them. Once in a while, these common tasks should hold our attention.

Give Me That Old College Tory!

John R. Strack

J OURNALS OF OPINION have been raising the dust lately with a rash of articles denouncing conservatism among college undergraduates. The phenomenon is something of a menace, they suggest; the conservative is out of place on a college campus. It is proper for a young man to be wild, rebellious, anarchistic. If one isn't a socialist at twenty, one's future is in doubt, and so on.

Nowhere in all this, however, has any college conservative taken pen in hand to announce that his is a noble and just cause, "one that is in the right because it has stood the test of time, and sound because it is rooted in our heritage."

A defense of what seems to be a much-maligned movement is in order; and what better man to do the job than a leading campus conservative himself? Hoping to gain an insight into the nature of collegiate conservatism, the writer, in a college way himself, recently visited just such a conservative. Now, once and for all, it is possible to quiet the fears of those who

387

eye with trepidation the future of the nation and its youth.

The audience was held in the headquarters of the Campus Watchbird & Safeguard Society, a group which aims to insure conformity between administration policy and student activity. The president of the club, a sophomore majoring in economics, motioned the writer into his office and waved him to a seat as he finished signing some papers. Shorter than most and rather stocky, he was wearing a carefully tailored and severely styled brown suit. From his vest he slipped out a cigar, offered it in a gesture of welcome and said in a tone of hearty camaraderie: "What can I do for you?"

When he was told that an explanation of the collegiate conservative's position was sought, he beamed, sat back in his leather chair, folded his hands across his stomach and said: "Well, now, I'm always glad to pass on to the students what little I've acquired in my time here. For it seems to me that the great hope of the future is the youth of the nation, and I shall consider worth-while whatever effort I can contribute to the encouragement of that youth. So, ask me your questions, and I will do my very best to provide you with the answers."

Q. *Well, briefly, what is collegiate conservatism?*

A. Collegiate conservatism, my friend, is not as some have said, the wave of the past; nor is it, as others have suggested, belief in the proposition, "What is, is right." Rather, it is a way of facing up squarely to the future of the problems of the past, a way of courageously confronting tomorrow's threat with yesterday's deterrent.

To the college man it means what Beatrice meant to Dante, what the essence means to the existence, or, in other words, what tracks mean to a train, only more respectable, of course.

Q. *What sort of student do you think becomes a conservative?*

388

A. One of my young freshman friends asked me that same question not too long ago, and I told him that if it isn't born in you, you'll never be a college Tory. One must have an almost inborn reverence for the venerable.

A serious turn of mind is not a handicap, nor is an impatience with ephemera; and an ability to dress tastefully doesn't hurt, either. The college conservative is a man who has learned early that if you go out walking on a dark night and watch the ground to avoid tripping, you won't see the stars, but you'll never trip, either. In sum, he doesn't go for this so-called "adventures-of-ideas" nonsense. He prefers to take his excitement in small, measured doses.

Q. *How, then, do you answer the charges of faculty people that the conservatism of the college man is an almost unnatural thing, that college people pick up the conservative banner, not because they stand for the same things great conservatives of the past stood for, but rather because they want to escape the risks involved in a normal life, preferring instead the almost bloodless security of a perfectly controlled environment?*

A. Well, let me repeat what I said at the dinner of the freshman Eternal Verities Club last month. The collegiate conservative does not answer the charges of his attackers; that is not his business. Instead, he looks to the erroneous motives of his assailants, to see what lies in their make-up that causes them to yelp like dogs at even the most established institutions.

We have been on campus almost a year now and know quite well what it means to face opposition. I think the best way to deal with opponents is to put them in their place, early. Now, who opposes us? Well, for one, the left-wingers and latter-day Bohemians do. Should this trouble us? No, everybody sees them for what they are. What we are concerned about is the effect they can have on the

389

innocent, for too many people are being led astray by the "liberals." We must handle the problem the way a patient father would treat an unruly son, showing him his error and chastising him when he seems in danger of worsening, but leaving the door open for him to return home when his brief infatuation with unreality has run its course.

Q. *Still, the arguments seem to carry weight with some people?*

A. Correct, my friend! The innocent *are* being duped and I feel it is my obligation to answer some of these wild charges—just for the record.

First of all, we respect the way things are. Right now it is winter; we like that. We like also the organization of the campus, the wonderful balance of power between the Registrar's office and the Public Relations Department. The campus is divided so well that one or the other of these can influence every sphere of school activity from the surveillance of the department chairmen to the operation of the cafeteria. Nothing can get out of hand.

We see things this way: in every generation, it is given to a few men to offset the influence of the so-called "progressives" who would do violence to what we hold sacred. Now this does not mean that we are intolerant, that we select only those ideas and recognize only those institutions which support our own. Nothing could be farther from the truth.

Conservatives are simply not capable of discriminating between views that are different from and those that are opposed to their own. In philosophy, for instance, we try to see the whole history of thought from Saint Thomas to Hegel as a *unity* and we *accept* the Hobbeses and the Kants along with the Burkes and Berkeleys. In religion we do tend to have an affinity for negative theology, and we do have a team of crackerjack natural lawyers, but we don't begrudge anyone else the right to hold legitimate differences of opinion with us—all they have to do is ask us.

Q. Ask you?

A. Yes, ask us. As I was telling my history professor the other day, that crew-cut, white-haired chap who claims to be a liberal—he's not really, you know; beneath that plaid sports coat beats the heart of a tired old man whose spirit cries out from the depths to be delivered from his current persuasion—anyway, as I was telling him, ours is a student view that tells time in terms of centuries. Events of today, like this new drive to double the size of the university in a "Great Leap Forward," must be seen against the background of the more realistic, original and age-old aims of the founders of the school. Remember, as our national director, Maxim Waterbug, a man in whom we place our deepest faith, has said: "Enthusiasm repressed is preferable to foolishness expressed."

In this regard, we do not oppose change for itself; rather, we oppose it because it tends to disturb

the way things are. Now we *want* to live with the future; we *accept* it as a fact of life, the same way we accept such things as the "curve" marking system and the inevitability of warfare.

But what we want from the future is a guarantee of the same order that prevailed in the past. If not, then we will impose order on it! If a teacher gives two A's a semester, we want some assurance that he will continue to do so. If he doesn't, we want to know why!

Man's great discovery in the universe is order. We see the stars wheeling about in their close-knit formations and imitate them when we put up our  fences and build our sewers. The triumph of man the maker is written in the remotest regions of the earth. From the Hill of Arizona to the teeming cities of Spain we see signs of the orderly pressure man has exerted on nature. The more order the better, say we. That's the way we like it.

Now if you want to see an example of the dynamic application of order to a situation, take my own case. I'm a sophomore now, and will graduate in 1963. After that I'll take an M.A. in Business Administration at Harvard. I'll spend two and a half years with IBM and leave for a higher salary in the insurance game. I will take my vacations in the mountains and will choose a wife whose tastes in such matters will be compatible with my own.

Q. *But where in the world would you ever expect to find such a wife?*

A. No problem there, my friend. Why Mount St. Marycliffe College, for instance, teems with girls whose training makes them perfect matches for young men who see things our way. I'll have no trouble finding a mate; together we will raise three children, two boys and a girl—in that order—and will have our home paid for by the time I am 36. I will retire at 58 and in my 60th year will complete work on my autobiography: *Looking Backward—the Story of My Life*, I'll call it. I will revise it just before I die to include my last seven years. I anticipate being outlived by my widow by six and one-half years, but just in case, will leave enough to support her comfortably for eight years.

I will bring my children up as rugged individualists, and will train them to look to their own resources in everything. I expect them to be paying half their way by their 15th year. With luck, I think nothing will go wrong.

I hope to worship my God, love my family, protect my home, respect my neighbors and uphold the law in such a manner that my friends as well as my enemies will remember me as a staunch and upright supporter of the Establishment. I have submitted this plan to National Headquarters for review. I am proud to say that our Regional Headquarters has already given it an "Honourable Mention."

Q. *What do the conservatives plan to do on campus this year?*

A. Well, first we hope to play a larger part in influencing student opinion to favor what we support. For one thing, we support the idea of noninterference in local matters, like this sit-in fuss over the so-called abuse of social minorities. We feel that the business of students is studying and minding their own business. We want to demonstrate that we do have a positive program; we are *for* maintaining the present size of the school; we will oppose the "Great Leap Forward." Incidentally, we may construct a bird sanctuary to keep our eagles in, and we hope especially to distribute copies of *The*

Mercurial American: How to be Conservatively Conscientious, our best-selling bible, you know. This is the story of how a rich Massachusetts family is carrying the traditions of old Salem into the modern publishing and political worlds.

• •

With that, conservatism's exponent pulled out his pocket watch, saw that it was late and said: "You'll have to excuse me now, I must get over to the cafeteria. After that, I've got a rally to organize. This week our group is protesting this Peace Corps nonsense. Nobody's come out against it yet, and I'm afraid the thing may catch on. Nice talking to you."

Thus dismissed, the writer strolled off, eyes glistening, pleased to know the college Tories could still yet defend themselves well.

The Left Side
of Paradise

William F. Gavin

A S EXECUTIVE CHAIRMAN of the Take Umbrage Committee of Young Americans' Hysterical Organization of Stevensonites (YAHOOS), I would like to take umbrage at the obviously slanted article, "Give Me That Old College Tory" (AM. 4/8). John Strack is undoubtedly a tool of the Interests and I have no doubt that the AMA, NAM and DAR were pleased no end by the space devoted to the neo-neo-Fascists and anti-freedom-of-riot groups which are infesting our otherwise dedicated outposts of the New Frontier.

I have long suspected that Fordham College was harboring people who enjoy wearing "carefully tailored and severely styled brown suits." Mr. Strack's enthusiasm for such anti-working-class mufti is his own business, but I'll bet my autographed picture of Arthur Miller that my own shredded T-shirt and humble khakis (suitable for sit-ins, cook-outs and, depending on the weather, small riots) are as worthy in the eyes of the emerging nations as his brown (I'll bet his shirts are the same color!) uniforms. Once the youth of America takes to wearing suits, can fascism be far behind?

Mr. Strack seems to laugh off the fact that conservatism (so called) is sweeping the college campus. We of the YAHOOS do not take such a bright view of reactionary movements. Only last week we held a meeting, followed by a rally, to decide what course of action should be taken against the conformists in our midst. After passing our usual resolution denouncing Trujillo, States' Rights, William F. Buckley, brown suits, IBM, bloodless security, Trujillo, Franco, the American Legion, Catholic War Veterans, Trujillo, and, finally, Trujillo, we decided that no positive action was needed at this time. But now the picture has changed. If AMERICA is seen to be a tool of the Fat Cats, we and our brother organization, Socialists Against Private Spending (SAPS) must act; you can readily see that Mr. Strack and his tailfin-loving friends are in for it.

I have chosen to reply to Mr. Strack's fawning interview with the president of the Campus Watchbird and Safeguard Society, by publishing my own interview with Alger T. Wunwurld, a junior majoring in Non-Conformity and a founder of YAHOOS. Mr. Wunwurld was gracious enough to devote a few minutes out

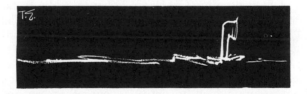

of his busy schedule (he pickets the pickets who are picketing the Russian Embassy on weekends, and devotes all of his spare time to writing letters of protest to the New York *Times*. Recently he broke his own record by writing a letter of protest against his own letter of protest, protesting the lack of protest on our campus!) to state his considered views on the state of campus liberalism today. Taller than most and rather thin (he recently went on a hunger strike protesting the lack of hunger strikes on our campus), Mr. Wunwurld gave his answers while walking a picket line protesting the—oh, the hell with it.

Q. *Mr. Wunwurld, what is collegiate liberalism?*

A. Liberalism is many things. It is being for, instead of against, things.

Q. *For? For what?*

A. It doesn't matter, so long as you're for something. Take me, for instance, I'm for Stanley Kramer movies, the World Federation, Freedom to Riot, Murray Kempton, Dorothy Day, Linus Pauling, Social Security, Job Security, Old-Age Security, Folk-Singing, Mort Sahl and SANE.

Q. *But doesn't your insistence on security leave you open to the same charge offered against campus conservatism? Do you not also prefer "the almost bloodless security of a perfectly controlled environment?" It would seem that your desire for more security would make you more of a conformist than your adversaries.*

A. This isn't true. I'm surprised you do not realize the difference between the conformist pattern of the young conservative and the dissenting voice of young Liberals. Only a few days ago, I said to a history professor: "I'm against McCarthyism, provincialism, clericalism and sin, in that order." "We need more men like you," he said. You can see that I'm not afraid to speak out against what I feel is wrong in the world.

Q. *Yes, but isn't it true that most history professors claim to be against the same things and that your dissent is, in reality, merely a playback of what they give to you?*

A. Not at all! This professor, for instance, was against McCarthyism, clericalism, provincialism and sin, in that order. You can see the difference. Plenty of room for honest disagreement between faculty and students.

Q. *Yes. Tell me, Mr. Wunwurld, what do young Liberals feel are the greatest problems facing America today?*

A. Conformity, for one thing. We *all* agree that conformity is bad.

Q. *Conformity? What kind of conformity?*

A. Never mind, what kind of conformity . . . just conformity. We go in big for "adventures of ideas."

Q. *What does that mean?*

A. Oh, you know . . . adventures . . . and ideas. We're all for freedom of thought, freedom of riot, that sort of thing. Adventure. . . . Ideas. . . .

Q. *You mentioned "freedom of riot." Just what do you mean by this?*

A. We feel it's every American's birthright to riot for what he believes in. Rioting is the backbone of our nation. Where would we be without the freedom to riot, guaranteed to us by our forefathers? We believe it is the right of every freedom-loving person to disrupt the United Nations and attack guards with chains and razors, although we do wish they would wait until Adlai is finished speaking. After all, isn't the United Nations for all people? There are some in this country who would deprive students of this fundamental right. A good riot is proof of nonconformity and that's what we're *for*. It's "adventures of ideas" in action. The reactionary press and the Big Boys in Washington have used smear tactics against those patriots who rioted against the HUAC in San Francisco. Is this the way to impress the emerging nations?

Q. *Mr. Wunwurld, it has been suggested that collegiate liberalism is an "unnatural thing," and that young Liberals, who demand more Federal aid for everything, are in reality trying "to escape the risks involved in a normal life." Is this true?*

A. Quite untrue. If properly applied, Federal aid to education can increase the risks involved in a normal life. Think of the healthy risks the parents of children attending private-parochial schools will have to take if aid is given only to public schools. On the other hand, if we do breach that historic wall separating Church and State, imagine the ingenious methods which will have to be devised by those who agree with the *New Republic*'s claim that the state should "draw people to its system of schools and away from centrifugal systems." You can readily see that Federal aid can involve enough risks for any 100-per-cent, anti-centrifugal American.

Q. *It has been rumored that contemporary collegiate liberalism has become dogmatic and is, in effect, illiberal. Is this true?*

A. I'm glad you asked me that. There seems to be a general misunderstanding of the views of YAHOOS, despite our efforts at free exchange of ideas. I suspect this misunderstanding is the deliberate and malicious work of the reactionary press and Madison Avenue. Let me clear up a few of the most confused ideas.

First, it is *not* true that we believe that social salvation is impossible outside the editorial page of the New York *Post*. We embrace all roads to understanding and progress. Only last week we sent out a task force to deliver copies of the *Nation* to underdeveloped areas that are infested with the blight of conservatism. Many YAHOOS swear by James Reston and Walter Lippmann, despite the fact that the writings of these gentlemen are found in what might be called areas of reactionary fall-out. Of course, it is true that we believe that all those who seek the one true road will read the New York *Post*.

399

Another distortion and, I feel, a truly shocking example of the power of the McCarthyites, is the vicious myth that YAHOOS apply a double standard in the realm of political morality. William F. Buckley, that unregenerate tool of the Interests and the Pentagon, in his *Up From Liberalism*, devoted an entire chapter to the Paul Hughes case. Buckley pretended to believe that the hiring of a con man by Joseph L. Rauh of the ADA to spy on Senator McCarthy was an example of gross hypocrisy, since Rauh had previously denounced the hiring of confidential informants in no uncertain terms. You can see for yourself that Buckley misses the point. Proof of this is the fact that I have never heard any YAHOOS or a single progressive professor question the motives of Joseph Rauh in this case, while it is a matter of record that McCarthyism is denounced regularly by students and faculty. If this so-called double standard was applied by YAHOOS, why wasn't the Hughes case discussed in Richard Rovere's scholarly, unbiased account of the McCarthy days, *Senator Joe McCarthy*? You can see that logically speaking, the YAHOOS position—a position distorted and twisted by perverse and medieval minds—is the only logical one. **Do** I make myself clear?

Q.   *Quite. What do YAHOOS plan to do on campus this year?*

A.   We are planning many things, despite the fact that "planning" is a dirty word in certain circles. We plan to have a music festival in which we hope Pete Seeger will sing and play, to be followed by a march on the United Nations building, where we will protest the fact that Pete Seeger is being denied the martyr's right to be in jail, where he was ordered when he was previously denied the right to refuse to deny that he had ever had been a Communist, a denial of which would have been ample proof that the right to deny is being forcibly denied patriotic citizens of this country.

Q. *Mr. Wunwurld, may I be the first to congratulate you on your old-fashioned spirit of protest?*

A. Thank you. Coming from one of the biggest YAHOOS and SAPS on campus, that is a compliment!

As Mr. Wunwurld strode off into the night, I was jostled by an oaf carrying a placard. This placard read "Hungary . . . Tibet . . . Poland . . . China. . . ."

I'm still wondering exactly what that sign was supposed to mean. After all, what is there to protest about in these countries? I cannot remember ever seeing YAHOOS carrying a sign like that.

♦

And Mary Sang

Snow on the gray crest of Mount Hermon wakes
My Son is a river in the waiting landscape
Out of my desert valley He has made a lake

My Son is the sea I am set to drift on
Everywhere I look He is the horizon
From and to Him I move Who is all directions

My Son is rain in Silo and I the first flower
I reach to Him and all my roots are sure
He is the storm and the early spring shower

My Son is a white alb of water falling
In the cup of my hands I have held Him flowing
Beneath the ferns and rock He is a small spring

My Son is a well deep in the unseen flesh
He is the water rising and the surface at rest
My children drink of Him and be refreshed

JAMES F. COTTER

What is "America"?

What is "America"?

Thurston N. Davis

THE QUEST for self-identity, for self-definition, is on occasions a useful and salutary undertaking for us all. Not only persons, but institutions also, can profitably ask themselves who or what they are and how they came to get that way. For a journal like AMERICA, a fiftieth birthday is as good a time as any to face the mirror of candid self-scrutiny and ask what we have been trying to be and do for half a century.

However, as Hippocrates once said, though art is long, life is short. There are so few left who can tell us, step by step for fifty years, how AMERICA happened to choose each of the paths it took at every forking of the roads. The last member of the first editorial staff, Fr. James J. Daly, died in 1954. The two survivors of our relatively early days are Fr. Gerald C. Treacy and Fr. J. Harding Fisher. Our sole hope for anything resembling full self-identification rests, therefore, with these two distinguished veterans—and, of course, with Fr. John LaFarge, now rounding out his 33rd intensely active year on the staff of this Review.

AMERICA's North Star in this voyage of self-discovery must be the original "editorial announcement" of April

17, 1909. In it our founder and first Editor, John J. Wynne, who was likewise the father of *The Catholic Encyclopedia,* spelled out the objectives and the modus operandi of the Review he was launching.

AMERICA was created "to meet the needs of the time." One such need was for "a review and conscientious criticism of the life and literature of the day." Other needs were for

> . . . a discussion of actual questions and a study of vital problems from the Christian standpoint, a record of religious progress, a defense of sound doctrine, an authoritative statement of the position of the Church in the thought and activity of modern life, a removal of traditional prejudice, a refutation of erroneous news, and a correction of misstatements about beliefs and practices which millions hold dearer than life.

To accomplish these ends, Father Wynne saw that a weekly journal would be required. The topics that demanded discussion were "too numerous, too frequent, and too urgent," he said, to be handled by a monthly magazine. He saw this need as "imperative." Father Wynne noted that the weekly diocesan press of those times did not attempt to chronicle events of secular interest or to discuss contemporary issues in the light of Christian principles. Many of them were "excellent in their way," he said, but they were "limited in the range of subjects and circumscribed in territory." The United States needed a national journal something like *The Tablet* of London.

What was Father Wynne's program for AMERICA? The new Review, he wrote, was to discuss

> . . . questions of the day affecting religion, morality, science and literature; give information and suggest principles that may help to the solution of the vital problems constantly thrust upon our people. These discussions will not be speculative nor academic, but practical and actual, with the invariable purpose of meeting some immediate need of truth, of creating interest in some social work or movement, of developing sound sentiment, and of exercising proper influence on public opinion.

Naturally, in its efforts to follow the course charted by its founder, AMERICA has tacked this way and that with the passing years, depending on who was at the wheel and who happened to be his first and second mates. After all, these plans and provisions of Father Wynne had to be translated into the weekly reality of ink on-paper by the men who year after year wrote and edited AMERICA's pages. Father Wynne made a brave and distinguished start, but his editorship (1909-1910) was not lengthy enough to give us grounds for judging how fully he would have succeeded in realizing the objectives he had plotted for AMERICA. His successor, the scholarly historian, preacher and former Jesuit provincial Thomas J. Campbell, Editor-in-Chief from 1910 to 1914, put no new or peculiarly personal stamp on the Review. But in the quiet years just before World War I, Father Campbell did far more than keep the franchise. The international coverage of those years was remarkably good, and Father Campbell's editorials were models of vigor and clarity.

In 1914 an extremely forceful personality came on the scene as AMERICA's third Editor. Richard Henry Tierney soon impressed the young Review with a polemic spirit, a readiness for controversy and a deep concern for the international responsibility of American Catholics. Father Tierney left the relatively quiet life of a seminary professor to become Editor. As his biographer notes, "he threw himself into the turmoil of a

405

journalistic career with a bounding energy and enthusiasm. . . . He had full scope to show whatever individuality and originality and power and personality he possessed"—and he had these qualities in abundance. He soon became, in his biographer's words, "the journalistic spokesman of the Catholic Church in the United States."

Through the years of the first World War and after, the vigor and energy of Father Tierney's rather imperial temperament dominated the pages of AMERICA and made themselves felt in the larger world outside. Some of the flavor of that period—Father Tierney arguing with President Woodrow Wilson and two Secretaries of State in defense of Mexican Catholics, his attack (he was a total abstainer) on the Prohibition Amendment, his drive for funds for the relief of Europe's starving millions after the war—is caught in the pages of his biography, *Richard Henry Tierney,* by Francis X. Talbot (America Press, 1930).

When Father Tierney's health broke beyond repair in 1925, Wilfrid Parsons succeeded him. There were changes of emphasis under the new Editor, but no lessening of liveliness and controversy. Father Parsons was always in the thick of things. During his time AMERICA stayed shoulder-deep in the continuing controversy over the persecution of Catholics in Mexico. The fateful events of the Al Smith campaign, the rise of fascism in Europe, the Great Depression, the disputes over Fr. Charles Coughlin and the beginnings of the New Deal are only some of the big strands that run through the fabric of Father Parsons' years as Editor.

What was AMERICA under Father Parsons? A penetrating answer can be found in an article published by *Social Order* in March, 1958 to commemorate Wilfrid Parsons' 71st birthday. Myles Connolly wrote from Hollywood to say:

> The AMERICA office, as presided over by Father Parsons, was a combination of employment office, embassy, information center, marriage bureau, Travelers' Aid Society and Number 10 Downing Street.

It was Father Parsons who chose and remodeled our present editorial residence, Campion House on West 108th Street in New York City. His good taste and practicality mark every corner of it, from the great crucifix over the main altar in our chapel to the fittings of the editorial board room. Father Parsons left the editorship in 1936, but his heart was always at Campion House and his pen ever at AMERICA's service. On the eve of his death in October, 1958, his arthritic fingers were still tapping out the lines of his weekly column, Washington Front—as sprightly and informed a bit of political reporting as could be found in the American press. At our fiftieth anniversary we salute a great priest-editor and a dear departed friend.

AMERICA got a new look under the next Editor, Francis X. Talbot. Father Talbot succeeded Father Parsons just as Francisco Franco was coming to power in Spain. An historian, a man of letters, a facile and wide-ranging mind, Father Talbot immediately impressed a new and distinctive stamp on the Review he edited. For one thing, AMERICA was given a fresh format. "Farewell to Old Style," said AMERICA on June 27, 1936. "From now on titles will be big and bold, . . . and every page will satisfy the esthetic sense." Heavy black rules and sans-serif type predominated, and the magazine took on that strong, emphatic look which many thought excessively stark and harsh. If so, the new, rather "nazoid" face of AMERICA certainly belied the mild, poetic and gracious Editor of those times.

During all these years, Paul L. Blakely, a Kentuckian and a vigorous States'-righter, wore out at least a gross of typewriter ribbons composing the editorials for AMERICA's pages. Father Blakely, associate editor of AMERICA for 29 years, was one of the most prolific journalists of his time. Through the entire middle period of AMERICA's life, the magazine was all but identified with the name of Paul Blakely—and with John Wiltbye, his pen name. Mention of Father Blakely recalls two other great names of our middle years, William I. Lonergan and Joseph C. Husslein. The work of Father Husslein, who was a member of the AMERICA staff from 1911 to 1927, is discussed in some detail elsewhere (p. 139) in this issue.

In 1944 the Second World War was drawing to a climax and a close. International issues were the order of the day; a new world was struggling to be born; the menacing shadow of Red imperialism was beginning to grow and grow over Eastern Europe and Asia. It was right, therefore, that one who had been among the very first to sound the warning against international communism, who was so widely acquainted with world affairs, so gifted a linguist and so well travelled, should direct the policies of AMERICA during these crucial years of reconstruction. He was John LaFarge.

Father LaFarge has little in common with warmongers or polemicists, and these were the days of the great world quest for peace. During this time, then, John LaFarge, the wise irenicist, made a most vital contribution. But his objective—and the objective of AMERICA—was always peace with justice. Therefore, when justice demanded it, the Editor of AMERICA spoke out unequivocally—against the shoddy in art, against the backsliders in social justice, against the hatemongers—in a word, against the blind who happened to be leading the blind of those particular years. Father LaFarge farsightedly opposed the policy of unconditional surrender for Germany at a time when such a stand was most unpopular. He had pioneered in the field of interracial justice, and of course these preoccupations were mirrored in AMERICA during his years as Editor. Probably no other single white American has worked so long and so arduously in the cause of the Negro here and everywhere. A long spell of ill health—today fortunately belied by his seventy-nine ripe years—forced his retirement from the editorship in 1948.

Robert Hartnett succeeded Father LaFarge. Large in body, mind and sympathies, Father Hartnett had been a superb teacher. His former pupils, remembering provocative classes in political science at the University of Detroit, were saddened to learn that the university was losing him. He brought to his editorial work the refinements of a scholar. None but those closely associated with him as colleagues will ever know how meticulously he worked, how many hours he carved out

408

of extremely busy days for reading and research, how faithful he was in scholarly correspondence, how exacting in demands on his collaborators—but primarily on himself.

To habits of scholarship Father Hartnett wedded the forceful, analytical and honest mind of a born debater. Perhaps the height of his debating career was the occasion when, before a large student audience at Yale, he took on Paul Blanshard. This was an era of great debates. The years during which this manly and priestly Jesuit presided over the policies of AMERICA were years when tension and controversy welled over in the United States. The infant United Nations made its first teetering steps on the world stage; Unesco was already under severe fire; China went Communist; the Korean War flared up and fizzled in a fire of frustrations; General MacArthur was relieved of his command; the Truman doctrine was formulated; Nato slowly gained ground in Europe; U. S. economic and technical aid programs were debated in the light of isolationist or internationalist preoccupations. Concern over Red espionage and the infiltration of Communists, later brought to a head in the notorious Rosenberg case, had set the stage for Sen. Joseph R. McCarthy. AMERICA under Father Hartnett turned to each of these problems as it arose, and the AMERICA record of those years, so obviously marked with the genius of its editor, is there for all to read. During the final months of his editorship in 1955, Father Hartnett set in motion detailed plans for a new format for AMERICA. The present dress of the Review dates from Oct. 1, 1955.

The foregoing account is necessarily most incomplete. For one thing, it discusses Editors-in-Chief as though they alone were responsible for AMERICA. Doubtless, each of these men gave a distinctive cachet to the Review, but their work could never have been done without the constant assistance of many lay collaborators and, above all, of teams of dedicated Jesuit associates. These fellow priests, both associate editors and business and circulation directors, have for fifty years given AMERICA unique strength in depth and unusual organizational stability. Elsewhere in this volume we list all their names. But it should be noted here that during the last fifteen or twenty years AMERICA could not have been published without able colleagues like the late Francis P. LeBuffe, Joseph A. Lennon, Cornelius E. Lynch, Joseph P. Carroll, Edward F. Clark, Joseph C. Mulhern, James P. Shea, Stephen J. Meany, the late Daniel M. O'Connell, Paul A. Reed, Joseph F. Mac-Farlane and Patrick H. Collins—men who labored in the office now directed by Clayton F. Nenno, who became treasurer of the America Press in 1958.

Associate and contributing editors who have given long and distinguished service to AMERICA during the last two decades are John A. Toomey, the late Gerard B. Donnelly, Edward A. Conway, Allen P. Farrell, the late Albert I. Whelan, Vincent P. McCorry and two long-time members of our present staff, Robert A. Graham and Vincent S. Kearney. For shorter terms during the same period AMERICA benefited from the presence on its editorial staff of Gordon George, William J. Gibbons, J. Gerard Mears, W. Eugene Shiels, Edward J. Duff and the late John P. Delaney. Still more briefly on our masthead during these years were the names of

FATHER WYNNE   FATHER CAMPBELL

William A. Donaghy, John Courtney Murray, J. Edward Coffey, Philip S. Land, Louis E. Sullivan, Richard E. Twohy, John J. Scanlon, Richard V. Lawlor, Daniel Fogarty, Thomas J. M. Burke, Joseph Small and Francis J. Tierney.

Charles Keenan, Managing Editor for many of his sixteen years on the staff, deserves special mention, because during these years—to paraphrase Parkman—not a line was turned or an apt parenthesis entered but Keenan led the way. A native Irishman, a man slight in bodily frame, Father Keenan loomed large in AMERICA's world as an editor's editor. Two other distinguished contemporary veterans of the present staff, Harold C. Gardiner and Benjamin L. Masse, assuredly deserve space in this issue for the articles in which they discuss AMERICA's contributions to the fields of literary criticism and Christian social thought. Finally, how can we ever express our gratitude to all the others—clergy and laity, here and abroad—who have given the fruit of their talents to our Review?

WHAT IS AMERICA? It is not merely the approximately 70,000 pages bound into the hundred volumes that now, the product of fifty years, span fifteen feet three inches on library shelves. Depending on how one appraises it, AMERICA is something more or less than the sum of those pages. It is what each of its editors and contributors, and all of us together, have made it. By our individual and collective failure or success we must measure the net gain or loss of fifty years.

Frankly, we are not ashamed. Of course, we could have done better. Even by the standards of an essay-writing age, there were probably too many general essays in our early volumes. In expressing its opinions, AMERICA has never laid claim to infallibility, nor does it do so now. However, there were moments when we spoke with too strident, too intransigent, too dogmatic a voice. There were other occasions when that voice sounded too cloistered, too timorous, too studied. But if at times mistakes were made in emphasis or attitude or expression, they were at least the mistakes of honesty. This same honesty compels us to say that the overwhelming part of our work has expressed exactly what we wanted to say as we wanted to say it.

AMERICA set out to be and still is an "opinion" magazine—a journal of Christian opinion. A journal of opinion is not a family magazine. It does not attempt to amuse, entertain, instruct or edify, though it may happen to do all these things at one time or another in the performance of its specific function. An opinion-journal exists to express self-consistent opinions, proposals and criticism and to foster discussion of them by competent minds. (An AMERICA department, State of the Question, is intended to create a forum for discussions of this nature.)

There is no better definition of a journal of opinion than the one formulated several years ago by Father Hartnett:

> It is a magazine which has a definite, coherent outlook in terms of which its editors and contributors analyze and reach judgments about current events and trends, especially in the social, economic, political, literary and (in some cases) religious fields. It addresses itself to a general readership, to those persons who, regardless of occupation or station in life, are interested in analyses terminating in judgments, based on a coherent outlook, about current events and trends in the fields mentioned.

A magazine of this nature, if it adheres to its principles, will not appeal to everyone; its circulation will remain relatively limited; it will never be a "popular" journal. In fact, if it is conscientiously performing its function of expressing ideas and opinions, it is likely at times to be exceedingly unpopular, at least in certain quarters.

Among the 310 publications that currently are members of the Catholic Press Association of the United States, strictly speaking only AMERICA and *Commonweal* can be described as weekly journals of opinion, though *Ave Maria* is now moving into this category. Outside the field of the Catholic press, the *New Republic, Christian Century, New Leader*, the *Nation*, the *Reporter*, and most recently the *National Review*— each with its own political or ideological point of view —likewise fit the definition. Of course, opinion journalism is not necessarily restricted to the field of the weeklies and biweeklies. The monthly *Atlantic* and *Harper's* in the secular field might be called opinion journals. Among Catholic publications there are the

Catholic World, published since 1865 by the Paulist Fathers, the *Voice of St. Jude* and the monthly of the Passionist Fathers, the *Sign*—all of which have made significant contributions to opinion journalism.

But has not the day of the opinion journal passed? Looking back thirty, forty or fifty years, it is possible to understand the role once played by these journals. A generation or so ago, the sociology of communication was vastly simpler. So were the dynamics of opinion-making. But there have been so many changes in the intervening years. In those earlier times public opinion was a much more limited thing than it is today. Opinion-makers then aimed their shafts at an elite, for it used to be sufficient to reach the leaders of society.

Today, however, the opinion-maker must reach everyone, for everyone weighs the same as everyone else in the egalitarian scales of the Gallup poll. Digest magazines bring capsulated opinion to everyone's bedside.

FATHER TIERNEY

FATHER PARSONS

Radios beam Fulton Lewis Jr. to our dinner tables. Three big national television networks project the images of opinion-molders like Edward R. Murrow into our living rooms.

But there were no Arthur Godfreys, Edward Murrows, TV networks or big book clubs when AMERICA was founded in 1909. People then would have stared in wild surmise at the phrase "communications media." There was advertising, of course, but no Madison

413

Avenue. The opinion-molding of the movie industry had not yet begun. In 1909, Henry Luce, an eleven-year-old boy at Hotchkiss School, had not so much as dreamed of his *Time-Life-Fortune* empire. There were lobbies in the Washington of 1909, but not the smoothly tooled, opinion-making lobbies of today. The Rockefeller brothers, either mere infants or still to be born, were not issuing—as they did in 1958—important opinion-making statements on education and the national security. Universities were for the most part just liberal arts colleges, not a great formative bloc of influence in the democratic process. The Federal Government of those days might well have wanted to influence public opinion, but it hadn't one-tenth of its present panoply of agencies and instrumentalities for doing so.

In this mid-century quest to establish once again the identity and definition of America, we do not discount these contemporary realities. It is quite true that today opinion-making has become the objective of all the above as well as of a score of organizations like the Fund for the Republic, but this development does not impair the usefulness of or diminish the necessity for the journal of opinion. There is no conflict between the opinion journal and, for example, the Fund for the Republic. The Fund operates on a high level in the field of mass adult education, and it approaches this legitimate and necessary work in a creative manner. But the area of the Fund's influence is distinct from that of the opinion magazine, and its occasional telecasts, releases and brochures, though powerful in the world of opinion, lack the constantly repeated impact of the weekly journal read at leisure by the thoughtful few.

In fact, the small-circulation journal of opinion has a more vital role to play than ever before. The relatively simple days of 1909 are gone. There are so many new problems, so many emerging issues, all of them intertwined one with the other in the most complicated ways. There is a plethora of opinion about each phase of every single problem. All these new questions, arising from the rapidly changing configuration of our times

414

and our society, demand analysis and discussion.

Vast and knotty international problems are cropping up from week to week and year to year. Again, U. S. society confronts in 1959 a problem which it was only commencing to recognize in 1909, that of religious pluralism. Today, all sorts of questions relating to intergroup tensions and to methods of trustworthy and fruitful communication between and among the diverse segments of our society have given public discussion of opinion a fresh framework of reference. Think, moreover, of the speed with which science and technology have developed in the last half-century. They raise hundreds of questions of the greatest urgency.

If they were here today, what would Fr. John Wynne and his first board of editors think of the long roster of topics—all of them pressing, perplexing and complex—that supply the grist for today's opinion journalism? They make quite a litany: space control and intercontinental ballistic missiles; hydrogen bomb tests and manned satellites to the moon; automation and the problem of the new leisure; a tangle of questions relating to urban life, suburbia, working mothers, slum clearance, city planning and juvenile delinquency; the mental-health menace, psychiatry, applied depth psychology and the new-fangled arts of persuasion. There is the new problem of how to control our vast abundance of foodstuffs; and all the related questions of an affluent society. There are new questions of public health and plans for voluntary health insurance. What about the "censorship" problem? What is beating the "beat" generation? Is the very meaning of "nature" changing? How are we to think of the bulging population of the world? With what alternatives can we confront the rising wave of neo-Malthusianism?

Father Wynne and his first AMERICA staff would have a difficult time picking and choosing today among these and a thousand other questions. Africa is in danger of slipping into the Red camp. Southeast Asia shudders under the shadow of the Red Chinese commune plan. How long can we count on the solid support of Latin America in the battle of the free world against tyranny? What are all these new freedoms—to read, to travel, to criticize, to know? To what extent is John Dewey's instrumentalism responsible for the alleged failure of education in the United States? Was the legal positivism of Justice Oliver Wendell Holmes a cause or a mere symptom of the declining health of American law? Father Wynne and his companions could take at least some measure of comfort from these last two questions. They would recognize the name of the younger Holmes. And in 1899—ten years before AMERICA—John Dewey had published *The School and Society.* But I suspect that most of these questions—along with conceptions like mass culture, baby sitting and supermarkets— would be as unintelligible to Father Wynne as the titles of a double-feature recently playing at our neighborhood theater: "The Blob" and "I Married a Monster from Outer Space."

FATHER TALBOT   FATHER LaFARGE

THIS IS ALL very well, say the devil's advocates at this point. We concede that you Jesuits of 1959 deal with a vastly greater proliferation of topics than did your predecessors. But is anybody reading you?

AMERICA goes directly to approximately 50,000 persons or libraries each week, and its circulation continues to grow. Moreover, we have good reason to believe that our readers are a most communicative lot, and so the magazine regularly gets passed along to a much larger group. Last year, when a friend gave us a check and asked that AMERICA be placed in 40 university or college libraries, we were hard put to find that many that didn't already have it. Early this year the chaplain at Leavenworth Penitentiary wrote: "The men here pounce on AMERICA as supplemental reading for their college courses. It is in constant circulation and demand." AMERICA is frequently consulted in the Congressional Library. The British Museum requested and got a complimentary subscription. AMERICA is clipped for dossiers used at State Department briefings. Moscow's *Literaturnaya Gazeta* is one of our subscribers. *Time* (May 28, 1956) backed into a compliment by saying that AMERICA "comes up to any secular standards." It is a special source of encouragement to us that so many editors, Catholic and non-Catholic, read AMERICA regularly. Many of them have been kind enough to tell us that, even in realms where they decide to differ from us on points of policy, they look to our Review for balanced Catholic opinion.

Newsweek and *Time* have cited AMERICA, we believe, more frequently of late than any other periodical of its type. AMERICA's editorials are frequently guest editorials not only in the Catholic press but also in the U. S. secular press and in the overseas edition of the New York *Herald-Tribune*. Our articles are sometimes "lifted" by foreign journals—often without a credit line. AMERICA is one of two Catholic periodicals on the tables of the reading rooms of the Harvard and Yale Clubs of New York City. The librarian at the National Press Club in Washington, we hear, recently posted a notice calling back missing copies of AMERICA needed for binding. Of course, AMERICA is indexed in *Readers' Guide to Periodical Literature* and in the *Catholic Periodical Index*.

Let us further clarify AMERICA's identity. It is a Catholic journal. At the top of our masthead, under the word AMERICA, we print "National Catholic Weekly Review." Again in the masthead, just above the names of the full-time and corresponding editors, we publish this legend each week: "Edited and published by the following Jesuit Fathers of the United States and Canada." This serves further to identify AMERICA as a

Catholic review published by Jesuit priests of North America. The reader who gives this legend a moment's consideration will immediately understand that the good ship AMERICA is not sailing under the flag of the Society of Jesus as such. Our Review is not the official publication, or an official publication, of the Jesuit order. AMERICA is no more and no less than it claims to be in this published self-identification. That is, it is a Catholic weekly review, the supervision of whose editorial content has been turned over to an Editor and a limited number of associate editors appointed to this work of weekly journalism by the provincial su-

FATHER HARTNETT   FATHER DAVIS

periors of the eleven provinces of the United States and Canada. (The French-Canadian Jesuits publish their own review, *Relations*.)

Allow me to repeat that AMERICA does not exist to publicize "the Jesuit line"—there is no such thing—on the temporal issues of our day. Individual Jesuits are free to, and often do, disagree with us; their letters, disputing some point of editorial policy, frequently appear in our Correspondence. Likewise, as is perfectly obvious, AMERICA does not pretend to be an "official" voice of the Church in any sense whatever. Despite the constant plaint of certain bigots that Catholic opinion is a frozen iceberg of conformity, any moderately informed person knows that apart from the defined dogmas of the Church and the principles of the natural moral law, Catholic opinion, especially on temporal issues, is variegated to the degree that one might almost say that its disarray is the most notable thing about it. All this is elementary, but is needs to be stated from time to time, and our fiftieth birthday is a good occasion on which to repeat it.

W HAT PRINCIPLES and policies actually define and characterize AMERICA? Obviously, AMERICA strives to be contemporary, to write and rewrite up to its deadline about strictly current events and contemporary trends. Its editors labor to stay "on top of the news"— ahead of it, if possible. But these are not distinctive traits. Any news magazine accomplishes this weekly feat and gives more extensive coverage than AMERICA.

AMERICA is specified by its concern for moral questions. This concern lies at the heart of what we may call the corporate personality of a journal like ours. Our preoccupation is with the moral hits or misses, the spiritual triumphs or failures of man in all the varied enterprises of the modern world. Almost every human event—from the closing of Little Rock Central High School to a sputnik racing into orbit—has certain definite moral and religious overtones. Our ears strain to catch these notes, from whatever source they come.

No one should look in AMERICA's pages for comment by neutral observers. We are not neutrals. We are deeply committed. Obviously, as Jesuit priest-journalists, we are committed to our holy faith. Secondly, we are committed to the moral law of God, as this law is promulgated through the universal forum of human conscience. We are committed, on a wide and varied field of subjects, to the principles enunciated by the Popes, the Vicars of Christ, and in the annual statements of the American hierarchy.

We conceive it to be a large part of our task to point out how, in concrete and specific cases, the universal teaching of the sovereign pontiffs and of the bishops on social, economic, scientific and cultural topics can be applied and implemented. It is not enough for us to compose disquisitions on the naked principles of justice

419

or on rights and obligations in the abstract. Our effort is to find out how, and according to what prudential judgments, these universal norms may be translated into laws, institutions and attitudes.

There are universally valid principles on racial justice. But how do fair-minded people go about changing attitudes regarding the desegregation of Virginia's high schools? What can American Catholics do toward helping Negroes into the ranks of white-collar employment? Anyone with a mere nodding acquaintance with the social encyclicals of recent Popes acknowledges the right of workingmen to form unions of their own choosing. But what of right-to-work laws? What remedies are there for the recent hardening of attitudes between labor and management? Is it true that congressional investigations of labor racketeering have gone so far that they are now harming the cause of labor itself? We address ourselves to problems such as these. No one will quarrel with the principle that college students should read deeply and widely under proper guidance. AMERICA does not devote space to this obvious major premise. We are in business to discuss whether a book like James Joyce's *Ulysses* is proper fare for a student in a Catholic college seminar, and under what conditions.

In these and a hundred other editorial concerns, AMERICA's attitude is and should be characterized by the widest and most universal of interests. Nothing, absolutely nothing, that concerns the good of the human person on any level of life should ever be outside our purview. The inspiring meditation on the Incarnation of Jesus Christ, found in the second week of the *Spiritual Exercises* of St. Ignatius, founder of the Society of Jesus, sets our editorial policy in this important respect. I wish there were space to quote pertinent passages from this meditation, but those interested may consult it for themselves. Let it be noted that it is in a meditation on the *Incarnation* that St. Ignatius gives us this amazing picture of universal human diversity, struggle and suffering.

It is no exaggeration to say, therefore, that every single line of type in AMERICA bears, or should bear, in some manner at least, on the meaning and consequences for daily life of the Incarnation of the Son of God. Catholicism draws its philosophy of life from the central reality of the Incarnation, from the fact that the Word became flesh and dwelt among us. The never-ending work of the Church is to elaborate, on every level of life and for every clime and time, the conclusions to be drawn from this shattering event. In that phase of the Church's deathless mission AMERICA has, we feel, a small but not insignificant part to play in the world of contemporary American life.

AMERICA is independent. As is obvious, of course, we acknowledge complete and unqualified dependence on the dogmas of our revealed religion and on the dictates of the moral law. But in the entire field of human affairs, wherever there is an area for prudential judgment, we gratefully recognize the immense freedom that we enjoy as editors. Our cordial relations with subscribers, advertisers, benefactors and that esteemed and valued group known as the AMERICA Associates are such as in no way to infringe on our editorial freedom. We ourselves act as censors of what we publish. There is absolutely no shadowy specter of "publishers" hanging over our shoulders, reading our galleys or inserting changes in our page proofs. Our considerable independence is made possible, of course, by the trust that our ecclesiastical superiors have been good enough to repose in our prudence. Without this leeway it would be almost impossible to publish AMERICA.

We are not beholden to any political party or any special interest. We take stands on public issues, but we do not endorse political candidates nor involve ourselves in partisan politics. We have never done so, and we have no intention of changing this policy. I think it is fair and true to say that not for one moment have we ever consciously deviated from the principle laid down in 1909 by John Wynne: "AMERICA will aim at becoming a representative exponent of Catholic thought and activity without bias or plea for special persons or parties" (AM. 1/17/00). Perhaps, on second thought, we have harbored one bias—a bias for the poor. But that would have been all right with Father

421

Wynne, since the poor are not "special persons" and they usually have no "special parties" to speak for them.

One final pillar of policy. AMERICA shuns an editorial attitude that is nagging, negative or contentious. We fail in this at times, to be sure, but our overarching concern is to write positively and constructively. We are fully aware of the terrible evils of the world. The twist in man's nature, product of original sin, is a cosmic tragedy to which we do not shut our eyes. We tend to look editorially, however, for what is promising and hopeful. We make no secret of the value we set on Christian optimism and on the habit of hope.

We are encouraged to pursue this policy by a statement that appears at the very beginning of the *Spiritual Exercises* of St. Ignatius Loyola, in a brief note that is called the Presupposition to the Exercises. In it St. Ignatius says: "Let it be presupposed that every good Christian is to be more ready to save [put a good construction on] his neighbor's proposition than to condemn it. If he cannot save it, let him inquire how his neighbor means it; and if he means it badly, let him correct him with charity." If in the future AMERICA should ever habitually set its course athwart the spirit of this Ignatian principle, it will in my opinion have ceased to justify its existence.

THE IDEAL of the Jesuit is not pure contemplation. His vocation is rather that of the contemplative in action. This double polarity of Jesuit life, certainly of the life of a Jesuit journalist, is symbolized for AMERICA's editors by the location of the house they live in. Campion House on West 108th Street in New York lies halfway between Upper Broadway and Riverside Drive. When we leave our residence to take a walk, we must choose between two courses. One is a contemplative stroll along the tree-lined, Old-World paths of Riverside Drive, with the Hudson River flowing quietly by at our side. The other is a brisk walk up or down the teeming and shabby sidewalks of mid-Manhattan's Broadway—a multilingual, interracial neighborhood to which all the nations of the earth seem to have sent delegates.

Like contemplation and action, these two city streets go unswervingly along, meeting only in the mind of the one who must wed them in a synthesis of the two things they represent. Perhaps at times in our fifty years we have walked too frequently on one rather than the other. But for the most part I believe we have mixed in just proportions the hustling immediacy of Broadway with the reflective quiet of the Drive. In the year 2009, when AMERICA celebrates its centenary, we can only guess where its editors will be residing. But wherever they are, you can take it for granted that two such paths as I have described will still meet at a cross-roads in their editorial sanctum.

Editorial

Our New Editorial Table

A MASTERPIECE WAS UNVEILED in our editorial office during the last week of March. It was delivered to our door with no fanfare and considerable incongruity when a rather battered truck jockeyed for space on our traffic-lined street and three husky men in old work clothes hopped out, swung open the doors and gently maneuvered the masterpiece into the editorial sanctum. The three "workmen" were two young Jesuit semi-narians from the house of philosophy at Shrub Oak, N. Y., and a Jesuit lay brother.

When the object of their maneuvering was stripped of its protective case and paper wrappings, a truly magnificent table gleamed out in the editorial board room. It is 12 feet long and of trapezoid shape. Under the satin sheen of the finish, the oaken grain has been matched in a lovely design; the angle-joints are so silky-smooth to the touch that they seem no more than penciled lines. Every ounce of its 400 beautifully designed and executed pounds bespeaks the craftsman whose loving care and pride in his work has produced this masterpiece of the cabinetmaker's art. And who was the artist-craftsman? He was the Jesuit lay brother.

Several days after we had rejoiced in our new table,

425

we read an address given at the sixth annual convention of the National Science Association at Denver. Dr. Howard E. Wilson, dean of education at the University of California at Los Angeles, delivered a severe indictment of one aspect of American culture and education. The problems that science teachers face, he charged, are problems that are deeply rooted in a "tradition of sloppy craftsmanship" on all levels of American education and work. This blight, he stated, extends from the garage mechanic who does a slipshod job of repairing a car to the would-be English teacher who does not bother about proper punctuation and good grammar. Yet our times demand precision as never before.

The successful career is, of course, a legitimate goal, but there is a deeper motivation to inspire worker or scholar to forswear "sloppy craftsmanship." It is the motive of proper pride in one's work. The unsung [What? After *this* tribute?] Jesuit lay brother to whom we are indebted labored for 175 long hours over his masterpiece. He was not struggling toward a successful career. But he was driven to achieve the perfection of his craftsmanship, because he knew the dignity of his work. He knew it because he realized that any legitimate product of the human hand and brain *is* human— and therefore Godlike. He knew, too, that any such human work has been sanctified by the Son of God— was He not known as the "son of the carpenter"?

The burden of inculcating such an esteem for the job at hand lies heavily on our whole educational establishment. But it lies primarily on the home. It is within those walls that children must first learn, not merely that "what's worth doing is worth doing well," but worth doing perfectly for their own sake as human persons, for the sake of the work and for the glory of God.

We shall sit around this lovely table twice every week, discussing and hammering out editorials and comments and the entire policy of this Review. We can only hope and literally pray that our work, in content and style, will reflect—week in and week out—the hatred of "sloppy craftsmanship" that shines out of every shimmering grain of our new board-room table.

Morals in the Market Place

MANY A CORPORATION EXECUTIVE probably rubbed his eyes in disbelief when he opened the September issue of *Fortune* and started reading Louis Finkelstein's "The Businessman's Moral Failure." Disdaining a soft approach to his audience, Rabbi Finkelstein, who is chancellor of Manhattan's Jewish Theological Seminary, starts his homily with this stern, uncompromising paragraph:

> If American businessmen are right in the way most of them now live, then all the wise men of the ages, all the prophets and the saints were fools. If the saints were not fools, the businessmen must be.

Dr. Finkelstein then proceeds to document his indictment with a litany of vivid detail. Not only do businessmen, he says, have a false view of their role in society—a view which emphasizes their place on the economic ladder rather than a concern for the civilization which they dominate. They tend in many lamentable ways to transgress in their daily actions the ethical laws that apply to their calling. The author instances crooked tax returns, bribery of purchasing agents, suborning of legislators and law-enforcement officials, use of wire-taps to learn the secrets of competitors, violations of building codes, spreading of false information by insiders in order to make a killing in the stock market, false advertising and a number of other malodorous practices. "It is impossible to conduct business in the U. S. today," a fast-rising young executive confided to Rabbi Finkelstein, "without breaking the law." Even if this is an exaggeration, the author comments, there is a great deal of skulduggery in the marketplace. The disturbing truth is, he says, that the U. S. businessman

> is preoccupied chiefly with gain, coasting on the spiritual momentum of the past, divorced from our sources of inspiration. He is a leading citizen of a largely hedonistic nation propelled by meaningless drives toward materialistic and frequently mean-ingless goals.

Only those who do not know Rabbi Finkelstein will imagine that his indictment springs from hostility to businessmen, or to our system of capitalistic enterprise. On the contrary, he has been led to criticize the American businessman because of his high regard for him and his role in our society. "The fate of the world," he asserts, "hangs on his decisions." Or again: "Today's crisis demands the businessman's leadership in the area of human behavior," for the world needs "ethical leadership from those it respects as supremely practical."

Nor is the Jewish leader's message merely negative. His program for placing ethics on the corporation agenda may differ in detail and emphasis from the Holy Father's teaching on the duties of businessmen, but it is in basic accord with it. Like Pope Pius XII, Rabbi Finkelstein is warring on the maxim "business is business." He is calling for the supremacy of ethics in business, as in all other phases of life. He would save our business civilization from moral decay by having businessmen concentrate, not only on profits, but also, and even more, on saving their immortal souls. The heartening thing is that the editors of a plush businessman's magazine like *Fortune* apparently agree with him.

Puerto Rican Pastoral

THE CATHOLICS of the mainland United States were quite unprepared to read in their morning papers that several Catholic bishops living under the American flag had, late in October, jointly forbidden their people to vote for a certain political party. Whatever the record may show elsewhere, such a prohibition is unprecedented in American Catholic history.

Up to the present, our American democratic system has unfailingly given the Church wide latitude and ample opportunity to voice its convictions on moral issues, to preach the Gospel and to take the means necessary to defend its liberty and well-being. Down through the years, as a result, the weight of Catholic leadership, like that of the Protestants, has been able to influence for good the course of national life. But a sharp line has invariably been drawn this side of the voting booth.

Thus, it has always seemed unnecessary, improper, even something of a profanation, for the authority of the Church to be extended, through the pulpit, to the point of a formal prohibition against voting for one particular party, or for an individual candidate. The record of the Catholic Church and of her priests has been without blemish in this respect.

For generations it has seemed evident to American Catholics that only the gravest peril to civil order, or an imminent threat to the Church's very existence, joined to a complete (and highly unlikely) breakdown of the democratic process itself, could be conceived as justifying the use of such an extreme measure.

Consequently, the U.S. Catholic remains profoundly confused and bewildered—not to say embarrassed—by the action of the bishops of Puerto Rico. In a joint pastoral read in all their churches on October 23, they forbade Catholics to vote, in that Commonwealth's November 8 elections, for the Popular Democratic Party headed by Gov. Luís Muñoz Marín.

Catholics on the mainland, together with most of their fellow Americans, are in a state of considerable ignorance concerning the true character of the PDP and the peculiar circumstances on the Island which led to the bishops' exceptional step. Perhaps, too, they do not understand the Latin temperament of the Puerto Rican political leaders who, while American citizens, rule the Commonwealth under their own constitution and who are heirs, not of the Anglo-Saxon traditions of New England, but of the culture and traditions of New Spain.

It may also be that, contrary to widespread belief, the pastoral did not impose a formal obligation in conscience, but was simply a way of articulating a longstanding grievance of the Church against the Puerto Rican powers-that-be. Any fair-minded observer in the States must also concede that Church leaders are in a better position to judge the moral issues involved than are outsiders who do not bear before God the responsibility of shepherds of souls in Puerto Rico.

Here on the mainland, few Catholics—and even fewer Protestants—are aware of the long struggle of the Church's leaders with Governor Muñoz and his party. That these grievances are important, and that they touch moral and religious issues, seems evident. In a land overwhelmingly Catholic, the governmental party has consistently advanced public policies offensive to the moral sense of Catholics and alien to the traditions of the Puerto Ricans. Passages in the party's own program, as published in the public press, have been interpreted as an open repudiation of this tradition and as an insult to the Church.

No Catholic should presume to pass definitive judgment on actions taken by responsible Church leaders in distant regions under circumstances difficult to evaluate. Obviously, however, American Catholics can and must decide whether such a course can be regarded as a model of what is proper for their own country. What is involved here is nothing less than the conception that we American Catholics entertain about the function of ecclesiastical authority within the American democratic system. Is the attempt to use Church authority to determine, by formal prohibition, how a man votes, an action that is necessary or desirable in this country?

It happens that in October U.S. Catholics found themselves with two examples to study and to choose between. During that very month, the French hierarchy issued a grave declaration on the moral problems posed by the continuing conflict between the Government of France and the Algerian nationalists. In France, the hierarchy's statement was met with wide attention and respect on the part of those to whom it was addressed. No Frenchman—not even a secularist or Freemason—challenged the right of moral leaders to offer guidance to their coreligionists at a time of widespread public confusion and anxiety.

In Puerto Rico, on the contrary, the bishops' letter stirred up a storm of controversy even in the churches in which it was read. Public officials denounced it as "incredible medieval interference in a political campaign." The difference in the reactions may stem, not from any divergence of viewpoint over the respective hierarchies' right to speak about public issues of moral import, but from a difference in the manner in which this vital mission was exercised.

The French document limits itself to setting forth guide lines by which French Catholics can form their consciences, and on the basis of which they can perform their civic duty with peace of mind. The Puerto Rican pastoral adds the suggestion of coercive action by ecclesiastical spokesmen. It does so, moreover, in that very area of civic life, the voting booth, which a democratic state necessarily surrounds with a maximum of privacy and independence. In this atmosphere the individual citizen is presumed to be able to come to a free decision—a choice enlightened, it may be, by the guidance of those he trusts and reveres, but a choice responsibly made by himself.

Catholics in the United States cannot but wonder about the nature of a situation which would persuade Church leaders to embark on a course of action so open to misinterpretation, not to say futility. It must indeed be a grave situation, for, in a healthy democracy, such a step as that taken by the Puerto Rican bishops can only be viewed as a profound disruption of normal political processes.

Under any circumstances now conceivable to the American Catholic citizen, there is no substitute for reliance on the established procedures by which the moral forces in a democracy are free to defend and promote the sound development of the common good and to secure, at the same time, the liberty and well-being of religion.

Standing outside the Puerto Rican discussions as they do, U.S. Catholics must regret the turn of events which now threatens to embitter that island Commonwealth with the fruits of prolonged civic dissension. Surely, too, they must regret the circumstances which have, in a sense, put the Church's prestige at stake in the coming Puerto Rican elections. More directly of concern to them, however, as it must be to all their fellow Americans, is the understanding of how this incident may be made to affect and influence events closer to home.

At this point and in this hour, history offers enlightenment and reassurance. It clearly affirms that American Catholics stand at one with their fellow citizens in loyal allegiance to those principles of conscientious citizenship which undergird our nation's attempt to preserve the structure of a free and democratic society. Indeed, the very difficulty that American Catholics feel in comprehending the pattern of events in Puerto

Rico testifies to their unquestioning conviction—inculcated, by the way, in the parochial schools now flourishing in this free country—that human dignity and the freedom of the Church have been and will remain best guaranteed by the wholehearted and unswerving support all Americans pledge to the democratic rules that lie at the core of our most cherished civic institutions.

◆

OUR LADY OF TAOS

Mary, narrow, lean as a bow,
Spreads her thin hands
From her hand-span waist
Gowned in green gauze
The brown hands of Indian
Women have wound, round-
Haloed in white wax roses,
Closed eyelids veiling black
Moon-down eyes, the ark
Of her womb a room
Beneath the tented pale-
Wrapped breast, the night-colored
Hair: the fair linened altar
At her feet, the neat, caped,
Plaited, tender leather lady,
Moon-shod monstrance mother,
Mounts her crescent as a
New moon rides the mountain's
Rim; old image of older mother,
Mother of oldest meaning—
All things are younger than she
Save her undying Son.

NANCY-LOU PATTERSON

433

Looming Moral Thunderheads

THE renowned classical scholar, Prof. Werner Jaeger, calls education by its Greek name, *paideia*. In its profoundest meaning *paideia* is the work of handing on the cultural heritage of our race from one age to the next. If so, then all of us—for in this school everyone is a teacher—must be concerned over the present general misunderstanding of what is meant by the natural moral law, and over the particular difficulty we encounter in bringing its true import home to a rapidly growing number of our physical and social scientists.

Today one is inclined to write the phrase "natural law" within quotation marks. For people by and large (many Catholics included) no longer know its significance. Speaking historically, this blackout of the natural moral law is a relatively recent development. For two centuries and more the tradition of the natural law flourished in our oldest American universities. One has only to examine the evidence in the archives of Harvard and Yale—the big folio sheets on which, each spring for decade after decade, there was imprinted that almost unvarying list of ethical "theses to be defended"

434

by senior-year students in their oral examinations. All the way from the 1630's to the 1850's and later, this scholastic tradition of ethical learning persisted without substantial deviation at Harvard, Yale and elsewhere. Then, about a century ago, simultaneously with the transforming process by which our older colleges passed from religious to secular control, the living tradition of the natural law was broken. Today few American secular educators even know that it once existed.

One who does honor the tradition, and who conceivably regrets its passing from the academic scene, wrote a letter that was published in the New York *Times* on December 14. John Herman Randall Jr., Woodbridge Professor of Philosophy at Columbia University, stated:

> In view of the justifiably great public interest in the recent statement of the Catholic bishops [that of November 25, 1959, relative to the promotion of artificial birth control], I should like to suggest that the level of discussion would be raised if it were focused on the central issue, the meaning of "the natural law" in Catholic thinking. This idea has so completely disappeared from non-Catholic thought about moral questions that I am afraid very few non-Catholics understand its meaning. . . . It would be immensely helpful in bringing together all sensitive men of good will if an informed Catholic teacher would try to explain to all of us just what Catholic thinkers mean by "the natural law" of morality.

We wrote to Professor Randall that same day, inviting him to take part with us, not in a debate, but in a common attempt to elucidate the question of the natural law itself and the problems raised by a widespread public failure to recognize its principles and appreciate the relevance of its applications. His gracious reply, indicating that he was at least tentatively willing to join us in such an exploration, has encouraged AMERICA to project plans for it in a coming issue.

Obviously, it is the birth-control question that has provoked this fresh and welcome interest in the natural law, with which, of course, the morality of artificial birth control is intimately connected. A number of Protestant leaders, some of them prominent in their advocacy of birth control as a solution to the so-called "population explosion," are entering the discussion of the natural law through the birth-control door. Many of them are concerned to explain to themselves and their followers how it is that official Protestant doctrine in this field has shifted so radically during the last two generations.

The scrupulously fair-minded dean of Union Theological Seminary in New York, John C. Bennett, recently noted that Protestant systems of ethics "have their day and lose force" and that "neither Luther nor Calvin can be said to be a guide on specific moral issues." As Dean Bennett remarked, in 1908 the Anglican Bishops were unequivocally opposed to contraception; opposed it again, though less vigorously, in 1920; gave permission for its "conscientious use" in 1930; and appeared as its advocates in 1958.

Commenting on Dean Bennett's statement, Robert Hoyt, editor of the Kansas City-St. Joseph *Reporter,* wrote with tact and insight in his December 25 issue:

> The absence of principles and laws in Protestant morality makes one wonder about the future. For example, at present there are influential secularist thinkers making the same arguments about homosexuality that their forebears were making earlier in the century about birth control. They are ridiculing the Christian teaching on sexual perversion as another example of the power of inherited inhibitions. . . . May we expect that Protestant leaders of thought, subjected to the pressure of this new ethical fashion, will gradually find it possible to regard acts of homosexuality as morally neutral? What limits are there in the methods of Protestant theology to prevent such an adaptation? For that matter, what action of any kind—no matter how shockingly evil or perverse it may appear to the Protestant conscience today—can be definitely and permanently outlawed as un-Christian, unnatural, inhuman?

On Fighting Communism

IF THE READER will turn to the "State of the Question" in this issue, he will see that, as we anticipated, the reprinting of Gerard E. Sherry's editorial on negative anticommunism provoked a lively response. By way of adding to the discussion, we should like to sketch in broad strokes the only kind of anticommunism which seems to us to be at the same time informed and Catholic.

At the risk of offending sophisticated readers, we begin by making the fundamental point that communism is both a movement of social protest and a conspiracy aimed at world conquest. As a form of social protest, it appeals to the victims of capitalist oppression and colonial exploitation, promising them a better and happier life in a society devoted to equality and justice. As a conspiracy, communism poses a double threat to the non-Communist world—the threat of direct, or military, aggression, and the threat of subversion, or indirect aggression.

It is scarcely necessary to argue that an intelligent and effective anti-Communist program must be well-rounded and all-embracing. It must be directed not only at blunting the appeal of communism as a social protest, but also at exposing and checking its conspiratorial drive for power. Such was the program sketched in 1937 by Pope Pius XI in *Divini Redemptoris*. Referring to the anti-Communist role of the Catholic press, the Pope wrote that "its foremost duty is to foster in various attractive ways an ever better understanding of social doctrine." In addition to this "foremost duty," the press should also, he explained, "supply accurate and complete information on the activity of the enemy and the means of resistance. . . ."

On the anti-Communist program thus outlined, there ought to be general agreement among Catholics. Some of our coreligionists, however, seem surprised—when it is called to their attention—at the priority the Pope gives to propagating the Church's social teaching. Clearly such Catholics are not well acquainted either with the social implications of their faith or with the state of affairs in the world today. They don't understand that even should Khrushchev call for baptism tomorrow, Catholics would still be obliged to strive zealously for social justice. They don't appreciate, either, the appeal of communism in underdeveloped countries, whose leaders are impressed both by the industrial progress of the Soviet Union and its official policy of racial equality. Such Catholics seem blind to the pressing necessity of demonstrating to the world, by our solicitude for justice here at home, the superiority of our way of life to the Communist way.

Whence it follows that a Catholic segregationist, for instance, no matter how loudly he may discourse against communism, is scarcely an anti-Communist at all. Like the Catholic who opposes foreign aid, he is, rather, an unwitting ally of the Kremlin. And the same is true of reactionary business men, of crooked labor leaders and of dishonest politicians.

But waging an anti-Communist war on the social front, though indispensable, is not enough. To the subversive activities of the Communist Fifth Column, it is only a partial answer; and to the Red Army, with its

438

missiles and bombs, it is no answer at all.

In dealing with communism as a conspiracy, two questions are of capital importance. One has to do with methods; the other with emphasis.

With regard to methods, anti-Communists differ on the importance to be attached to democratic procedures in combating the Communist plot. Some would cut constitutional corners to wage a more effective fight. Others insist on giving Communists the full benefit of all the rights and immunities they are sworn to destroy. Unless communism is fought with clean democratic hands, say the constitutionalists, we risk being infected by the very totalitarian poison we abhor. That this danger is not negligible appears today from the persistent effort in some quarters to turn "liberal" into a smear word and to stigmatize as "socialistic" all proposals for social reform. The blurring of ideological lines is not the least of the temptations which earnest anti-Communists must stoutly resist.

There remains the question of emphasis. In some areas of the world the threat of indirect Communist aggression is obviously greater than the danger of direct aggression. This is true of a country like Indonesia, as it is true of the Middle East and Africa generally. The opposite, however, is true of the United States, since over the past decade it has become agonizingly clear that far and away the bigger threat to our security is Soviet power as symbolized in the Red Army. That doesn't mean that our domestic Communists, no matter how discredited they are right now, can be safely ignored. On the contrary, they must be watched and exposed and checked at every point (and, of course, our security agencies must be ever alert to Soviet espionage). But it would be a fatal error, we believe, so to concentrate on the Soviet Fifth Column in our midst as to miss the greater threat from abroad.

That is the reason this Review has consistently supported big appropriations for defense and mutual security. That is why we have frowned on proposals for tax relief. That is why we have approved all sorts of foreign commitments—from the Truman Doctrine on

Greece and Turkey, through the Marshall Plan, to Nato and Seato. We have thought—and still think—that the only hope of stopping further Communist expansion by force—and the only hope for peace in the short run—is to maintain a clear margin of military superiority over the Soviet Union.

We have been disturbed, consequently, by the preoccupation of some Catholics with our domestic Communists. Too often, we note, this preoccupation is accompanied by coolness toward Nato, by hostility toward foreign aid, by complaints about taxes, by opposition to desegregation, even by defeatist fears of Communist infiltration of Catholic schools and other Catholic organizations. In short, we are led sadly to suspect that some of the anticommunism in Catholic circles is no more than a form of escapism from the anxieties and frustrations of a harsh and turbulent world.

We are fighting for our lives against a shrewd and implacable foe. We cannot afford to make mistakes. If with our knowledge of communism we combine a realistic appraisal of existing situations; if, in addition, we never lose sight of the lessons of justice and charity read to us by all the recent Popes, we should be able, with God's help, to fight successfully, and with much greater unity, for values we hold dearer than life itself.

Woman and the Common Good

IN MOST pagan societies womankind has an inferior status. She is the first of the male's chattels and a social cipher. It is the Gospel that ultimately powers the trend to give woman the suffrage, free her from civic disabilities and allow her to enter careers where she may use her talents for the upbuilding of the common good.

Despite female emancipation, the male still harbors the illusion that woman is the "lesser man." He fosters a stereotype of the female which emphasizes her lower intelligence, emotional instability and general ineptitude to engage in activity that transcends home economics. Obviously it is to the advantage of the lordly male to perpetuate this stereotype and to encourage its submissive acceptance among women. In this way the male employs tradition and prejudice to prevent the distaff element from encroaching on such typically male preserves as architecture and engineering, government and politics. One of the great pities of the modern world is that women so generally subscribe to the stereotype created for them by the dominant male, and thereby hinder their own achievement of status as well as hamper their power to contribute significantly to the betterment of the whole human family.

Somebody ought to inform June's girl graduates that the masculine posture of superiority finds cold comfort in the Church's attitude toward woman's place in society today, even though the Church deplores the philosophy of individualism that atomizes society and tends to eject woman from the home and make her a mere economic competitor of the male. Woman needs no Magna Charta of her rights and duties in the modern world: all the elements of her status and vocation were repeatedly emphasized in the many addresses of Pope Pius XII. Perhaps it will be good to summarize some of the basic positions of the papacy right here.

1. Male and female enjoy an absolute equality in personal dignity and value. Man and woman both enjoy a common temporal destiny, one toward which all history moves, and which was indicated in the command that Adam and Eve received together: "Increase and multiply, and fill the earth and subdue it and rule over it. . . ." Scripture does not say that men shall do the ruling while women do the multiplying.

2. Because of this shared goal, there is no field of human activity which must remain closed to woman; her horizons reach out to the regions of politics, labor, the arts, sports. Granted that the primary function and sublimest mission of woman is motherhood, still her personal perfection can be attained in other ways. Not all women are called to motherhood, neither is it possible for all to find their maximum personal development in the atmosphere of the home. "Woman's place is in the home" is a neat generality that reflects the ideal pattern of a perfect human society. It is not a precept of natural or divine law, least of all in an era of social upheaval that tends to debase woman, destroy the family and endanger the common good of all the world.

3. Despite the fact that woman's supreme throne is the home, modern conditions specifically demand female participation in public life, both social and political. It is the right and the duty of woman to take an active part in the movements of the day. The fate of

human relations is at stake everywhere. The mission of woman is urgent, to cooperate with men for the total good of country and society generally.

The paramount principle of social and political action is always the achievement of the common good, not the preservation of historically male prerogatives. So far, unfortunately, the male of the race has not distinguished himself in securing the welfare of humanity. It is not impossible that history may yet see women assuming a dominant role in the direction and control of community activity, and with the Church's blessing, too. Today, of course, this is still a "man's world." It will be a long while before the male crawls into the doghouse and growls: "Move over, Rover."

◆

IDYLL

Peace flowed along the River Tyne,
As Bede, the boy, sang in his stall,
Peace brought by angels drawing near
To serve the hours canonical.

Culling from out the Fathers' mead
The flowers of truth, the quiety Bede
Himself became a meadow rare:
The flock of Christ had pasture there.

Then lest his children learn a lie,
He taught them still, Death standing by,
Closing in peace and God's pure praise.
The lovely idyll of his days.

SISTER M. PAULINUS, I.H.M.

443

The Still, Small Voice

MORALITY MAY SEEM to be expiring, but it is not yet dead. Proof of its vitality lies in the fact that the word "conscience" is being heard again in the land.

For most of us, conscience probably means the still, small voice of the inward judge who levies praise or blame on our past actions. Conscience in this sense is a source of peace or remorse—a soft pillow or a gnawing worm.

But in the strict sense, conscience is not an assessor of dead choices but an imperative guide. It is nothing else than the final judgment I make on the moral quality of an action I am here and now on the point of performing. In the more crucial moments of the last stages of deliberation, it gives out with the "must" or "must not" that are the characteristic notes of inescapable duty. Conscience in this sense is the most personal and private experience in the world, and yet it puts on us the absolute necessity of obeying, for the moral judgment of conscience is nothing else, ultimately, than the will of God as known to us here and now.

We must remember that the judgment called con-science is not an irrational impulse that wells up from the Id, nor a censorious delivery of some mysterious Superego. The judgment of conscience is a work of human reason, and in fact it can be expressed as the conclusion of a syllogism. This, in truth, is precisely what links conscience to the realm of morality: the major of the syllogism is a general moral precept and its minor applies the precept to the contemplated choice of the will. In this sense conscience is not really private at all. The majors that preside over our choices are drawn from the authority of parents or the state or the Church, or even from our knowledge of natural law or divine revelation. But in every case, if they pos-sess any genuine binding power over the human will, that power must be derived from God, who alone has the final right to put shackles upon the liberty of the human will.

Forming the judgment of conscience is not always easy. We are subject to error and lie under the influ-ence of passion. Sometimes, too, we find that our moral judgments are complicated by conflicting loyalties. St. Thomas More faced such a decision when he had to choose between obeying Henry VIII and what he conceived to be the more pressing claims of God. Many are starting to ask now what a Catholic Presi-dent would do, if he faced a choice between "national interests" and "Church doctrine."

When such perplexities arise, every man is ultimately bound to do the will of God as he sees it. If he acts with complete sincerity, he commits no moral fault, even though his choice is objectively in error.

That sincerity before God is not enough for those who pose a religious test for Catholic candidates. For them, conscience is dandy, so long as it is not a Cath-olic conscience. In this case they remain distrustful and suspicious, like Aaron in *Titus Andronicus:*

". . . for I know thou art religious,
And hast a thing within thee called conscience,
With twenty popish tricks and ceremonies,
Which I have seen thee careful to observe. . . ."

445

Southerners as Americans

A MERICANS coming back from Europe always express the feeling that while abroad they "discovered" America. In much the same way, the first earthlings to return from another planet will likely have a lot to tell us, not so much about possible non-earthlings as about ourselves. A stay away from home is always a revelation.

Many a Southerner, too, has found that living in another part of the country has helped him find out what it means to be a Southerner. As an instance of this, nine respected Southern scholars—born, reared, largely trained in the South, but now doing research or teaching in the North—have just published a provocative, hard-hitting analysis called *The Southerner as American* (University of North Carolina Press, $5).

Mere latitude apart, what is the South, and what is a Southerner, anyway? The most important fact about the Southerner is that he is an American. But an American with a difference. He is pictured as different in accent, though even that is a varying and vanishing trait. The main difference lies in his history, true or fanciful; he is a man burdened with that history. After the Civil War, many Southerners, having tasted that un-American thing, defeat, escaped into a warm dream, a sentimental myth, the tragic Lost Cause. Today a new, revisionist school of Southern historians has for some years been demythologizing the fictitious South.

This stock-taking by the revisionists has already yielded meaningful results. It is now clear that the Old South was not really old, nor was it uniformly idyllic: the ante-bellum mansions of Natchez, Mobile and Charleston were few and far between, and three-fourths of all white Southerners owned no slaves. Slavery was not the paternalistic or neutral thing it was once held to be; indeed the great majority of slaves did not remain loyal. Nor was the corruption of Reconstruction entirely the fault of one race or region; it continued unabated even under restored white supremacy. Jim-Crowism, in the form we have known, is a recent, not a venerable social myth and usage. And research has quite exploded the concept of a monolithic South, together with the notion that the Civil War happened because there were "two civilizations"; the war came very largely because of that "peculiar institution"—the fact that the South possessed slavery and slavery was found profitable in the South.

The tragedy of Southern history—and vestiges of this regrettably remain—has lain in the inner conflict of true, self-reliant, freedom-loving Americans doing violence to their best selves by depriving fellow human beings of much of their humanity. The Southerner has been deeply Christian, and it has taken soul-wrenching casuistry to reconcile Christian charity and justice with the "peculiar institution." Many a Southerner—and Northerner—has thought it, but few have been so blunt as the politician who proclaimed: "I believe more in the purity of the Anglo-Saxon race than I do in the principles of democracy."

Out of this inner paradox of the South, and partly because of it, has come a sort of victory of pen over sword—not in cold historiography, but in the literary word. From Poe to Faulkner and Flannery O'Connor, the Southern writer shows a more-than-American tragic grandeur. Even the lowly of Yoknapatawpha County, Faulkner's Bunrens and McCaslins, have dignity and stature. Here is an awareness of the local, the regional, the highly individual. Southern characters, in and out of books, have time to be neighborly and personal. Surely these are precious human values in a largely homogenized, undifferentiated America.

The new upsurge of racism and race-baiting dema-goguery seems only a noisy death gasp. There is a New South intent on being its true self, fully American, Christian, authentically human. It may take some time for this South to win out.

◆

PURE DESERT

The more one runs in the spiritual
life the less tired one gets.—PADRE PIO

This is pure Gobi desert, you declare.
I see, past sandstorms (of exaggeration)
And rage of flesh at ghostly motivation,
Pink health invade your prayer.

Pure desert, you complain, though now you walk
Who once had shuffled through the arid miles.
Sighting a day of flight, I shelve my smiles
And share your pilgrim talk.

All true ascesis as a desert lies,
Hot wind, hot sand, no water and no way.
The ego agonizes through each day.
Freedom is when it dies.

I coax you onward: soon first breeze of bliss,
Soon sun that scorches cooled to sun that warms.
Your youth will dance when shady lanes lock arms
With each green oasis.

JESSICA POWERS

448

Patient, Doctor, Human Life

A BISHOP of the Church of England recently made headlines by telling British medical men that they have no moral obligation to keep elderly persons alive by "extraordinary means."

We have no quarrel with Doctor Mortimer, who is Bishop of Exeter: most of his quoted remarks could have been drawn from Catholic manuals of morality. But if the press thought his comments newsworthy, then it is time to emphasize the most basic principles that bind both patient and physician in the preservation of human life.

We all recognize that man has a positive duty to protect life when it is threatened by injury, disease or the debilities of advanced age. For although life is not a supreme good, to be maintained at all costs, it is certainly the fundamental natural good that makes possible the achievement of every other valuable goal. The root question, then, is this: what is the minimal required standard of responsible stewardship for man when his life, which is God's exclusive property, is threatened by any of the many ills that assail our mortal frame?

Catholic moralists agree that we are bound to take *ordinary* means to preserve life, but that *extraordinary* means, generally speaking, are not obligatory. The crux of the problem lies in the meaning and application of these two terms.

Probably most moralists would agree that ordinary means are those which lie at hand and are in common use among doctors and surgeons. In the average well-organized community of today, therefore, it would seem reasonable to regard intravenous feeding, blood transfusions or injections of insulin as ordinary means of preserving life. We might say the same thing of using the oxygen tent or of a growing number of routine operations and amputations, even major ones.

Extraordinary means are usually regarded as those which entail great hardship, suffering or expense beyond that which men would prudently consider proper for a serious undertaking, according to the state of the individual in question. That too may be called extraordinary which offers no solid hope of success or utility. Examples of such extraordinary procedures would be the continuation of a costly treatment after a patient has fallen into a terminal coma, as well as some of the more drastic operations on the heart and brain. No one would doubt that Fred Snite, who spent more than 18 years in an iron lung, at a probable cost of a million dollars, took an extraordinary and nonobligatory means of keeping himself alive.

We simply do not have any neat formula for applying these sound norms of convenience and utility. Hence, in particular cases, it is often difficult to impose a clear duty of employing a specific means of preserving life. Fortunately, when life is at stake, the human instinct is to seek help over and above the hazy minimum that morality rigorously demands.

The medical man's duties are stricter than those of his patient. He must not only do the minimum to which the patient is bound, but also do whatever the patient reasonably requests as well as what professional standards require. It is by close adherence to a certain moral rigorism that medical men maintain the reputation for setting a high value on life. It is by never yielding to defeatism that yesterday's extraordinary means of preserving life become today's commonplace, and thus contribute immeasurably to the enrichment of humanity and the nobility of medical art.

The Day

Which the Lord Has Made

IN A SERIES of dramatic flashbacks on Good Friday, the
Church unrolls her memories of the suffering and
death of Jesus. The recollections are bitter indeed.

On Holy Saturday, the Church is mute. Here is the
one day in the year that has no service of its own. The
altar of sacrifice is denuded and the tabernacle stands
empty. The Church tarries numbly at the tomb and
wills to know the void that is in her life when Christ is
gone.

But neither Friday's grief nor the Sabbath loneliness
is marked by despair. The grief is softened by sudden
thrills of hope, the sense of abandonment is lightened
by throbs of expectation, up to the moment when the
Church, as if unable to repress her remembrance of the
hour when death was swallowed up in victory, anticipates the rapture of Easter in the vibrant gladness of its
Vigil.

Now indeed her joy is great, as the flood of triumphant happiness transforms the ugliness of yesterday's
defeat. The whistle of the lash and the thud of the hammer are blotted out by the merry clangor of bells and
the thunder of the Alleluia. The purple flecks of Christ's

451

blood redden gloriously in the light of the paschal candle. The poignancy of Good Friday's Reproaches yields to the almost extravagant fervor of the Exsultet. In a kind of inspired daring, the Church presumes to rationalize the abysmal blunder of Adam, the ultimate folly of his prevarication in Eden: "How truly necessary was that sin of Adam, which was blotted out by the death of Christ! How fortunate was that sin which occasioned for us such a redeemer!"

Is the Church justified in these unrestrained outbursts of joy? We can answer that question by posing another. What is the central fact in the long narrative of human hopes and fears? What is the axis of revolution about which the world of history turns? It is not the emergence of language, nor the discovery of fire, nor the invention of the wheel. It is the event that took place between the darkness and the dawn of the day "which the Lord has made." "On this day, God, you overcame death through your only-begotten Son, and unlocked for us the gate to everlasting life."

To the eyes of faith, human history could never again be a story of damning sin and crushing guilt, of wrenching grief and farewells forever, after that moment when the portals of hell were shaken and the grave gaped and the One who bore our iniquities burst His cerements in lightnings of glory. For with the same thrust of omnipotence that exalted the name of Jesus above every name that is in heaven or on earth, grace began to rush like a torrent over the wastelands of mankind's sorrow and despair. The earth began to flower with the hope that men could live in newness of life, a "holy and sanctified people." Even more, the resurrection of the Prince of Life was an earnest of their own regeneration from the burnt ruins of the world. History is not closed by the mordant kiss of the worm. It opens on the day when this corruptible body puts on incorruption, and mortality robes itself in everlastingness.

"I am risen, and am still with you, alleluia." We must not forget that Christ is our hope, the author of our faith and the cause of our joy, even in these sad days when the "sky grows darker yet, and the sea rises higher."

A Great Inaugural

IN EDITORIALIZING on President Kennedy's Inaugural Address, one is tempted to string together a dozen sparkling quotations and let it go at that. For the materials from which this relatively brief address was fashioned are the heady stuff that history is made of. As Lincoln assuaged the spiritual anguish of a war-sundered nation in his Second Inaugural, as in his First Inaugural Franklin Roosevelt stirred to new hope and enterprise a generation mired in the despondency of economic depression, so did John F. Kennedy rally an apprehensive people to face with confident energy the seemingly insoluble problems of our times.

To some readers this comparison with deathless documents of the past may seem farfetched. After all, nothing the President said on January 20, as a bitter cold wind swept over snow-covered Capitol Hill, has not been said before. The contemporary challenge to freedom is plain for everybody to see. The rise of Castro in Cuba has dramatized the Communist threat to the Western Hemisphere. The anarchy in the Congo, which the Soviet Union and its stooges are recklessly exploiting, has raised the specter of newly independent nations trading 19th-century colonialism for 20th-century slavery. The civil war in Laos provides only the most recent reminder that our unity with "those old allies whose cultural and spiritual origins we share" is less firm than it ought to be. And as background for all this, we have been living for more than a decade now with the awful and frustrating knowledge that while "man holds in his mortal hands the power to abolish all forms of human poverty," he also holds in the same fragile vessels the ability to destroy all human life.

What was there, then, about this Inaugural which in the opinion of some raised it far above pedestrian levels and destined it, perhaps, to a high place in history?

It was written in excellent modern style—lean, terse, evocative. It was rich in literary, biblical and historical overtones. To old friends and ambitious foes, to our neighbors in the hemisphere and the masses of people in underdeveloped lands everywhere, it said with precision, and with feeling deep but disciplined, exactly what should have been said. And the address was studded with quotable gems: "Let us never negotiate out of fear. But let us never fear to negotiate"; "For only when our arms are sufficient beyond doubt can we be certain beyond doubt that they will never be employed"; "If a free society cannot help the many who are poor, it cannot save the few who are rich"; "And so, my fellow Americans: ask not what your country can do for you—ask what you can do for your country."

But it was none of these merits that touched the Inaugural with greatness. What gave it distinction is suggested perhaps by a line from one of our American poets. "Each crisis," wrote John Greenleaf Whittier, "brings its word and deed." Early in the address, one sensed that President Kennedy was bringing the word that answered the need of our troubled age. Harking back to our revolutionary past, he sounded his keynote:

Let the word go forth from this time and place, to friend and foe alike, that the torch has been passed to a new generation of Americans—born in this century, tempered by war, disciplined by a cold and bitter peace, proud of our ancient heritage —and unwilling to witness or permit the slow undoing of those human rights to which this nation has always been committed, and to which we are committed today.

And what did that torch signify? It signaled that this generation is called, as other generations were called before it, to testify to its loyalty to the American dream:

Now the trumpet summons us again—not as a call to bear arms, though arms we need—not as a call to battle, though embattled we are—but a call to bear the burden of a long twilight struggle, year in and year out, "rejoicing in hope, patient in tribulation"—a struggle against the common enemies of man: tyranny, poverty, disease and war itself.

Is this challenge one to deplore, to be fearful of, to shrink from?

In the long history of the world, only a few generations have been granted the role of defending freedom in its hour of maximum danger. I do not shrink from this responsibility. I welcome it. I do not believe that any of us would exchange places with any other people or any other generation. The energy, the faith and the devotion which we bring to this endeavor will light our country and all who serve it—and the glow from that fire can truly light the world.

Surely, that was the word today's crisis called for—a word that scorned cynicism and routed despair, that soared above sterile caution and the diffidence that sires defeat, that found challenge exhilarating, a word that echoed the ringing words Shakespeare put into the mouth of King Henry V on the eve of Agincourt:

From this day to the ending of the world,
But we in it shall be rememberéd . . . .
And gentlemen in England now a-bed
Shall think themselves accurs'd they were not here,
And hold their manhoods cheap whiles any speaks
That fought with us upon Saint Crispin's Day.

So it was that the word spoken by the President—a man who for all his youth has endured pain and lived with death—wedding American idealism and American power, summoned free men to a grand crusade.

From the word born of the crisis to the deed it de-

455

mands the gap is wide and treacherous. This, too, the President knows. Counting "a good conscience our only sure reward," he concluded, "let us go forth to lead the land we love, asking His blessing and His help, but knowing that here on earth God's work must truly be our own."

So it must be. Without God's help the deed will not be done. With it, we cannot fail.

Catholics and Others

JUST A WEEK AGO everyone who could manage it hustled off to a TV set to see the show: the historic and gripping solemnities that surrounded the Inaugural of the dynamic young man now formally empowered to speak for 180 million Americans. From Florida to Alaska, from Maine clear out to Hawaii, from Capitol Hill all the way down to 1600 Pennsylvania Avenue, from the shining top hat on his head to the last burst of band music outside the White House—it was a red, white and blue day! On such a day there are no political parties in America—there are only Americans. And all together we prayed that the new President would be equal to the gigantic tasks ahead.

Almost all our Presidents have had, each of them, a distinctive style, a special stamp—something they gave us to remember them by. In a thousand ways President Kennedy will impress his name and face and decisions on his time and on the conciousness of children yet unborn. But what will be the singular Kennedy cachet?

Washington is the man who crossed the Delaware and fathered his country; Jefferson is a wigged Virginian squire from Monticello; Lincoln freed the slaves and spoke briefly and tersely at Gettysburg; Teddy

Roosevelt charged up San Juan Hill; Coolidge was a silent man and chose not to run in 1928; a second Roosevelt smiled, smoked cigarettes in a long holder, talked admirably over the radio and put through the New Deal. The popular imagination is limited; it cuts a man down to a few sharply defined lines of truth or caricature, drawn with the brush of sentiment or pride or partisan pique.

How will Mr. Kennedy be remembered? For those of us who grew to adulthood with him here in the United States, who in 1928 were around to observe and now remain to remember what happened, who watched fascinated last fall as history proved that 1928 would not repeat itself, the answer is obvious. He will be remembered as a Catholic and as the first of his Catholic kind to be elected to the Presidency. (Strange, isn't it, that so few people could confidently tell you the precise religious affiliations of any former President, including Mr. Eisenhower.)

On November 8, 1960, with the election of John F. Kennedy, the full first-class citizenship of U.S. Catholics was at long last ratified. How do these now fully enfranchised Catholics regard their coreligionist in the White House? They respect him as their President, and they can't deny—whether they voted for him or not— a certain natural pride that he happens to be a Catholic. But there is no gloating in the Catholic attitude. Catholics look for no special preferments or special favors, and they will get none. They would be disappointed in the President if he ran things any other way. But they do want justice, particularly with respect to their schools, and they want the President and the Congress and the country at large to think hard about giving it to them.

With Mr. Kennedy in the White House, things are bound to be different for American Catholics. We shall all have fresh confidence, new courage, fewer resentments, and an easier feeling about shouldering our share of the common day-to-day work of America. Look for this in hundreds of ways—and we sincerely hope that you will not look in vain. With it now clearly established that our country does not accord prior rights

458

to Anglo-Saxon Protestants, you can expect to find Catholics turning up in all sorts of places where, formerly, nursing real or partially imagined resentments, they never quite felt at home: on all the citizens' committees that heretofore they frequently seemed to shun —committes to clear slums, organize municipal orchestras, build new wings on public libraries, raise money for the Red Cross, and all the rest. We shall be surprised if, from now on, Catholics don't take a more active and constructive interest in the public schools—to which they choose not to send their children.

In a word, it isn't hard to think that with a Catholic President in Washington, we Catholics might even call a halt to our old game of mutually excommunicating each other as "liberals" or "conservatives," and form a new citizens' front called COAU—Catholics and Other Americans United. It's time we did.

Values in the Space Race

W HY CONQUER SPACE? Practical minds often brush off the mastery of this forbidding environment as if it were no more than cosmic boondoggling or a game of "follow the leader" in which we or the Russians will be the first to encounter slimy little green men on Martian meadows.

From the short-range viewpoint, the current race into the unknown is politically and psychologically imperative. From the long-range point of view, man's triumph over gravity, no less than the invention of the wheel or of writing, puts humanity on the threshold of a new era of history that will enlarge our horizons immeasurably.

● In our own century, the development of the airplane profoundly influenced the strategy of warfare. The only assumption compatible with our security is that the future of military power is inextricably linked to the art which is the junction point of today's most advanced technology. That art is rocketry. Russia got there "fustest with the mostest," and her presumptive lead creates a threat of control which must not be disregarded. Control of space holds a completely unexplored potential for control of the globe. To cite just one possibility: the nation which puts in orbit one deliverable megaton bomb, as Sen. Stuart Symington has observed, "literally holds the sword of Damocles over its adversaries."

As long, therefore, as we live in a world without law, we must achieve that slender measure of safety which comes from wedding military science to the exploration of an environment which may provide war with its most dangerous dimension. We must either control space or be able to counter control by another power; in any event, we must master its difficulties.

● United States technological prestige has been appreciably dimmed by Russian successes in space. The USSR is not only enjoying its new role as the paladin of science; it is determined to play it boldly for every propaganda value that may accrue. The Soviets have literally hitched their wagon to the stars. After all, is there a better place to display one's prowess than the open skies?

Prestige is defined as the power to command esteem. It is a prime psychological value rooted in brilliance of achievement. France and Red China are seeking prestige diligently: so are Russia and the United States. Why? Because prestige is the basis of world leadership.

In today's world, justifiably or not, space feats have become the primary symbol of world leadership and a criterion of ideological superiority. Russia uses every conquest of space as a compelling argument for the Communist way of life. Since the United States aspires to remain the leader of the free world and a strong locus of values for the uncommitted nations, the implications are clear. If we are in the race for space, we must not lag on the laps; technological prestige tends to create political power, just as scientific competence tends to produce military supremacy.

We dare not neglect psychology, therefore, any more than we dare neglect physics. The world is no longer content to bow to affluence or be captivated by grants-in-aid or noble slogans. But it is powerfully drawn by a technology that springs from a way of life which promises to end want, injustice and inequality every-

461

where.

● Of course, the basic and enduring justification for the thrust into space lies in its ability to increase scientific knowledge and contribute to the welfare of mankind. As Wernher von Braun said on April 28: "This is a period of dynamic evolution which is pushing outward the horizons of human kowledge with explosive force." The impulse toward that evolution comes from nothing more mysterious than human curiosity. But the desire to know is a tendency of our nature as radical as the instincts of self-preservation and procreation. In God's design, it is what makes possible the material and spiritual advancement of the human species. Space may or may not be the last frontier, but it is certainly an endless frontier, whose margins fade forever as we move. Despite its formidable challenges, history counsels no other attitude toward it than optimistic expectation.

Let's Stamp Out Summitry·

D IPLOMACY is the theory and practice of international negotiations. Its traditional formalities are the etiquette of nations. Its institutionalized processes are designed to preserve civilized communication between sovereign powers, even when the most delicate issues are at stake.

Centuries of experience have taught the West that diplomacy is most effective when it is exercised by specialists in foreign affairs. Its pursuit demands the skill of the artist, the single-mindedness of the career man and—increasingly today—the intimate knowledge of details and their interrelations that characterizes the practical scientist. Effective diplomacy is not the province of the amateur, however powerful, prestigious or presumptuous he may be. This is especially true in the era when the portfolios of the Big Powers are loaded with plutonium and the heavy fears of all mankind brood over the conference tables.

SHEER SHOWMANSHIP

Yet the ascendancy of the high-level amateur over the professionally competent is the pattern of Big Power diplomacy that has become ever more dominant from the "horse-trading" days of Teheran and Yalta to the obscene "heavyweight match" that was scheduled for May 16 in Paris.

What judgment should the ordinary citizen pass on these jet-assisted take-offs whereby Mr. Eisenhower seeks to give Chancellor Adeuauer periodic assurance that we will not scuttle Berlin, or Charles de Gaulle hopes to attain a little bit more status with Prime Minister Macmillan? What are we to think of the swashbuckling summitry whereby Nikita Khrushchev strives to slug out international problems of the most vital nature in a fantastic brand of "diplomacy" of his own contriving, while accredited foreign ministers watch the water buckets in the corner of the prize ring?

What we have seen of personal summitry so far is hardly encouraging. It diverts the time and energy of the highest officials of government into channels of futility. It produces neither fruitful agreements among allies nor genuine relaxation of tension anywhere. It is a hasty process whereby the weightiest matters receive cursory discussion and immature decision. In the words of Sir Ivone Kirkpatrick, who was once United Kingdom High Commissioner for Germany: "So far as I am aware, no summit meeting has had any result except to create in the minds of all the participants a sense of weary and angry frustration."

Some of the perils of summitry, particularly as they affect the President of the United States, were explored in the April issue of *Foreign Affairs* by Dean Rusk, former Deputy Under Secretary of State. The basic obstacle to effective diplomatic communication on the Presidential level is that it is difficult or impossible for the President to give proper preparation to the business of international negotiation without serious neglect of his political and constitutional duties at home. Thus the constant climbing of summits adds intolerable burdens to the discharge of an office that many already consider an impossible task. According to Mr. Rusk, "the process needs time, patience and precision, three resources which are not found in abundance at the highest political level." Moreover, the raw confrontation of heads of government involves added tensions because every summit, in a sense, is a court of last appeal, where the costs of miscalculation are multiplied by the seriousness of the issues and the power of the men who take part.

If inadequate preparation, limitations of time and accidents of personality make personal summitry a hazardous gamble even in tête-à-têtes among leaders who are allies, the psychological barriers to meaningful diplomatic dialogue are infinitely higher in the staged summit. These affairs tend to be as stylized as a ballet, as elaborate as a Hollywood colossal and as public as life in a goldfish bowl. The expanded conference table is flanked by cameramen, messengers and translators. The principals are backed up by battalions of experts with their position papers and a mania for writing *ad hoc* memos. The press hovers in the wings, anxious to score beats and put calculated leaks on the wire. Under such circumstances a summit is not a meeting of minds but an exchange of propaganda-slanted speeches. There is no genuine dialogue on issues, no reasoned harmonizing of differences, but only an artificial debate. Is this the way to pursue effective communication between nations? It makes as much sense as if Rome and the separated churches of the East were to stage a council of reunion in the piazza in front of St. Peter's.

PSYCHOLOGICAL WARFARE

The gravest risks of personal summitry are inevitably associated with confrontations of East and West, whether they take place in the intimacy of Camp David or in the Elysée Palace. For here, because of the very nature of Soviet political strategy, true communication becomes fundamentally impossible. Soviet diplomacy is not intended to be a form of communication, but a tool of political and psychological warfare. As Dean Acheson pointed out just a year ago, its aim is not rational debate but propaganda. Hence its lifeblood is publicity. The conference table is turned into a grandstand. The forms of negotiation are prostituted to project images of power that are meant to deceive, divide and overawe. And in this clash of the megaphones, the odds favor the Soviet Union. The free world lies open to the sound and fury, but the Iron Curtain rises too high to permit an effective counteroffensive.

The supreme example of the perils of personal summitry was just given to us in Paris. In what was patently the Biggest Show on Earth, the Russian lion-tamer took the abashed American cat by the scruff of its neck, just to teach it a lesson—in Soviet diplomacy. Such obscene buffoonery, disgraceful at any time in international relations, is intolerable in what may be the twilight of oncoming nuclear darkness. Its utter irony was that it made impossible, at least temporarily, civilized negotiation between the United States and Russia: the rupture of rational dialogue was done in the very name of the art whose core is communication. No further argument is needed to bring in question the whole trend to summitry that we see today.

The policy of personal summitry needs immediate reappraisal in the light of the recent debacle in Paris. At best, it is comparatively ineffective. At worst, it envenoms international relations. If such a reappraisal restores to honor the old-fashioned diplomacy with its high style and painstaking attention to detail, the world will be the gainer. Let's return the portfolios to the foreign ministers, when we are dealing with our friends. As for the assumption that, in dealing with the Soviet Union, it does no good to talk with anyone but the boss, that depends on how firmly we resist his yen for summitry. He will give Gromyko a fat portfolio when he really wants to negotiate.

A last note of warning. Khrushchev, the incurable optimist, is ready to let bygones be bygones. He is already looking forward to a new summit. He is currently giving us a time of grace in which to regret our iniquities, nurse our fears and elect a President "who can negotiate." Whoever that President may be, let him remember the indignity his office met in Paris. Let him remember too that he already lies under an ultimatum; he has been served notice that he must capitulate on Berlin or suffer the consequences of his obdurate refusal to "negotiate" in the interests of peace.

Second Thought

Second Thought

The Threshold Moment

IT HAS BEEN five years since Msgr. John Tracy Ellis of the Catholic University of America delivered his paper on the condition of American Catholic intellectual life. The substantial and spontaneously affirmative response of American Catholics to Monsignor Ellis's indictment indicated that many were not complacent about the quality of Catholic education or about its product.

Indeed, most of those who would have disputed his thesis held their fire prudently for a year or two. And when they did speak out, it was not to directly dispute the thesis, because at its critical points it was indisputable. Rather the lament was over "Catholic breast-beating." Apparently public *mea culpas* were more scandalous than the conditions that occasioned them. Likewise, the critics did not challenge the validity of Monsignor Ellis's evidence. They questioned his methodology. Thus, although Monsignor Ellis had devoted only two paragraphs of his 200-paragraph paper to *Who's Who in America* and the lack of Catholic representation therein, one journalistic critic launched what was an ostensibly massive attack by questioning the relevance of that single piece of evidence.

I mention these negative tactics, not to refute them at this late date, but primarily to emphasize how sporadic and ineffectual they were and how pathetic they were by comparison with the realistic and rueful acknowledgment of our intellectual disabilities by the many hundreds of Catholics—teachers, school administrators and concerned nonacademic people—in whom Monsignor Ellis's paper had struck a sensitive nerve. Almost as one, the realists agreed that some remedial action was indicated.

All of which, however, brings me to my central question. It is this: "What has been done to remedy the situation Monsignor Ellis described in 1955?"

If Monsignor Ellis had been only half-accurate in his evaluation of American Catholic intellectual life, obviously it would take more than five years to correct the deficiencies he had found and which he had so painstakingly, charitably and candidly listed. Impatient as I tend to be in these circumstances, I am prepared to concede that a five-year reform program would have been an impossibly foreshortened crash program. But I am wondering what has been done in terms of reform and whether or not the future is promising.

I am wondering for three reasons:

1. Monsignor Ellis's indictment was a gravely serious one.

2. Initial response, particularly on the Catholic college and university campus, was one of reform enthusiasm.

3. Talks I have had in the last four or five months with lay, clerical and religious intellectuals, while they have not left me uneasy about the future, have not reassured me either.

Perhaps we ought to reread Monsignor Ellis's paper to recall just how serious were the defects he had discovered in his scrutiny of the intellectual life of American Catholics.

He spoke then of the "paucity of scholars of distinction" in Catholic higher education; of the "perpetuation of mediocrity"; of the "senseless duplication of effort and the wasteful proliferation that have robbed Catholic graduate schools of the hope of superior achievement"; of the failure of American Catholics to recognize "scholarly accomplishments" in fellow Catholics and to "lend them their support"; of the "vocationalism and anti-intellectualism" present in the schools; of the "pervading spirit of separatism" in both American Catholics at large and the intellectuals; of the "failure" of American Catholics "to have measured up to their responsibilities to the incomparable tradition of Catholic learning"; and of our "failure to produce national leaders and to exercise commanding influence in intellectual circles."

The first response on the Catholic college and university campuses was indeed gratifying. Monsignor Ellis's paper became not only the principal topic of discussion among professors over coffee; it was an object of formal departmental and interdepartmental study and analysis in our schools. Significantly, the few denials of its realism and accuracy did not come from the schools.

I suspect we will never know with anything like scientific accuracy which educational reforms and corrections can be traced to Monsignor Ellis's study. I don't think that kind of information is particularly important. What is important is whether there have been reforms; whether that first, instinctive, enthusiastic affirmation that greeted Monsignor Ellis's statement has spent itself fruitlessly and has now dwindled with no discernible effect or promise.

The kind of evidence that would be reassuring here is not limited to the easily recognizable action, such as the elimination of wasteful duplication of efforts; conspicuous recognition and reward by the Catholic community of scholarship; palpable attempts to discourage Catholic separatism; insistence on newer, more rigorous academic requirements.

Talking with Catholic intellectuals can yield evidence of a less obvious, but perhaps just as important, kind. Has there been a shift in the attitude, a change in the atmosphere, so far as the academic and nonacademic Catholic community is concerned? Is there not only remedial desire and pressure, but also the freedom to translate that desire and direct that pressure into action?

There seems to be no clear-cut answer to these latter

470

questions. We seem to be on a threshold at the moment. We can go forward, but we can also go back. Our schools, I think, hold the key to that larger question that goes beyond the schools, the question of whether the future of the Church in America will be shaped and determined in its essential features by the forces of reaction, separatism and the siege mentality, or by the forces of progressiveness, openness and intellectual respectability and influence.   DONALD McDONALD

◆

In Our Town

In our town the Lion's Club remembers
Christmas with a crèche of strong black timbers—
larger this year with tiny bulbs that flick
like city lights. And right there by the track
where the New York Central makes it tremor
four times a day. So nice to see a glimmer
of something new on the long gray ride home.
Everyone's heard of it and they come
from up the line to see the crib, and smile
that we thought of it. They say it's all
we're known for, only borough in the valley
that has no tree, or bright red lights strung willy
nilly across Main Street.
                    But we only nod
and share a secret by the small straw bed.

MARY ANN MAC NEIL

471

Washington Front

Washington Front

TWENTY YEARS AGO the Swedish scholar Gunnar Myrdal argued that the Negro in America was caught in a vicious circle. The Negro, he said, was the victim of prejudice, which made it difficult for him to secure adequate income, housing, education or even the right to vote. However, a break in any part of the circle would modify all of the factors arrayed against him. Such a change could create an entirely new arrangement of forces—this time favorable to the Negro. If, in other words, the Negro's opportunities for education or voting increased, then each of the other factors would also take a turn for the better. These changes would make it easier for Negroes and whites to work together. Prejudice would be reduced, thus making it still easier for Negroes to get jobs, education, housing and the right to vote. Whether the first crack in the circle resulted from improved conditions of life for Negroes or from the lessening of prejudice made no difference. The effect in any case would be cumulative.

The Myrdal analysis looks remarkably good in retrospect. In 1940, in spite of some degree of equality in unemployment relief and pressure from the Supreme Court to force equality before the law, no one could contemplate the Negro's position in society with much

473

optimism. The war brought fair employment practices. Negroes moved to the industrial centers of the North, increasing their political power. White primaries were outlawed. The dire-predictions of the fearful failed to materialize. Prejudice was reduced.

Under the Truman Administration the armed forces were largely integrated. The place of the Negro in many of the industrial unions was made secure. In 1948 the two major parties stated their support of political, social and economic equality. The support of the parties for civil rights was in part the product of their desire for votes from the increasing Negro electorate.

The developments of the 1940's made it easier for the Supreme Court to outlaw segregated schools in 1954. During the six years following the court's decision 12 per cent of the schools of the border and Southern States were desegregated. The number of Negro voters in the South has continued to climb—aided by the work of the Attorney General's office and the Commission on Civil Rights. Congress passed two civil-rights laws. Today only 16 Senators can be counted on to vote against civil rights.

There seems to be a kind of geometric rate of change in the status of the Negro. More was accomplished in the 1950's than in the preceding five decades, but the break in the circle in the 1940's was essential for this development. Aided by the cumulative effect of the work of the last two decades, the Kennedy Administration should be able to make really great strides toward racial equality. HOWARD PENNIMAN

Comment

Comment

Inaugural Prayers

A religious people are delighted that their government, in its Presidential Inaugurations, formally acknowledges the place of God in the business of the state.

But we deferentially observe that the four prayers offered at the Capitol on Jan. 20 were too wordy for an already cold and gusty day. God was addressed bilingually, several times reminded of the calendar and pluralistically praised, thanked and besought. Secularists must have felt that organized religion was seizing a golden opportunity to turn a civil event into a specific act of worship. A continuance of this trend could wind up by making obnoxious the whole concept of recognizing the Almighty from the public platform.

Public prayer is difficult for men, even when they profess one faith in a common sanctuary. When the occasion is civil and the petitioners are of many faiths, due regard must be shown for all circumstances. Perforce, the theme of the prayer must reflect some common denominator of commitment. For all that, the pious should not be sermonized, neither should the indifferent be wearied, nor even the atheist needlessly provoked.

Ideally, public prayer on civil occasions should sound only the essential notes of praise and repentance, thanksgiving and petition. It should do so with directness, simplicity and a quick amen. It will lose nothing if its sense and sentiment echo the familiar themes of the Judeo-Christian tradition.

In January, 1965, esteemed clergymen will do well to limit Inaugural prayers to one or two minutes at most. The norm of public prayer is not fullness of utterance but sincere engagement of the people's heart.

Here is a model. The occasion was public, the need desperate, but the message brief: "Lord, remember me when thou comest into thy kingdom." God listened, and that's what counts.

476

Up Mehitabel!

Are we missing out on the finer—and furrier—things. of life? With International Cat Week starting on Nov. 6, isn't it time to improve our cat I.Q. a bit?

Do you realize that there is no Cat Explosion, despite tabby's addiction to casual motherhood? The cat population has been stable since the Great Depression. America boasts only 21 million cats. Any judgment to the contrary is based on wanton extrapolation from the cat density of a few jungle areas in urban centers.

Do you know that 50 per cent of our cats have to scrounge for fodder on Skat Row? And that only four per cent live in Park-Avenue-type pads? Are you aware that even pampered pets have to form milklines at the icebox or do cockroach patrol in the pantry?

Do you know that a well-fed feline has an overkill potential of 13 rats per night, and that the mouse potential defies calculation?

The Golden Age of cats came in ancient Egypt; those were the days when a cat had a better chance of being mummified than a statesman. Since then the cat has suffered a colossal loss of puss: although Castro touched most bases in his UN oratorathon, the Maximum Leader forgot to press a resolution against catocide.

Another point. The family with a cat is a family without delinquents. Cats build character. How can kiddies go wrong, if all their time is taken up in feeding, grooming and scratching the Velvet Menace?

Cats have a program: what this country needs is a good five-cent fish head with catnip dressing. Beyond that, all the cat asks is freedom to roam and tighter dog-walking ordinances. The independent cat scorns the dog as a fawning satellite who will bootlick his master for a synthetic bone.

Beware the cat. It's just waiting for some Karl Manx to rise and mew: "Mousers of the world, unite. You have nothing to lose but your bells and bows."

Seven Against Space

On April 9, with a happy flourish, the Space Administration czar unmasked the seven heroes who will become our first bona fide "space rats." A composite sketch of Captain Icarus shows that he is a military test pilot of 34 and packs 164 lbs. of splendidly coordinated grit into a moderate frame. He totes an egghead I.Q. and has a strong yen to become a rocket jockey. Moreover, he is highly motivated (his wife is keen on lobbing him into orbit:

any connection?) and, sure enough, he is a believing *Protestant*.

Ay, there's the rub! No good for Administration spokesmen to say "That's just the way the rocket tumbles." We predict:

1. Some bilious editorial apologete will darkly ask: "How does it happen that from the volunteers among our 80,000 military pilots, not one Catholic filtered through the Space Curtain?"

2. Some communion breakfast orator will harangue the Knights: "Instead of asking where are our Catholic Salks and Einsteins, let us ask where are our Catholic astronauts? Columbus et al. were of our faith. Are the *Niñas, Pintas* and *Santa Marías* of the cosmic seas to be piloted solely by heretic helmsmen?"

3. A Catholic educator will demand a look-see at the 566 "Who am I?" questions used in screening the fledgling spacemen. Were those questions slanted to put a Catholic profile in a poor light?

4. Inevitably some aspiring politico will stir our bile to a boil by observing that a space team without a Catholic is like an All-American team without a Notre Dame player: serves us right if the Commies beat us to the moon!

But in all seriousness, our prayers go with these brave men in their arduous training and their perilous flight. It frightens most of us, even in quiet reverie, to think of the day when the first of the seven thunders into the blue, spins about the earth dizzily, and then endures the fiery buffeting of re-entry. We hope indeed that our first Magellan of the empyrean may have God for his co-pilot.

Wesley Rides the Rockets?

Zealous John Wesley, mounted on a horse, rode 250,000 miles in evangelizing England and Ireland to Methodism. His form of the Gospel, transferred to America, was well adapted to lay preaching and frontier life.

Is Methodism getting ready for space? Are astronauts, doubling as preachers, going to ride the rocket circuit to the frontiers of Alpha Centauri?

Celebrating the 175th anniversary of U. S. Methodism, Washington Bishop G. Bromley Oxnam recently asked if the seminaries were preparing missionaries to share our riches with the cosmos.

Hmmm. . . . We hope *all* the sects will move slowly before men begin to contaminate the planets with the scandal of disunity, or transfer the confusing ecumenical dialogue to the stars, or show the universe a Word whose pearly luster is muddied by private interpretation.

Christ gave no command to evangelize the Outsiders; His Gospel is for earthmen. Are we to preach repentance to those who perhaps never sinned, or to baptize the unredeemed? Would Bishop Oxnam demand faith of those who never got a revelation, or whip up dogmatic squabbles among those who never doubted, or encourage sectarian fragmentation among those who perchance were never divided? The thought is gruesome.

What's more, those stellar jaunts will be centuries long. Pity the devoted pilot-preacher. The tree of U. S. Methodism, during the last 175 years, produced 22 branches. What will the faithful astronaut come home to, after preaching near Epsilon Eridani? One awful thought—he may make his earthfall only to find there is no such thing as Methodism; since he went out on circuit, the ecumenical problem has been solved, and all the earthbound Methodists have "come home to Mother," the Mother Church, that is! Could be!

Let Bishop Oxnam take thought. Cosmic Christianity is more than· a challenge. It's a caution.

Queen of Spacemen

In the catalog of saints there are patrons galore. Brewers and weavers have them; so do wet nurses, plasterers and gravediggers. Travelers call on St. Christopher and aviators invoke the Little Flower. But so far as we know, there is no official patron of astronauts. We have protectors against hernia, snakes and earthquakes. But we have no intercessor against the perils of blast-off, orbiting or re-entry, no guardian of voyagers through the "deserts of vast eternity" that lie beneath the moon, the sun and the distant stars.

The rocket launchers may adopt St. Barbara as their own. The communications engineers in the blockhouses already have St. Gabriel for their advocate. But neither of these, nor even the many patrons of mariners will suffice for the pioneers who navigate in the bowels of space ships. These men need a patron who is powerful on earth and in the heavens, too.

One of our readers suggests that in selecting such an advocate no better choice could be made than Mary Assumed into Heaven. After death our Lady was in a very physical sense borne through outer space, and surely she may be called queen of the starry deeps as well as Queen of Heaven.

We like the suggestion that Mary be chosen patron of the astronauts who will soon be probing the solar system. Certainly there is nothing to hinder devout spacemen from seeking confirmation of the choice by the Holy See.

Dormididactics

You never heard that word before? No wonder; we just invented it. We're giving a name to the latest of the black arts—the art of teaching students while they are asleep.

Yes, there are psychologists who think we are wasting one-third of our lives by wandering aimlessly in the land of ·Nod. Why dream, when precious sack-time can be used to absorb a packaged course in sociology or history? What this country needs is an efficient mechanical tutor to drill us in facts and figures while we are knitting up the "ravell'd sleave of care."

We dread to think of successful assaults on the citadel of sleep. If the experimenters are lucky, the next generation will not be troubled by teacher or classroom shortages. Live teaching will be supplanted by nocturnal imbibition of knowledge at the hands of a patient, fully transistorized pedagogue. We can even do away with the schools; surely the taxpayer will welcome the day when a bedside squawk-box will give us all the benefits of a university. As for students, why should they spend four years in college? We can see the day when they hibernate for three months with a mechanical educator and emerge with a bachelor's degree.

Somehow, though, it doesn't seem right to invade the realm of Morpheus with packaged pedagogy, nor to make liars of poets who used to call sleep by such pretty names as the balm of woe, sore labor's bath or the prisoner's release. As for Macbeth, now he can rest in peace. 'Twas not Macbeth did murder sleep; 'twas the experimental psychologists.

The pay-off, from *Tiny Tim's Garden of Verses*:
"Now I lay me down at ease,
To learn in sleep my A B C's;
If I should die in bed alone,
Please, Lord, switch off
my Dormiphone."

To Bind in Conscience

Political discussion often uses the expression "Catholics are bound in conscience" as though it were an indication of some specifically Romanist syndrome. Let us scotch this error once and for all. Let us see what it means to bind in conscience.

Any dictionary gives us a clue. Literally, to bind is to tie or fasten tightly; and thus we bind sheaves, books, madmen and prisoners. The root idea in binding is to impose some form of constraint or immobility.

But for centuries our language has sanctioned figurative usage of the word "bind." We speak of binding ourselves by oath or contract. In the strictest sense, we speak of being bound by law, conceived as an externally imposed guide of choice and action. All these metaphorical uses connote some limitation on our freedom of movement. They imply a constraint that operates not by brute force but by moral influence.

Doesn't our experience of the "still small voice" of conscience bear this out? It is conscience that makes us aware of obligations. It is conscience that prods us to do our duty. It is conscience that makes us chafe under the yoke of law.

It is conscience that issues categorical imperatives, those unique forms of necessity which do not physically compel, yet which cannot be resisted without guilt, remorse and the awareness that we have deviated from rectitude.

To bind in conscience, then, is simply to put one under the moral necessity of pursuing a determinate course of action. To be thus bound, of course, is not a uniquely Catholic experience; it is the common lot of all men who are in bondage to Duty, the "Stern Daughter of the Voice of God."

⚜

Anyone for Steno?

If you can believe the "Help Wanted" columns, a girl doesn't just go out and get herself a job these days; she is lured into employment by promises of Sutton Place glamor, easygoing bosses, slow dictation and the "opty" to meet what abbreviated ad lingo calls "impt. ppl." Here are some samples from a single page of the Oct. 16 N. Y. *Times:*

• SECY. Unique exciting posit for brite gal who cannot stand routine. Aver skls ok. No pressure no tests.

• SECY. "Sutton Place." If you can't afford the rental live vicariously & at least work in this posh atmos. Lovely office lovely view lovely people who say "thank you."

• SECY. This is the opty you've been waiting for! Wk with nicest boss in NY. Meet NY's top echelon! Non pressure,

inform atmos.

• SECY. This furniture showroom is fab! What a lovely place to start your career! Loads of pub contact! Non pressure! Lt steno ok.

• SECY-RECEPT "Plush." Meet and greet, take a few letters, be pleasant & cheerful & a fine job in a fine firm is yrs! Coffee on the house.

• SECY. Plus bonus. Easygoing boss who's not around much & all benefits. Be yr own boss.

• SECY—JRS. Your opty to enter the stimulating, fascinating, exciting Busn. World. These are positions for brite gals in highly attractive offices staffed by young men on their way up to exec and sales mgmt.

What, no wedding rings on the house?

⚜

Science

SCIENCE

The Tunguska Mystery

It isn't every day that I can wrap up astronomy, flying saucers and Soviet science in one package. But I have the chance now, thanks to a weird event that took place in the upper valley of the Podkamennaya Tunguska, in Central Siberia, on June 30, 1908.

It was a cloudless morning. Precisely at seven o'clock a fiery object, blazing like the sun, swept into the earth's atmosphere from the south and exploded with a cataclysmic crash in the marshy scrub forest of the Siberian wilderness. What followed was awe-inspiring.

For a hundred miles around the big blast, water gushed from the ground or was swept in sheets from the lakes and rivers. The earth shook and concentric ridges formed on its surface. At 250 miles a black mushroom cloud was seen boiling up into the stratosphere. In selected directions, it appears, fences were toppled at a distance of 250 miles. Horses and men were bowled over at Kansk, 450 miles from the blast. The sound of the blast was heard 600 miles away and at the same distance (in Irkutsk) disturbances in the earth's magnetic field were recorded.

Other effects were noted on an even wider scale. The Tunguska object produced air waves that circled the globe twice. The seismic shock of the crash was registered in Washington, D.C. The atomized debris of the explosion, rising some sixty miles into the air, created luminescent clouds which gave much of Asia and Europe a series of "white nights."

It took many years for scientists to bring together the few facts I have cited. The Tunguska phenomenon happened in one of the most remote and uninhabited areas on earth. It was not until 1921 that a Russian expert on meteors began to gather eyewitness accounts of the catastrophe. Even he was not able to penetrate the actual site of the blast until 1927. The first aerial survey of the region took place almost thirty years after the original event. In fact, it was not until 1958 that the USSR sent a large and well-fitted expedition into the bleak tundra area.

What was the nature of the object that hurtled from the sky back in 1908? The conventional scientific explanation is that a stone or nickel-iron meteorite, moving at many miles a second and weighing possibly a million tons, intercepted the earth in its flight and hit the scrub forest with a terrific jolt of blast and fire. We are all aware, of course, that the earth sweeps up many meteors in its orbit and that some of them are quite hefty.

The meteoritic theory immediately runs into grave difficulties. Searching has uncovered no meteor fragment anywhere in the impact area or beyond it. No trace of a crater has ever been found at the site of the explosion. Are we to believe that a large and solid mass was so completely vaporized by heat that it left no trace of itself?

Another puzzle: at the very center of the Tunguska explosion a forest of about 700-800 square miles was entirely destroyed, with the charred trees stretching radially outward along the ground for from 20 to 40 miles. Yet in the middle of this devastated area, in a circle about one mile across, the detonation left the trees erect, but with their crowns and branches shorn. How could a meteor have produced such an effect?

A second, but rather beatnik, explanation of the Tunguska affair has been popular in semi-scientific organs of the Soviet press for the last fifteen years. Its originator was science writer A. N. Kazantsev, who argues that the explosion was caused by the disintegration of a spaceship that unwisely entered our atmosphere. Conceivably, the alien vehicle was made of "anti-matter," and quite inevitably vanished in a burst of sheer energy when it invaded the environment of the earth. Kazantsev's theory enjoyed new popularity in 1959, after inspired amateurs investigated the Tunguska territory and reported unusual radioactivity at the site of the explosion.

This curious report brought about a new official expedition to the explosion site in 1960, under the leadership of Dr. V. G. Fesenkov. This study did not confirm the reports on radioactivity, but did result in a third theory about the Tunguska object.

Dr. Fesenkov thinks that what struck the earth in 1908 was a small comet, several miles in diameter and weighing perhaps a million tons. Since comets are presumably composed of dust and frozen gases (they are often called "flying gravel pits"), such a theory could explain why the explosion left no crater and why the object disappeared without a trace. It simply bored into the atmosphere at 25 miles per second and vaporized when its kinetic energy turned into heat as a result of friction. Dr. Fesenkov feels that his cometary theory is supported by the fact that the object approached the earth from a direction opposite to that of the earth's motion around the sun, whereas meteorites always orbit the sun in the same sense as the earth.

We must conclude, I believe, that the Tunguska mystery still clamors for a satisfactory explanation. There is no doubt Soviet scientists will continue to give the matter considerable attention.

Celestial visitors like the Tunguska object were once quite common, when the solar system was young. But objects of a comparable magnitude probably do not strike the earth now more than once in a hundred thousand years. Thank heaven! If the Tunguska object had arrived just four hours and 17 minutes earlier, it would have landed on St. Petersburg (now Leningrad). The results of such a calamity may be left to the imagination.

L. C. McHugh

Melting the Snowman

Last September the conqueror of Mount Everest, Sir Edmund Hillary, set out from Katmandu on a search for the Abominable Snowman. Four months later, in the London *Sunday Times*, he said: "In spite of our efforts, the *yeti* still remains a very real part of the mythology and tradition of the Himalayan peoples and it is undoubtedly in the field of mythology that it rightly belongs."

For many centuries, all along the rugged flanks of the vast Himalayan range, there have been popular legends of a hairy wild man that dwells in the timberline forests and inaccessible icy caves of the world's loftiest mountains. Reports of these strange creatures began to reach the West at the turn of this century. By 1953, I would say, the Snowman had achieved a status like that of the Flying Saucers: there was lots of "evidence" for its existence, if you didn't ask too much concreteness and took an optimistic view of the value of human testimony. Its mewing or yelping cry was heard by various mountain people. It was fleetingly sighted in Tibet and in the northern parts of Burma, Sikkim and Nepal. Respectable explorers—British, American and even Russian—saw and photographed its tracks amid the eternal snows. Some even examined specimens of its scalp, which is sometimes preserved as a relic in Tibetan monasteries. There were even reports that mummified *yeti* were to be seen in some of these remote lamaseries.

Admittedly, none of this evidence took the scientific world by storm, although it did arouse considerable speculation. Any reader who wants to bone up on the lore of the. Snowman can consult pp. 127-183 of *On the Track of Unknown Animals*, by Bernard Heuvelmans (surely one of the most fascinating books ever seen in print). There one will find all the reasons why, if the *yeti* really exists, it is so hard to track down, describe with exactness or even classify as to type. I mention this last point because it is important to realize that legend provides for several kinds of Snowmen. The broad classifications run thus:

1. *Nyalmo*, a real giant that can attain a height of 16 feet. This is the most elusive of all Snowmen, found only above an altitude of 16,000 feet. *Nyalmo* may be only a fiction derived from the native Tibetan principle that "the higher you go, the bigger the *yeti*."

2. *Rimi*, probably identified with the Sherpan *chuteh* and the Tibetan *metoh-kangmi* or "filthy man of the snow." This hairy ginger-and-black monster, sometimes eight feet tall, has long arms, bowed legs and a somewhat human face. Like all Snowmen, it has a high conical skull. It is normally vegetarian and tends to travel in groups at the 13,000-foot level.

3. *Yeti*, the garden variety of the species, also known as *yeh-teh*, *mi-teh* or *rackshi bompo*. The *yeti* proper flourishes on the flanks of Everest. It is never larger than a small man. In most respects it looks like a small edition of the *chuteh;* however, it is usually reddish in color and relishes human flesh.

4. Hillary, finally, allows for the *thelma*, a *yeti* no more than two feet high which (being the smallest) fittingly lives down in the jungle. It has human features and likes to pile sticks and stones in little mounds.

While *nyalmo* and *thelma* are almost

surely fictional, descriptions of the *chuteh* and common *yeti* are so widespread, persistent, detailed and supported by various kinds of indirect evidence that some respectable investigators feel there must be some anthropoid reality behind the *yeti* story. Heuvelmans, in the book I mentioned above (published by Hill & Wang, 1959), believes that the evidence for the Snowman is no worse than that which has admitted many a fossil or living species to the halls of paleontology or zoology.

Accordingly, Heuvelmans thinks that a giant anthropoid biped really dwells in the Himalaya. It is a primate, omnivorous, with a flat face and a conical skull; it reaches a height of from five to eight feet, depending on age, sex or geographical range. He gives it the name of *Dinanthropoides nivalis,* "the terrible anthropoid of the snows."

No doubt Sir Edmund views this nomenclature wryly. The *yeti* tracks he found turned out to be those of foxes and wild dogs, distorted by the heat of the sun. His *chuteh* skins, when examined, proved to be the pelts of the very rare Tibetan blue bear. The scalps which were shown to him as relics of the *yeti* were given special study; the scientific conclusion was that they were not scalps at all, but were molded from the skin of a rare member of the goat family.

Sir Edmund, of course, would be the first to admit that his fruitless search and explaining away of purported evidence do not actually disprove the existence of the rumored Snowman. It is very difficult to prove that something which might possibly exist, does not exist in point of fact. It may be assumed that the search for the elusive *yeti*, like the effort to establish the reality of Flying Saucers, will continue to engage the energies of those who have the resources and opportunity for such bizarre forms of research.

Here's a final fillip to disturb the unbelievers. Since 1934 the Dutch paleontologist von Koenigswald has discovered several molars, almost human in character, amid the fossil curiosities of Hong Kong and Cantonese apothecary shops. The peculiarity of these teeth is that they have five or six times the bulk of human molars. Whatever wore them, back in Pleistocene times, was probably 11 to 13 feet high. In any event, there is growing evidence that a huge ape called *Gigantopithecus,* inhabited Kwangsi Province in South China, half a million years ago. How would you go about proving that this big primate had no cousins that still survive in the relatively unknown stretches of the mighty Himalayan mountains?

L. C. McHugh

Hummingbirds

Crawford H. Greenewalt is not only president of E. I. du Pont de Nemours & Company at a staggering $600,000 per year; he is also a distinguished amateur ornithologist.

Since 1953 Mr. Greenewalt has lugged 250 pounds of photographic gear more than 100,000 miles in a dedicated effort to make a portrait gallery of hummingbirds in "living color." Sixty-seven plates of 57 species, together with new material on the iridescent coloration and flight techniques of what the naturalist Audubon called a "glittering fragment of the rainbow," form the substance of Greenewalt's 250-page book, *Hummingbirds,* which was issued by Doubleday & Co. during November.

I have not seen Mr. Greenewalt's book, already hailed as a "classic of natural history," because its list price is $22.50. But happily, the "poor man's Greenewalt" can be found on pages 658-679 of the November issue of the *National Geographic.* There you will see not only a summary of Mr. Greenewalt's studies on hummingbirds, but also 23 magnificent color reproductions of his work, some of them photographed at speeds of one 30-millionth of a second.

British and Continental lyric poets, who often pay tribute to the thrush, the skylark, etc., have generally ignored the glittering glory of the hummingbirds. Why? The answer is simple. They were not familiar with these flying sapphires, topazes and amethysts. Hummingbirds are the boast of the Western Hemisphere; they range the mountain and the plain, the forest and the desert, from Alaska to Tierra del Fuego. They are found nowhere else.

This does not mean that hummingbirds are rare. It is true that only one species (the ruby-throat) is found east of the Mississippi, and even in our West there are no more than eleven breeding species. But over their entire range, the suborder of the Trochilidae counts at least 319 species: hummingbirds will be found wherever and whenever flowers bloom in the Western world. If hummingbirds seem rare to us, that is because they are not much given to

migration, and most of the species inhabit a band that is five degrees wide on either side of the Equator. The center of the world, from the viewpoint of a hummingbird, lies in Ecuador, which is home for half the known species.

Everybody knows that hummingbirds are tiny. One Cuban species of *zum-zum* is only 2¼ inches long. The giant of hummingbirds is the Andean *Patagona gigas;* its 8½ inches sound impressive until you see how much of it goes into tail and bill, and learn that the whole bird weighs less than an ounce. Our American ruby-throat, in fighting form, weighs in at 150 to the pound. Smallest of the known hummingbirds is the South American *Calliphlox amethystina,* which can balance a dime on the scales but is a lot more colorful.

No description of hummingbirds can overlook the fact that these living gems have an extraordinary power plant that enables them to expand energy faster than any other warm-blooded animal. To hover in the air like a helicopter, a hummingbird must burn up fuel, per unit of weight, ten times faster than a man running nine miles per hour. Even when it sinks into torpor (a state like hibernation) at dusk of a day on which it has not fed well, a hummingbird uses up its reserve energy at the rate of a man taking his morning walk.

◆

Such prodigious living explains why hummingbirds eat 50 to 60 meals a day and take most of their food on board in the form of sugar (nectar from flowers), which is high in food value and quickly convertible into energy. To absorb enough calories to live the normal life of a hummingbird, a 170-pound man would have to eat 370 pounds of boiled potatoes every day. To hover like a hummingbird, a man would have to perspire at the rate of 100 pounds of water per hour, just to keep his skin temperature below the boiling point of water. That's really living it up!

Some of the most remarkable facts about hummingbirds involve their flight techniques. Some species can beat their wings 80 times a second. All of them have wings that develop power on the upbeat as well as the downbeat. Most remarkable of all, hummingbirds have a "reverse gear" that enables them to fly backward almost as well as they can fly forward.

But not even the hummingbirds have everything! Their legs are so poorly developed that they cannot walk about at all. Worst of all, their spectral beauty is not matched by their vocal performance. Hummingbirds are not songbirds. Mr. Greenewalt found only one species whose operatic capabilities would excite the envy of even the lowly English sparrow.

Lack of a musical repertory does not lead to neurotic withdrawal in a hummingbird. The little bird is both bold and curious, disdainful of man's presence and extraordinarily pugnacious. It can afford to manifest these traits because it is faster than a rapier in the face of danger.

Too bad more poets cannot observe the flashing gems whose Indian names signify such things as "tresses of the daystar," and whose scientific nomenclature translates into expressions like "fiery topaz" and "golden torch." But our own priest-poet of the South, John Banister Tabb, who perhaps knew only the ruby-throat, paid it this lovely tribute:

A flash of harmless lightning,
  A mist of rainbow dyes,
The burnished sunbeams brightening,
    ing,
From flower to flower he flies.

L. C. McHugh

The Hurricane

The hurricane is a meteorological monster spawned in tropical seas. Our very name for these rampaging atmospheric vortices probably comes from an old Carib word that meant "evil spirit" or "storm god." But for the scientist, the Atlantic hurricane is identical with the Asiatic typhoon and the Australian willy-willy. The only difference is that the homegrown variety gets its start in the Gulf of Mexico, the Caribbean Sea, or more often in the southern stretches of the North Atlantic.

Everybody knows that the Weather Bureau nowadays christens every newly born Atlantic hurricane with a girl's name. The reason for this is neither humor nor sentiment, but the need for a short, easily pronounced identification tag in emergency situations. When Ione was zooming in on Washington, D. C., back in 1955, the local weather number was dialed 395,000 times in a single day.

Names like Donna and Ethel also point up the fact that every hurricane has an individuality of its own. When we talk of these storms in a general way, we have to use broad averages to cover their characteristics.

Hence, in general, we can say that the hurricane season runs from May to October; outside these six months, hurricanes are rare. As for frequency, Atlantic hurricanes have averaged about eight per year since 1890. Their span of life runs around nine days, although a duration of five weeks is not unknown. Even the size of hurricanes varies considerably, but usually a fully developed storm stretches over a diameter of about 400 miles.

◆

Only a hardy meteorologist would assert that he knows for sure how a hurricane is born, and perhaps we are foolish to outline the most popular current theory in a few sentences. But it would go like this: Somewhere on the ocean, a low-pressure mass of hot, moist air, squeezed by opposing trade winds, starts rising up a heat chimney in a spiral movement. As the air rises, it cools. As it cools, the water vapor condenses and heat is given off. The heat further warms the rotating air, which becomes even lighter and rises more swiftly. Meanwhile, more hot, moist air swirls in at the bottom of the chimney—as much as a million tons a second. Thus moisture and heat feed the growing monster at an increasing pace; in a few days it has turned into a symmetrical storm of wind, cloud and rain—a low-pressure heat engine that tends to sweep west and then veer north like a steam roller out of control.

Fortunately, every hurricane has a suicidal mania. Its northern motion deprives it of new heat. If it rolls over land, there is no water to feed it. Both factors cause the hurricane to dissipate its energy budget in a few days.

The most unique part of a hurricane is its eye, a central area about 14 miles in diameter where the wind is no more than a breeze and the sun shines fitfully through filmy clouds. Seen from the eye, the inner edge of the hurricane looks like a curtain of cloud that may rise seven miles high. This circular surface represents the region where the centrifugal force of the spiraling winds just balances the pressure that tends to force the air inward toward the center of the storm.

The chief destructive forces that are associated with the hurricane are tides, floods and winds.

Historically, the most dangerous spot to tarry in during a hurricane is a coastal area. As the hurricane moves toward the land, it drives a mighty surge before it, one that acts like a tidal wave. The Galveston hurricane of 1900 killed 6,000 people; most of them drowned in a tidal surge.

On land, a like danger to life lies in flooding. A big hurricane can deposit from three to six inches of rain over a million square miles. In limited areas, the total rainfall commonly reaches 20 or 30 inches. It was flooding that caused most of the deaths and damage done by Diane in 1955.

As for property damage, most of it is due to wind. Most hurricanes attain wind velocities of at least 100 miles per hour sometime during their life. Back in 1955, Janet touched 175 miles an hour at Chetumal, Mexico, before the anemometer disintegrated. Engineers who studied the effects of a big wind that swept the Florida Keys in 1935 estimated that its gusts must have reached 200-225 miles per hour.

Obviously, every hurricane is a titan of energy, even though its mechanical efficiency is abysmally low. An average hurricane expends heat energy at the rate of 16 trillion kilowatt-hours per day; that's 8,000 times the total electrical energy generated each day in the United States. The energy budget of a respectable hurricane is equivalent to 500,000 atom bombs every 24 hours.

Such a fantastic quantity of energy shows how hard it will be to kill a hurricane, unless we can find some weakness in these giants that will enable us to trigger off a self-destructive chain reaction while they are still at sea. It has even been suggested that the way to nip a hurricane in the bud is to pop off an atom bomb in its eye while it is still a baby. The trouble with that suggestion is that we don't know enough about hurricanes to estimate the effect. The bomb might dissipate the baby's strength. On the other hand, it might only rouse the baby's fury and make it squall more lustily.

L. C. McHugh

Vanishing Wildlife

At one time the passenger pigeon nested in North America in larger numbers than any other vertebrate land animal on record. The 19th-century naturalist Audubon once observed a single flock of these birds that contained more than one *billion* individuals. But the pigeon was a tasty morsel and reckless slaughter made it disappear from the countryside by 1906. The last known specimen died in captivity in 1914.

Only last-minute conservation efforts saved the American bison from a like extinction. This huge beast once roamed over most of the United States and Canada. As many as 60 million buffaloes provided the Indians with food, clothing, shelter and fuel. Neither strength nor numbers saved the buffalo herds from wanton destruction. In 1889 there were only 541 bison left in the United States and even fewer in Canada.

What befell the passenger pigeon and the bison is part of the pattern of extermination that often follows man wherever he hunts or fishes, clears the forest or plows the plain.

It is possible that Stone Age man played a part in wiping out the woolly mammoth and the giant sloth. It is very probable that the European lion and the North African elephant did not wholly disappear until Roman times. Altogether, it is estimated that 106 species of mammals (not to mention birds) have become extinct since the time of Christ. Wildlife conservationists think that as many as 600 animal species are in peril of extermination at the present time.

What is most distressing about the world's vanishing wildlife is that the rate of extinction is rising. About one species a year is disappearing from the earth right now. Man's careless waste ot one of the earth's most precious and irreplaceable resources runs on apace in field and forest, in lake and river and sea—whenever the God-appointed steward of lower nature invades the biosphere and exploits it for his utility and profit.

Although nature was unusually lavish in endowing the United States with resources, her prodigality was more than matched by the human tendency to squander them. It was not until 1908 that we realized our timber was not inexhaustible. It took the dust-bowl days to impress us with the necessity of conserving the soil. Even today we are grossly ignorant of the extent to which much of our wildlife is being depleted.

There are perhaps 50 species of animal life that are seriously threatened in the United States today. The California condor, our largest soaring bird, numbers no more than 60 individuals. Over the vast stretch between Alaska and Florida, there are probably no more than 1,500 bald eagles. It is doubtful that there are 800 grizzly bears in the entire country. Coyotes and bighorn sheep are growing rare. The flamingo, the snowy egret and the great white heron are vanishing.

Sometimes, of course, conservation methods help a species to stage a comeback. The trumpeter swan was thought to be extinct in 1900. They now number 700. Strong public interest is an even more potent savior of wildlife. The whooping crane was reduced to 14 specimens in 1938, but numbers 38 today, thanks to the unusual publicity that attends its every migration.

The ethics of man's religious and social life is well developed. It is hardly a century since man began to realize his natural duties as the husbandman of the earth's resources. It is only today that he is developing a rudimentary conscience to guide him in his attitudes toward earth's flora and fauna. We are just beginning to see that good stewardship may imply more than wise economic use of our abundant riches. The good life of man on earth embraces elements of beauty and companionship with nature, too.

A population explosion may someday blow most of our wildlife into outer space. But does anyone want to hasten the day when there will be no violets by mossy stones and no lark to sing at heaven's gate as "Phoebus 'gins arise"?

L. C. McHugh

Poems

The Human Condition

En route through space
And ice Siberia-bound
The bus stops now and then
Time out prisoners
Bethlehem town Last
Time to look around
Be back in fifteen minutes
All of you
And I mean
Fifteen minutes

Move along murderer
Move along jade
Fifteen minutes of sheep
Cattle stable hay
Not much here to take away
Where is endless sleepless
Night in day out of time
Out of space
Past all frozen sun

Lift him up Cain
Let every eye see
You kiss his frail fingers
He will not shrink away

Give him your apple Eve
It belongs to him
Drink the wine his mother
Pours to the ruddy brim
Drink   For on your journey
The sun itself grows dim

All aboard now
Time shouts again
All aboard flesh souls sin God
With you amen

Sister Mary Aquin, b.v.m.

FALCON

On the other side of our hill
spring comes sooner;
the frost is gone by noon there
and winter with clenched teeth singing
and April chill—
on the other side of our heart.

Warm on the smooth rocks our faces
search high in the light,
homesick and pale for the sight
of a bird on the shouldering sky
where the huge air races—
on the other side of our heart.

The valley is silent and still
and the grasstops unstirred
where the ants work, but one arrogant bird
has found where the air is and mastered it
over our hill—
on the other side of our heart.

Downwinging the stream he floats,
tight wing sliding
perfect and black there, riding
his wind; and the grace of his flight
stops breath in our throats—
on the other side of our heart.

Children wild and white, debating
ironeyed of love,
we follow a taut wing above—
small heart in a towering pride—
while we sit waiting—
on the other side of our hill.

ROBERT B. BROCK

"O" Antiphons

"O" Antiphons vocal again on tongue
enjoin us: take to the attic rung
and from the cedar trunk bring out
hillside, a stable, and devout
ceramics—shepherds and kings
in apprehensive poses;
open the cardboard cases where
upon the kapok whiteness, stare
Mary and St. Joseph; and shed
cocoons of cotton from a carven head
of the cradled Child.
                          On strings
our youngest one disposes
tissue wings upon a tree
and squeals a challenge: "You'll see
the Infant smile if you bend low
enough."
              So eagerly we go
the depths of creeping things.
Here faith discloses
no paltriness about our wooden crèche
wherein, upon the straw, we have
the Child, whose curve of lip
transcends the clay of flesh.

SISTER MARY HONORA, O.S.F.